Separation Methods in Proteomics

Separation Methods in Proteomics

edited by

Gary B. Smejkal
Alexander Lazarev

CRC Press
Taylor & Francis Group
Boca Raton London New York

CRC Press is an imprint of the
Taylor & Francis Group, an **informa** business

CRC Press
Taylor & Francis Group
6000 Broken Sound Parkway NW, Suite 300
Boca Raton, FL 33487-2742

First issued in paperback 2019

ISBN-13: 978-0-8247-2699-7 (hbk)
ISBN-13: 978-0-367-39157-7 (pbk)

Library of Congress Cataloging-in-Publication Data

Catalog record is available from the Library of Congress

Visit the Taylor & Francis Web site at
http://www.taylorandfrancis.com

and the CRC Press Web site at
http://www.crcpress.com

Preface: Evolutionary Nature of the Proteomic Revolution

Gary B. Smejkal and Alexander V. Lazarev

For the past decade we have witnessed success and broad adoption of proteomic techniques, fueled by rapid innovations in mass spectrometry and bioinformatics. Proteomics is currently applied to the discovery of new protein biomarkers of disease,[1,2] toxicity,[3] and drug efficacy.[4] Recently, numerous attempts were undertaken to use proteomic analysis workflow in early diagnostics.[5] Proteomics, in conjunction with gene expression and metabolic profiling, forms the foundation of a systems biology approach to understanding and modeling fundamental mechanisms of life, essentially driving biology closer to the realm of exact sciences.[6]

Regardless of its wide adoption, proteomics is a constantly evolving field. It is interesting to note, that while instrument development has received enormous attention in recent years, the separation techniques used for analysis of proteins have remained fairly conservative. However, the debates on which separation technique is best suitable for proteomic analysis continue to stir the scientific community.

There are generally three distinct directions in proteomics. The first approach, the historical two-dimensional gel electrophoresis (2DE) approach, well regarded for its resolving power, massively parallel separation, quantitative nature, and instant visualization of thousands of protein species, including distinct posttranslationally modified protein isoforms, has been criticized for the labor-intensive process and the lack of reproducibility.

The second approach, based on variations of multidimensional chromatography, apparently is easier to automate, although analysis typically takes longer due to the serial nature of the separation process.

The third emerging approach calls for using various microarray technologies for protein profiling, primarily driven by the diagnostic factions of proteomics.[5,7,8] Because this approach does not typically utilize any separation techniques at the time of analysis, it will not be covered in this book.

There are several permutations of the first and second approach being reported from time to time. These methods generally combine the chromatography and electrophoresis steps, either an ion-exchange chromatography fractionation followed by a one-dimensional sodium dodecyl sulfate polyacrylamide gel electrophoresis (SDS-PAGE) or isoelectric focusing followed by a reverse phase chromatography.[9–12]

Separation steps are typically coupled with either rapid matrix-assisted laser-desorbtion ionization-time of flight (MALDI-TOF) peptide mass fingerprinting (PMF) or more versatile electrospray ionization-tandem mass spectrometry (ESI-MS/MS) protein identification.[8]

It is now becoming apparent that all of the approaches described above are optimized for particular applications or user preferences and should be considered complimentary to each other. As high-end tandem instruments capable of accurate mass determination and data-dependent analysis of posttranslational modifications become affordable, diversity of liquid chromatography-tandem mass spectrometry (LC-MS/MS) techniques allow automated data acquisition and confident protein identification from complex samples. On the other hand, considering the wide adoption of proteomic techniques by biochemists, electrophoresis is still arguably the most frequently used method of protein separation to date.

"2D OR NOT 2D?"

That was the question posed in Shakespearian satire by Fey and Larsen in a *Current Opinions in Chemical Biology* review article published in 2001.[13] This marked a period that seemingly began a movement away from 2DE, or at least an attempt by some researchers to avoid using it.

However, as one of the major contributors to the field of protein separation, Pier Giorgio Righetti, has recently illustrated in his review of the history of electrophoresis,[14] this technology is a prolific area of research and remains an integral part of proteomic analyses today and most likely for some time to come.

In the past, the technique was viewed as procedurally complex, highly irreproducible, and incapable of eliciting all of the protein constituents of a proteome, particularly low abundance and membrane proteins. For years, there were concerns expressed by many groups that subpopulations of proteins, such as hydrophobic membrane proteins, possibly clinically significant biomarkers, or targets of therapeutic treatment, might not be accurately represented in the two-dimensional (2D) array. This criticism has been largely dismissed by recent critical advances in 2DE, a technique 30 years old, but still in its formative years. For example, the implementation of neutral pH buffer systems has made possible the commercial mass production of stable polyacrylamide gels making the second dimension of electrophoresis both reproducible and procedurally benign. Likewise, the refinement of immobilized pH gradients has eliminated the problems inherent to carrier ampholyte-driven isoelectric focusing (IEF) and because the pH gradient is insusceptible to drift, proteins are resolved reproducibly along the x-axes of the 2D array.

Some efforts have been made to increase the capacity of 2DE for resolving the entire protein content of cells. The synthesis of linear sulfobetaine surfactants has facilitated the isolation of even the most recalcitrant membranes proteins, and it is this segment of the cellular proteome that is often thought to represent the majority of potential drug targets. Zwitterionic detergents have replaced the nonionic ones used earlier for protein solubilization. It is the increased stringency of these new surface reactive agents that not only enables the isolation of membrane associated proteins, but ensures their solubility is maintained during the first dimension of electrophoresis.

It is the purpose of this volume, nearly two years in the making, to provide a comprehensive overview of the separation methods currently employed in proteomic analyses.

Regardless of the chosen analytical approach, a common concern of missing the low abundance components of the proteome still remains prevalent in proteomics. Hence, the need for complexity reduction prior to proteomic analysis remains the most popular paradigm at the moment.

The first section of this book deals with sample preparation. It includes several descriptions of immunological depletion of high-abundance proteins, specifically from plasma and sera, but applicable to any species for which there are available antibodies. As communicated in subsequent chapters, the complexity of proteomes can be diminished by fractionation into smaller, less complicated "subproteomes" that can be analyzed individually. Ion-exchange chromatography, free flow electrophoresis (FFE), and solution-phase IEF are all such "divide and conquer" strategies for the isolation of disparate fractions of defined isoelectric point (pl) intervals from complex samples.

Several applications to specific areas of research are described in the following chapters. These chapters provide valuable insights, often derived empirically, for several representative samples. For example, the extraction of proteins from myelinated tissues such as the optic nerve as described by Bhattacharya et al. (Chapter 9), indeed a challenge in terms of sample preparation, will prove invaluable to the researcher trying to elucidate proteins from recalcitrant samples that are resistant to usual solubilization schema.

Reduction and alkylation of protein disulfides is reviewed in Chapter 12 by Bai et al. who prescribe new reagents that alkylates nearly 100% of the cysteines. Further, this chapter explains the rationale of reduction and alkylation before IEF, further improving the reproducibility of 2DE separation.

In recent years, rapid development of novel, more affordable mass spectrometry instrumentation capable of producing data of high-mass accuracy at great resolution, has strengthened the idea to alleviate the need for high-quality separations prior to analysis, shifting the burden of separation onto a mass spectrometer.

Several comprehensive reviews have been written recently on using automated liquid-phase separation methods.[15–21] We allocate three chapters of this book to cover some of the new trends in chromatographic separations coupled with mass spectrometry, such as a utility of accurate mass determination to enable identification of proteins in complex mixtures, creative methods combining classical top-down and bottom-up approaches to assure reliable automated protein identification, and the novel, promising area of high-resolution monolithic capillary columns.

Many approaches have been developed to challenge the most famous claim of 2D gels, an ability to quantitatively assess protein expression, by development of mass spectrometric methods for relative[22–24] or absolute[25] quantification of differentially expressed proteins. However, most successful efforts on automated LC-MS/MS methods are still being pursued by a fairly small community of technology champions and frequently are applicable to the samples of relatively low complexity. It is also important to note that one of the biggest adoption barriers for quantitative LC-MS/MS and multiple permutations of classical Multidimensional Protein Identification Technology (MudPIT) techniques has been related to the bottlenecks in data analyses such as limitations of commercially available software tools to handle reliable automated data-dependent data acquisition, protein identification, and validation of the results thus obtained applicable for the analysis of complex biological samples.[26]

Recent advancements in functional proteomics call for shifting the experimental efforts to the detection of protein interactions. Conventional approaches to protein isolation and analysis portrays proteins in their "naked," unnatural form and completely ignores protein-DNA, protein-lipid, and protein-carbohydrate interactions of biological relevance. It is this growing interest in protein interactions that will be the impetus for the development of new technologies capable of analyzing proteins in their native state. Accordingly, two chapters in this book discuss the application of nuclear magnetic resonance (NMR) for the analysis of such interactions.

We truly believe that proteomics is yet to bring its best results. The "Proteomic Revolution" all of us have been waiting for is finally over. Proteomics is now widely accepted as one of the essential techniques for biomarker discovery. Now it is the time for the evolutionary process: selection of the best, most successful technologies that can deliver high-quality knowledge in our quest for understanding life.

LITERATURE CITED

1. Hale, J.E., Gelfanova, V., Ludwig, J.R., Knierman, M.D., Application of proteomics for discovery of protein biomarkers, *Brief Funct. Genomic Proteomic.,* 2, 185–193, 2003.
2. Gao, J., Garulacan, L.A., Storm, S.M., Opiteck, G.J., Dubaquie, Y., Hefta, S.A., Dambach, D.M., and Dongre, A.R., Biomarker discovery in biological fluids, *Methods.,* 35, 291–302, 2005.
3. Wetmore, B.A. and Merrick, B.A., Toxicoproteomics: proteomics applied to toxicology and pathology, *Toxicol. Pathol.,* 32, 619–642, 2004.
4. Walgren, J.L. and Thompson, D.C., Application of proteomic technologies in the drug development process, *Toxicol. Lett.,* 149, 377–385, 2004.
5. Steel, L.F., Haab, B.B., and Hanash, S.M., Methods of comparative proteomic profiling for disease diagnostics, *J. Chromatogr. B.,* 815, 275–284, 2005.
6. de Hoog, C.L. and Mann, M., Proteomics, *Annu. Rev. Genomics Hum. Genet.,* 5, 267–293, 2004.
7. Angenendt, P., Progress in protein and antibody microarray technology, *Drug Discov. Today,* 10, 503–511, 2005.
8. Charboneau, L., Scott, H., Chen, T., Winters, M., Petricoin, E.F., III, Liotta, L.A., and Paweletz, C.P., Utility of reverse phase protein arrays: applications to signalling pathways and human body arrays, *Brief Funct. Genomic Proteomic.,* 1, 305–315, 2002.
9. Essader, A.S., Cargile, B.J., Bundy, J.L., and Stephenson, J.L., Jr., A comparison of immobilized pH gradient isoelectric focusing and strong-cation-exchange chromatography as a first dimension in shotgun proteomics, *Proteomics.,* 5, 24–34, 2005.
10. Tu, C.J., Dai, J., Li, S.J., Sheng, Q.H., Deng, W.J., Xia, Q.C., and Zeng, R., High-sensitivity analysis of human plasma proteome by immobilized isoelectric focusing fractionation coupled to mass spectrometry identification, *J. Proteome Res.,* 4, 1265–1273, 2005.

11. Ogorzalek Loo, R.R., Yam, L., Loo, J.A., and Schumaker, V.N., Virtual two-dimensional gel electrophoresis of high-density lipoproteins, *Electrophoresis.,* 25, 2384–2391, 2004.
12. Wu, S., Tang, X.T., Siems, W.F., and Bruce, J.E., A hybrid LC-gel-MS method for proteomics research and its application to protease functional pathway mapping, *J. Chromatogr. B. Analyt. Technol. Biomed. Life Sci.,* 822, 98–111, 2005.
13. Fey, S.J. and Larsen, P.M., 2D or not 2D. Two-dimensional gel electrophoresis, *Curr. Opin. Chem. Biol.,* 5, 26–33, 2001.
14. Righetti, P.G., Electrophoresis: the march of pennies, the march of dimes, *J. Chromatogr. A.,* 1079, 24–40, 2005.
15. Smith, R.D., Trends in mass spectrometry instrumentation for proteomics, *Trends in Biotechnol.,* 20, S3–S7, 2002.
16. Zhou, M., Conrads, T.P., and Veenstra, T.D., Proteomics approaches to biomarker detection, *Brief Funct. Genomic Proteomic.,* 4, 69–75, 2005.
17. Wang, H. and Hanash, S., Multi-dimensional liquid phase based separations in proteomics, *J. Chromatogr. B.,* 787, 11–18, 2003.
18. Delahunty, C. and Yates, J.R., III, Protein identification using 2D-LC-MS/MS, *Methods.,* 35, 248–255, 2005.
19. Romijn, E.P., Krijgsveld, J., and Heck, A.J., Recent liquid chromatographic-(tandem) mass spectrometric applications in proteomics, *J. Chromatogr. A.,* 1000, 589–608, 2003.
20. Wysocki, V.H., Resing, K.A., Zhang, Q., and Cheng, G., Mass spectrometry of peptides and proteins, *Methods.,* 35, 211–222, 2005.
21. Righetti, P.G., Castagna, A., Antonucci, F., Piubelli, C., Cecconi, D., Campostrini, N., Zanusso, G., and Monaco, S., The proteome: anno Domini 2002, *Clin. Chem. Lab. Med.,* 41,425–438, 2003.
22. Ong, S.E., Foster, L.J., and Mann, M., Mass spectrometric-based approaches in quantitative proteomics, *Methods.,* 29,124–133, 2003.
23. Julka, S. and Regnier, F.E., Recent advancements in differential proteomics based on stable isotope coding, *Brief Funct. Genomic Proteomic.,* 4, 158–177, 2005.
24. Righetti, P.G., Campostrini, N., Pascali, J., Hamdan, M., and Astner, H., Quantitative proteomics: a review of different methodologies, *Eur. J. Mass Spectrom.,* 10, 335–348, 2004.
25. Kirkpatrick, D.S., Gerber, S.A., and Gygi, S.P., The absolute quantification strategy: a general procedure for the quantification of proteins and post-translational modifications, *Methods.,* 35, 265–273, 2005.
26. Johnson, R.S., Davis, M.T., Taylor, J.A., and Patterson, S.D., Informatics for protein identification by mass spectrometry. *Methods,* 35, 223–236, 2005.

About the Editors

Gary Smejkal is Applications Scientist at Pressure BioSciences where he has applied that company's novel pressure-cycling technology (PCT) for sample preparation in proteomics and biomarker discovery. Formerly from Proteome Systems, he played an integral role in the continuing development of the multicompartment electrolyzer (MCE) for solution phase isoelectrofocusing (IEF) and two-dimensional gel electrophoresis (2DGE) with focus on sample preparation and protein-staining methodologies. He has designed and presented 2DGE training courses at several major academic institutions, including the Harvard School of Public Health, and at government affiliates such as the U.S. Food and Drug Administration. He serves on the Editorial Advisory Boards of Expert Reviews in Proteomics, Expert Opinion in Drug Discovery, and the Japanese Journal of Electrophoresis.

Previously, he worked 14 years at the Cleveland Clinic Foundation Lerner Research Institute and Cleveland State University Department of Chemistry, both in basic and clinical research, where he specialized in the electrophoresis of high molecular-weight plasma proteins. He has over 40 publications in the field of protein electrophoresis. At the Cleveland Clinic Foundation Department of Clinical Pathology, in collaboration with Case Western Reserve University Department of Biomedical Engineering, he developed a rapid electrophoretic assay for the clinical analysis of von Willebrand Disease phenotypes.

Dr. Alexander Lazarev is a Director of New Technology Development at Proteome Systems, Inc. He joined Proteome Systems at the commencement of the company's North American subsidiary in September, 2001, to work on the research and development of consumable products and applications for high-throughput proteomic research and discovery of protein biomarkers of disease and toxicity. Primary focus of his work is research and development (R&D) and technology scouting in the areas of protein separation, analysis of proteins and their posttranslational modifications, mass spectrometry, and bioinformatics.

Prior to Proteome Systems, he was Senior Research Scientist at the Proteomics R&D Division of Genomic Solutions, Inc. His responsibilities there included proteomic contract services, technical training and support of robotics, mass spectrometry, and informatics solutions for proteomic research.

Alexander has over 10 years of experience in development of analytical methods, related to drug metabolism, separation, purification, and identification of proteins, oligosaccharides, and small molecules, metabolic profiling, and proteomics. He has been invited to speak at a number of high-profile scientific seminars and symposia and serves as an ad hoc Executive Board member of The Central New England Chromatography Council (CNECC).

About the Editors

Contributors

Birte Aggeler-Schulenberg
Invitrogen
Eugene, Oregon

Suresh P. Annangudi
Cleveland Clinic Foundation
Cleveland, Ohio

Mikaela Antonovici
John Buhler Research Centre
Winnipeg, Manitoba, Canada

Fengju Bai
Indiana University School of Medicine
Indianapolis, Indiana

Jerome Bailey
Agilent Technologies, Inc.
Wilmington, Delaware

James Behnke
Pressure Biosciences
West Bridgewater, Massachusetts

Scott J. Berger
Waters Corporation
Milford, Massachusetts

Sanjoy K. Bhattacharya
Cole Eye Institute (i31)
Cleveland Clinic Foundation
Cleveland, Ohio

Vera L. Bonilha
Cleveland Clinic Foundation
Cleveland, Ohio

Barry Boyes
Agilent Technologies, Inc.
Wilmington, Delaware

Mary Grace Brubacher
Bio-Rad Laboratories
Hercules, California

Elena Chernokalskaya
Millipore Corporation
Danvers, Massachusetts

Steven A. Cohen
Waters Corporation
Milford, Massachusetts

John S. Crabb
Cleveland Clinic Foundation
Cleveland, Ohio

John W. Crabb
Cleveland Clinic Foundation
Cleveland, Ohio

Kumar Dasuri
John Buhler Research Centre
Winnipeg, Manitoba, Canada

Bruce Dawson
The Biotechnology Application Centre
(BAC BV)
Naarden, The Netherlands

Christoph Eckerskorn
BD Diagnostics
Franklin Lakes, New Jersey

Hani El-Gabalawy
John Buhler Research Centre
Winnipeg, Manitoba, Canada

Xiangming Fang
Genway Biotech
San Diego, California

Jerald S. Feitelson
Genway Biotech
San Diego, California

Catherine Fenselau
University of Maryland
College Park, Maryland and
Greenebaum Cancer Center
University of Maryland School of Medicine
Baltimore, Maryland

Susan Francis-McIntyre
Michael Barber Centre for Mass Spectrometry
University of Manchester
Manchester, England

Simon J. Gaskell
Michael Barber Centre for Mass Spectrometry
University of Manchester
Manchester, England

Heidi Geiser
University of New Hampshire
Center for Structural Biology
Durham, New Hampshire

Marcia Goldfarb
Anatek-EP
Portland, Maine

Eva Golenko
Perkin Elmer Life and Analytical Sciences
Wellseley, Massachusetts

Xiaorong Gu
Cleveland Clinic Foundation
Cleveland, Ohio

Sara Gutierrez
Millipore Corporation
Danvers, Massachusetts

Qiuhong He
Magnetic Resonance Research Center
University of Pittsburgh Medical Center
Pittsburgh, Pennsylvania

Tim Hermans
The Biotechnology Application Centre (BAC BV)
Naarden, The Netherlands

Douglas Hinerfeld
University of Massachusetts Medical School
Shrewsbury, Massachusetts

Joe G. Hollyfield
Cleveland Clinic Foundation
Cleveland, Ohio

Lei Huang
Genway Biotech
San Diego, California

Alexander R. Ivanov
Harvard NIEHS Center for Environmental Health Proteomics Facility
Harvard School of Public Health
Boston, Massachusetts

David Innamorati
Charles River Laboratories
Worcester, Massachusetts

Jon M. Jacobs
Pacific Northwest National Laboratory
Richland, Washington

Niclas G. Karlsson
Proteome Systems Ltd.
North Ryde, Sydney, Australia

Ira S. Krull
Northeastern University
Boston, Massachusetts

Askar Kuchumov
BD Diagnostics
Franklin Lakes, New Jersey

Nathan P. Lawrence
Pressure BioSciences, Inc.
West Bridgewater, Massachusetts

Alexander V. Lazarev
Proteome Systems, Inc.
Woburn, Massachusetts

Jack T. Leonard
U.S. Genomics, Inc.
Woburn, Massachusetts

Chunqin Li
Pressure Biosciences
West Bridgewater, Massachusetts

Sheng Liu
Indiana University School
of Medicine
Indianapolis, Indiana

Mary F. Lopez
Perkin Elmer Life and Analytical
Sciences
Welseley, Massachusetts

James Martosella
Agilent Technologies, Inc.
Wilmington, Delaware

Kevin M. Millea
Waters Corporation
Milford, Massachusetts

Ingrid Miller
Protein-, Peptide-, and
Immunochemistry Group,
Institute of Medical Chemistry
University of Veterinary
Medicine
Vienna, Austria

Jonathan Minden
Carnegie Mellon University
Pittsburgh, Pennsylvania

Gordon Nicol
Agilent Technologies, Inc.
Wilmington, Delaware

Nicolle H. Packer
Proteome Systems Ltd.
North Ryde, Sydney, Australia.

Aran Paulus
BioRad
Hercules, California

John Pirro
Charles River Laboratories
Worcester, Massachusetts

Aldo M. Pitt
Millipore Corporation
Danvers, Massachusetts

Malcolm G. Pluskal
Consulting Services
Acton, Massachusetts

Anton Posch
Research Institute for
Occupational Medicine
Bochum, Germany

Phani Kumar Pullela
Chemical Proteomics Facility at
Marquette
Marquette University
Milwaukee, Wisconsin

Amir Rahbar
University of Maryland
College Park, Maryland

Vernon Reinhold
University of New Hampshire
Durham, New Hampshire

Contents

Part V *Related Techniques*

Part I

Sample Preparation

1 Applications of Pressure Cycling Technology (PCT) in Proteomics

Feng Tao, James Behnke, Chunqin Li, Calvin Saravis, Richard T. Schumacher, and Nathan P. Lawrence

CONTENTS

1.1 INTRODUCTION

It was announced in 2001 that the human genome had been sequenced, thus heralding a new era for biological research.[1,2] However, the sequenced genome didn't provide information about protein species in cells and tissues. Consequently, focus then shifted to the intricate interactions and roles of proteins in cells—that is, to a new field called proteomics. *Proteomics* by definition is the systematic determination of protein sequence, quantitation, modification state, interaction partners, activity, subcellular localization, and structure in a given cell type at a particular time.[3] There are an estimated 30,000 to 40,000 human genes. However, the old canonical definition of "one gene, one protein" is no longer valid. One gene does not necessarily produce one protein. Many genes are made up of a series of substructures, or exons, that can be combined in different ways to give rise to a diverse series of similar proteins that have distinctly different functions. Once proteins are made, many of them can be modified by phosphate, carbohydrates, or lipids with different structural configurations, thus adding to the complexity to understanding the action and relationship of proteins. A comparable study of the protein version of the genome, the proteome, must deal with approximately 30,000 genes that can be rearranged to yield the more than 800,000 proteins, which can then be modified with over 300 different submolecular entities.

Proteomics must define which proteins are produced in a certain type of cell at a specific time, how they are modified, where they reside, in what form they associate *in vivo,* and most importantly, how the protein or protein complexes function. A key to the success of proteomic analysis is to instantaneously release all the proteins from cells or tissues while retaining their native configurations and biological activities.

The field of systems biology, which includes proteomics, is now rapidly advancing through basic biological research, pharmaceutical development, and clinical diagnostics—all helped by progress in genome sequencing and bioinformatics. To this end, improved instrumentation is needed to facilitate rapid data generation in proteomics studies. Many texts about proteomics often give a cursory overview of the preparatory steps in analyses, although the efficient release of proteins from cells and tissues is a critically important first step in most analytical processes (regardless of whether studies involve human or animal physiology, medicine, agriculture, or the environment) because the quantity and quality of extracted material can profoundly affect the success of the downstream applications. Many publications devoted to proteomics focus on the advancement of analytical tools, such as two-dimensional (2D) gel analysis, mass spectrophotometry, and data mining, and on the need for continued improvement and development in these key areas. Rarely is there discussion of the need for improvements in methods of releasing proteins from cells and tissues. An important, underemphasized area that needs improvement is sample preparation, the first step toward analysis, which is often the bottleneck that limits rapid discovery in proteomic research. Many laboratories still rely on manual methods to release proteins, such as mortar and pestle grinding, or on semiautomated methods, such as bead beating, rotor-stator homogenization (Polytron™), and sonication. The limitations of these and other current techniques, as well as their inherent problems, are discussed later in this chapter. The major requirements for new instruments used for sample preparation include (*a*) solubilizing all cellular proteins, naturally occurring protein complexes, or both; (*b*) preventing formation of improper aggregates; (*c*) removing contaminants; (*d*) quenching proteolytic enzymatic activities; and (*e*) achieving reproducibility.

This chapter focuses on a new sample preparation system developed by Pressure BioSciences Inc. (PBI), the Pressure Cycling Technology Sample Preparation System (PCT SPS). This system releases proteins from cells and tissues based on PBI's proprietary Pressure Cycling Technology (PCT).[4] The PCT SPS cycles pressure between ambient and 35 kpsi (235 MPa) to control biomolecular interactions while releasing proteins from cells and tissues. This chapter describes the PCT SPS, the potential mechanisms of action of PCT, and the effects of pressure on proteins, presents data derived from the PCT SPS, and compares these data to data derived from other protein-release methods. Additional requirements may be the selective extraction of subset proteins, such as abundant protein-depleted samples.

1.2 SAMPLE PREPARATION BY THE PCT SPS COMPARED TO TRADITIONAL METHODS

Currently many laboratories use manual grinding with a mortar and pestle, sometimes at liquid nitrogen temperature, to release cellular components.[5] Other current mechanical or chemical methods include sonication, bead beating, French press processing, freezer milling, rotor-stator homogenization, enzymatic digestion, and chemical dissolution—all of which have inherent limitations. These limitations include safety issues, potential for contamination, extended process time, lack of automation, and poor reproducibility or applicability to only a narrow range of tissue types.[6] Sample preparation methods also carry the risk, to varying degrees, that they will perturb important naturally occurring complexes or relationships among the constituent molecules, making interpretation of data more difficult.

The PCT SPS comprises a pressure-generating instrument (Barocycler™ NEP2017 or NEP3229) and single-use, fully enclosed sample processing tubes (PULSE™ Tubes).[7] Both Barocycler instruments (Figure 1.1) are capable of rapidly cycling hydrostatic pressure between ambient and up to 35,000 psi (235 MPa). Samples to be processed are placed inside the PULSE Tubes, the PULSE Tubes are placed inside the pressure chamber of the Barocycler instrument, and cycles of pressure are applied. Up to six PULSE Tubes can be processed simultaneously in the two reaction chambers of the Barocycler NEP2017; up to three PULSE Tubes can be processed in the one reaction chamber of the Barocycler NEP3229. The ability to run multiple PULSE Tubes simultaneously allows the user to run assay controls with every sample preparation. The processing temperature can be controlled using an external circulating water bath, allowing for sample processing between 4 and 37°C.

The PULSE Tube, (Figure 1.2), is a single-use device designed for containing and processing the sample, and its unique design also allows for the transmission of pressure to the sample from the Barocycler instrument. It consists of a cylindrical body open at both ends, and in the middle of the Tube is a perforated lysis disk. The lysis disk is an intergral of the Tube body formed during the molding process. It separates the PULSE Tube into two compartments, each with a specific function. Raw tissue samples are loaded into the sample chamber or ram end of the Tube. The ram, a polypropylene disk fitted with two O-rings, is inserted after the sample is loaded to seal the sample chamber. Even with liquid samples, such as cell suspensions, the ram completely seals material in the Tube. The opposite end comprises the fluid retention chamber into which processing buffers are typically introduced and from which extracted material is withdrawn. This end of the Tube is sealed with a threaded cap that also is fitted with two O-rings.

Figure 1.3A shows the workflow of PCT SPS. The design of the PULSE Tube permits both collection and storage before proceeding to sample processing. Samples may be processed at the convenience of the researcher. After processing, the researcher may store the sample in the PULSE Tube for later analysis or continue with downstream applications. To initiate sample preparation, the researcher loads the assembled PULSE Tubes containing samples into the tube holder. When the sample, ram, lysis buffer, and cap are in place, the assembled PULSE Tubes are

A

B

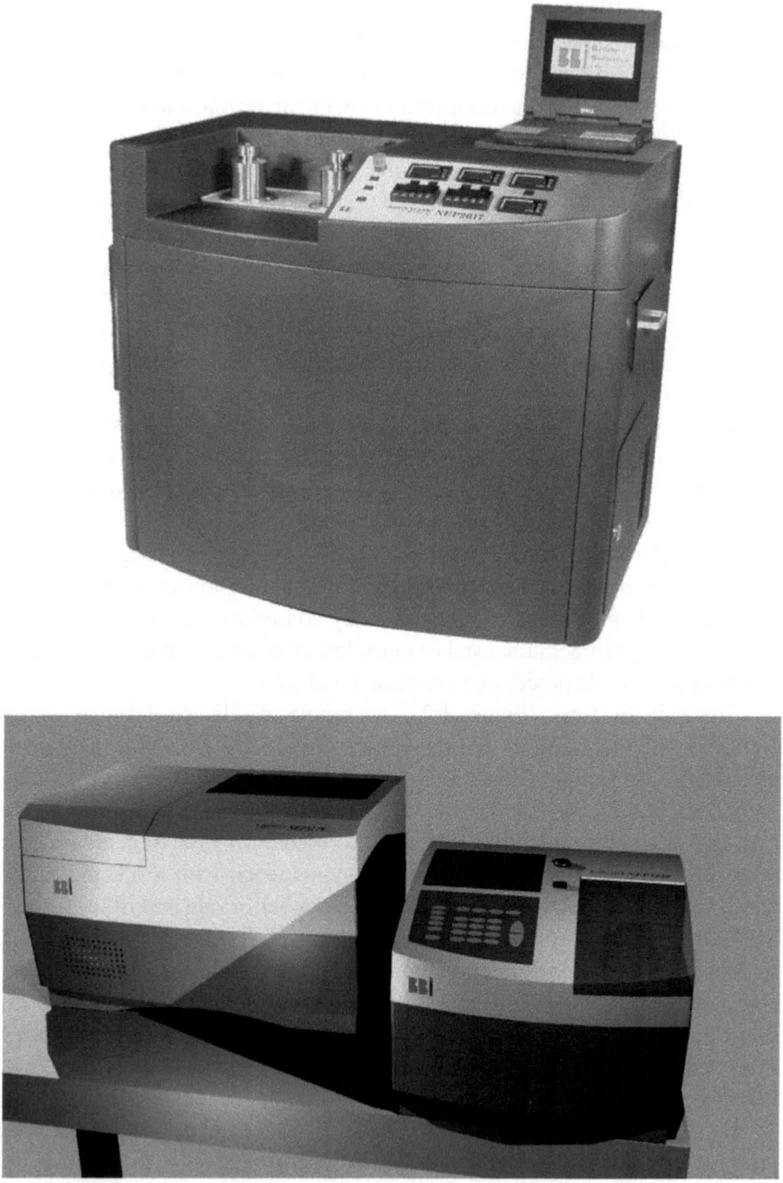

FIGURE 1.1 (A) Barocycler NEP2017; (B) Barocycler NEP3229.

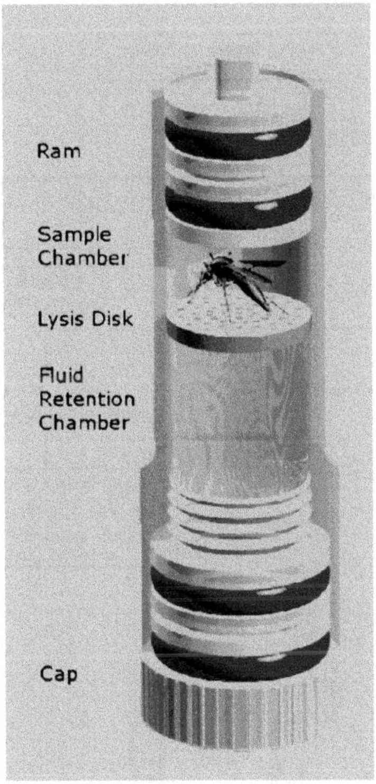

FIGURE 1.2 An illustration of a PULSE Tube.

loaded into a tube holder. The holder is then placed into the pressure chamber, which contains sample chamber fluid and has been equilibrated to a desired temperature. In operation (Figure 1.3B), pressure is applied in the chamber, and as the process moves toward equilibrium, the pressure difference between the fluid in the pressure chamber and the interior of the Tube forces the ram to move toward the lysis disk, ultimately compressing the sample against the disk, which in turn forces the material through the lysis disk holes and into the fluid retention chamber. During this process, the material is broken apart and mixes with the buffers that were placed in the PULSE Tube. When pressure is released, the ram rapidly retracts to near its starting point until pressure is again added. This action is repeated with additional pressure cycles. Cycle numbers and times may vary with the material being processed, but five 1-minute cycles at high pressure with 15-second ambient intervals are usually sufficient to release proteins from many cell suspensions and solid tissues. Chaotropic agents, buffers, or other chemicals such as detergents can be added to the sample processing chamber and used as lysis solutions in the PCT SPS.

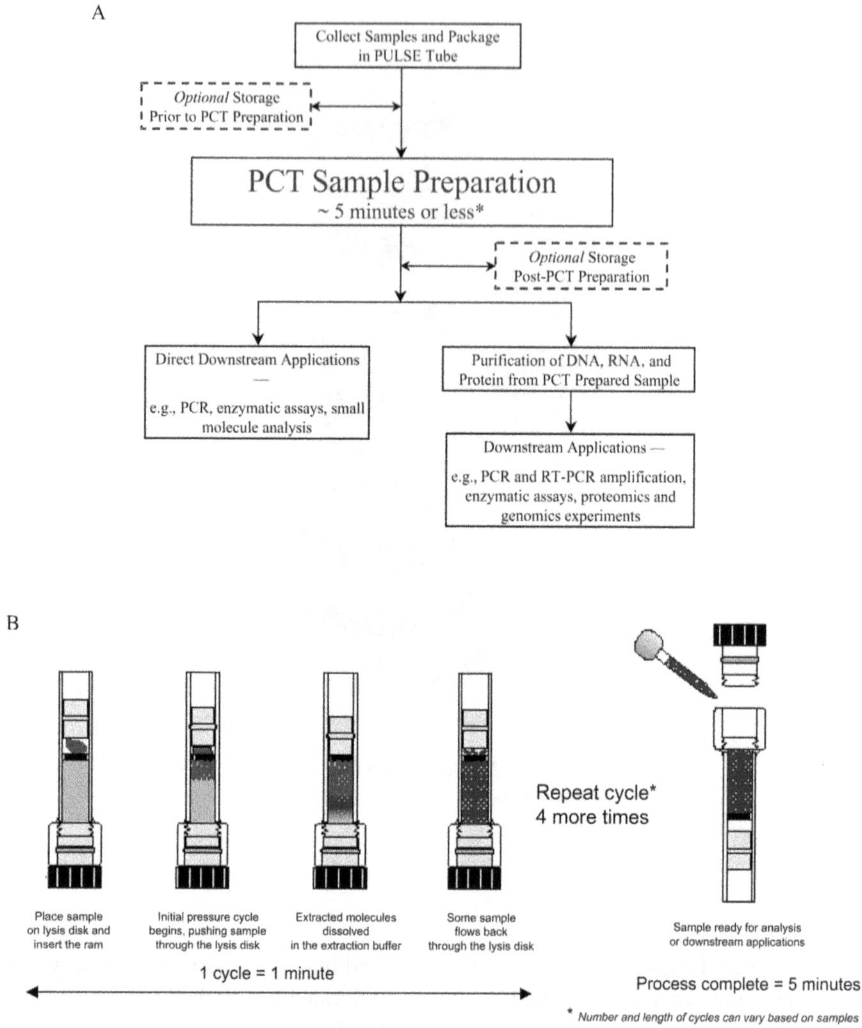

FIGURE 1.3 (A) The PCT SPS process; (B) one cycle of sample processing in a PULSE Tube.

1.3 POTENTIAL MECHANISMS OF ACTION OF PCT

The PCT SPS has proven to be effective in the release of biomolecules from a wide variety of samples (Table 1.1). To this end, the mechanisms of action of PCT are being extensively studied. One must consider that PCT extraction is a multifactorial process with multiple variables that may significantly contribute to the outcome of the process. Variables include pressure, temperature, buffers used, sample type, targeted complexes, and the design of the processing device. Five hypothetical mechanisms

TABLE 1.1 Samples Extracted for Total Protein by the PCT SPS

Sample Group	Examples	Applications
Animal tissues	Soft: rat liver, brain, kidney, lung	ELISA, SDS-PAGE, Western blot, 2D gel
	Hard: rat and chicken skeletal muscle	Western blot, SDS-PAGE
Plant tissues	Soft: corn sprout	SDS-PAGE, 2D gel
	Hard: apple seed	SDS-PAGE, 2D gel
Insects, small organisms	Mosquito	SDS-PAGE
Blood	Blood, blood spot	SDS-PAGE
Microbes	Yeast, *C. elegans*	SDS-PAGE, 2D gel
	Mycobacteria sp., bacteria/spores	SDS-PAGE, 2D gel

have been proposed for PCT extraction in PULSE Tubes: (*a*) rapid pressure changes that cause damage in cells and tissues in part by pressure shock waves or bubble collapse; (*b*) pressure-mediated segmentation of tissue from larger into smaller pieces with the lysis disk and ram movement playing important roles; (*c*) solvation changes that in turn cause the dissociation of macromolecular complexes and the diffusion of molecules in a high-pressure environment during pressure cycles; (*d*) possible phase changes during pressure changes (similar to a freeze-thaw process under atmospheric conditions), which result in protein release; and (*e*) solvent exchange across the cell membrane during pressure changes or solvent penetration into cellular compartments upon pressurization, which in turn induces sample extraction. We propose that the most important mechanism of action relates to the bubble collapse hypothesis, which is currently being investigated at PBI in collaboration with gas and fluid physicists. However, it is possible, and even likely, that two or more of these mechanism are working in concert.

1.4 PCT AND PROTEIN STABILITY

The behavior of biosystems under high hydrostatic pressure is governed by Le Chatelier's principle, which predicts that pressure favors processes accompanied by negative volume changes and that, conversely, pressure inhibits processes accompanied by positive volume changes. As with general chemical reactions, a pressure-mediated reaction is governed by the Gibbs free energy change (ΔG), which is a function of pressure and temperature, for example, $d(\Delta G) = (\Delta V)dp - (\Delta S)dT$. When temperature is constant, volume changes determine the signs of free energy change or equilibrium of reactions under pressure. In cases of proteins and other biopolymers, because the packing of the atoms is not always perfect, Kauzmann proposed that the partial molar volume (V_i) of a protein in solution is composed of three components: atoms, cavities, and hydration (Equation 1.1).[8] Pressure and temperature give rise to changes in cavities and hydration, but not to volume change of atoms. Thus, at constant temperature, the derivative of the free energy change with respect to pressure in either protein folding or macromolecule ligand binding is the volume change of the reaction (Equation 1.2). Or, if pressure is applied to a system in equilibrium, the system will shift toward the side that occupies the smallest volume.

$$V_i = V_{\text{atoms}} + V_{\text{cavities}} + V_{\text{hydration}} \tag{1.1}$$

$$d(\Delta G / dp)_T = (\Delta V)_T \tag{1.2}$$

Volume change (compressibility) studies on proteins show the hydration properties of these biopolymers.[9] It is difficult to predict the sign of ΔV with respect to a specific protein or protein complex under pressure. As with heating, a protein pressure-temperature phase diagram (typically ellipse shaped) predicts that protein denaturation occurs at elevated pressure. Unlike thermodenaturation, the kinetics of pressure denaturation is dependent on pressure range, rate of compression and decompression, and the time period of exposure.

For many applications in proteomics, native protein structures and their enzymatic activities must be preserved for analyses. In theory, the native and biologically active conformation of a protein may be preserved only when the sign of volume changes (ΔV_i) of molecules between low and high pressures is negative. Since pressure changes in the PCT SPS can be more rapid and uniform than temperature changes in similar reactions, it may be possible that proteins retain native conformation and activity. Under typical PCT SPS conditions, when the pressure is at 35 kpsi (235 MPa) or less, no damage or breakage of covalent bonds is expected. To obtain native or reversibly folded proteins, or to preserve large macromolecular complexes under PCT conditions, determining the effects of different solvents, pressure, and temperature is required to yield a negative sign for ΔV. For example, proteins are generally intact or thermally stable at temperatures near 20 to 25°C. However, under pressure, such a generally favorable temperature may be shifted to a different range. Some compounds, such as naturally occurring osmolytes, are known to destabilize denatured forms of proteins and to protect proteins from denaturation. These agents may be incorporated in the PCT extraction process.[10] Also, finding the balanced value of pressure and cycling conditions, which not only preserves protein structure and activity, but also favors extraction, is desirable.

1.5 PCT AND THE ROLE OF EXTRACTION BUFFERS

Chemical composition of the extraction solution used in the PCT SPS can be crucial in achieving sufficient extraction and preservation of protein conformation and activity. For example, in the extraction and purification of proteins carrying a charge, the ionic strength of the solutions plays an important role in the disruption of the sample and solubility of protein molecules. Studies in the ionization of various compounds indicate that pressure effects of buffers are related to the volume of the ionized compound, which is defined as the difference in partial molar volumes of the products and reactants of the buffer's ionization reaction.[11] For example, carboxylic acids have negative ionization volumes, whereas bases, such as amines, generally have positive ionization volumes. The volume change comes primarily from solvent electrostriction, that is, the tendency of a polar solvent to adopt a higher density configuration around an ion.[12] Elevating the hydrostatic pressure lowers the free energy of ion solubilization and perturbs the equilibrium toward the ionized form. Buffers with high charge density have larger absolute ionization volumes.[13] The presence of internal hydrogen bonding is also associated with large absolute ionization volumes. Based on these and other studies, compounds that pose large ionization volumes and low barrier hydrogen bonds may be evaluated for extraction. Exploiting the interplay between pressure and hydrophobicity may lead to additional significant protein extraction methods.

Hydrophobic interactions are usually assumed to be a force responsible for the low solubility of nonpolar substances in water. These interactions are the primary driving force of protein folding and the dominant physical and chemical contributor of membrane proteins anchored in lipid bilayers. Pressure has also been used to effectively disaggregate and refold proteins from aggregates and inclusion bodies, resulting in high yields of native protein.[14]

It is hypothesized that PCT may affect protein extraction in the following ways: (*a*) any transition that results in deprotonation to form charged residues is favored at high pressures; in other words, dissociation of chemical species into their respective ions contributes to an overall reduction in system volume due to electrostrictions; (*b*) the exposure of hydrophobic groups to water is also highly favored under pressure; and (*c*) pressure-induced weakening of the hydrophobic effect, or hydration of buried hydrophobic residues, substantially contributes to dissolution of protein hydrophobic core or aggregates at high pressures. Based on these rationales, extraction by pressure is perhaps superior to thermally based extraction processes, since highly hydrophobic proteins, for example, membrane proteins, can be more soluble rather than sticky under pressure conditions. In similar situations, salt bridges or ionic interactions, found in native proteins (although rare), can be exposed by high pressure to the solvent and the state of electrostriction altered in favor of denaturation.

1.6 EXAMPLES OF PROTEIN EXTRACTION USING THE PCT SPS

Proteins have been extracted from a variety of animals, plants, and microbes using the PCT SPS (See Table 1.1). The pressure cycling conditions were typically five 1-minute cycles at 35 kpsi (235 MPa) at 4°C or 20°C using the Barocycler NEP2017. Samples were extracted in PULSE Tubes containing a variety of lysis buffers. Positive controls using either mortar and pestle grinding or sonication were prepared in parallel with the PCT-treated samples. Negative controls were obtained by incubating samples in identical buffers to the pressure-treated samples, but without pressure treatment.

Data shown in this section illustrate some of the more significant research findings using the PCT SPS. Specifically, examples are shown that demonstrate that PCT SPS treatment can maintain protein antigenicity, protein structure, and protein activity using the analytical tests described. Data are also shown that demonstrate that the PCT SPS can efficiently release proteins from a wide range of samples and that the released proteins are suitable for 2D gel electrophoresis analysis. Also discussed are the intrinsic advantages of the PCT SPS that make this method superior to other extraction methods currently available.

The SDS-PAGE and Western blot data (Figure 1.4) show that although the sample was subjected to 35 kpsi (235 MPa) the specific protein nitric oxide synthase retained its immunoreactivity.

Several published examples show proteins that retain their bioactivity under high-pressure conditions.[15–18] For example, HIV reverse transcriptase activities are retained even after 30 cycles of pressure treatment (Figure 1.5). These data, in association with the published examples, indicate that proteins may maintain both structure and function following PCT, even at ultra-high-pressure levels.

In other experiments, the PCT SPS was proven to efficiently release proteins from a wide variety of samples. Figure 1.6 shows proteins released from chicken

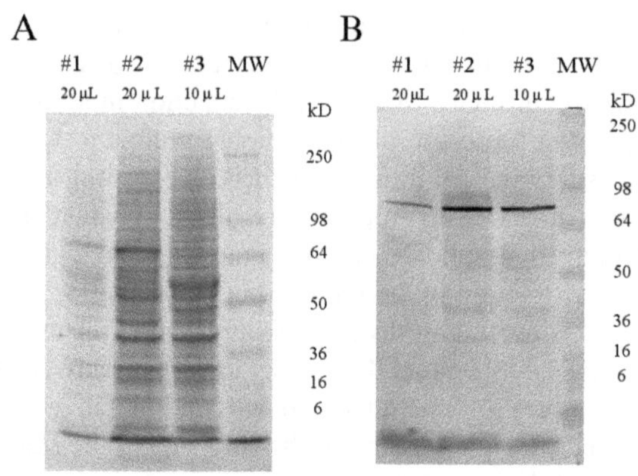

FIGURE 1.4 (A) Western blot of rat brain protein extracted by PCT, five cycles, 1 min/cycle, −25°C in phosphate-buffered saline (PBS) (lane 2) and by mortar and pestle grinding in liquid nitrogen followed by Omni homogenization (lane 3). A negative control sample (shown in Lane 1) was obtained under similar conditions (applying no cycles of pressure). Crude lysate sodium dodecyl sulfate–polyacrylamide gel electrophoresis (SDS-PAGE) is shown; (B) Western blot of protein extracts shown in part A. Primary antibody: universal monoclonal anti-nitric oxide synthase (mouse); secondary antibody: anti-Mouse IgM (μ-Chain specific) alkaline phosphatase conjugate.

muscle. Similar yields of proteins were obtained using both the PCT SPS and mortar and pestle grinding methods. Although the majority of the protein species in PCT extracts are similar to those obtained by the positive control methods, there are some distinguishable differences in the proteins with both high and low molecular weights. The significance and potential implications of these findings are discussed at the end of this chapter.

In several cases, proteins released by the PCT SPS were further evaluated by one-dimensional (1D) and 2D gel electrophoretic analyses. A comparison of proteins released from the nematode *Caenorhabditis elegans* by the PCT SPS versus sonication indicated that the PCT SPS yielded 37% more protein. 2D electrophoresis gels showed that more high molecular weight proteins (possibly due to decreased proteolysis), as well as more high pH proteins, were released by PCT than by sonication. Further, analysis of gel spots from the PCT-treated nematodes showed membrane proteins including ATP synthase subunits and channel proteins. Proteins released from *C. elegans* by PCT were analyzed by a hybrid ion trap/TOF MS instrument (MALDI-QIT-TOF). This method successfully discerned structural and glycol-linkage motifs.[19]

In independent studies by outside laboratories of Proteome Systems (Woburn, Mass.) using PBI's PCT SPS, the efficiency of the system to extract proteins from

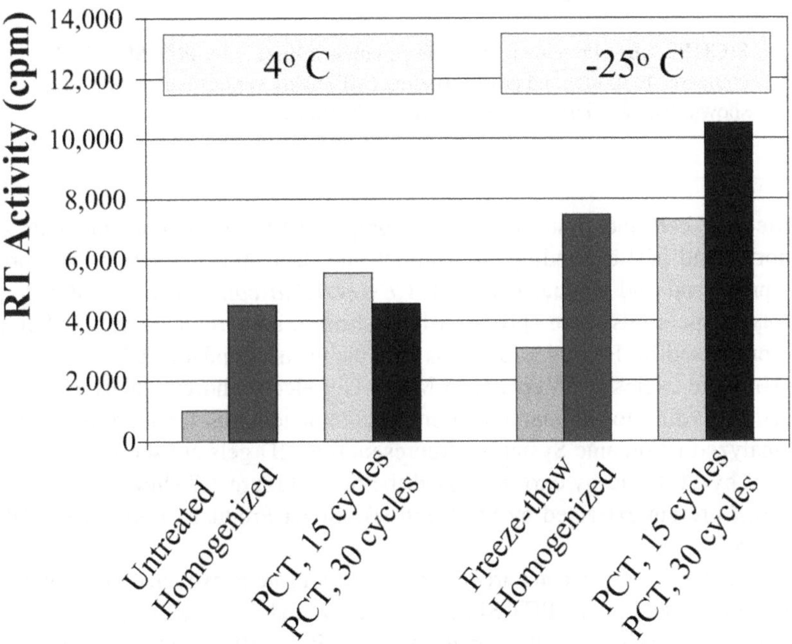

FIGURE 1.5 Reverse transcriptase (RT) activity from HIV-1 infected H9 cells. Cells were treated by PCT or dounced at 4°C or −25°C. HIV-1 reverse transcriptase activity was measured by a standard RT assay.

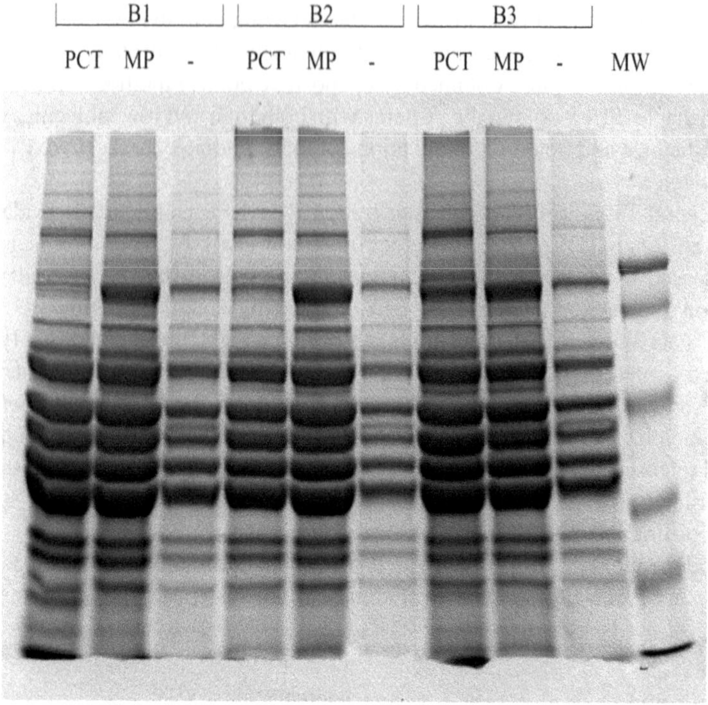

FIGURE 1.6 Proteins released from chicken muscle by PCT SPS (PCT) compared to mortar and pestle grinding (MP) using various buffers (B1–B3) shown in a 12% Tris-glycine reducing SDS-PAGE.

a variety of cell and tissue types was compared to conventional methods such as mortar and pestle grinding, sonication, and bead mill extraction. The organisms processed and evaluated included *Escherichia coli,* rat liver and brain, *C. elegans,* apple seeds, corn sprouts, whole shrimp, and whole zebrafish. Samples were processed by Proteome Systems Inc. using the standard PCT conditions of five 1-minute cycles at 35 kpsi (235 MPa). Gel electrophoretic analysis was performed according to the manufacturer's recommendations for the type of organism analyzed (Proteome Systems). Representative 2D gels are shown for proteins released by PCT and by currently used methods. Figure 1.7 shows a comparison of *E. coli* protein extracted by bead mill (BioSpect Products, Bartlesville, Okla.) and by PCT.

Figure 1.8 shows a comparison of rat liver protein extracted by mortar and pestle grinding and by the PCT SPS, followed by analysis on a 2D gel. Scientists at Proteome Systems reported that, in all cases, the PCT SPS was at least equivalent to the comparison methods in terms of yield and quality. In some cases, such as with *E. coli,* PCT may have yielded more proteins than the bead mill or sonication methods. However, for some researchers, the features of the PCT SPS, particularly its

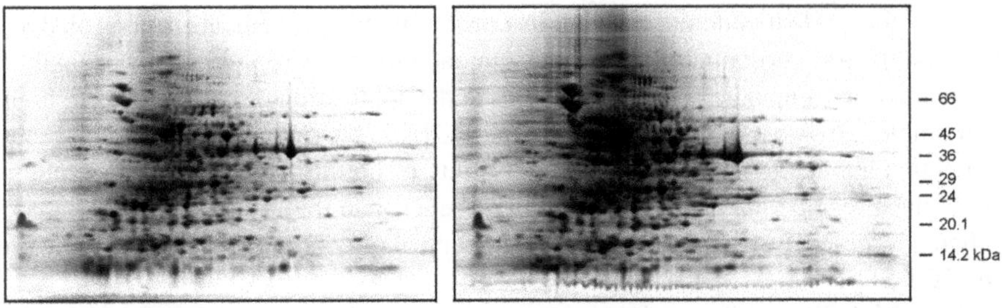

FIGURE 1.7 2D gel analysis comparing bead-mill (1800 oscillations/min, three 1-minute cycles) (left) and the PCT SPS (ten 20-second cycles) (right) for the extraction of proteins from reconstituted lyophilized *E. coli* cells. IPGs were pH 3–10 (Courtesy of Proteome Systems).

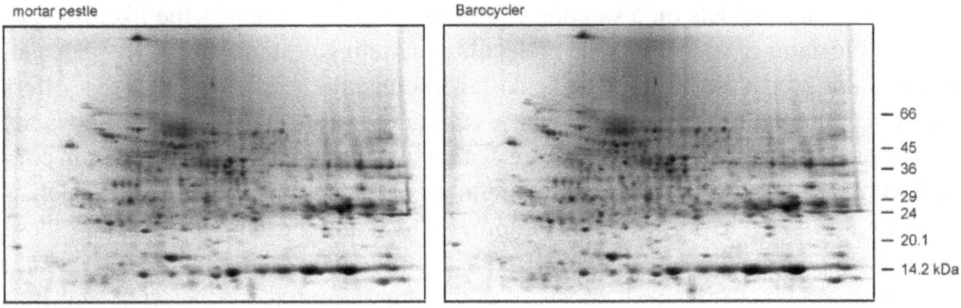

FIGURE 1.8 2D gel analysis comparing mortar and pestle grinding (left) and the PCT SPS (right) for the extraction of proteins from rat liver (Courtesy of Proteome Systems).

ease-of-use and the safety factor of full containment in the PULSE Tube (particularly for bacteria), make the PCT SPS a more desirable method for the extraction of proteins for 2D electrophoresis gel analysis. The features of the PCT SPS are discussed further in the following section.

1.7 CONCLUSION

Physical and chemical properties of proteins are determined by their primary sequences, secondary and tertiary structures, posttranslational modifications, and bound ligands. Differences among proteins may be subtle in their sequence or structure, but their physical and chemical properties, as well as their function, can be significantly different due to variations in their posttranslational modifications. Protein extraction and purification that best retains protein functionality and structure *in vivo* can be highly challenging, especially when a large protein population is assessed.

In an era of high-volume, high-throughput biochemistry and biology, sample preparation remains problematic, especially for hard-to-lyse materials. The shortcomings of traditional methods are well known; for example, many such methods have not been entirely amenable to automation, with its concomitant improvements in control, documentation, and reproducibility. A new approach to the disruption of tissues and to the extraction of nucleic acids and proteins is the application of rapidly cycled, high and ambient hydrostatic pressure using the PCT SPS. Using the PCT SPS, protein extraction applications, including a wide variety of liquids, suspensions, and solids, have given results that are comparable to or better than traditional methods. Safety, efficiency, ease-of-use, reproducibility, and control of conditions are improved with the PCT SPS.

The PCT SPS exhibits unique properties compared to other high-pressure applications. First, and most significantly, this system uses cyclic pressure instead of static pressure. Pressure increase from ambient and high levels, such as 35 kpsi (235 MPa), can be generated in seconds, and decompression occurs in a fraction of a second. Incubation periods, which can be as short as a few seconds, at each pressure level also may be varied. Manipulations of these parameters allow for much greater control of pressure impacts on a sample. Data show that PCT reduces the likelihood of irreversible damage to many labile molecules, including proteins and RNA.[6] Second, samples are processed in a specially designed container (PULSE Tube) that offers additional, unique mechanisms to release cellular contents. Third, PCT employs not only mechanical forces, that is, pressure at a set temperature, but it also works in concert with various solvents or buffers. A number of parameters can be varied, such as pressure levels, cycle profiles, temperature, buffers, and the sample container; these options are generally not available with other sample preparation methods.

Using PCT, several mechanisms of action may account for the release of biomolecules, including proteins and small molecules from cells and tissues. It is probable that microbubble collapse has the most profound effect in the process. It is also likely that two or more mechanisms are acting in concert. Further studies are being conducted to increase our understanding of these mechanisms, which in turn should lead to a better understanding of the extraction process using PCT. Ultimately, this increased understanding of the mechanisms of action of PCT could lead to the release of better quality and quantity of biomolecules, and perhaps to the release of biomolecules that better reflect the true nature of these biomolecules *in vivo*.

The PCT SPS exhibits unique and advantageous features compared to other currently available methods. For example, PCT does not expose proteins to the physical shearing that is inherent in bead beating or by processing in a French press. In contrast, the PCT SPS applies cycles of pressure to open the cells in order release their contents. As pressure is released, the proteins are more likely to resume their native configuration. Experiments by PBI scientists (data not shown) indicate that the repetitive application of very short pulses of high pressure is more effective in releasing cellular contents than one continuous application of pressure. For example, 1, 3, and 5 cycles of pressure resulted in 20, 40, and 75% expected yield of RNA extracted from rat liver (F. Tao unpublished results, 2005). More significantly, the biological activity of enzymes and other proteins released by PCT retain greater functionality than proteins released by continuous pressure, other physical means, harsh chemicals, and other currently available sample preparation methods. One good example is the HIV-1 protease data shown in Figure 1.5. Data from experiments conducted by PBI scientists and independent laboratories show that the PCT SPS releases at least as many species of proteins as conventional extraction methods do. On the other hand, distinct protein species were found in the PCT extracts, in particular, in the high molecular weight species. This phenomenon is potentially very important and exciting, since proteins and protein complexes released by PCT may not be released by other available methods, thus offering the potential for new discoveries that may not be possible without the PCT SPS. And because the PCT SPS is instrument based and employs unique mechanisms to release proteins, it has excellent potential for development as an extraction system that complements or replaces traditional extraction methods for proteomic studies.

REFERENCES

1. Pandey, A. and Mann, M., Proteomics to study genes and genomes, *Nature*, 45, 837–846, 2000.
2. Cho, A. and Normile, D., Nobel prize in chemistry: Mastering macromolecules, *Science*, 298, 527–528, 2002.
3. Liebler, D.C., *Introduction to Proteomics: Tools for the New Biology*, Human Press, Totowa, N.J., 2002, pp. 8–10.
4. Schumacher, R.T., Manak, M., Garrett, P., Miller, W., Lawrence, N., and Tao, F., An automated sample preparation solution for nucleic acid and protein extraction from cells and tissues, *Am. Lab.*, 34, 38–43, 2002.
5. Johnson, B., Breaking up isn't hard to do: A cacophony of sonicators, cell bombs and grinders, *The Scientist*, 12, 23, 1998.
6. Salusbry, T., Disruption, in *Protein Purification Methods: A Practical Approach*, Harris, E.L.V. and Angal, S., Eds., IRL Press, New York, 1994, pp. 87–97.
7. Garrett, P.E., Tao, F., Lawrence, N.P., Ji, J., Schumacher, R.T., and Manak, M., Tired of the same old grind in the new genomics and proteomics era? *TARGETS*, 1, 156–162, 2002.
8. Kauzmann, W., Some factors in the interpretation of protein denaturation, *Adv. Protein Chem.*, 14, 1–63, 1959.
9. Taulier, N. and Chalikian, T.V., Compressibility of protein transitions, *Biochim. Biophys. Acta*, 1595, 48–70, 2002.
10. Bolen, D.W. and Baskakov, I.V., The osmophobic effect: Natural selection of a thermodynamic force in protein folding, *J. Mol. Biol.*, 310, 955–963, 2001.
11. Hess, R.A. and Reinhard, L.A., Unusual properties of highly charged buffers: Large ionization volumes and low barrier hydrogen bonds, *JACS*, 121, 9867–9870, 1999.
12. Nernst, W. and Drude, P.Z., Uber Elektrostriktion durtch freie lonen, *Phys. Chem.*, 15, 79–85, 1894.
13. Kitamnura, Y. and Itoh, T., Reaction volume of protonic ionization for buffering agents: prediction of pressure dependence of pH and pOH, *J. Solution Chem.*, 15, 715–725, 1987.
14. Estevez, L., Degraeve, P., Espeillac, S., and Lemay, P., High-pressure induced recovery of L-galactosidases from immunoadsorbents: Stability of antigens and antibodies. Comparison with usual elution procedures, *Biotechnol. Lett.*, 22, 1319–1329, 2000.
15. Cheung, C.Y., Green, D.J., Litt, G.J., and Laugharn, J.A., Jr., High-pressure-mediated dissociation of immune complexes demonstrated in model systems, *Clin. Chem.*, 44, 299–303, 1998.

16. Green, D.J., Litt, G.J., and Laugharn, J.A., Jr., Use of high pressure to accelerate antibody: Antigen binding kinetics demonstrated in an HIV-1 p24:anti-HIV-1 p24 assay, *Clin. Chem.*, 341–342, 1998.
17. Rudd, E.A., Reversible inhibition of lambda exonuclease with high pressure, *Biochem. Biophys. Res. Commun.*, 230, 140–142, 1997.
18. Remaley, A., Hess, R., Fischer, S., Sampson, M., and Manak, M., Pre-analytical sterilization of serum by pressure cycling treatment: A novel procedure to prevent laboratory acquired infections, *Clin. Chem.*, 46 (Suppl. 6) A39, 2000.
19. Geiser, H.A., Hanneman, A., Rosa, J.C., and Reinhold, V.N., HTP Proteome-Glycome Analysis in *Caenorhabditis elegans*, poster presented at the Annual Conference of the Society for Glycobioslogy, Boston, November 9–12, 2002.

2 Applications of Ion-Exchange (IEX) Chromatography to Reduce Sample Complexity Prior to Two-Dimensional Gel Electrophoresis (2DE)

Malcolm G. Pluskal, Eva Golenko, and Mary F. Lopez

CONTENTS

2.1 INTRODUCTION

After completion of the human genome, proteomics has rapidly gained momentum as the next major undertaking in bioscience.[1] The complexity of the proteome of an organism presents a significant analytical challenge[2] and has stimulated interest in approaches to reduce sample complexity prior to high-resolution analysis. The challenge in these complex samples is twofold: (*a*) depletion of high-abundance proteins, such as albumin or immunoglobulin (IgG) in human plasma,[3] and (*b*) increasing the concentration of low-abundance proteins above the threshold of sensitivity of the analytical method[4] employed in the project. Current methods of analysis depend heavily on two-dimensional gel electrophoresis (2DE), and multidimensional high-performance liquid chromatography (HPLC) linked directly to mass spectrometry is also gaining some interest.[5,6] Sample complexity reduction has been attempted with the following technologies: (*a*) biospecific affinity ligands, such as mimetic blue dyes and immunochemical reagents,[3] (*b*) detergent extraction leading to fractionation based on solubility,[7] (*c*) ion-exchange resolution under nondenaturing conditions,[4] (*d*) size exclusion by ultrafiltration[8] or gel permeation chromatography, (*e*) preparative electrophoresis by SDS-PAGE or isoelectric focusing (IEF),[9] and (*f*) multidimensional IEX linked to reverse-phase chromatography.[5,6]

In this chapter, we emphasize the application of IEX chromatography in small volume (<1 ml) centrifugal format devices as a prefractionation step prior to 2DE analysis.

2.2 IEX CHROMATOGRAPHY BASICS

2.2.1 Theory

IEX is the reversible adsorption of charged molecules to an immobilized group on a suitable solid phase that bears the opposite charge.[9] In practice, the process employs the following steps: (*a*) equilibration of the IEX surface with a suitable counterion (anion or cation) and pH so the molecule of interest will be strongly adsorbed onto the surface; (*b*) binding of the sample and displacement of the bound counterion (molecules with the same charge as the surface will be unretained by the solid phase); (*c*) elution of the retained material by increasing the counterion concentration or changing the pH of the system to reduce the strength of the adsorption interaction with solid phase (during this step molecules are resolved on the basis of their charge state at the pH of the elution buffer); (*d*) recycling the IEX surface back to the equilibration buffer conditions in step 1 for the next sample.

IEX is a powerful technique used to resolve molecules on the basis of their charge state that can be influenced by (*a*) the molecules' primary sequence, which can be used to estimate a theoretic pI, or isoelectric point, for the structure;[4] (*b*) their native tertiary state conformation in solution; (*c*) their association with other proteins to form quaternary complexes; and (*d*) their association with other macromolecules, such as nucleic-acid and carbohydrate structures, with charged glycan residues, such as sialic acid content. The resolving power of a charge-based separation, such as IEF in 2DE or nondenaturing IEX chromatography, is an invaluable tool when applied to sample complexity reduction in proteomics.

2.2.2 IEX Chromatography Media Available

IEX chromatography technology is based on two distinct classes of functional groups on a range of solid supports: (*a*) anion exchangers such as diethylaminoethyl (DEAE), quaternary aminoethyl (QAE), quaternary ammonium (Q); and (*b*) cation exchangers such as carboxymethyl (CM), sulfopropyl (SP), and methyl sulfonate (S). Within these two categories are further groupings of weak (exchangeable proton having a distinct pKa in the normal pH range for proteins) and strong (fixed formal charge or very extreme pKa out of the normal pH range) exchanger surfaces. The above range of chemistries offers a wide range of IEX options for sample preparation of proteins and nucleic acids within the pH range of 3 to 10. Beyond this range, proteins and nucleic acids are not stable and will undergo hydrolysis of amino acid side chains and nucleic-acid bases leading to fragmentation of the primary structure.

The above chemistries have been immobilized on a range of solid supports. Bead- or particle-based media based on silica, controlled pore glass, regenerated cellulose, agarose, dextran, cross-linked acrylate, and, recently, zirconium, europium, and alumina metallic supports are one type. In most cases these supports are porous, but in the case of DNA, nonporous beads offer some unique properties for resolution of these large molecules. These supports are available in a range of pore sizes and configurations offering flexible IEX capacities and flow rates. (For a review of commercial supports see[10].) Membrane-based media are made from regenerated cellulose or polymeric microporous materials and offer high flow, efficient mass transfer onto the chromatography surface, and comparable capacities to the aforementioned beaded media.[11,12] Chromatography beds based on continuous monolithic polymerization with a macroporous inner structure[13] offer a singular path through the pores of this media, enabling rapid mass transfer at high flow rates. The so-called cast in-place expanded bed—beaded media with hybrid properties of the IEX capacity of particles and the high flow rates of a microporous membrane structure—has found recent application with reverse-phase media in a microvolume sample preparation format.[14]

2.2.3 Illustration of IEX Chromatography Separation Correlated with 2DE

Charge-based separation forms the basis for protein separation by IEX and IEF with carrier ampholyte media or immobilized pH gradient (IPG) technologies. In a recent comparison[4] between these two modes of charge-based separation there appeared to be good correlation between retention by an anion IEX surface and the proteins position in the IEF pH gradient. An example of such as correlation is shown in Figure 2.1 for a mammalian cell lysate resolved on regenerated cellulose based DEAE anion exchange surface.

The incremental salt elution fractions, when analyzed on pH 3–10 carrier ampholyte IEF gels, showed progressively more acidic fractions in the resulting 2DE pattern. This clearly illustrates that for most proteins, their charge-based separation on an anion exchange surface reflects their pI in a urea-based IEF separation. However, as previously reported,[4] some components of the proteome do not correlate well and presumably reflect the subunits of protein complexes resolved in the nondenaturing IEX process reflecting their aggregate pI. In the presence of urea, these complexes would be denatured and the individual components would then separate according to their individual pI states. These data clearly support the view that protein prefractionation

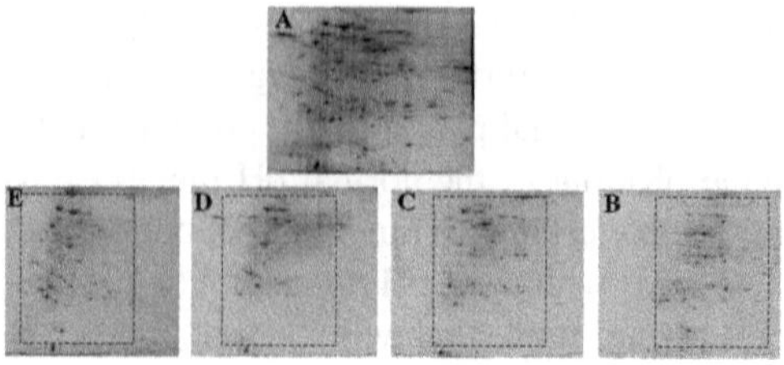

FIGURE 2.1 2DE analysis pattern of whole cell lysate proteins retained by regenerated cellulose membrane modified with DEAE chemistry. Subsets of proteins (0.1 to 0.2 mg in 20 mm NH_4 CH^3COO^- buffer pH 8.5) derived from WT2 rat fibroblast cells grown in culture were loaded onto IEX membranes in a centrifugal device format (Ultrafree™ MC, Millipore Corp., Billerica, Mass.) and spun at 500 xg for 2 minutes. The flow through and wash fraction (0.2 ml above buffer) was not analyzed by 2DE in this experiment and contained basic pI proteins. (A) A sample of the unfractionated lysate is shown for comparison resolved on a pH 3–10 IEF gel as described.[15] Retained material was then sequentially eluted with increasing concentrations of 20 mM NH_4 CH^3COO^- buffer pH 8.5, as follows; (B) 100 mM; (C) 200 mM; (D) 300 mM; (E) 400 mM. The data is displayed in a backward orientation to depict the progression of proteins from acid (left) to basic (right) regions of the 2DE gel pattern. A box surrounds the location of where the bulk of the proteins in the sample are found in the 2DE pattern.

by IEX chromatography can be used to reduce sample complexity by a process that correlates directly with the first dimension of 2DE. This allows isolation of enriched fractions that can then be loaded on appropriate narrow-pH-range IEF gels to improve resolution and the presence of lower abundance proteins in 2DE analysis.

2.3 APPLICATIONS OF IEX CHROMATOGRAPHY PRIOR TO 2DE ANALYSIS

2.3.1 Fractionation of Human Plasma

Human plasma, or serum, is a complex sample that is receiving considerable attention in proteomic research. Human plasma is a major challenge for sample complexity reduction processes because the top six abundant proteins (albumin [54.3%], IgG [16.6%,] α-1-antitrypsin [3.8%], IgA [3.4%], transferrin [3.3%], and haptoglobulin [2.9%]) account for 85% of the total protein mass of plasma[3] and because the total dynamic range of protein concentration spans ten orders of magnitude[16,17] (most analytical methods only have a dynamic range of four orders). Clearly plasma prefractionation is necessary to access the full dynamic range of protein abundance in plasma.

In some preliminary experiments, IEX prefractionation of plasma was investigated with some high-capacity polymeric-based membrane ion-exchange surfaces (Mustang™ Q and S membranes in a Nanosep™ device format from Pall Life Sciences, East Hill, N.Y.). Plasma was diluted into 20 mM Tris HCl buffer at pH 8–8.5 (Mustang Q) and 20 mM Na CH^3COO^- pH 4.0 or 20 mM MES buffer pH 5.5 (Mustang S) and loaded (up to 0.075 ml of plasma diluted to 0.2 ml final volume) onto the Nanosep devices (bed volume 0.008–0.016 ml with a binding capacity of

0.4 to 0.9 mg protein). Flow through and small increments of NaCl in the above buffers were then passed through the device by centrifugation at 500 xg for 2 minutes and the filtrate was collected. After acetone precipitation (see Figure 2.2) the samples were then resolved on precast one-dimensional (1D) sodium dodecyl sulfate–polyacrylamide gel electrophoresis (SDS-PAGE) gels and stained with colloidal Coomassie Blue™. A modified SDS sample buffer was used (no reducing agent) to improve the SDS-PAGE resolution between albumin and the IgG molecule. In the absence of a reducing agent, IgG remains as an intact 150 kDa molecular weight complex and can be easily visualized above the strong albumin band. Reducing the IgG complex yields 55 and 15 kDa bands, the former being hard to resolve from the overloaded albumin. The resulting gel images are reproduced in Figure 2.2.

In this study, the resolution between albumin and IgG was optimized on both anion and cation IEX surfaces. At pH 8.0 (Figure 2.2A), IgG was not retained by the Mustang Q surface but was retained and eluted with 50 mM NaCl in the loading buffer at pH 8.5 (Figure 2.2B). Albumin retained at pH 8.0 and eluted between 100 and 150 mM NaCl. Closer examination of other lower abundance proteins bands shows some degree of fractionation between the two loading pH levels. The cation-exchange surface at pH 4.0 showed most of the protein content of the plasma was not well retained (Figure 2.2C). In contrast, at pH 5.5, albumin was not retained, and IgG eluted over a 25 to 100 mM NaCl range, reflecting the polyclonal nature of IgG in plasma (Figure 2.2D). In combination, these two IEX chemistries evaluated in a simple centrifugal format offer the potential to achieve prefractionation of plasma without first depleting abundant proteins such as albumin and IgG. However with albumin and IgG depleted plasma as a starting point, IEX fractionation has the potential to improve resolution and detection of lower abundance proteins.

2.3.2 Impact on IEX Prefractionation on Resolution in 2DE

Prefractionation of complex protein mixtures with subsequent separation on 2DE gels is the preferred approach for many proteomics applications.[18,19] The objective of many differential display studies in proteomics is the comparison between the levels of proteins in experimental and control samples and the determination of quantitative and qualitative differences in individual protein expression. In many cases this is carried out by 2DE resolution linked to mass spectrometry analysis. Reducing the complexity of whole cell lysates improves the detection of low-abundance, extremely acidic or basic, or very low molecular weight proteins. In addition, fractionation of areas where many proteins are crowded in the 2DE pattern can reduce the number of proteins that comigrate in these gels. This can prevent lower abundance proteins from being obscured by higher abundance species.[20,21]

An example of the application of membrane-based IEX fractionation (VIVASCIENCE Inc., Edgewood, N.J.) of a Jurkat T lymphocyte total cell lysate on Q-A and S-A centrifugal spin devices is summarized in Figure 2.3. The cell lysate was resolved into acidic (Figure 2.3B) and basic (Figure 2.3C) fractions following the manufacturer's directions. Fractionation of this cell lysate resulted in a significant reduction

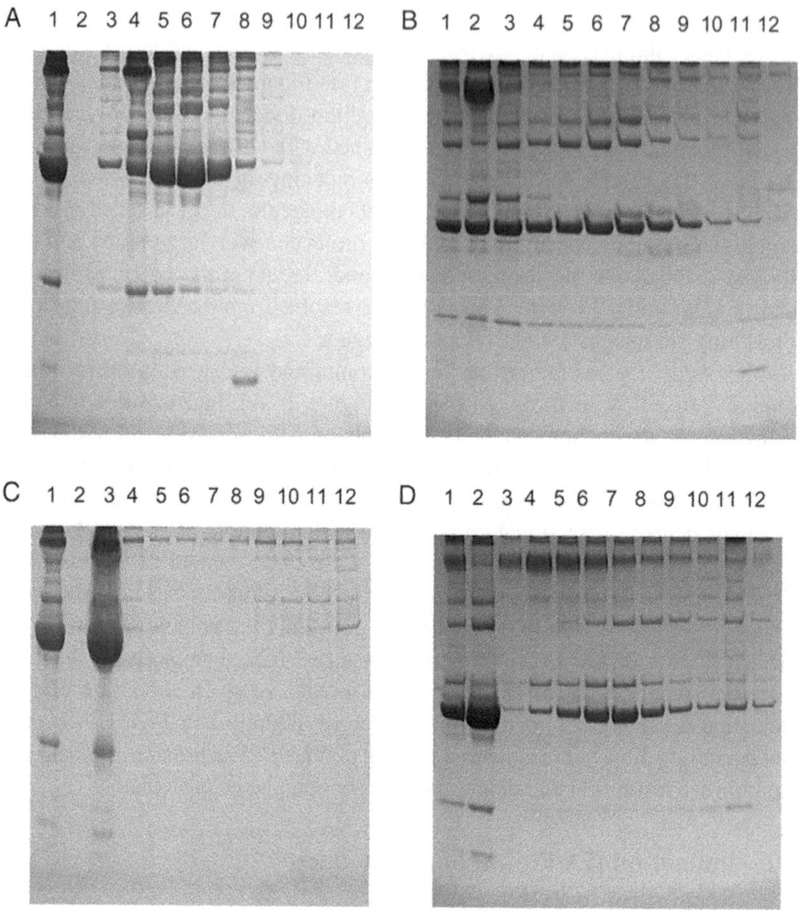

FIGURE 2.2 Methods development for fractionation of human plasma on membrane based IEX surfaces. Human plasma (0.075 ml) was diluted with buffer to a final volume of 0.2 ml and loaded onto the Mustang Q or S membrane based IEX Nanosep devices (Pall Corp.). Centrifugation at 500*g* for 2 minutes was used to process liquid through these membranes. After one recycle, the flow through was retained (FT). The device was then washed with 0.2 ml of loading buffer followed by sequentially increasing concentrations of NaCl up to 1.0 M in 0.2 ml volumes to elute the retained proteins (EL). The resulting fractions were all precipitated with four volumes of cold (−20°C) acetone, followed by 1D SDS-PAGE analysis and colloidal Coomassie blue staining. (A) Loaded at pH 8.0 on Mustang Q. Lane 1, unfractionated lysate; lane 2, FT + wash; lane 3, EL + 25 mM; lane 4, EL + 50 mM; lane 5, EL + 75 mM; lane 6, EL + 100 mM; Lane 7, EL + 125 mm; Lane 8, EL +150 mM; Lane 9, EL + 175 mm; Lane 10, EL + 200 mM; Lane 11, EL + 300 mM; Lane 12, 1000 mM; (B) loaded at pH 8.5 on Mustang Q. Lane 1, unfractionated lysate; lane 2, blank; lane 3, FT + wash; lane 4, EL + 50 mM; lane 5, EL + 100 mM; lane 6, EL + 150 mM; lane 7, EL + 200 mM; lane 8, EL + 300 mM; lane 9, EL + 400 mM; lane 10, EL + 500 mM; lane 11, EL + 750 mM; lane 12, EL + 1000 mM; (C) loaded at pH 4.0 on Mustang S following the elution pattern of B; (D) loaded at pH 5.5 on Mustang S following the elution pattern of A.

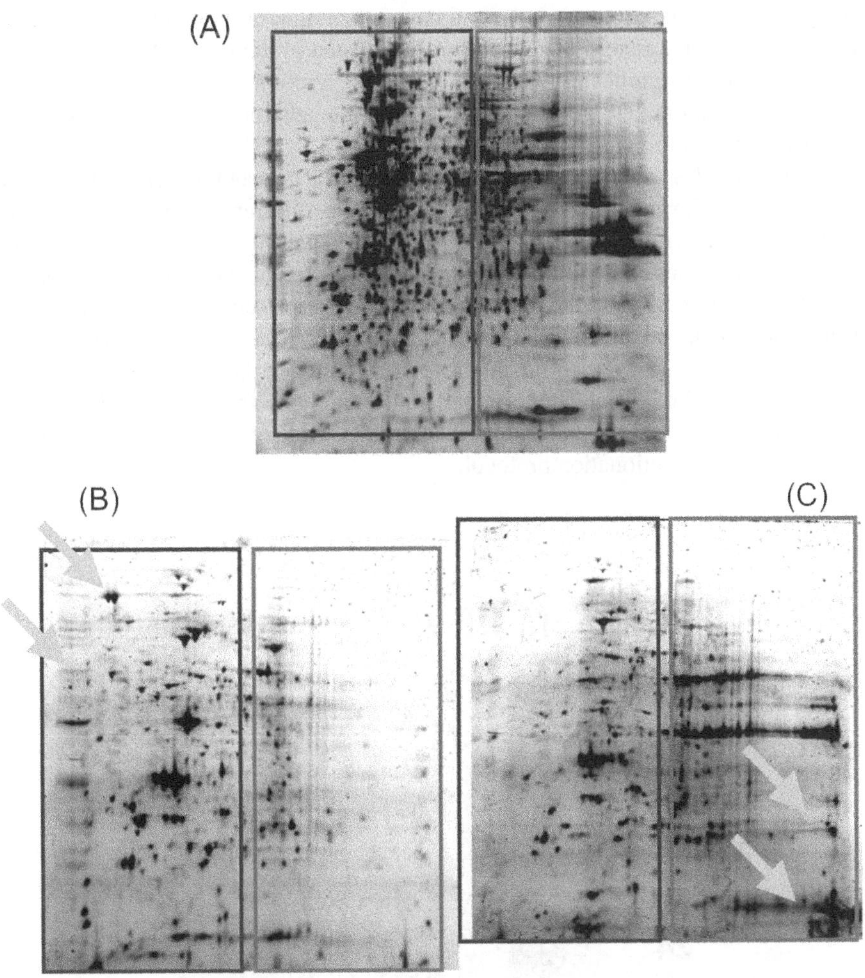

FIGURE 2.3 2DE analysis of whole cell lysate fractionated on membrane based IEX surfaces. Jurkat T lymphocyte cells were grown in Roswell Park Memorial Institute (RPMI) media formulation 1640 10% FBS, 4 mM L-glutamine to log phase and harvested. Cells were pelleted by centrifugation and washed three times with RPMI containing no bovine serum. The cells were homogenized in a tissue homogenizer in 2 ml nondenaturing extraction buffer containing 12.5 mM Tris/acetate, pH 8.2, 4% octyl-beta-glucopyranoside, mammalian protease inhibitor cocktail (Sigma). 20 ml RNAse A 1 mg/ml and 20 ml DNAse I 62u/ml (Invitrogen) were added per 1 ml extraction buffer, and the homogenate was centrifuged at 12,000 rpm for 10 minutes. The supernatant was aliquoted and stored frozen at –80°C. 2D gels were run essentially as described.[22] All reagents and precast gels were from Genomic Solutions except IPG strips, which were from Amersham. Before rehydration of IPG strips, an aliquot of Jurkat T protein was TCA or acetone precipitated and resuspended in Thiourea buffer. 2D gels were stained with SYPRO© Ruby or silver stain. The resulting 2DE gels are shown: (A) unfractionated; (B) acidic fractions; and (C) basic fractions. The blue-colored and red-colored boxes represent the acidic and basic ranges of protein separation on 2D gels, respectively.

of complexity of the protein pattern and better resolution of individual protein spots as compared to the protein pattern for unfractionated proteins. In addition, some of the proteins were significantly enriched during the fractionation, resulting in clearly defined spots on the 2DE gels not detectable on the gels for unfractionated samples (see green arrows in Figure 2.3B and Figure 2.3C).

These samples were analyzed with ID SDS-PAGE gels and by densitometry, and the results are shown in Figure 2.4. To quantify the differences among unfractionated (UF), acidic (A), and basic (B), the following analysis was performed, and the results are summarized in Table 2.1. The molecular weight standards run on the gel were first used to estimate the mass values of the test proteins. The net intensity of each band was then calculated as the total integrated intensity for the band minus background divided by a total net intensity of all bands in the corresponding lane to estimate the level of enrichment of a particular protein by fractionation. The results of these experiments show that a number of proteins were enriched by the IEX-based chromatography fractionation protocol.

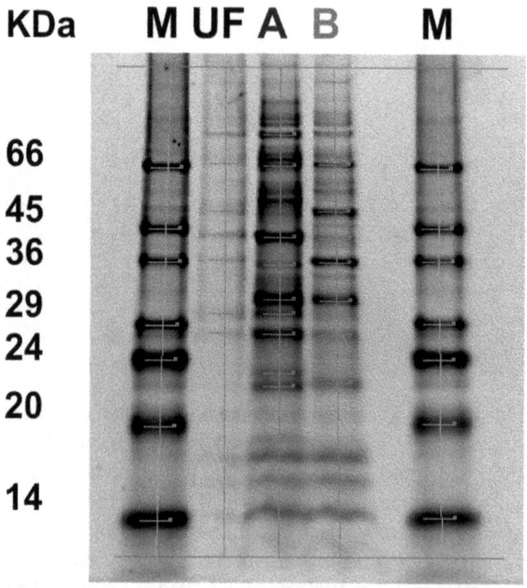

FIGURE 2.4 1D- SDS-PAGE analysis of cell lysate fractionated on membrane-based IEX surfaces. Unfractionated (UF), acidic (A), and basic (B) Jurkat T protein fractions represented on the 2DE gel images in Figure 2.3 were also run on a 1D SDS-PAGE gel. The lanes labeled M are the molecular weight standards. The level of enrichment of some of the proteins was quantified using Kodak Digital Science™ 1D Image Analysis Software.

TABLE 2.1 Gel Densitometry Quantification of IEX-Fractionated Jurkat T Whole Cell Lysate

M (kDa)	UF (%)	A (%)	B (%)
96.0	—	—	0.1
77.5	0.1	0.5	0.3
67.1	0.0	0.5	0.7
51.0	0.1	—	1.2
46.6	—	—	0.2
42.7	0.3	1.7	0.4
36.0	0.1	0.2	1.6
31.8	—	1.9	1.6
30.2	0.3	0.5	—
27.6	0.2	1.2	0.5
23.2	—	0.3	—
22.5	—	0.6	0.6

2.4 SUMMARY

Sample complexity reduction is an important first step in the sample preparation process for proteomics. IEX chromatography offers a flexible technology platform for resolving proteins on the basis of their molecular charge state. The process can be readily correlated with the first dimension, IEF of high-resolution 2DE analysis where proteins are resolved based on their isoelectric point, or pI. Proteins can be resolved by anion- or cation-exchange IEX chemistries leading to fractionation of complex samples, such as human plasma and whole cell lysates. This prefractionation step can lead to increased resolution of the 2DE gel and improved identification of lower abundance proteins.

REFERENCES

1. Huber, L.A., Is proteomics heading in the wrong direction? *Nature Rev. Mol. Cell Biol.*, 4, 74–80, 2003.
2. Klein, J.B. and Thongboonkerd, V., Overview of proteomics, *Nephrol. Basel, Karger,* 141, 1–10, 2004.
3. Bailey, J., Zhang, K., Zolotarjova, N., Nicol, G., and Szafranski, C., Removing high abundance proteins from serum, *Genet. Eng. News*, 23, 32–37, 2003.
4. Butt, A., Davidson, M.D., Smith, G.J., Young, J.A., Gaskell, S.J., Oliver, S.G., and Beynon, R.J., Chromatographic separations as a prelude to two-dimensional electrophoresis in proteomics analysis, *Proteomics*, 1, 42–53, 2001.
5. Wagner, K., Racaityte, K., Unger, K.K., Miliotis, T., Edholm, L.E., Bischoff, R., and Marko-Varga, G., Protein mapping by two-dimensional high performance liquid chromatography, *J. Chromatogr. A*, 893, 293–305, 2000.

6. Opiteck, G.J., Ramirez, S.M., Jorgenson, J.W., and Moseley, M.A., III, Comprehensive two-dimensional high performance liquid chromatography for the isolation of over expressed proteins and proteome mapping, *Anal. Biochem.*, 258, 349–361, 1998.
7. Rabilloud, T., Blisnick, T., Heller, M., Luche, S., Aebersold, R., Lunardi, J., and Braun-Breton, C., Analysis of membrane proteins by two dimensional electrophoresis: Comparison of the proteins extracted from normal and *Plasmodium falciparum*-infected erythrocyte ghosts, *Electrophoresis*, 20, 3603–3610, 1999.
8. Harper, R.G., Workman, S.R., Schuetzner, S., Timperman, A.T., and Sutton, J.N., Low-molecular weight human serum proteome using ultrafiltration, isoelectric focusing, and mass spectrometry, *Electrophoresis*, 25, 1299–1306, 2004.
9. GE Healthcare (formerly Amersham-Pharmacia). *Ion Exchange Chromatography: Principles and Methods*, Lit. Ref. 18–114–21, 1979.
10. Levison, P.R., Mumford, C., Streater, M., Brandt-Nielsen, A., Pathirana, N.D., and Badger, S.E., Performance comparison of low-pressure ion-exchange chromatography media for protein separation, *J. Chromatogr. A*, 760, 151–158, 1997.
11. Gottschalk, U., Fischer-Fruenholz, S., and Reif, O., Membrane adsorbers: A cutting edge process technology at the threshold, *BioProcess Int.*, 56–65, 2004.
12. Gomme, P., Forstch, V., Gilgen, S., McCann, K., Hurt, B., Schonmann, C., Bertolini, J., and Hodler, G., IgG Purification using IEX Filter Technology, paper presented at Plasma Product Biotechnology 2003, Curacao, Netherland Antilles, 2003, abstracts pp. 32–34.
13. Strancar, A., Monolithic supports for chromatography, *Genet. Eng. News*, 23, 50–51, 2003.
14. Pluskal, M.G., Microscale sample preparation, *Nat. Biotech.*, 18, 104–105, 2000.
15. Patton, W.F., Pluskal, M.G., Skea, W.M., Buecker, J.L., Lopez, M.F., Zimmermann, R., Belanger, L.M., and Hatch, P., Development of a dedicated two-dimensional gel electrophoresis system that provides optimal pattern reproducibility and polypeptide resolution, *BioTechniques*, 8, 518–527, 1990.
16. Anderson, N.L. and Anderson, N.G., The human plasma proteome: History, character and diagnostic prospects, *Mol. Cellular, Proteomics*, 11, 845–867, 2002.
17. Anderson, N.L., Polanski, M., Pieper, R., Gatlin, T., Tirumalai. R.S., Conrads, T.P., Veenstra, T.D., Adkins, J.N., Pounds, J.G., Fagan, R., and Lobley, A., The human plasma proteome: A nonredundant list developed by combination of four separate sources, *Mol. Cellular, Proteomics* 4, 311–326, 2004.
18. Hochstrasser, D.F., Sanchez, J.C., and Appel, R.D., Proteomics and its trends facing nature's complexity, *Proteomics*, 2, 807–812, 2002.
19. Pandey, A. and Mann, M., Proteomics to study genes and genomes, *Nature*, 405, 837–846, 2000.
20. Cordwell, S.J., Nouwens, A.S., Verrils, N.M., Basseal, D.J., and Walsh, B.J., Sub proteomics based upon protein cellular location and relative solubilities in conjunction with composite two-dimensional electrophoresis gels, *Electrophoresis*, 21, 1094–1103, 2000.
21. Lopez, M.F. and Melov, S., Applied proteomics: Mitochondrial proteins and effect on function, *Circ. Res.*, 90, 380–389, 2002.
22. Tonge, R., Shaw, J., Middleton, B., Rowlinson, R., Rayner, S., Young, J., Pognan, F., Hawkins, E., Currie, I., and Davison, M., Validation and development of fluorescence two dimensional differential gel electrophoresis proteomics technology, *Proteomics* 1, 377–396, 2001.

3 Separations in Proteomics: Use of Camelid Antibody Fragments in the Depletion and Enrichment of Human Plasma Proteins for Proteomics Applications

Mark ten Haaft, Pim Hermans, and Bruce Dawson

CONTENTS

3.1 INTRODUCTION

When the bulk of the human genome was deciphered in 2000, the scientific community quickly realized that we could draw few meaningful conclusions from the vast amount of data generated by the project. The event triggered the dawn of proteomics—the study of how proteins interact with each other and other molecules in metabolic pathways. The three billion nucleic acid base pairs identified in the human genome are believed to make up roughly 40,000 genes, which after full cotranslational and posttranslational modifications are believed to total in the millions of proteins. Unlike DNA and RNA, protein activity is based on molecular structure. While there have been countless attempts to predict protein activity with supercomputers, these efforts have produced little useful results due to the complex structure of proteins.

The single largest proteomics market is in the field of differential proteomics, where samples of serum from diseased and nondiseased populations are compared and contrasted to search for differences in protein levels. These proteins become diagnostic markers of disease, targets for clinical therapeutic intervention, or therapeutics themselves. More than half the human plasma proteome comprises housekeeping proteins such as human serum albumin (HSA) and immune gamma globulins (IgGs). These proteins have no relevance to disease state and mask the presence of important proteins. As the analytical tools for proteomic analysis offer increasingly improved sensitivity, the ability to differentiate important proteins from housekeeping proteins is become increasingly important and difficult. Ian Humphrey Smith of the Human Proteome Organization said in 2001 "... solving this problem will require designing affinity reagents, or molecules that capture certain classes of proteins, to screen out the high abundance proteins found in cells before analysis ..."[1]

Scientists have struggled with technologies such as size-exclusion chromatography and isoelectric focusing to deplete samples of HSA and IgGs, yet these techniques are not specific to the unwanted molecules, and important proteins are removed along with the depleted fractions. Some attempts have been made at standard products based on dye-based ligands and conventional bioprocessing media such as Protein-A, but the poor specificities, low capture capacities, and high costs have stalled these products in the market. We have developed a technology based on affinity chromatography that uses naturally occurring camelid single chain antibody (VHH) fragments as ligands. The application areas for this technology are numerous (e.g., the purification of proteins from different sources).[2,3] The camelid single domain antibodies can solve the problems in proteomics by providing high-affinity, high-specificity binders that can remove over 95% of these proteins. Unlike antibody reagents, VHH fragments can be easily manufactured and are very stable. Unlike conventional affinity reagents such as blue dye and Protein-A, these molecules offer much better specificity and higher capacities. Unlike antibody fragments such as Fabs and scFvs, VHHs are designed by nature and do not suffer the same stability and hydrophobic agglomeration problems. Finally, unlike peptides, VHHs offer multiple binding sites for multiple epitope recognition and better capacity, and they are far less expensive.

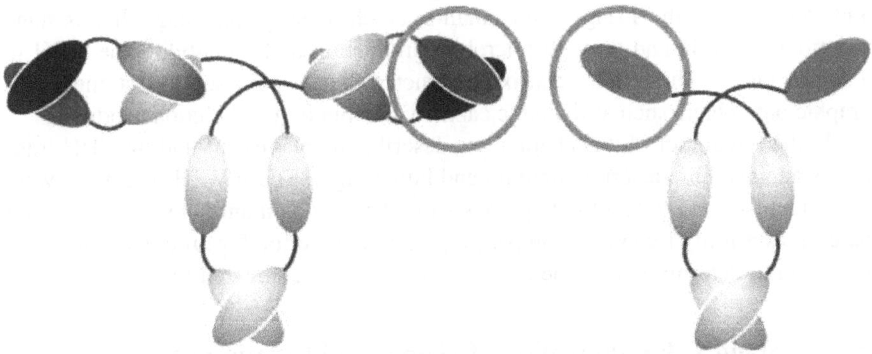

FIGURE 3.1 The difference between classical and single domain camelid antibodies. The difference between classical antibodies (left) and heavy chain antibodies (right). The binding domain is encircled.

3.2 HEAVY-CHAIN ANTIBODIES FROM THE CAMELID FAMILY

The serum of animals of the camelid family contains a unique type of antibodies devoid of light chain. This type of antibodies was discovered in 1993 at the University of Brussels.[4] Only a fraction of the immune repertoire of the animals of the camelid family comprises the heavy chain antibodies. The other fraction are normal classical antibodies. The heavy chains of these so-called heavy-chain antibodies bind their antigen by one single domain, the variable domain of the heavy immunoglobulin chain, referred to as VHH.[5,6] With a molecular weight of approximately 12 kDa, the VHH domain is the smallest known intact binding fragment derived from a functional immunoglobulin (Figure 3.1). VHHs show homology with the variable domain of heavy chains of the human VHIII family. The VHHs obtained from an immunized camel, dromedary, or llama have a number of advantages compared to the Fab, Fv, or scFv fragments derived from other mammals. One major advantage is that only one domain is cloned and expressed to generate an active binding fragment. This makes effective production in microorganisms such as *Saccharomyces cerevisiae*[7,8] possible.

3.3 ROADMAP TO LLAMA ANTIBODIES

The route to acquire the best antibodies for the task, such as in finding highly specific binders to abundant and irrelevant proteins in serum, is always the same and can be divided in three stages. The first stage is the immunization of the llama and the construction of the library. In this stage the immune repertoire of the llama after immunization is copied to a large microbial library. The second stage is the screening stage. In this stage we screen the library for several antibodies that can perform according to set specifications like specificity, binding, elution conditions, stability, and so forth. The third stage is the small-scale chromatography stage. In this stage we produce the ligand through microbial production and immobilize the VHH to a solid support. The affinity matrix constructed this way is used to test chromatographic parameters such as dynamic capacity and binding and elution conditions.

In the remainder of this chapter, we describe the process of finding VHH fragments against human serum albumin and human IgG. These VHH fragments were subsequently used to test the hypothesis that heavy-chain antibody fragments can be used as affinity ligands in sample preparation depletion for proteomics applications and to determine the benefits of this technology compared to other methods.

3.3.1 Stage 1: Immunization of Llamas and Creation of the VHH Library

Two llamas were immunized using standard procedures.[8] One llama was immunized with HSA, and the other llama was immunized with human Fc. Immune response of the llamas to these antigens was checked at regular intervals, using a small serum sample from the immunized animal. When the immune response reached a plateau, a larger blood sample was taken from the llama. From the peripheral blood lymphocytes, the mRNA was isolated. Using polymerase chain reaction (PCR) techniques, the VHH encoding fragments were amplified.[8] The DNA was cloned to the yeast *Saccharomyces cerevisiae*,[8] creating a VHH library. The size of both libraries was 1E + 7.

3.3.2 Stage 2: Screening for Antigen-Binding VHH Fragments

Screening for VHH fragments that could bind to the target of interest was done in the following way. The library was plated out on agar plates in a dilution high enough that single colonies were expected. Colonies were transferred to a non–protein binding 96-well plate. The colonies were grown in these plates and induced to produce VHH. The VHH-containing supernatant was used to check if the produced VHH could bind to the antigen using an enzyme-linked immunosorbent assay (ELISA). The positive clones from this ELISA were produced at small scale in shake flasks. For quick screening purposes, the VHH was purified from the supernatant using a Superdex 75 gelfiltration column after removal of the biomass.

3.3.3 Stage 2: Screening for the Ultimate Ligand

The selected VHH fragments were screened for several parameters to see if they could be used as ligands. In this case the parameters tested were affinity and specificity. Testing was performed using surface plasmon resonance (SPR). This technology can be used to accurately measure the antibody antigen interaction. For the anti-HSA and anti-human IgG ligands that were chosen for the experiments described in this chapter, the affinity was in the nanomolar range, as can be seen in Table 3.1. One of the important aspects of an anti human IgG ligand is that the ligand binds to all subclasses of human IgG. Table 3.2 shows binding of the anti human IgG ligand to all subclasses

TABLE 3.1 Affinity Constants for the Anti-Human IgG and Anti-HSA Ligands

Affinity Constants for the Anti-Human IgG Ligand	
ka (1/Ms)	3.2E + 05
kd (l/s)	3.7E − 03
KA (l/M)	8.6E + 08
KD (M)	1.2E − 09
Affinity Constants for the Anti-Human Serum Albumin Ligand	
ka (l/Ms)	7.1E + 05
kd (l/s)	5.0E − 03
KA (1/M)	1.4E + 08
KD (M)	7.1E − 09

TABLE 3.2 Subclass Specificity

	Response			
	IgG1	IgG2	IgG3	IgG4
hIgG binder	118.3	71.1	37.8	39.2
Control	6.1	7	5.3	8.3

of IgG as determined by SPR on a BiaCore 3000. Also crossreactivity of the HSA ligand to serum albumin from other species was tested. As can be seen from the results (Figure 3.2) there is no cross-reactivity of the HSA ligand to rat and rabbit serum albumin; however the HSA ligand does bind to mouse serum albumin. For the human IgG ligand, the crossreactivity to IgGs from other species was tested. As can be seen from the results (Figure 3.3) no cross-reactivity to mouse, bovine, or goat IgG was found.

3.3.4 Stage 3: Ligand Production

Ligands are produced by BAC's proprietary host system *Saccharomyces cervisiae*. The DNA sequence encoding the ligand is integrated in the genome of the yeast. Subsequently the ligand is produced by fed-batch fermentation.[7,8] After fermentation, the biomass is removed by microfiltration. The cell-free material is concentrated by means of ultrafiltration. The ligand is then purified using ion-exchange chromatography. This whole downstream processing route results in low-colored product, and the ligand has a protein purity of higher then 95%, as determined by sodium dodecyl sulfate–polyacrylamide gel electrophoresis (SDS-PAGE).

3.3.5 Stage 3: Coupling to Matrices

The ligand can be coupled to different matrices using several coupling chemistries. In the examples shown, ligand was coupled to a matrix using *N*-hydroxysuccinimide

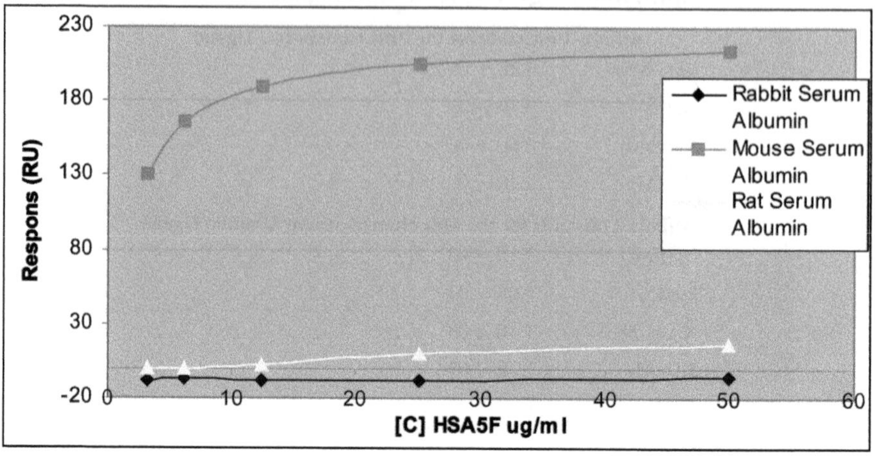

FIGURE 3.2 On a BiaCore 3000, the crossreactivity of the HSA ligand to serum albumin from other sources is tested. As can be seen, there is no cross-reactivity of the HSA ligand to rat and rabbit serum albumin; however, the HSA ligand does bind to mouse serum albumin.

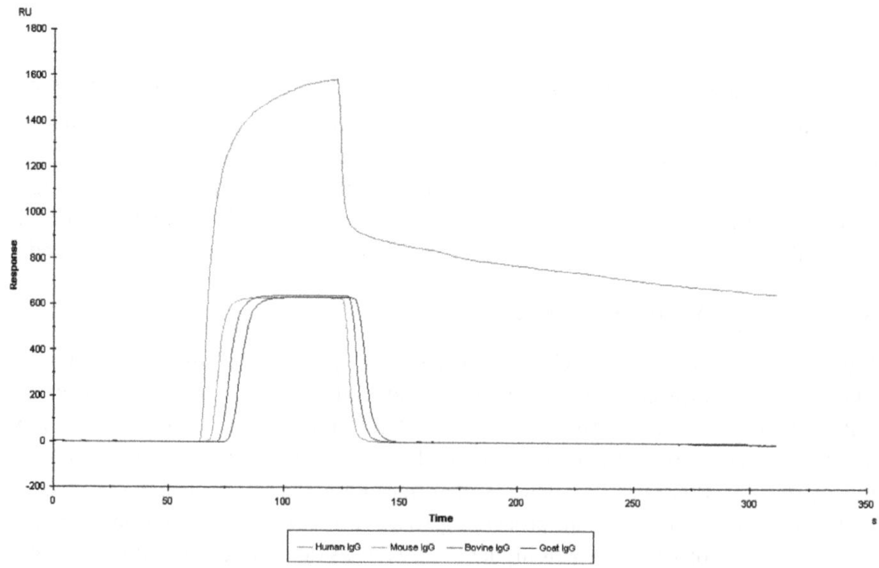

FIGURE 3.3 The cross-reactivity of the human IgG ligand to IgGs from other species was tested on a BiaCore 3000. As can be seen from the results, no cross-reactivity to mouse, bovine, or goat IgG was found.

(NHS) coupling chemistry. Ligands containing primary amino groups couple directly to the active ester of NHS to form a chemically very stable amide linkage.[9] Base matrix sepharose (Amersham Biosciences) was used because it is known for its low nonspecific binding. The coupling procedure used is as follows.

After purification, the anti-human IgG ligands were dialyzed to NHS coupling buffer, 0.1 M HEPES, pH 8.0. The following procedure was used for coupling of the ligands to NHS sepharose. Prior to coupling of the ligand to NHS, the matrix was washed with cold demineralized water acidified with acetic acid to pH 3. Then the matrix was washed twice with NHS-coupling buffer. The washed matrix was mixed with the ligand solution and left overnight at 4°C head over head or 1 hour at room temperature. Subsequently the gel material was filtered over a sintered glass filter, and the nonreacted groups of the gel material were blocked with Tris (0.1 M, pH 8.0) for 1 hour at room temperature. The coupled medium was washed using alternate low and high pH (3×10 column volumes phosphate-buffered saline (PBS) pH 2 and 3×10 column volumes PBS pH 7.4). The coupled gel material was now ready to use. Using the nonbound fraction, the coupling efficiency was determined by looking at the protein pattern on SDS-PAGE of the coupling solution before and after coupling.

The dynamic capacity of the affinity matrices was determined on an AKTA explorer 100 (Amersham Biosciences). Column volume that was used for these tests was 400 μl. Conditions for testing were the following: equilibration buffer PBS, pH 7.4 (Roche), flow 150 cm/hr; elution buffer PBS with an adjusted pH to 2.1. The low pH in the elution ensures full elution of the proteins of interest from the affinity column. The eluted fractions are immediately neutralized with 2 M Tris buffer. As sample for these experiments, pure HSA (Sigma) and pure human IgG (Sigma) was used. The dynamic capacity was determined using peak integration of the elution peak. For the HSA affinity matrix, typical dynamic capacities fell in the range of 8 to 10 milligrams HSA per milliliter affinity matrix of a settled matrix bed. For the human IgG affinity matrix typical dynamic capacities fall in the range of 13 to 15 milligrams human IgG per milliliter of affinity matrix of a settled matrix bed.

3.3.6 Stage 3: Depletion of HSA and Human IgG from Human Plasma

The affinity matrix was tested using an Akta explorer 100 for efficiency of the depletion and nonspecific binding of the affinity matrix. Onto a small affinity column (400 μl) a sample of human plasma was loaded, and the same chromatographic procedure as described previously was used. The flow-through fractions and elution fractions (Figure 3.4) were collected and used for SDS-PAGE analyses. SDS-PAGE was performed on NOVEX Tris glycine 4–20% gels according to the supplier's protocol. Figure 3.5 shows the starting material, flow through (depleted fractions), and the eluted fractions.

Using a combined approach and thus depleting a sample of human plasma from HSA and human IgG, a 2D gel electrophoresis experiment was performed. Figure 3.6A and 3.6B show the nondepleted and depleted 2D gels of human plasma. The 2D gel electrophoresis was performed as described by Klooster et al.[10]

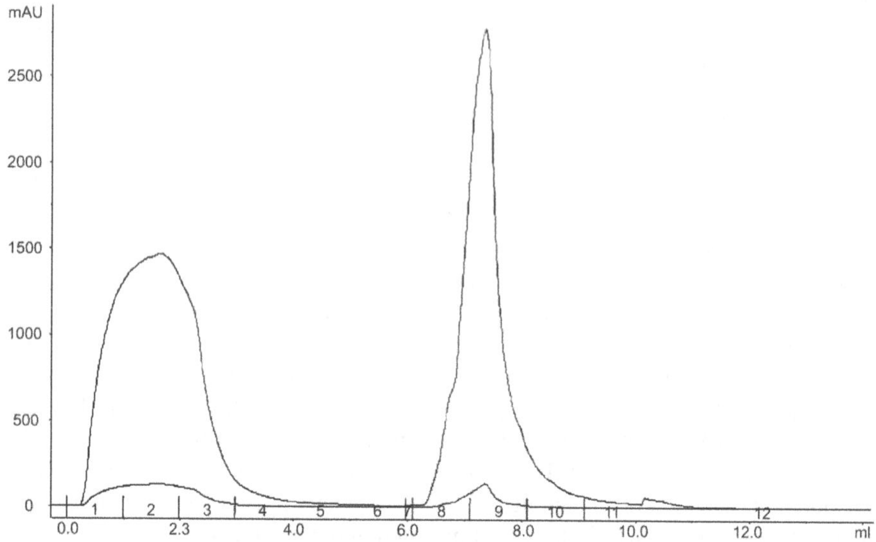

FIGURE 3.4 Depletion of a HSA from human plasma. The lower optical density line shows the optical density at 280 nm, the higher optical density line is the optical density at 214 nm. At time point 0, a sample of human plasma is injected onto the column. The HSA-depleted fraction is the flow through of the column (fractions 1, 2, and 3). After washing out of the flow through, the column is eluted, the depleted HSA is collected (fractions 8, 9, and 10), and the eluted fractions are immediately neutralized with 2 M Tris buffer.

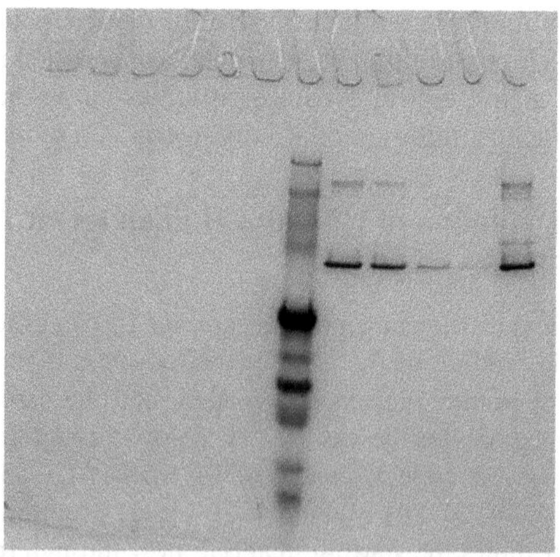

FIGURE 3.5 From left to right: 1, fraction 1 flow through; 2, fraction 2 flow through; 3, fraction 3 flow through; 4, fraction 4 flow through; 5, fraction 5 flow through; 6, fraction 6 flow through; 7, marker; 8, fraction 8 elution; 9, fraction 9 elution; 10, fraction 10 elution; 11, fraction 11 elution; 12, starting material—diluted human plasma. The flow-through and elution fractions are the fractions of the chromatography experiment of Figure 3.4.

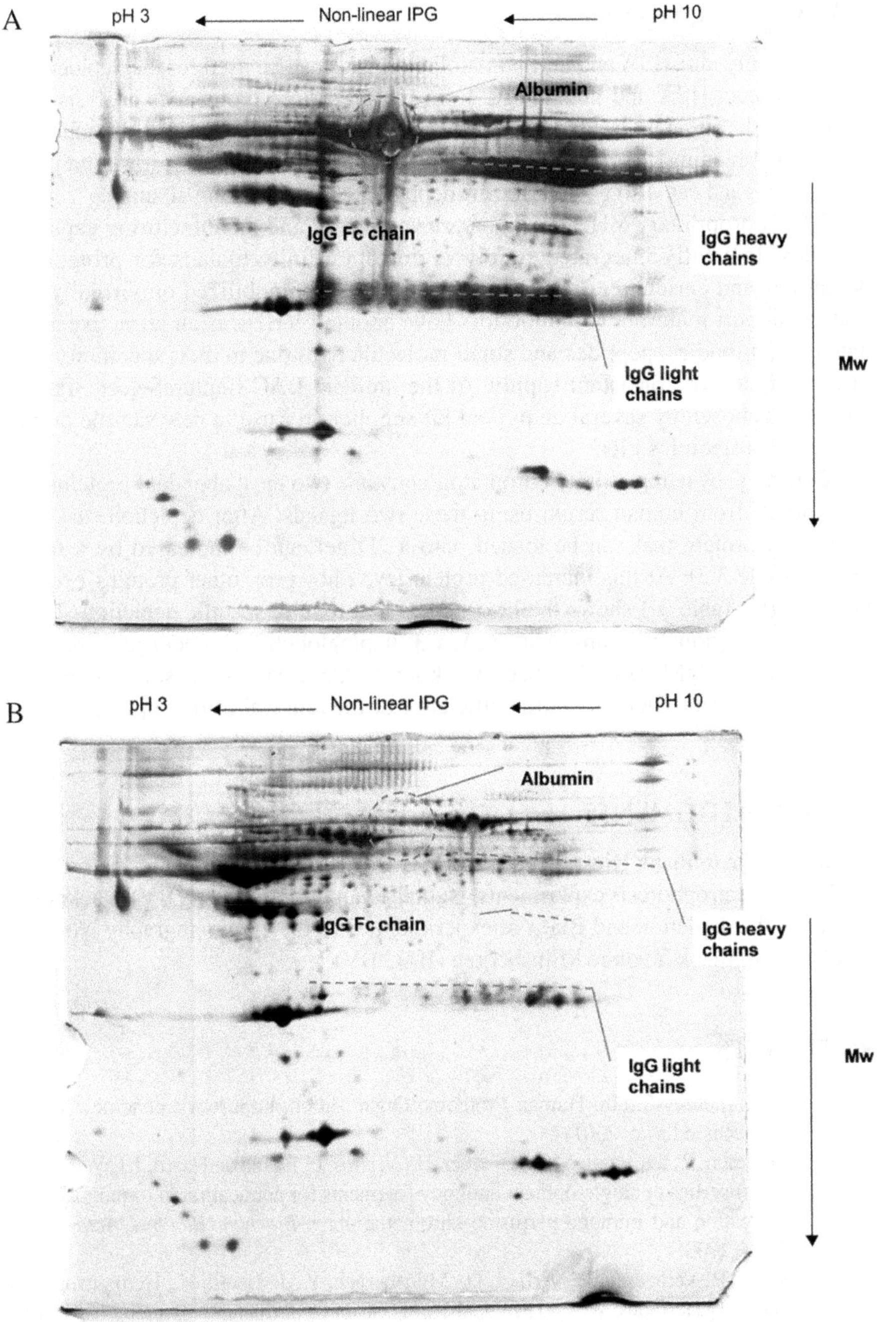

FIGURE 3.6 (A) 2D gel of nondepleted human plasma; (B) 2D gel of human plasma after depletion of the HSA and human IgG using BAC's anti-HSA and antihuman IgG ligands.

3.4 CONCLUSIONS

BAC's affinity anti-HSA and anti-human IgG ligands can be used for the depletion of CaptureSelect® HSA and human IgG from plasma before proteomics analysis. The removal of the target molecules is very efficient, more than 95% can easily be achieved. Nonspecific binding of other proteins to either the ligand or the matrix was found. The anti-HSA ligand can also be used for the depletion of mouse serum albumin.

BAC's proprietary VHH discovery, development and manufacturing capabilities provide a fully integrated supply of custom affinity ligands for proteomics depletions and enrichment applications that can be immobilized on virtually any matrix support material. Collaborators have adopted VHHs as an attractive alternative to antibodies, peptides and small molecule dyes due to their specificity, stability and ease of consistent supply. At the moment BAC CaptureSelect® ligands have been chosen by several analytical kit suppliers for use in new sample preparation for proteomics kits.

Currently, by using affinity chromatography, the two most abundant proteins can be depleted from human serum using these two ligands. After depletion, the total amount of protein that can be loaded onto a 2D gel can be increased by a factor of 2.3 (Table 3.3). At this increased protein level, however, other proteins become problematic. Table 3.3 shows the next targets that require specific depletion. These include macroglobulin, transferrin, IgA, C3, haptoglobulin, α-a acid glycoprotein, fibrinogen, and IgM. BAC has been working on these targets for some time now, and we are currently looking at the efficiency of the removal of these proteins from human serum.

ACKNOWLEDGMENTS

We would like to thank Rinse Klooster (MCB University of Utrecht) for performing the 2D gel electrophoresis experiments. Nathalie van Egmond (BAC BV) is acknowledged for the depletion and BiaCore experiments. For the chromatography work we would like to thank Yvonne Mijnsbergen (BAC BV).

REFERENCES

1. Ian Humphrey-Smith, Human Proteome Organization, http://www.genomeweb.com (last accessed May 2001)
2. Verheesen, P., ten Haaft, M.R., Lindner, N., Verrips, C.T., and de Haard, J.J.W., Beneficial properties of single-domain antibody fragments for application in immunoaffinity purification and immuno-perfusion chromatography, *Biochim. Biophys. Acta*, 1624, 21–28, 2003
3. Oranje, P., Verheesen, P., Verbart, D., Mijnsbergen, Y., de Haard, H., Hermans, P., ten Haaft, M., Hermens, W., and Meulenberg, J., Scalable purification of adeno-associated virus serotypes by immunoaffinity chromatography using llama-single domain antibody fragments, 2005, submitted for publication.
4. Hamers-Casterman, C., Atarhouch, T., Muyldermans, S., Robinson, G., Hamers, C., Songa, E.B., Bendahman, N., Hamers, R., Naturally occurring antibodies devoid

TABLE 3.3 The Abundant Proteins in Human Plasma

	Increased Relevant Protein Load on Gels as f(depletion efficiency)					
Plasma Protein	**mg/ml**	**mg/0.1ml**	**Depleted at 95.0%**	**Cumulative Removed**	**Remaining**	**Protein Load Multiplier Increase**
Albumin	40.0	4.00	3.80	3.80	4.80	1.8
IgG	11.0	1.10	1.05	4.85	3.76	2.3
Macroglobulin	3.0	0.30	0.29	5.13	3.47	2.5
Transferrin	2.8	0.28	0.27	5.40	3.20	2.7
AAT	2.4	0.24	0.23	5.62	2.98	2.9
IgA	2.4	0.24	0.23	5.85	2.75	3.1
C3	1.7	0.17	0.16	6.01	2.59	3.3
Haptoglobin	1.1	0.11	0.10	6.11	2.49	3.5
LDL	1.1	0.11	0.10	6.21	2.39	3.6
IgM	1.1	0.11	0.10	6.31	2.29	3.8
Alpha-a Acid Glycoprotein	1.0	0.10	0.10	6.41	2.19	3.9
ACT	0.4	0.04	0.04	6.44	2.16	4.0
Ceruloplasmin	0.3	0.03	0.03	6.47	2.13	4.0
C4	0.3	0.03	0.03	6.50	2.10	4.1
C1 Inhibitor	0.2	0.02	0.02	6.52	2.08	4.1
Others (from study)	—	1.74	0.00	6.52	2.08	4.1
Total Protein Load	—	8.60	6.52	—	—	—

of light chains, *Nature,* 363, 446–448, 1993; Muyldermans, S., Single domain camel antibodies: Current status, *Rev. Mol. Biotechnol.*, 74, 277–302, 2001.

5. Muyldermans, S., Atarhouch, T., Saldanha, J., Barbosa, J.A.R.G., and Hamers, R., Sequence and structure of VH domain from naturally occurring camel heavy chain immunoglobulins lacking light chains, *Protein Eng.*, 7, 1129–1135, 1994.
6. Muyldermans, S., Single domain camel antibodies: Current status, *Rev. Molec. Biotechnol.,* 74, 277–302, 2001.
7. Thomassen, Y.E., et al., Large-scale production of VHH antibody fragments by *Saccharomyces cerevisiae*, *Enzyme Microb. Technol.,* 30, 273–278, 2002.
8. Frenken, L.G., van der Linden, R.H., Hermans, P.W., Bos, J.W., Ruuls, R.C., de Geus, B., and Verrips, C.T., Isolation of antigen specific llama VHH antibody fragments and their high level secretion by *Saccharomyces cerevisiae, J. Biotechnol.*, 78, 11 –21, 2000.
9. Hermanson, G., *Bioconjugate Techniques*, Academic Press, San Diego, 1993.
10. Klooster, R., Hermans, P., ten Haaft, M., Improved removal and isolation of blood proteins with Camelid antibodies, 2005, submitted for publication.

4 Novel Plasma Protein Separation Strategy Using Multiple Avian IgY Antibodies for Proteomic Analysis

Sun W. Tam, Lei Huang, Douglas Hinerfeld, David Innamorati, Xiangming Fang, Wei-Wei Zhang, John Pirro, and Jerald S. Feitelson

CONTENTS

4.1 INTRODUCTION

By studying human plasma or serum, a number of diagnostic markers and therapeutic targets were discovered.[1,2] Markers help to understand diseases such as cancer, hemophilia, osteoarthritis and cardiovascular diseases.[3–5] Normally, many disease processes are correlated with quantitative and physiological changes of proteins in body fluids.[6,7] Proteins from damaged tissues can enter into the bloodstream as leakage markers.[8–19] In general, plasma can contain enzymes, cytokines, chemokines, soluble receptors, transport proteins, peptides, hormones, blood-clotting factors, and other soluble proteins. Well over 3,700 distinct plasma proteins can be identified using mass spectrometry and other methods.[11] However, only a handful of plasma proteins are now routinely used in clinical diagnoses or prognoses of diseases.[12] This may be due to our lack of knowledge of many yet uncharacterized plasma proteins, their posttranslational modifications and relevant physiological functions.

The (approximately fourteen) most abundant plasma proteins constitute about 95% of the protein mass of plasma.[9] However, the vast majority of proteins that may be biomarkers for health and disease in various organs are present in much lower abundance.[11,12] Plasma proteins can be present in a dynamic range of 10^{10} from concentration at mg/ml, such as albumin (accounting for 50 to 75% of total plasma protein), to pg/ml for some interleukins. Such an enormous dynamic range of protein concentrations represents a nearly insurmountable problem for detection and quantitation in a single assay. The unusually high abundance of albumin and IgG in serum can significantly affect the resolution and sensitivity of many techniques. For instance, plasma proteins resolved by 2D gel electrophoresis become crowded in the range from 45 to 80 kDa and pI of 4.5 to 6. The removal of highly abundant proteins, such as albumin, from serum and plasma is absolutely necessary for the detection of low-abundance proteins by 2D electrophoresis, ICAT™, or Multidimensional Protein Identification Technology (MUDPIT) proteomic methods.[13] In most cases, plasma or serum samples also must be diluted to detect proteins of interest. However, dilution often is not the best approach since it further lowers the concentration of low-abundance proteins, often reducing them below the limits of assay detection. There is an unmet need to selectively remove high-abundance proteins from plasma and to enrich low-abundance proteins in plasma to make the study or measurement of low-abundance proteins technically more feasible, particularly in multiplex settings.

Two types of competing products are currently on the market. Chemical dye-based methods for removing serum albumin (HSA), such as Cibacron Blue, have excellent binding capacity and are relatively inexpensive. However, these chemical affinity products are highly nonspecific. Scientists cannot afford to lose one seventh of their sample when mining for valuable protein biomarkers. Several companies sell Goat IgG-based immunodepletion products against human serum albumin. The use of monoclonal antibodies against HSA as the immunoaffinity depletion reagent was far more efficient and specific than the dye-based resins for human samples.[13] The use of affinity-purified polyclonal antibodies (pAbs) to specifically remove albumin or IgG was successful to enrich low-abundance proteins from human plasma samples.[14] This approach has much better specificity than the chemical affinity methods but still artificially captures many nontarget proteins due to the Fc region (the stem) of IgG molecules. This region is known to bind complement proteins, rheumatoid factors, human anti-mouse antibodies (HAMA), and multivalent Fc binding proteins, such as IgM.

More importantly, specificity is usually limited to human antigens, since the host producing IgG (goat, rabbit, mouse, etc.) is genetically closely related to humans and thus cannot recognize many conserved epitopes on the human protein that are also present in the antibody-producing species. The search for novel biomarkers is heavily dependent on the analysis of the plasma proteome from mammalian model organisms, such as rodents. The removal of albumin from nonhuman mammalian serum and plasma has largely relied on the binding of albumin to Cibacron Blue, such as the Montage™ albumin Depletion Kit (Millipore Inc., Bedford, Mass.). As mentioned, this method lacks specificity, thus removing many potentially important proteins from the sample.

In this chapter, we describe a novel method for the specific removal of albumin from the serum and plasma of human, rat, and several other mammalian species.[15] Avian IgY antibodies raised against HSA, transferrin, fibrinogen, IgA, IgM or IgG, which are covalently linked through their Fc regions to microbead carriers (Seppro™, are coupled with high-throughput liquid chromatography for selective removal of albumin and other abundant plasma proteins. IgY is the functional equivalent of IgG. From an evolutionary perspective, IgY antibodies are considered to be the ancestor of mammalian IgG and IgE antibodies.[16] One of the important advantages of using IgY is that there is an enhanced immunogenicity against conserved mammalian proteins due to the phylogenetic distance between donor and recipient organisms.[17] This enhanced immunogenicity makes production of antibodies against conserved mammalian proteins generally more successful in chickens than in other mammals. In addition, IgY antibodies tend to recognize the same protein in a number of mammalian species, making them more widely applicable as immunoaffinity separation reagents and for drug target discovery.[18] Avian IgY antibodies with high avidity against bacterial or human proteins have been developed.[19–21] The avian immune response was also shown to be persistent, very low quantities of antigen were required to obtain high and long-lasting IgY titers in the yolk from immunized hens.[17,22] This platform allows for very specific and high-throughput depletion of albumin and other abundant proteins from serum and plasma of multiple mammalian species utilizing a high-throughput liquid chromatography system, such as the Applied Biosystems Vision protein workstation.

4.2 MATERIALS AND METHODS

4.2.1 Coupling of Affinity-Purified IgY to UltraLink® Hydrazide Gel

Coupling of affinity purified IgY antibodies to UltraLink Hydrazide beads was performed essentially as described in the manufacturer's protocol (Pierce Biotechnology Inc., Rockford, Ill.). The antibody conjugation scheme employed was based on hydrazide chemistry resulting in the IgY antibodies oriented with the $F(ab')_2$ antigen-binding regions facing outward and the Fc regions of the antibodies covalently coupled via their carbohydrate residues through a 23 atom spacer arm to the solid phase material. This strategy allows maximum efficiency of antigen binding, in contrast to previous random orientation methods using ε-amines in lysine for coupling to the solid phase.

4.2.2 Spin Column Method for Removal of Serum and Plasma Abundant Protein

Human serum samples (Sigma Cat. H-1388) were centrifuged at 14,000 rpm for 5 minutes in an Eppendorf microfuge to remove insoluble materials. Appropriately diluted samples were loaded onto a 1.5 ml spin column (Pierce Biotechnology Inc.) containing GenWay Biotech's Anti-HSA or MIXED6 IgY gels. After incubation for 15 minutes with occasional mixing, the column was centrifuged at 8,000 rpm for 8 seconds. The flow-through samples were collected for analysis by one-dimensional (1D) sodium dodecyl sulfate–polyacrylamide gel electrophoresis (SDS-PAGE). Proteins captured by the IgY gels were removed with stripping buffer (0.1M Glycine-HCl, pH 2.5) and analyzed in parallel.

4.2.3 Albumin Depletion with Liquid Chromatography Workstation

Human, rat, and mouse sera were supplied by Bioreclamation, Inc. (Hicksville, N.Y.). Sera were centrifuged at 21,000*g* for 10 minutes at 4°C to remove insoluble material. All immunodepletion of serum was performed on the Applied Biosystems Vision™ Workstation liquid chromatography system. For HSA separation, serum was diluted 1:10 in Tris-HCl-buffered saline (TBS) and injected at 0.1 ml/min onto a column containing Seppro products: anti-HSA IgY antibodies linked to UltraLink Hydrazide beads equilibrated in TBS. The flow-through (depleted) fraction was collected. The column was rinsed with TBS containing 0.05% Tween-20 until no further protein was detected in the flow-through material, and then bound proteins were eluted with 100 mM Glycine-HCl, pH 2.5, at 1 ml/min for retrieval of albumin and associated material. The column was neutralized with 100 mM Tris-HCl, pH 8.0, and then reequilibrated in TBS prior to application of subsequent samples. Depletion using the Montage kit was performed as described in the manufacturer's protocol.

4.2.4 One-Dimensional Electrophoresis

Following depletion, samples were concentrated on Amicon Ultra 4 10kDa cutoff ultracentrifugation columns (Millipore Inc.). Protein quantitation of the concentrated sample used the Bradford dye-binding assay. Four micrograms of the concentrated samples were reduced with 50 mM DTT and subjected to electrophoresis on 4–12% Bis-Tris SDS NuPage gels (Invitrogen, Carlsbad, Calif.) according to the manufacturer's protocol.

4.2.5 Two-Dimensional Electrophoresis

Prior to isoelectric focusing (IEF), samples were acetone precipitated and solubilized in 40 mM Tris, 7 M urea, 2 M thiourea and 2% CHAPS, reduced with tri-butylphosphine, and alkylated with 10 mM acrylamide for 90 minutes at room temperature. Following a second round of acetone precipitation, the pellet was solubilized in 7 M urea, 2 M thiourea, and 2% CHAPS, and 75 μg protein were subjected to IEF on 11 cm pH 3–10 immobilized pH gradient (IPG) strips (Proteome Systems, Sydney, NSW, Australia). Following IEF, IPG strips were equilibrated in 6 M urea, 2% SDS, 50 mM Tris-acetate buffer (pH 7.0), and 0.01% bromophenol blue and subjected to SDS-PAGE on 6–15% Gel Chips™ (Proteome Systems). All gels were stained in Sypro® Ruby (Molecular Probes, Eugene, Oreg.) and imaged by a charge-coupled device camera on a fluorescent imager (Alpha Innotech, San Leandro, Calif.).

4.2.6 Protein Digestion and MALDI Analysis

Protein spots were automatically detected and excised using the Xcise apparatus (Shimadzu Biotech, Japan). Gel pieces were washed twice with 150 μL 25 mM ammonium bicarbonate, pH 8.2, 50% v/v acetonitrile (ACN), dehydrated by the addition of 100% can, and air dried. Trypsin (Promega, Madison, Wisc.) in 2 mM ammonium bicarbonate (20 μg/μL) was added to each gel piece and incubated at 30°C for 16 hours. The peptides were extracted by sonication. The peptide solution was automatically desalted and concentrated using ZipTips from Millipore (Bedford, Mass.) on the Xcise apparatus and spotted onto the Axima (Kratos, Manchester, U.K.) MALDI target plate. Peptide mass fingerprints of tryptic peptides were generated by matrix assisted laser desorption and ionization-time-of-flight-mass spectrometry (MALDI-TOF-MS) using an AximaCFR (Kratos).

4.2.7 Bioinformatic Database Search

All spectra were automatically analyzed by the BioinformatIQ integrated suite of bioinformatics tools from Proteome Systems. Protein identifications were assigned by comparing peak lists to a database containing theoretical tryptic digests of National Center for Biotechnology Information (NCBI) and Swiss Prot sequence databases. Protein identification was evaluated based on percent coverage, Molecular Weight Search (MOWSE) score, number of peptide matches, peak intensity, and match of pI and molecular weight with the location of the protein on the 2D gel.

4.3 RESULTS

4.3.1 Cibacron Blue Depletion of Rat Albumin Removed Many Nontarget Proteins

Rat serum was depleted with the Millipore Montage kit to remove albumin, according to manufacturer protocol. Two-dimensional gel analyses were performed with the predepleted, postdepleted, and albumin-associated proteins. Based on the results shown in Figure 4.1, the postdepleted sample showed some removal of the albumin and some enhancement of lower abundant proteins. However, it was clear that many proteins were nonspecifically removed in the albumin fraction. Hence, the Cibacron blue dye does not selectively isolate the albumin-associated proteins but rather has strong nonspecific effects. Important proteins can be lost using this methodology.

4.3.2 Structural Comparisons of IgG and IgY and Their Covalent Coupling to Microbead Carriers

Despite the similarities between IgY and IgG antibodies, there are some profound differences in their chemical structures (Figure 4.2A).The IgY heavy chain is longer (65–70 kDa, versus ~50 kDa), and the IgY light chain is shorter (19–21 kDa versus 22–23 kDa).[22,23] The greater molecular mass of IgY is due to an increased number of

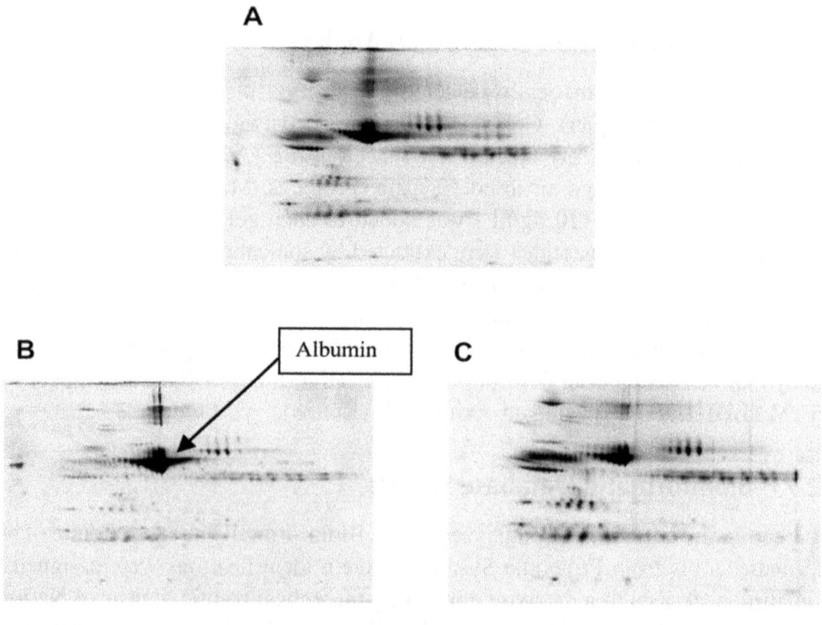

FIGURE 4.1 Cibacron Blue–based albumin depletion of rat serum. 2D electrophoresis of rat serum albumin depleted using the Montage kit. The large number of associated proteins in the albumin fraction indicates the lack of specificity of the resin-based depletion kit. (A) Predepletion Rat Serum; (B) albumin-associated fraction; (C) albumin-depleted fraction.

heavy-chain constant domains and an extra pair of carbohydrate chains in its Fc region. The latter structural feature is important for the efficiency of oriented covalent coupling of IgY via hydrazide chemistry to solid surfaces. In addition, the hinge region of IgY is shorter and less flexible compared to that of mammalian IgG. IgY is a more hydrophobic molecule than IgG,[24] which matches the lipid-rich environment of the egg yolk. The amino acid sequence and posttranslational differences between the two types of antibodies determine their biochemical features and immunological functions.

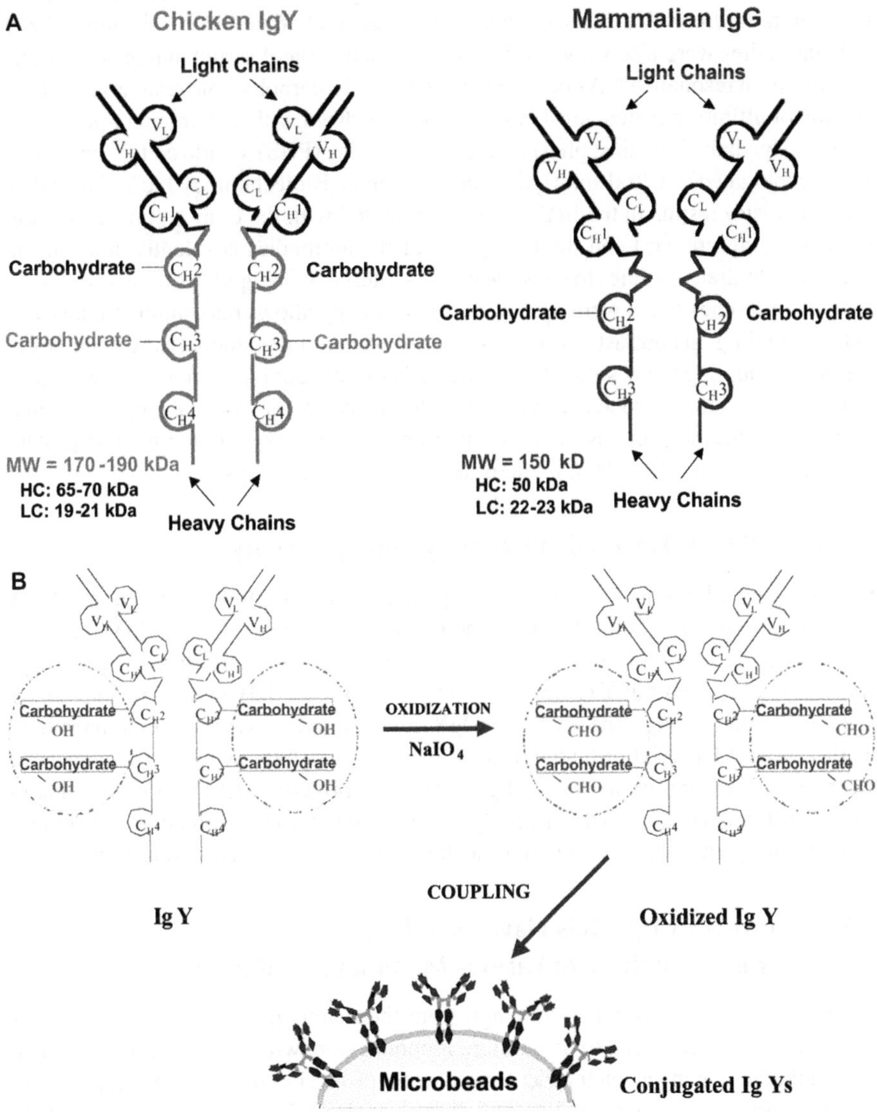

FIGURE 4.2 (A) Structures of IgY and IgG; (B) coupling of affinity-purified IgY to UltraLink Hydrazide Gel.

IgY antibodies react neither with mammalian IgG nor IgM,[25] nor with human anti-mouse IgG antibodies (HAMA),[26,27] nor do they bind to the rheumatoid factor (RF), which is an anti-immunoglobulin autoantibody found in many different disease organisms.[28] IgG molecules often give false positive results by interaction with RF in immunoassays.[24] The lack of cross-reactivity between IgY and IgG can be utilized in many ways to reduce unwanted reactions in assays in which anti-IgG antibodies are used. IgY does not activate mammalian complement factors,[29] which also helps to reduce the assay interference by complement factors in mammalian serum samples.[30]

IgY was shown to have good stability after being applied to latex microspheres, due to its more hydrophobic surface compared to mammalian antibodies.[31] Immobilized IgY antibodies were also shown to improve the detection of serum antigens with surface plasmon resonance.[30] As depicted in Figure 4.2B, carbohydrate residues on the Fc portion of affinity-purified anti-HSA IgY were oxidized with sodium meta-periodate, followed by dialysis against phosphate-buffered saline (PBS). Oxidized IgY was covalently conjugated to UltraLink Hydrazide Gel (Pierce Biotechnology Inc.). The hydrazide chemistry results in the IgY antibodies oriented with the $F(ab')_2$ antigen-binding regions facing outward and the Fc regions of the antibodies covalently coupled via their carbohydrate residues to the solid-phase material. Conjugation efficiency and the ratio of antibodies to gel were optimized. This strategy allows maximum efficiency of antigen binding, in contrast to previous random orientation methods using ε-amines in lysine for coupling to the solid phase. Oriented covalent conjugation of IgY antibodies to 60 μm diameter microbead carriers provides a convenient way to capture and separate ligand-specific proteins in several platforms, such as spin columns, fast protein liquid chromatography (FPLC), microtiter plates, and microtips.

4.3.3 IgY Gels Have High Capacity and Specificity

Comparison of binding specificity of polyclonal, antigen affinity purified anti-HSA IgY and monoclonal anti-HSA IgG showed higher capacity for the IgY antibody (Figure 4.3). Higher molecular weight albumin-associated protein bands were also significantly reduced with IgY (lanes AA1 and AA2 in Figure 4.3), indicating both higher specificity and binding capacity for the IgY. IgY antibodies were shown to have high avidity against bacterial or human proteins.[20] A further study using surface enhanced laser desorption ionization time-of-flight mass spectrometry (SELDI-TOF MS) analysis of anti-human recombinant insulin IgY compared to the corresponding IgG demonstrated that IgY had stronger binding capability and higher detection sensitivity.[32]

4.3.4 Anti-HSA IgY Gels Have Excellent Cross-Reactivity to Various Mammalian Albumins

Because chickens diverged so long ago from the mammalian clan, human proteins injected into chickens tend to be far more immunogenic, with the resulting polyclonal IgY antibodies having a much wider array of recognized epitopes. Thus it was predicted (Figure 4.4A) and experimentally verified that anti-HSA IgY will immunoprecipitate serum albumins from many mammalian species, such as mouse, rat, pig, and goat.

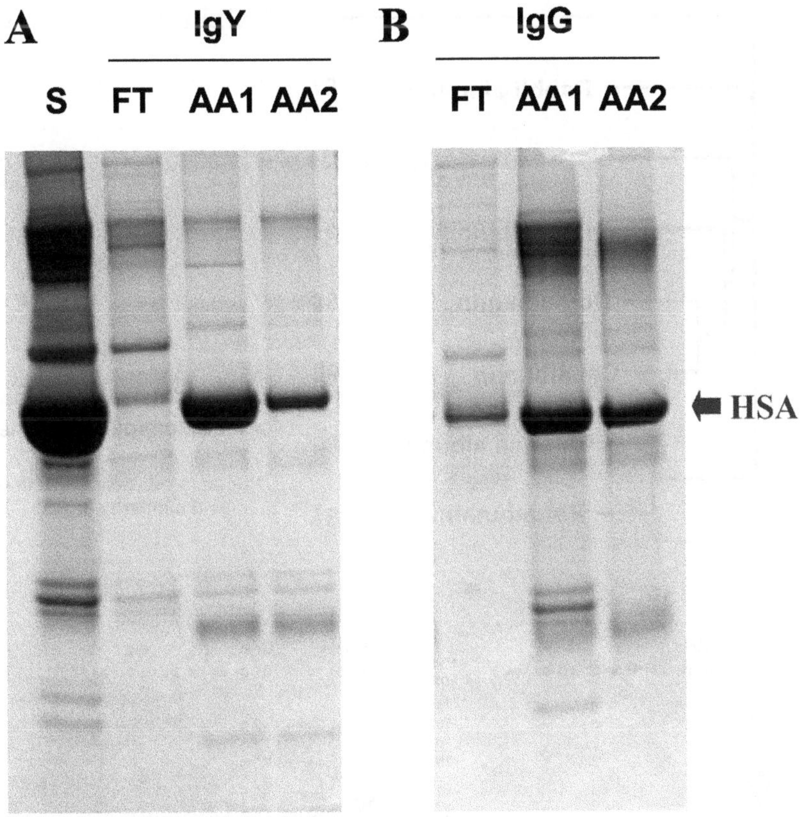

FIGURE 4.3 IgY has higher specificity than IgG for serum depletion. (A) 1D SDS-PAGE using GenWay Biotech's anti-HSA IgY covalently bound to Pierce's Ultra-Link Hydrazide microbeads; (B) 1D SDS-PAGE using Sigma's monoclonal anti-HSA IgG coupled to Protein G beads. S, undepleted human serum; FT, flow through after two rounds depletion; AA1, albumin-associated protein after first binding; AA2, albumin-associated protein after second binding. (Data courtesy of Ciphergen Biosystems, Inc.)

As shown in Figure 4.4B, the anti-HSA IgY gel is effective at depleting serum albumins from these species. Corresponding anti-HSA IgG products do not effectively remove albumin from nonhuman species (Table 4.1).

4.3.5 IgYs Cross-React and Are Also Effective at Separating IgG, Fibrinogen, and Transferrin from Human and Rat Plasma

Similar to results with albumin, cross-species reactivity was confirmed using anti-human IgG-Fc IgY gels to selectively remove IgG from rat and mouse serum (Figure 4.4C).

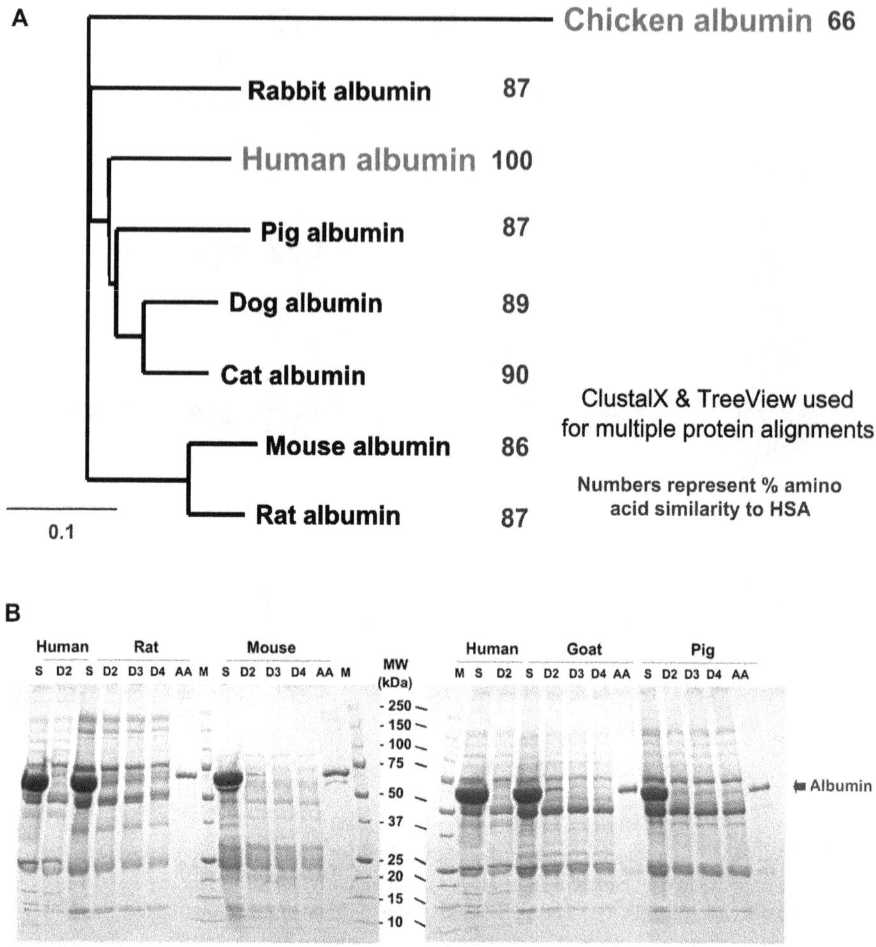

FIGURE 4.4 (Continued)

Figure 4.5 extends these results to include fibrinogen and transferrin. Therefore we conclude that IgY gels are effective at separating many plasma proteins from a variety of mammalian species.

4.3.6 Six Abundant Plasma Proteins Are Effectively Removed by MIXED6 IgY Gels

To test the feasibility of simultaneous removal of several of the most abundant plasma proteins, a physical mixture of Seppro™ products was made and used for immunoaffinity separation in a spin column format. In the first spin column, anti-HSA IgY gel

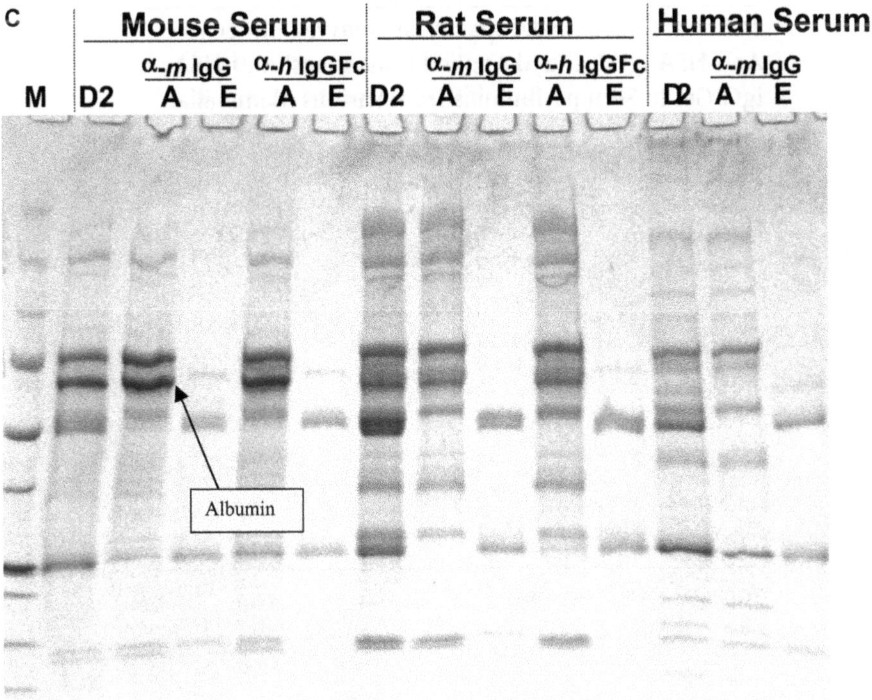

FIGURE 4.4 (A) Genetic parallax: clustering of mammalian serum albumins; (B) GenWay Biotech's anti-HSA IgY gel completely and specifically depletes albumin from rat and pig serum after two rounds of batch depletion (25 μl each) and mouse and goat serum after three rounds. MW, molecular weight marker; S, serum before depletion (1:10 dilution); D2, unbound material after two rounds of anti-HSA depletion; D3, unbound material after three rounds of anti-HSA depletion; D4, unbound material after four rounds of anti-HSA depletion; AA, albumin-associated proteins, recovered after stripping PBS-rinsed column with 100 mM Glycine (pH 2.5). It appears that this batch of mouse serum was partially proteolytically degraded. A 4 to 20% gradient SDS-PAGE was used; (C) anti-mouse-IgG and anti-human-IgG-Fc IgY Gels remove IgG from mouse, rat, and human sera. α-m IgG, chicken anti-mouse IgG IgY coupled to microbeads; α-h IgGFc, chicken anti-human IgG-Fc IgY coupled to microbeads; M, molecular weight marker; D2, after two HSA depletions; A, IgG depletion of D2; E, eluted bound protein from IgY gel. A 4 to 20% gradient SDS-PAGE was used.

removed 85 to 90% of the albumin in the sample (lane E1, Figure 4.6A). The flow-through sample from this column was applied to a second MIXED6 spin column containing a physical mixture of 6 IgY gels: anti-HSA (to remove residual albumin), anti-IgG-Fc, anti-transferrin, anti-fibrinogen, anti-IgA, and anti-IgM. Lane D of Figure 4.6A showed complete removal of all six proteins but only detected in the eluted fraction, as seen in lane E2. The samples were then applied in a 2D gel analysis (see Figure 4.6B). Figure 4.6A shows the position of the six proteins as circled

TABLE 4.1 Relative Capacity of GenWay Biotech's Anti-HSA IgY Gel and Applied Biosystems' Anti-HSA IgG Gel to Serum Albumins from Various Mammalian Species

Serum Albumin Species	ABI's Anti-HSA IgG Gel (%)	GenWay's Anti-HSA IgY Gel (%)
Human	100	100
Cow	12	59
Goat	0.5	25
Mouse	34	65
Pig	25	49
Rabbit	15	—
Rat	37	60
Sheep	0.5	—
Dog	—	21

Note: One round of depletion.

versus the sample after MIXED6 depletion (Figure 4.6B.b). The albumin associated fraction (Figure 4.6B.c) and the MIXED6 depleted fraction (Figure 4.6B.d) are also analyzed. The procedure showed specific and effective depletion of the sample, allowing appearance of low-abundance proteins previously undetectable without depletion. The protein identity of the proteins from MIXED6 were then identified by careful MALDI analysis by cutting out the gel spots followed by trypsin digestion and peptide mass fingerprinting. The Mowse scores for all the identified proteins are all significant within confident level (see Figure 4.6C). The data showed excellent and specific isolation of all the target proteins, as revealed by a 1D gel, a 2D gel, and mass spectrometry methods, respectively. These data indicate that the MIXED6 IgY antibodies can be used to specifically deplete abundant human proteins to enhance the detection of low-abundance proteins in plasma proteome.

4.3.7 High-Throughput Proteomic Sample Processing Is Facilitated by Use of FPLC Techniques

In order to extend the use of the IgY antibodies as a plasma depletion reagent, the conjugated antibody beads, including antibodies for albumin, IgG, and transferrin, were packed into small columns for the ABI Vision workstation. As shown in Figure 4.7A, all three proteins were removed from the starting material and eluted accordingly. The depletion and elution can be semiautomatically operated with the workstation by monitoring the ultraviolet absorbance, pH, flow resistance, and flow rate of the column chromatography. In Figure 4.7B, even after 21 cycles of

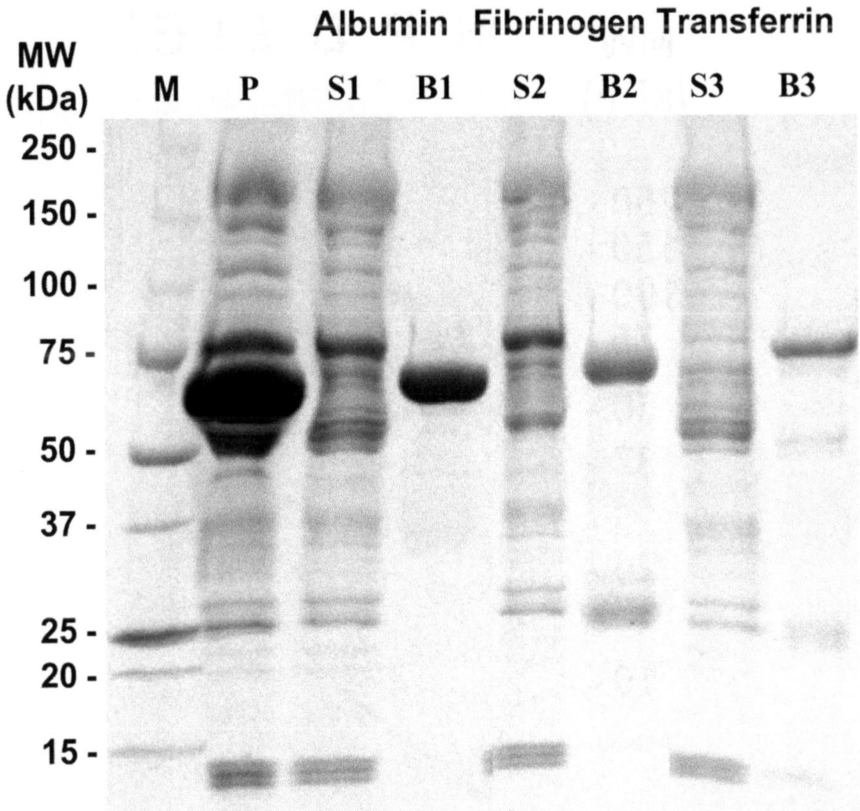

FIGURE 4.5 Separation of abundant rat plasma proteins using three specific anti-human plasma protein IgY gels. M, molecular weight marker; P, 1:8 diluted citrated rat plasma pre-depletion; S1, P after two batch depletions on an anti-HSA gel; B1, eluted bound protein from anti-HSA gel; S2, P after two batch depletions on an antifibrinogen gel; B2, eluted bound protein from anti-fibrinogen gel; S3, P after two batch depletions on an anti-transferrin gel; B3, eluted bound protein from anti-transferrin gel. A 4 to 20% gradient SDS-PAGE was used.

depletion, the quality of the flow-through material and eluted proteins stayed the same. This indicated that our workstation approach can be developed into a high-throughput format for plasma depletion. This approach will enhance our ability to process and study large amount of clinical or animal study samples in an efficient manner.

4.4 DISCUSSION

The examination of the human serum and plasma proteome and those of other mammalian species is becoming important, due to their clinical accessibility. Plasma is

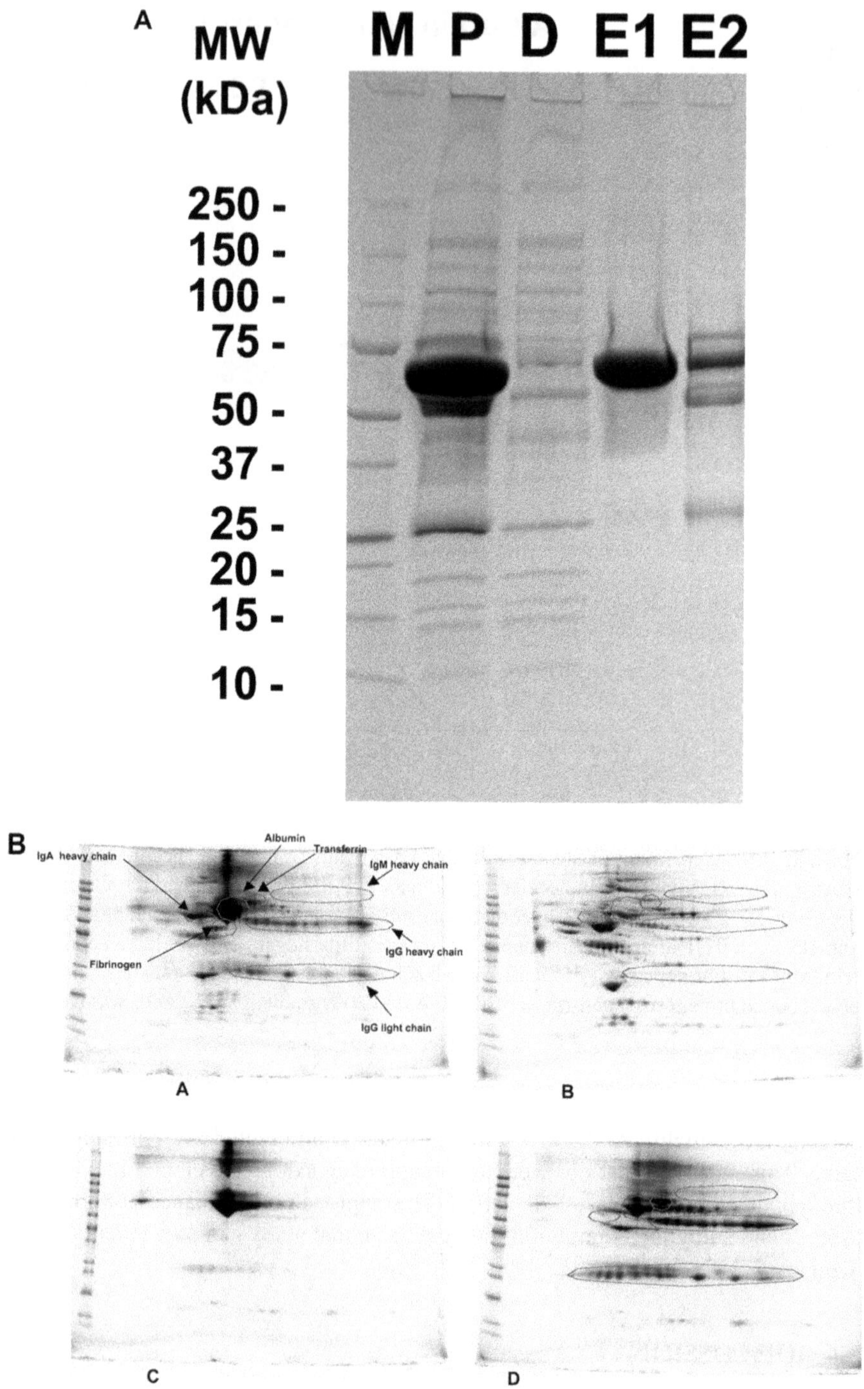

FIGURE 4.6 (Continued)

C

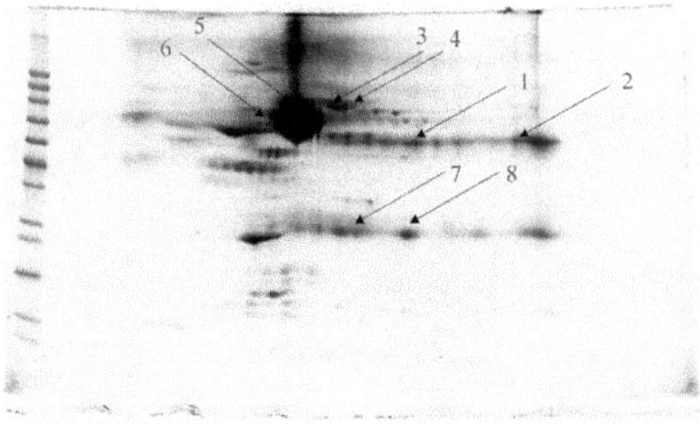

Spot #	Protein Name	Accession #	MW (kD)	pI	# of Peptides	Coverage	Mowse Score
1	Ig gamma-1 chain C region	P01857	36.1	8.46	11	43.63%	161
2	Ig gamma-1 chain C region	P01857	36.1	8.46	9	36.36%	104
3	Transferrin	P02787	77	6.87	28	45.85%	395
4	Transferrin	P02787	77	6.87	43	52.29%	662
5	Serum albumin precursor	P02768	69.3	5.92	24	50.74%	361
6	Serum albumin precursor	P02768	69.3	5.92	8	14.61%	86
7	Ig kappa chain C region	P01834	11.6	5.6	5	80.19%	54
8	Ig kappa chain C region	P01835	11.6	5.6	5	80.19%	72

FIGURE 4.6 (A) Mixed-6 IgY depletion of human plasma. 1D electrophoresis of human plasma depleted samples using the MIXED6 IgY gel in a two-step spin column format, with an anti-HSA antibody spin column followed by a spin column filled with anti-HSA and 5 other antibodies. M, molecular weight marker; P, 1:8 dilution citrated human plasma before depletion; D, P after depletion with MIXED6 IgY gels (two-column system) containing α HSA, α-IgG, α-fibrinogen, α-transferrin, α-IgA, and α-IgM. E1, eluted bound proteins from α-HSA gel; E2, eluted bound proteins from MIXED6 gel. A 4 to 20% gradient SDS-PAGE was used; (B) 2D gel analysis (pI 3–10) of MIXED6 IgY depletion of human plasma. Protein (75 μl) from each sample shown in Figure 4.6A was reduced and alkylated prior to separation in 2D gels. A, human plasma before depletion; B, human plasma after removal of serum albumin, transferrin, fibrinogen, IgA, IgG, and IgM; C, eluted bound proteins to α-HSA IgY gel; D, eluted bound proteins to MIXED6 IgY gel. The absence of associated proteins in the albumin fraction (C) from human plasma is indicative of the high specificity of the antibody for albumin; (C) MALDI-MS identification of depleted plasma proteins. All spectra were automatically analyzed by the BioinformatIQ integrated suite of bioinformatics tools from Proteome Systems. Protein identifications were assigned by comparing peak lists to a database containing theoretical tryptic digests of NCBI and Swiss Prot sequence databases. The identification of the protein is evaluated based on percent coverage, MOWSE score, number of peptide matches, peak intensity, and match of pI and molecular weight with the location of the protein on the 2D gel.

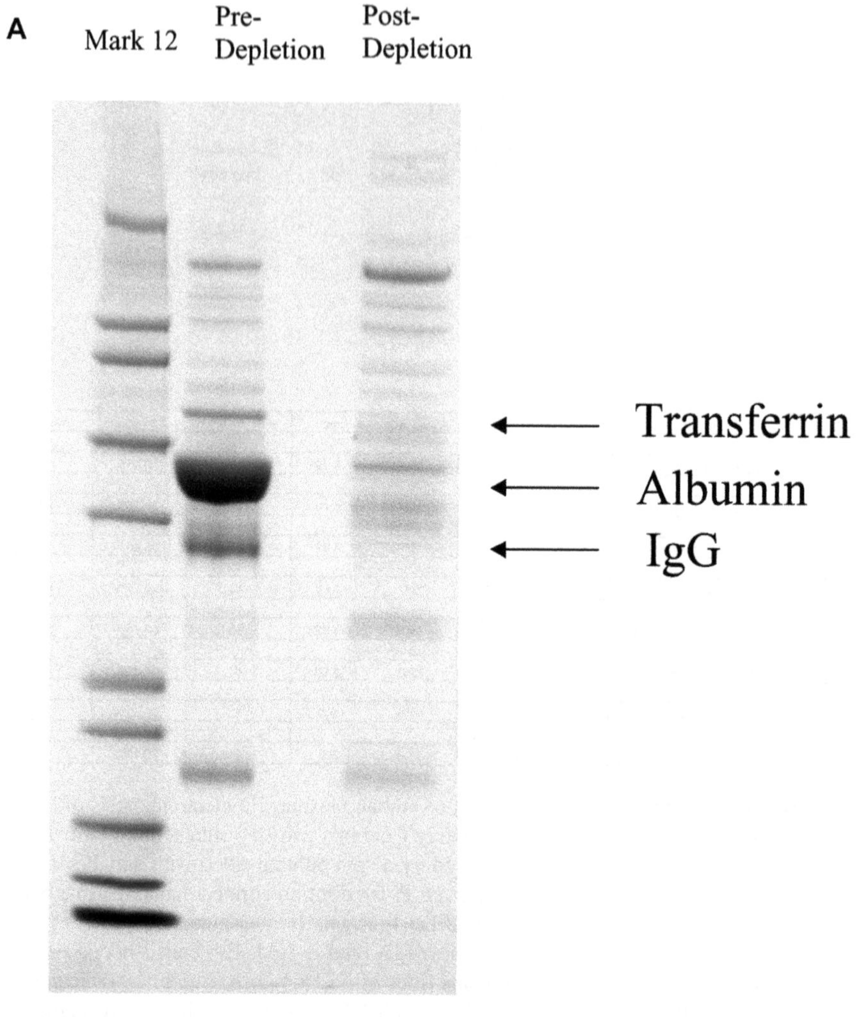

FIGURE 4.7 (Continued)

by far the most readily accessible mammalian tissue from human or animal subjects, allowing many studies for patient monitoring and animal model testing. Traditionally, blood is by far the most abundant tissue stored for subsequent analyses. Serum and plasma are believed to contain many yet uncharacterized posttranslationally modified forms of proteins or peptides, but in low abundance. These modified forms potentially reflect disease states and thus may be important as diagnostic markers, prognostic markers, clinical trial biomarkers, or drug targets.

The most abundant plasma proteins, roughly 12 proteins, make up about 93% of the protein mass of plasma.[9] However, the vast majority of proteins that may be

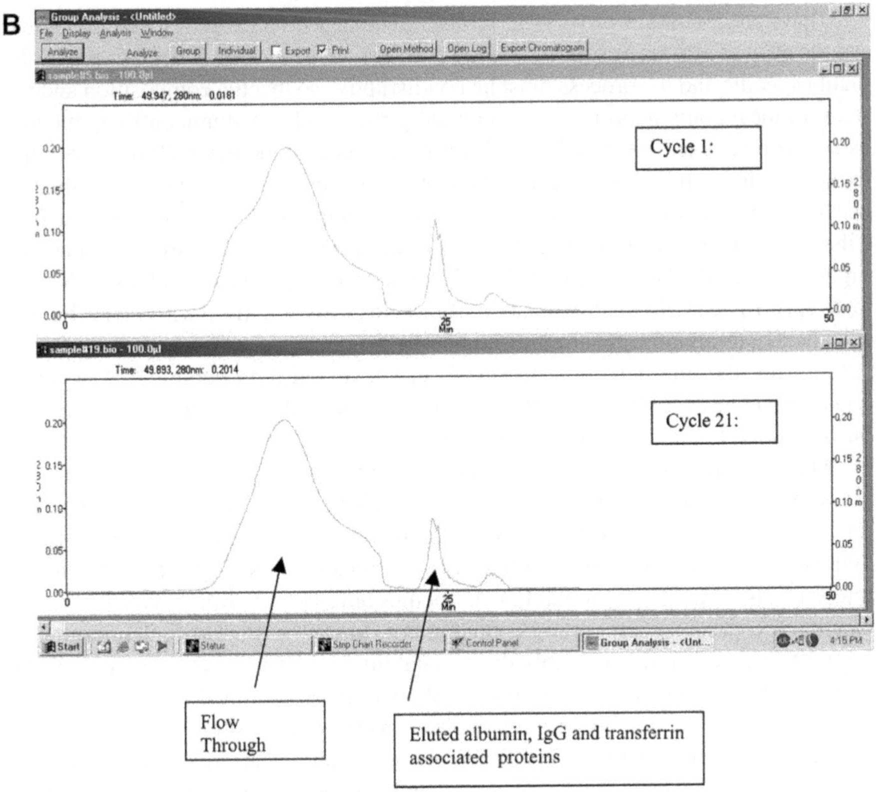

FIGURE 4.7 High-throughput automated depletion of human serum albumin using cross-reactive anti-human serum albumin IgY antibody and ABI Vision Workstation. (A) One-dimensional electrophoresis of human serum albumin-, transferrin-, and IgG-depleted samples using anti-human serum albumin (HSA), transferrin, and IgG IgY gel with the ABI Vision Workstation. The complete removal of albumin, transferrin, and IgG in the flow-through sample indicates good protein specificity and binding affinity of the IgY antibodies; (B) chromatography of human serum albumin, IgG, and transferrin depletion with Seppro gels and ABI Vision Workstation. The data indicate that the columns can effectively be reused at least 21 times with no loss of specificity or capacity.

biomarkers for health and disease in various organs are present in much lower quantities. Without proper depletion methodologies to specifically remove highly abundant proteins, many low-abundance proteins are not detectable with even the most sensitive mass spectrometer. This challenge is further complicated by the huge dynamic range (10^{10}) of plasma protein concentration. Current instrumentation, such as protein microarrays, liquid chromatography–mass spectrometry (LC-MS), or 2D gels, can only resolve about 10^3 to 10^4 distinct proteins.[33] High-abundance plasma proteins are also the main cause of assay background or assay noise, particularly in multiplex setting.[9]

There are a number of ways to remove abundant proteins in a given biological sample such as plasma or serum.[34,35] However, the challenge is that depletion must be highly specific and the process must be nondisruptive so that the composition and the status of the remaining proteins are maximally preserved. An immunoaffinity method using antibodies is best suited for this purpose, since antibodies with high specificity and avidity can be obtained and the interaction between antigens and antibodies occur under physiological conditions.[36,37] Unlike immobilized textile dyes, such as Cibacron Blue–based resins that nonspecifically remove albumin from plasma (see Figure 4.1), antibodies of either IgG or IgY can selectively remove albumin. Recent mass spectrometry (MS) identification of proteins that bind to Cibacron Blue but not specifically to albumin resulted in 60 unique protein species.[38] Another challenge is to ensure that the depletion process is effective and reproducible.[39] A specific, quantitative, and reliable method is required. Immunoaffinity capture using highly purified polyclonal antibodies is the best choice.

Mammalian IgG antibodies from rodents, rabbits, goats, or other mammals have cross-reactivity with human plasma proteins such as complements, rheumatoid factors, Fc receptors, and human anti-mouse antibodies (HAMA). Using IgG to deplete human plasma proteins may result in nonspecifically removing those proteins of interest.[32] In contrast, polyclonal IgY antibodies have been shown to have less cross-reactivity to these human protein contaminants. Since the protein depletion process requires large quantities of antibodies, scalability of the production of polyclonal IgG antibodies is challenging. However, the nature of continual and accumulative production of IgY via collecting eggs yolks can effectively overcome this difficulty. In addition to being suitable for large-scale and economical production processes, IgY antibodies also have higher capacity than IgGs for some antigens, such as serum albumins (see Table 4.1).

Anti-HSA IgY gels were used to deplete HSA from human serum samples. The specificity and capacity of depletion were assessed. It was found that anti-HSA IgY-gel selectively and effectively removed HSA from human serum samples. The capacity assay showed that anti-HSA IgY gel worked best when the molar ratio between IgY and HSA was at 1:1 and dilution of the human serum samples was at a range of 1:5 to 1:10.[32]

Proteomics studies use various types of animal models to determine mechanism, pharmacokinetics, toxicity, and preclinical efficacy studies. It is useful and convenient to analyze human samples and those of animal models using the same assay reagents and methods. Therefore, it will be ideal if the same antibodies used for human sample depletion can also be used for animal samples. Due to the extensive homology of family proteins between human and other mammalian species (examples shown in Figure 4.4), IgG antibodies against human abundant plasma proteins raised in animals, such as rabbits and goats, do not have strong cross-reactivity to the same proteins from other mammalian species, such as rats and mice (Table 4.1). For example, it was found that anti-HSA IgG antibodies were not effective in removing rat or mouse albumin.

In contrast, due to their genetic diversity, common sequences of human and other mammalian proteins are less homologous to those of chicken. IgY antibodies against these proteins have a broader recognition spectrum among mammalian species (see Figure 4.4A). Anti-HSA IgY antibody gel was applied to remove albumin from the serum samples of mouse, rat, goat, and pig (Figure 4.4B). Anti-HSA IgY gel depleted the albumins in those four samples of mammalian species with a practical efficiency comparable to that of removing human albumin. This illustrates the potential of IgY for cross-species usage in abundant, highly conserved proteins for sample preparation in proteomics assays.

The specific removal of abundant proteins from plasma across different species can significantly increase the relative protein concentration of low-abundance proteins being studied and thus enable their detection and quantitation. This approach will greatly enhance the detection sensitivity and ability to characterize low-abundance proteins from plasma using subsequent proteomics analyses such as 2D gel MS or LC-MS. Our study illustrates a novel method for the selective removal of albumin and other abundant proteins from serum and plasma of human, rat, mouse, goat, cow, or pig with avian IgY antibodies. These affinity-purified polyclonal IgY antibodies will be applied to meet the needs of protein separation, quantification, process development, and standardization. The advantage of this separation scheme compared to other methods is that multiple abundant proteins can be simultaneously removed (see Figures 4.5 and 4.6). Anti-HSA, anti-fibrinogen, anti-transferrin, and anti-IgG-Fc IgY antibodies have a high affinity and specificity for their corresponding targets from multiple mammalian species. This was confirmed by rigorous MS identification of predepleted and postdepleted proteins from 2D gels (Figure 4.6B and Figure 4.6C). Indeed, we are currently developing a MIXED12 Sepproproduct that is expected to selectively remove more than 95% of abundant plasma proteins. Use of this material will allow the investigator to load at least 20 times more low-abundance protein in a subsequent fractionation system (2D gel, LC, etc.) prior to MS analyses, significantly improving detection of rare biomarkers.

These antigen affinity purified polyclonal chicken antibodies provide a significant advantage over other available immunoaffinity methods that use murine monoclonal antibodies, since the latter rely on single epitopes for protein capture. In disease conditions, different posttranslational modifications can occur and alter the binding epitope of the antigen. By using polyclonal antibodies, one will have a better chance to capture various modified forms of disease-related proteins. Other immunoaffinity-based approaches that are dependent on goat-, rabbit-, or mouse-produced antibodies lack reactivity across multiple mammalian species and thus are typically limited to human samples.

Often, the albumin-associated fractions are excluded from further studies. In our study[40] and those done by other groups[41] albumin-associated protein complexes under normal or diseased conditions contain many uncharacterized proteins. The extremely high specificity of anti-HSA IgY gels permits effective implementation of Albuminomics™, largely avoiding artifactual capture and analysis of nonphysiologically bound proteins due to their association with the Fc portion of IgG or to nonspecific Cibacron Blue–type dyes.

In the multiple-column format permitted by the high-throughput liquid chromatography apparatus Vision™ LC system, it is feasible to separately elute each of the associated proteins for individual characterization, such as albumin- or IgG-associated proteins. The system allows the simultaneous monitoring of flow rate, pressure, pH, and protein elution. In this format, the IgY antibody gels can be regenerated more than 21 times (Figure 4.7B) and beyond (undocumented data) to process the same sample or multiple samples after proper washing and recycling of the column. The high reusability of the column enables multiple rounds of depletion of the same sample to achieve complete removal of undesirable proteins at a significant scale. The LC column format is also amenable to automation. To analyze clinical samples, speed and reproducibility are critical for the accumulation of statistically reliable information.

This novel technical platform, incorporating Seppro products with the high-throughput Vision™ LC system, can be exploited for the depletion of many abundant proteins from serum and plasma, therefore enabling enhanced detection of low-abundance proteins in proteomic studies. In this study, HSA, IgA, IgG, IgM, transferrin, and fibrinogen were specifically and simultaneously removed (Figure 4.6A). Normally, serum is collected from plasma through clotting and the removal of fibrinogen. Using an anti-fibrinogen IgY antibody, plasma can be directly analyzed without activation of many proteolytic enzymes that occur during clotting. By further separating other abundant proteins from plasma with IgY antibody conjugates, one can advance our current understanding of low-abundance plasma proteins in disease progression and intervention. This technology will advance the discovery and validation of biomarkers, drug targets, and therapeutic proteins and antibodies.

ACKNOWLEDGMENTS

We are grateful for the support from Jennifer Rutherford and Mai-Loan Nguyen in MS analysis. Dr. Jim Jersey has given productive comments on the manuscript.

REFERENCES

1. Sinz, A., Bantscheff, M., Mikkat, S., Ringel, B., Drynda, S., Kekow, J., Thiesen, H.J., and Glocker, M.O., Mass spectrometric proteome analyses of synovial fluids and plasmas from patients suffering from rheumatoid arthritis and comparison to reactive arthritis or osteoarthritis, *Electrophoresis*, 23, 3445–3456, 2002.
2. Imam-Sghiouar, N., Laude-Lemaire, I., Labas, V., Pflieger, D., Le Caer, J.P., Caron, M., Nabias, D.K., and Joubert-Caron, R., Subproteomics analysis of phosphorylated proteins: Application to the study of B-lymphoblasts from a patient with Scott syndrome, *Proteomics*, 2, 828–838, 2002.
3. Vejda, S., Posovszky, C., Zelzer, S., Peter, B., Bayer, E., Gelbmann, D., Schulte-Hermann, R., and Gerner, C., Plasma from cancer patients featuring a characteristic protein composition mediates protection against apoptosis, *Mol. Cell. Proteomics*, 1, 387–393, 2002.
4. Qin, L.X. and Tang, Z.Y., The prognostic molecular markers in hepatocellular carcinoma. *World J. Gastroenterol.*, 8, 385–392, 2002.
5. Johnson, P.J., A framework for the molecular classification of circulating tumor markers, *Ann. N.Y. Acad. Sci.*, 945, 8–21, 2001.

6. Tissot, J.D., Vu, D.H., Aubert, V., Schneider, P., Vuadens, F., Crettaz, D., and Duchosal, M.A., The immunoglobulinopathies: From physiopathology to diagnosis, *Proteomics*, 2, 813–824, 2002.
7. Hutter, G. and Sinha, P., Proteomics for studying cancer cells and the development of chemoresistance, *Proteomics*, 1, 1233–1248, 2001.
8. Lathrop, J.T., Anderson, N.L., Anderson, N.G., and Hammond, D.J., Therapeutic potential of the plasma proteome, *Curr. Opin. Mol. Ther.*, 5, 250–257, 2003.
9. Anderson, N.L., and Anderson, N.G., The human plasma proteome: History, character, and diagnostic prospects, *Mol. Cell. Proteomics*, 1, 845–867, 2002.
10. Ruepp, S.U., Tonge, R.P., Shaw, J., Wallis, N., and Pognan, F., Genomics and proteomics analysis of acetaminophen toxicity in mouse liver, *Toxicol. Sci.*, 65, 135–150, 2002.
11. Pieper, R., Gatlin, C.L., Makusky, A.J., Russo, P.S., Schatz, C.R., Miller, S.S., Su, Q., McGrath, A.M., Estock, M.A., Parmar, P.P., Zhao, M., Huang, S.T., Zhou, J., Wang, F., Esquer-Blasco, R., Anderson, N.L., Taylor, J., and Steiner, S., The human serum proteome: Display of nearly 3700 chromatographically separated protein spots on two-dimensional electrophoresis gels and identification of 325 distinct proteins, *Proteomics*, 3, 1345–1364, 2003.
12. Anderson, N.L., Polanski, M., Pieper, R., Gatlin. T., Tirumalai, R.S., Conrads, T.P., Veenstra, T.D., Adkins, J.N., Pounds, J.G., Fagan, R., and Lobley, A., The human plasma proteome, *Mol. Cell. Proteomics*, 3, 311–326, 2004.
13. Steel, L.F., Trotter, M.G., Nakajima, P.B., Mattu, T.S., Gonye, G., and Block, T., Efficient and specific removal of albumin from human serum samples, *Mol. Cell. Proteomics*, 2, 262–270, 2003.
14. Pieper, R., Su, Q., Gatlin, C.L., Huang, S.-T., Anderson, N.L., and Steiner, S., Multi-component immunoaffinity subtraction chromatography: An innovative step towards a comprehensive survey of the human plasma proteome, *Proteomics*, 3, 422–432, 2003.
15. Hinerfeld, D., Innamorati, D., Pirro, J., and Tam, S.W., Serum plasma depletion with chicken immunoglobulin Y antibodies for proteomic analysis from multiple Mammalian species. *J. Biomol. Tech.*, 15, 184–190, 2004.
16. Warr, G.W., Magor, K.E., and Higgins, D.A., IgY: Clues to the origins of modern antibodies, *Immunol. Today*, 16, 392–398, 1995.
17. Gassmann, M., Thommes, P., Weiser, T., and Hubscher, U., Efficient production of chicken egg yolk antibodies against a conserved mammalian protein, *FASEB J.*, 4, 2528–2532, 1990.
18. Zhang, W.-W., The use of gene-specific IgY antibodies for drug target discovery, *Drug Discovery Today,* 8, 364–371, 2003.
19. Ikemori, Y., Peralta, R.C., Kuroki, M., Yokoyama, H., and Kodama, Y., Research note: Avidity of chicken yolk antibodies to enterotoxigenic *Escherichia coli* fimbriae, *Poultry Sci.*, 72, 2361–2365, 1993.
20. Lemamy, G.J., Roger, P., Mani, J.C., Robert, M., Rochefort, H., and Brouillet, J.P., High-affinity antibodies from hen's-egg yolks against human mannose-6-phosphate/insulin-like growth-factor-II receptor (M6P/IGFII-R): Characterization and potential use in clinical cancer studies, *Int. J. Cancer*, 80, 896–902, 1999.
21. GenWay Biotech, http://www.genwaybio.com, (last accessed August 2005).
22. Hatta, H., Tsuda, K., Akachi, S., Kim M., and Yamamoto, T., Oral passive immunization effect of anti-human rotavirus IgY and its behavior against proteolytic enzymes, *Biosci. Biotech. Biochem.*, 57, 450–454, 1993.
23. Sun, S., Mo, W., Ji, Y., and Liu, S., Use of nitrocellulose films for affinity-directed mass spectrometry for the analysis of antibody/antigen interactions, *Rapid Commun. Mass. Spectrom.*, 15, 708–712, 2001.

24. Davalos-Pantoja, L., Ortega-Vinuesa, J.L., Bastos-Gonzalez, D., and Hidalgo-Alvarez, R., A comparative study between the adsorption of IgY and IgG on latex particles, *Biomater. Sci. Polym. Ed.*, 11, 657–673, 2000.
25. Hadge, D. and Ambrosius, H., Evolution of low molecular weight immunoglobulins. IV. IgY-like immunoglobulins of birds, reptiles and amphibians, precursors of mammalian IgA, *Mol. Immun.*, 21, 699–707, 1984.
26. Larsson, A. and Mellerstedt, H., Chicken antibodies: A tool to avoid interference by human anti-mouse antibodies in ELISA after in vivo treatment with murine monoclonal antibodies, *Hybridoma*, 11, 33–39, 1992.
27. Carlander, D., Stålberg, J., and Larsson, A., Chicken antibodies: A clinical chemistry perspective, *Ups. J. Med. Sci.*, 104, 179–189, 1999.
28. Larsson, A., Karlsson-Parra, A., and Sjoquist, J., Use of chicken antibodies in enzyme immunoassays to avoid interference by rheumatoid factors, *J. Clin. Chem.*, 37, 411–414, 1991.
29. Carlander, D., Kollberg, H., Wejaker, P.E., and Larsson, A., Peroral immunotherapy with yolk antibodies for the prevention and treatment of enteric infections. *Immunol. Res.*, 21, 1–6, 2000.
30. Vikinge, T.P., Askendal, A., Liedberg, B., Lindahl, T., and Tengvall, P., Immobilized chicken antibodies improve the detection of serum antigens with surface plasmon resonance (SPR), *Biosens. Bioelectron.*, 13, 1257–1262, 1998.
31. Davalos-Pantoja, L., Ortega-Vinuesa, J.L., Bastos-Gonzalez, D., and Hidalgo-Alvarez, R., Colloidal stability of IgG- and IgY-coated latex microspheres. *Colloids Surf B. Biointerfaces*, 20, 165–175, 2001.
32. Fang, X., Curran, K.W., Huang, L., Xiao, W., Strauss, W., Harvie, G., Feitelson, J.S., and Zhang, W.-W., Polyclonal gene-specific igy antibodies for proteomics and abundant plasma protein depletion, *Front. Pharma. Biotech.*, 4, 1–23, 2003.
33. Tam, S.W., Wiese, R., Lee, S., Gilmore, J., and Kumble, K.D., Simultaneous analysis of eight human Th1/Th2 cytokines using microarrays, *J. Immunol. Methods,* 261, 157–165, 2002.
34. Phillips, T.M. and Dickens, B.F., *Affinity and Immunoaffinity Purification Techniques*, Eaton Publishing, BioTechniques Books, Natick, MA, 2000.
35. Winzor, D.J., Quantitative affinity chromatography, *J. Biochem. Biophys. Methods*, 49, 99–121, 2001.
36. Sisson, T.H. and Castor, C.W., An improved method for immobilizing IgG antibodies on protein A-agarose, *J. Immunol. Methods*, 127, 215–220, 1990.
37. Pieper, R., Su, Q., Gatlin, C.L., Huang, S.T., Anderson, N.L., and Steiner, S., Multi-component immunoaffinity subtraction chromatography: An innovative step towards a comprehensive survey of the human plasma proteome, *Proteomics*, 3, 422–432, 2003.
38. Bailey, J., Zhang, K., Zolotarjova, N., Nicol, G., and Szafranski, C., Removing high-abundance proteins from serum, *Gen. Eng. News*, 23, 32–36, 2003.
39. Lollo, B.A., Harvey, S., Liao, J., and Stevens, A.C., Improved two-dimensional gel electrophoresis representation of serum proteins by using ProtoClear, *Electrophoresis*, 20, 854–859, 2000.
40. Tam, S.W. and Hinerfeld, D., unpublished results, 2004.
41. Zhou, M., Lucas, D.A., Chan, K.C., Issaq., H.J., Petricoin, E.F., III, Liotta, L.A., Veenstra, T.D., and Conrads, T.P., An investigation into the human serum "interactome," *Electrophoresis*, 25, 1289–1298, 2004.

5 Immunoaffinity Depletion of High-Abundant Proteins for Proteomic Sample Preparation

Nina Zolotarjova, Barry Boyes, James Martosella, Liang-Sheng Yang, Gordon Nicol, Kelly Zhang, Cory Szafranski, and Jerome Bailey

CONTENTS

5.1 INTRODUCTION

Human serum is frequently used in proteomics analysis for the discovery of new disease markers, drug targets, or studying protein expression patterns. This body fluid represents the most complex sample of the human proteome, composed of homeostatic blood proteins as well as tissue leakage proteins.[1] Difficulties arise in the proteomic analysis of serum due to the extreme concentration range of target proteins over 10 to 12 orders of magnitude. High-abundant proteins such as albumin, IgG, transferrin, haptoglobin, IgA, and alpha1-anti-trypsin represent up to 85% of the total protein mass in serum (Figure 5.1). These major protein constituents interfere with identification and characterization of important moderate- and low-abundant proteins by limiting the dynamic range of mass spectral and electrophoretic analysis. During protein isolation, separation, and analysis, these six proteins often mask the detection of the more important low-abundant proteins that are of high interest as biomarkers of disease or drug targets. In one- and two-dimensional gel electrophoresis (1DGE and 2DGE), for example, the spots or bands due to these six highly abundant proteins, as well as their fragments, often overlap or completely mask large regions of the gel, making detection of the myriad low-abundant proteins very difficult, if not impossible. Moreover, proteomic analysis methods commonly include an electrophoretic or chromatographic separation step which, of course, has a finite mass loading tolerance. The presence of a large quantity of high-abundant proteins limits the mass load of targeted proteins that can be initially sampled by these separation methods.

Mass spectrometry (MS), or combined liquid chromatography–mass spectrometry (LC-/MS), analyses enable identification of proteins by tandem MS analysis of peptide fragments. These identities can be obtained by database matching of peptide MS/MS spectra to previously known protein sequence information, or by *de novo* sequence determination of novel proteins. The few high-abundant proteins described above interfere in this analysis when the signals from their peptide fragments dominate the mass spectrum. One way to alleviate the inherent problems of high-abundant protein interference in proteomic studies is to deplete these proteins as completely as possible, while not removing the proteins of interest.

Removal of high-abundant proteins from serum or plasma has been accomplished by many methods, ranging through the use of various solid-phase and liquid-liquid extraction or precipitation methods. Solid-phase extraction methods have had the most success and have taken a variety of forms, including lectin affinity chromatography,[2] controlled pore membranes separations,[3] textile dye-modified LC media (for example, Cibacron Blue),[4] and affinity media based on antibody-antigen interactions.[5] However, many techniques inefficiently remove only one or at most two proteins and have poor specificity.[6,7] Ion exchange and controlled pore size membranes are used to fractionate proteins based on either charge or molecular dimensions. Since proteomic samples often contain many proteins of similar charge and size, these devices are prone to specificity and capacity concerns. The available membrane materials do not appear to have the requisite efficiency for targeted protein removal and leave significant quantities of interfering high-abundant proteins behind in samples.[3] A well-known method to remove albumin from serum and plasma samples uses dye affinity chromatography

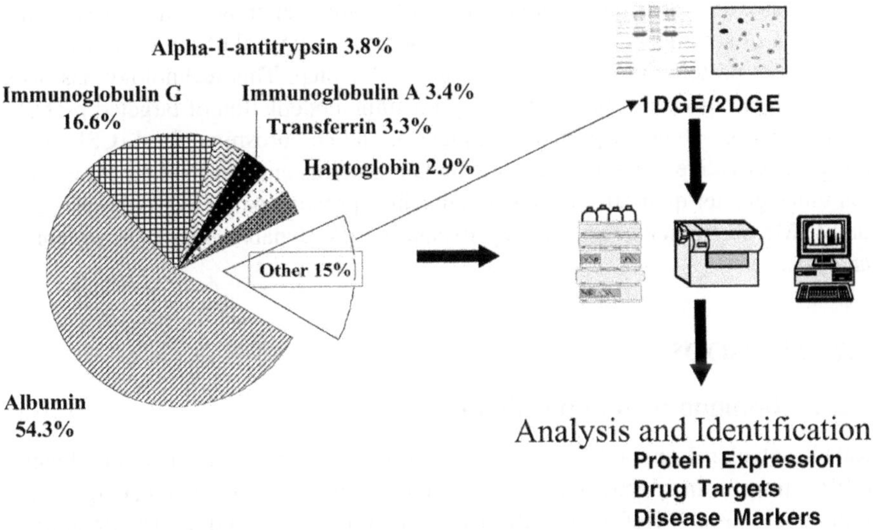

FIGURE 5.1 Composition of proteins in human serum. The protein composition is schematically depicted based on mass abundance in normal human serum. The six high-abundant proteins removed by the immunoaffinity column comprise approximately 85% of the total protein mass in human serum.

resins modified with Cibacron Blue or with related versions of this chlorotriazine dye. Previous studies have shown that the removal of albumin on Cibacron Blue is incomplete, and that many other lower abundant proteins are removed by this approach as well, either through interaction with the media base material or with albumin itself.[7] The poor selectivity of Cibacron Blue resins may result in the removal of proteins of interest through absorptive interactions that have both hydrophobic and ion-exchange characteristics and are thus difficult to optimize for specific removal of targeted high-abundant proteins. Commonly used affinity LC techniques for the removal of serum immunoglobulins are based on the use of Protein A– or Protein G–modified media.[8,9] These materials have also been further modified with antibodies to eliminate albumin or additional members of the immunoglobulin family, or both.[10] Conventional affinity materials prepared in this way typically remove only one protein at a time, and even in devices that claim removal of more than one protein (e.g., albumin and IgG), the removal is often incomplete, and the devices are rather short-lived (offering only one to several reuses per device).

Inefficient protein removal and the lack of specificity and reusability of these solid phase sample preparative devices has prompted the development of more efficient and specific protein removal methods.[11] Ideally, a depletion column should possess little or no binding of nontargeted proteins and preferentially should bind several high-abundant proteins simultaneously. Antibody-based selection strategies can offer the desired characteristics, since immunoaffinity separations are known to be capable of exquisite specificity and acceptable sample capacity.[12,13]

We have developed an immunoaffinity LC column approach that allows for the reduction of serum sample complexity by selective removal of the six most abundant serum proteins in a single chromatographic step. This technology has been demonstrated to exhibit practical utility for immunodepletion of targeted proteins present in high abundance in serum, plasma, and cerebrospinal fluid (CSF) samples. The effective removal of high-abundant proteins enables loading of seven to ten times greater quantities of lower abundance proteins onto electrophoresis gels or LC-MS instrumentation, thereby expanding the dynamic range of proteomic analysis.

5.2 METHODS

5.2.1 Immunodepletion Column

Rabbit polyclonal antibodies to six major serum proteins—human serum albumin (HSA), transferrin, alpha1-anti-trypsin, haptoglobin, immunoglobulin A (IgA), and immunoglobulin G (IgG) were affinity purified on corresponding protein antigen columns.[14] The resulting affinity-purified antibodies were covalently coupled to porous beads via their Fc region and cross-linked. Spatially controlled cross-linking of the antibodies resulted in preferential orientation of the antibody binding sites away from the solid-phase surface, supporting maximum binding capacity of targeted proteins.

Each affinity resin was analyzed to determine its binding capacity. The six affinity resins were packed separately into 2.1×30 mm columns, which permit binding capacities to be measured by serial injections of purified antigens. Column capacities were considered saturated when 10% of the total injected antigen was detected in the flow-through peak (absorbance at 280 nm). A final mix of the resins was made based on the relative abundance of each high-abundant protein in serum, combined with the measured individual resin capacity for a targeted protein. The mix was adjusted to allow for simultaneous removal of the six high-abundant proteins in human serum during one column run. In practice, the column is loaded with serum at about 90% of its effective maximum capacity in order to prevent column overload that can result from individual differences in protein concentration between samples.

The immunoaffinity column requires a proprietary two-buffer system (Buffer A and Buffer B) for operation. The two buffers provide the means to separate low-abundant proteins from high-abundant proteins and regenerate the column, all in about 20–30 minutes per injection. Buffers A and B are optimized to minimize coadsorption of nontargeted proteins to the column packing and to ensure reproducibility of column performance and long column lifetime. Buffer A (Agilent Technologies) is a salt-containing neutral buffer (pH 7.4) used for loading, washing, and reequilibrating the column. Buffer B (Agilent Technologies) is a low pH urea buffer used for eluting the bound high-abundant proteins from the column. Serum samples are injected onto the column and the high-abundant proteins are simultaneously removed as low-abundant proteins pass through in the flow-through fraction. After collecting the low-abundant proteins and washing the column, the bound proteins are eluted with Buffer B and the column is reequilibrated with Buffer A.

Through this series of sample loading, washing, collection, and reequilibration, multiple serum samples can be processed and depleted. After each pass of serum through the column, flow-through fractions containing low-abundant proteins can be pooled and concentrated for downstream postaffinity processing. The methods outlined below demonstrate a typical workflow for immunodepleting a given quantity of human serum samples. Similar procedures have been employed successfully for cerebrospinal fluid, amniotic fluid, and for urine analysis.

5.2.2 Sample Collection and Processing

Human serum was obtained from a healthy female volunteer. The blood was collected in a BD vacutainer tube with SST gel and clot activator (Becton Dickinson). Shortly after clot formation, the sample was centrifuged at 1000*g* for 15 minutes. The serum was removed, aliquoted, and stored at −70°C. Prior to injection on the immunodepletion system, serum samples were thawed and diluted five times with Buffer A, then "Complete" protease inhibitors cocktail (Roche) was added. The sample was transferred to a 0.22 μm pore size spin tube (Agilent Technologies) for removal of particulates by centrifugation at 16,000*g* for 1 minute at room temperature. The prepared samples were maintained at 4°C, either in the refrigerator or in the temperature-controlled autosampler stage of the Agilent 1100 LC system.

5.2.3 Chromatographic Set-Up

Analysis of proteomic serum samples often requires either high throughput of a number of different samples, or serial processing of larger volumes (pooling) of a single serum sample in order to bring low-abundant proteins of interest into detectable levels. Therefore, it is desirable to process serum samples with minimal workflow variability to ensure the precision and reproducibility of results. A fully automated LC station for sample processing ensures run-to-run consistency and increases the total throughput. High-performance liquid chromatography was performed on an automated Agilent 1100 LC system (Chemstation A.10.01 software) using a standard column configuration at ambient temperature. The LC station components included a binary pump, degasser, solvent cabinet (containing Buffer A and Buffer B), autosampler (300 μl loop) with thermostat, diode-array detector (280 nm) with 6 mm flow cell, and an analytical scale thermostatted fraction collector. The LC system setup allowed automated separations of multiple samples and fulfilled necessary requirements such as monitoring, precise fraction collection, injection, and cooling of injected and collected samples.

5.2.4 Immunoaffinity Separation

High-abundant protein removal from crude diluted human serum was performed on a 4.6 × 50 mm immunodepletion column (Agilent Technologies). The recommended column capacity range is between 15 and 20 μl of nondiluted human sera. A larger size 4.6 × 100 mm column (Agilent Technologies) exhibits a serum capacity of about 40 μl. Larger internal diameter columns have also been prepared and tested, with serum loading capacities that are proportional to the volume of packing material employed. Capacity is defined as the amount of undiluted serum that can be loaded on the column such that 99% of the targeted high-abundant proteins are removed for at least 200 injections on a particular column. After sample preparation and dilution, 90 μl (18 μl sera diluted 5× with Buffer A) of sample was injected onto the 4.6 × 50 mm column in 100% Buffer A at a flow rate of 0.25 ml/min for 9.0 minutes (Figure 5.2). After collection of the flow-through fraction, the column was washed and the bound proteins eluted with 100% Buffer B at a flow rate of 1.0 ml/min for 3.5 minutes. Afterward, the column was regenerated by equilibrating it with Buffer A (0% B) for 7.5 minutes for a total run cycle of 20 minutes.

Fraction collection of flow-through proteins was time controlled and corresponded to the ultraviolet (UV) 280 nm absorbance of the eluting proteins. The flow-through fraction was collected into a 1.5 ml tube (Sarstedt) and cooled to 4°C using the thermostatted fraction collector. Bound fractions were collected for analyzing the specificity of the immunodepletion procedure.

5.2.5 Analysis of Flow-Through Fractions by Enzyme-Linked Immunosorbent Assay (ELISA)

Standard sandwich ELISAs[14] were used to determine the completeness of removal of targeted proteins from human serum. Briefly, assay plates were coated with 50 μl of serum samples, flow-through fractions, or purified serum proteins diluted from 1:100 to 1:50,000 in Buffer A. The assay plates were incubated overnight at 4°C then washed with standard phosphate-buffered saline. Nonspecific binding sites were blocked with 200 μl of blocker solution (Bio-Rad) for 2 hours at room temperature.

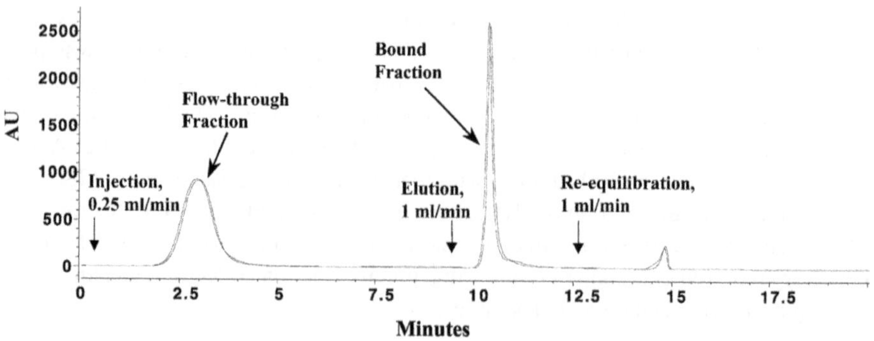

FIGURE 5.2 Chromatograms for the affinity removal of high-abundant proteins from human serum. Diluted human serum (90 μl) was injected on a 4.6 × 50 mm immunoaffinity column (0.25 ml/min) and a flow-through peak (2–4.5 minutes) containing the immunodepleted serum proteins was collected. The column was washed with Buffer A and the targeted high-abundant proteins were eluted with Buffer B (1.0 ml/min). The column was reequilibrated in Buffer A for subsequent injections. The chromatograms show runs 20 and 200 were identical as observed by the overlays—indicating high reproducibility and reusability.

The plate wells were washed with PBS before 100 μl of affinity-purified rabbit anti-human antigen antibodies (see section 5.2.1.) were added at 1:1000 dilutions or 1 μg/ml in blocker solution. After incubation for 2 hours on a shaker at room temperature, the washing procedure was repeated, then the detection antibody, horseradish peroxidase-conjugated goat anti-rabbit IgG antibody (Sigma), was added for a 1-hour incubation. Finally, the wells were washed as before, and 100 μl of 3,3′, 5,5′-tetramethylbenzidine (TMB) liquid substrate (Sigma) was added. The absorbance was measured after 15 minutes at 655 nm in an ELISA plate reader and plotted versus the standard concentration. Standard curves were fitted to a four-parameter equation by nonlinear regression analysis.[15] All samples were analyzed in duplicate or triplicate, and multiple sample dilutions were compared with standard curves to determine the concentrations of targeted proteins in crude serum and immunodepleted flow-through fractions.

5.2.6 Processing of Depleted Serum and Bound Fractions for Down-Stream Analysis

Flow-through fractions or bound fractions from several injections were pooled and buffer-exchanged in 4 ml spin concentrators with 5 kDa molecular weight cutoffs (Agilent Technologies). The sample was centrifuged at 7500*g* for 20 minutes at 4°C. Buffer was exchanged into 20 mM Tris-HCl, pH 7.4 by three rounds of addition of the buffer, with centrifugation for 20 minutes each time. The concentrated samples were aliquoted and stored at −70°C until analysis. Protein concentrations were analyzed using a BCA protein assay kit (Pierce).

5.2.7 One-Dimensional (1D) and Two-Dimensional (2D) Gel Electrophoresis

1D SDS-PAGE analysis was carried out using Invitrogen Tris-glycine precast gels (4–20% acrylamide, 10 wells, 1 mm) according to the manufacturer's protocol. For 2D electrophoresis, samples were prepared by mixing 250 μg of proteins with 185 μl of rehydration buffer containing 8 M urea, 2% 3-[(3-Cholamidopropyl)dimethylammonio]-1-propanesulfonate (CHAPS), 2% ampholytes, and 20 mM dithiothreitol. Samples were applied on 11 cm immobilized pH gradient (IPG) and pH 3–10 nonlinear strips (Bio-Rad) and processed according to the manufacturer's instructions. The second dimension was carried out on 8 to 16% precast Tris-glycine gels. Proteins were visualized by Coomassie Blue staining with GelCode Blue (Pierce).

5.2.8 Liquid Chromatography–Tandem Mass Spectrometry (LC-MS/MS)

For analysis of the specificity of the immunodepletion column, the bound fraction was resolved by 1D SDS-PAGE. After staining with the Coomassie Blue, protein bands were cut and destained, and the proteins were reduced, alkylated, and digested with trypsin using an in-gel trypsin digestion kit from Agilent Technologies. The peptides were extracted and analyzed by LC-MS/MS on an Agilent 1100 MSD Trap SL. Results were processed by Spectrum Mill software (Agilent Technologies).

Crude human sera and the flow-through fractions from the immunodepletion column, as well as from a Cibacron Blue albumin removal column, were digested with trypsin and analyzed by 2D LC-MS/MS on an Agilent 1100 MSD Trap SL. A strong cation exchange (SCX) column (Polysulfoethyl A, 1 × 50 mm, 5 μm, 300 Å, PolyLC) was used for the first dimension, and a reversed-phase (RP) column (Zorbax 300SB-C18, 5 μm, 0.3 × 150 mm, Agilent Technologies) provided the second dimension. The main flow rate was 0.1 ml/min using 3% acetonitrile/0.1% formic acid. Solvent A was 0.1% formic acid and solvent B was 100% acetonitrile/0.1% formic acid. The flow rate for the reverse-phase separation was 6 μl/min. Injection of 40 μl of a salt solution was used to elute peptides from the SCX column. Peptides eluted with the salt solution from the SCX column were trapped and desalted on a Zorbax 300SB-C18, 5 μm, 5 × 0.3 mm column (Agilent Technologies). The column was then switched inline with the flow for the analytical RP column, and the peptides were eluted off the trap column and onto the RP column. A separate RP run was performed for each salt slice. Higher concentrations of salt were repeated to ensure complete removal of peptides from the SCX column prior to loading the column with the next sample. The gradient for the reverse phase and salt slices is as follows: 5 minutes, 10% B; 90 minutes, 60% B; 92 minutes, 90% B; 97 minutes, 90% B; 100 minutes, 10% B. Salt slices: 10 mM, 25 mM, 50 mM, 75 mM, 100 mM, 200 mM, 500 mM, 1 M, and 2.5 M KCl.

5.2.9 Cibacron Blue Column

We compared the specificity of depletion with the immunoaffinity column to the specificity of Cibacron Blue-modified resin. For these experiments, a HiTrap Blue column (1 ml, Amersham Pharmacia Biotech) containing Cibacron Blue ligand covalently attached to highly cross linked agarose gel was employed. A variety of loading and elution conditions have been examined, and those described here are based on the conditions recommended by the manufacturer. Alterations in the ionic strength of the loading mobile phase can be used to improve the specificity of the separation, but this must be balanced by the requirement of complete removal of HSA from human serum.

To perform HSA depletion, 125 μl of serum were diluted four times with the loading buffer. Sample was loaded onto the column and the column was washed with five column volumes of the loading buffer—50 mM K_2HPO_4, pH 7.0. Elution buffer (50 mM K_2HPO_4, 1.5 M KCl, pH 7.0) was used for the collection of the bound fraction. Both the flow-through and the bound fractions from several runs were collected, pooled, and buffer exchanged into the immunoaffinity column loading Buffer A.

5.2.9.1 Analysis of Cibacron Blue Bound Fraction

The Cibacron Blue bound fraction (1.0 mg) was loaded on an anti-HSA immunoaffinity column (4.6 × 50 mm). The flow-through fractions from several injections were collected, concentrated, and analyzed by 1D SDS-PAGE and LC-MS/MS.

5.2.9.2 Analysis of Cibacron Blue Flow-Through Fraction

The Cibacron Blue flow-through fraction was reduced, alkylated, digested with trypsin, and analyzed by 2D LC-MS/MS on an Agilent 1100 MSD Trap SL. The data was analyzed with Spectrum Mill software. Similar analyses were performed on sera and the flow-through fraction from an anti-HSA affinity column.

5.3 RESULTS

5.3.1 Immunoaffinity Column Performance and Reproducibility

Figure 5.2 represents a typical chromatogram for a run on a 4.6 × 50 mm immunoaffinity column. To test the reproducibility of the separation, a specific serum sample from a normal healthy donor was injected repeatedly, with fraction collection of the flow-through and bound fractions. The column was stable and robust for at least 200 injections of serum samples, as illustrated by the chromatograms shown in Figure 5.2, which demonstrates the overlay of retention time and peak area for runs 20 and 200. The removal of high-abundant proteins was consistent and reproducible with no apparent antibody leakage or loss of column capacity (as measured with standard proteins during the course of the stability test experiments). The two column buffers used during loading, elution, and equilibration provided the desired robustness and contributed to the long column lifetime. The maintenance of column loading capacity indicates that the affinity binding sites are fully recovered, and thus the selected proteins are completely desorbed and solubilized by the elution Buffer B.

Figure 5.3 shows 1D gel electrophoresis data for the crude serum, flow-through, and bound fractions from the immunoaffinity column. Equal amounts of protein were loaded in each lane. Results show that high-abundant proteins in serum (lane 2) are clearly removed and are not visible in the flow-through fraction (lane 3). Also, low-abundant proteins that were not visible in the serum before depletion became visible in the flow-through fraction after removal of the high-abundant proteins. This could easily be observed using 2D gel electrophoresis. Figure 5.4 shows the protein pattern of human serum before (A) and after immunodepletion (B). Circles indicate the areas where the targeted high-abundant proteins reside. The depletion of high-abundant proteins unmasks the low-abundant proteins due to the substantial removal of protein mass from the sample. More than 85% of total protein was depleted after a single pass of serum through the immunoaffinity column. This enabled a large increase in low-abundant protein mass loading onto the gel (up to ten times). As a result, low-abundant protein fractions become enriched and more easily detectable on the gel, making the protein spots more amenable to quantitation and MS identification.

5.3.2 Completeness of the Depletion of High-Abundant Proteins

Standard sandwich ELISA assays were performed with a crude serum sample and the flow-through fraction from the immunodepletion column. As shown in Table 5.1,

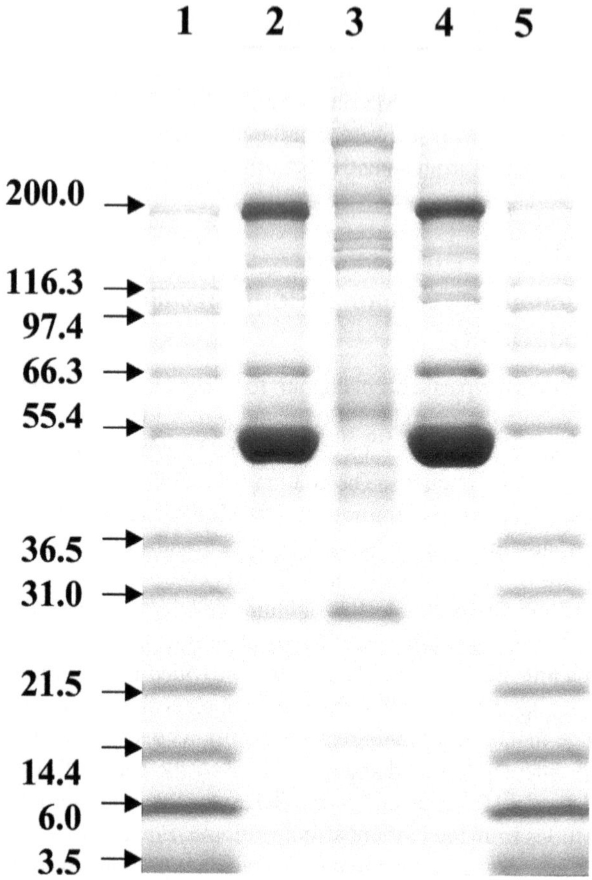

FIGURE 5.3 1D SDS gel electrophoresis of human serum protein fractions from an immunoaffinity column. An equal amount (10 μg) of crude serum (lane 2), flow-through (lane 3), and bound fractions (lane 4) were separated by 4 to 20% SDS-PAGE under nonreducing conditions. Lanes 1 and 5 are the molecular weight standards (Mark12) from Invitrogen. The proteins were stained with Coomassie Blue dye. Based on the protein assay of the flow-through fraction, 85% of total protein was removed from the crude serum.

more than 99% of the targeted high-abundant proteins were removed from the serum sample in a single pass on the immunodepletion column. This result indicates a high degree of efficiency in the performance of the immunodepletion column and is consistent with the qualitative impressions that result from the examination of gel electropherograms of flow-through fractions (Figure 5.3 and Figure 5.4), which lack any indication of the selected proteins. Gel electrophoretic analysis of bound fractions, with band selection, tryptic digestion, and MS analysis indicates the presence of multiple proteolytic fragments of the targeted high-abundant proteins, particularly

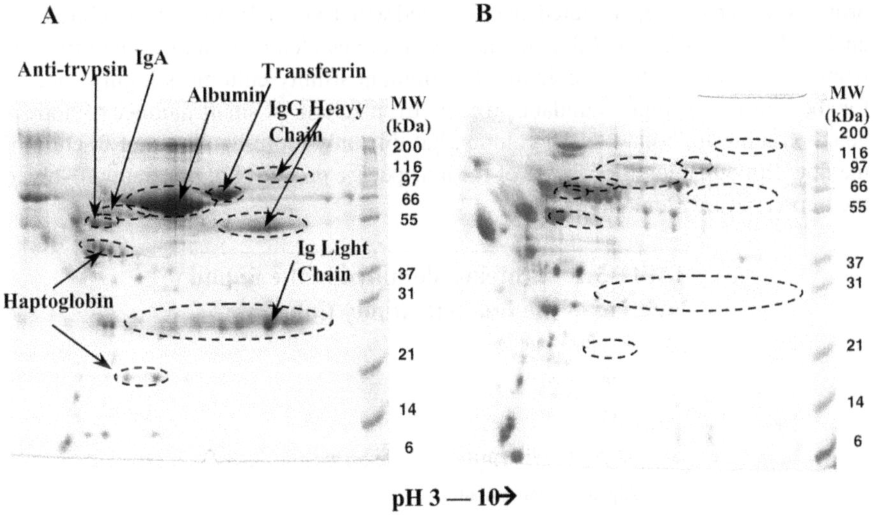

FIGURE 5.4 2D gel electrophoresis of human serum before and after removal of high-abundant proteins—250 μg of total protein was loaded on each gel. (A) Human serum before depletion. The targeted high-abundant proteins are circled; (B) human serum after depletion of the six targeted high-abundant proteins. The positions of the removed proteins are circled. Molecular weight standards (Mark12, Invitrogen). Proteins were visualized by staining with Coomassie Blue.

TABLE 5.1 Depletion Efficiency of High-Abundant Proteins from Human Serum

High-Abundant Proteins	Serum (18 μl) (μg) Mean ± SD (*n* = 3)	Flow through (μg) Mean ± SD (*n* = 3)	Depletion(%)
HSA	760 ± 60	5.05 ± 0.01	99.34
Transferrin	37 ± 17	0.24 ± 0.04	99.36
IgG	230 ± 22	0.86 ± 0.03	99.63
IgA	56 ± 5.5	0.23 ± 0.01	99.59
Haptoglobin	31 ± 8.7	0.21 ± 0.01	99.31
Alpha-1-antitrypsin	65 ± 5.7	0.74 ± 0.05	98.86

Note: Sandwich ELISA immunoassays were used to measure the concentration of high-abundant proteins in crude serum and the flow-through fraction after immunodepletion.

truncated versions of HSA. These fragments are efficiently bound by the polyclonal antibody selectors.

5.3.3 Specificity of the Immunodepletion

For analysis of immunodepletion column specificity, bound fractions were resolved on 1D gels (Figure 5.3, lane 4). The gel lane was cut into multiple sections, and the proteins were reduced, alkylated, and digested with trypsin. Peptides were eluted and analyzed by LC-MS/MS. Table 5.2 displays proteins identified in the bound fraction from the immunodepletion column. The immunoaffinity column is highly specific for the removal of high-abundant proteins from serum. A small number of nontargeted proteins are bound to the column, but this only represents a small fraction of the flow-through quantities. None of the nonspecific proteins bound quantitatively to the immunoaffinity column.

TABLE 5.2 Proteins Identified in the Bound Fraction of the Immunoaffinity Column

1.	**Serum Albumin**
2.	**Transferrin**
3.	**IgG**
4.	**Alpha-1-antitrypsin**
5.	Alpha-2-macroglobulin
6.	**Haptoglobin**
7.	Complement C3
8.	**IgA**
9.	Complement C4
10.	Apolipoprotein A-1

Note: Proteins retained by the immunoaffinity column were eluted and resolved with SDS-PAGE. Gel lanes were sliced and processed for identification by tryptic digestion and MALDI/MS or LC-MS/MS. Bold type indicates proteins targeted for removal by the immunoaffinity column.

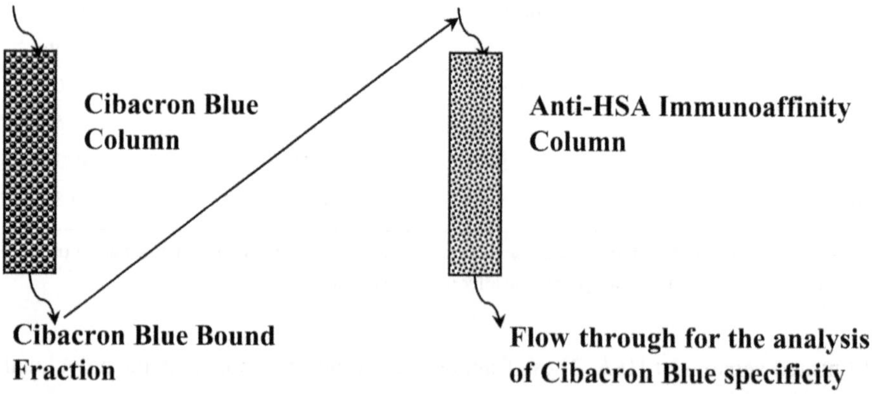

FIGURE 5.5 Experimental setup for the identification of proteins bound to the Cibacron Blue column. The Cibacron Blue bound fraction (1 mg) was loaded on an anti-HSA immunoaffinity column (4.6 × 50 mm). The flow-through fractions from several injections were collected and processed for further analysis by SDS-PAGE and LC-MS/MS.

We compared the specificity of the immunoaffinity column to the specificity of a Cibacron Blue dye column that is often used for the depletion of albumin from serum samples. A human serum sample was loaded onto a Cibacron Blue column, and the bound fraction was eluted after column washing. This fraction was depleted of albumin by passing it through an immunoaffinity column containing the material specific for the removal of albumin (see Figure 5.5 for the experimental setup). By removing HSA we are able to determine which proteins, besides albumin, bind to the Cibacron Blue resin. The flow-through fraction from the immunoaffinity column then represented all of the proteins nonspecifically retained by the Cibacron Blue column but not by the immunoaffinity depletion column. Analysis by SDS-PAGE revealed the presence of multiple proteins that were nonspecifically bound to the Cibacron Blue column (Figure 5.6). The identities of these proteins were confirmed by in-gel tryptic digestion followed by LC-MS/MS analysis. Tryptic digestion of this fraction with subsequent analysis of the resulting peptides by 2D LC-MS/MS was also conducted.

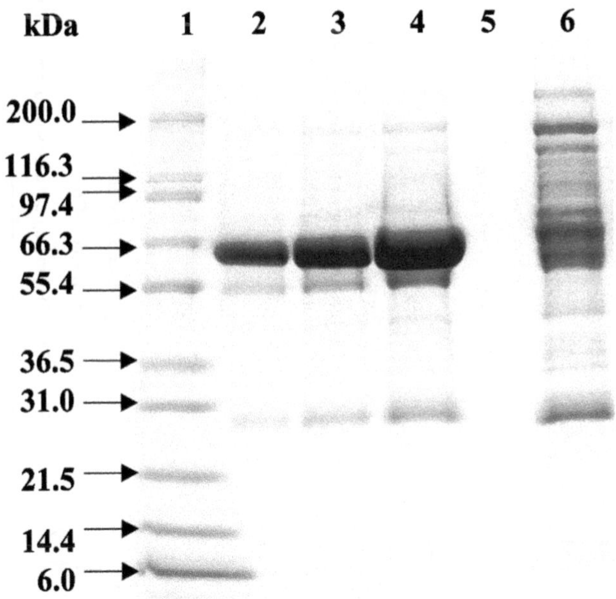

FIGURE 5.6 1D gel electrophoresis of a Cibacron Blue–bound fraction. The Cibacron Blue–bound fraction was separated by 4–20% SDS-PAGE under nonreducing conditions. Lane 1, Mark12 standards (Invitrogen); lanes 2–4, Cibacron Blue bound fraction; lane 2, 2 μg; lane 3, 4 μg; lane 4, 6 μg; lane 5, empty; lane 6, flow-through fraction from an anti-HSA immunoaffinity column. The Cibacron Blue–bound fraction was depleted of albumin and concentrated. A total of 7 μg protein was loaded on the gel lane.

TABLE 5.3 Proteins Identified in Bound Fraction of Cibacron Blue Column

1. Alpha-2-macroglobulin	19. ITIH4	36. Angiotensinogen
2. Complement C3	20. C1 inhibitor	37. Hypothetical protein XP_227784
3. Complement C4	21. Complement C7	38. Complement C5
4. Ceruloplasmin	22. Complement 9	39. ATP synthase F0 subunit 6
5. Alpha-1-antitrypsin	23. Ig-gamma chain	40. Vitronectin precursor (S-protein)
6. Serum albumin	24. Alpha-2-antiplasmin	41. Embryonic leucine zipper kinase
7. Apolipoprotein A-1	25. Clusterin	42. Cul-3
8. Antithrombin III variant	26. SNC66 protein	43. Unnamed protein product
9. Hemopexin	27. Peptidoglycan recognition protein L	44. Coagulation factor X
10. Complement factor B	28. Insulin-like growth factor binding protein complex acid labile chain	45. Complement C1s component
11. Kininogen		46. Paraoxonase/arylesterase
12. Ig-mu chain	29. Ig-lambda light chain	47. TFNR
13. Gelsolin	30. Haptoglobin-related protein	48. Glycosylphosphatidylinositol phospholipase D
14. Ig-kappa light chain	31. Alpha-1-microglycoprotein	49. Plasminogen
15. Heparin cofactor II	32. Afamin precursor	50. KIAA1926 protein
16. Serum amyloid P-component	33. Ig-alpha heavy chain	51. Serum aryldialkylphosphatase
17. ITIH1	34. Complement C6	52. KIAA1461 protein
18. ITIH2	35. ATP-binding cassette subfamily A member 9	

Note: The proteins retained by a Cibacron Blue column were eluted and depleted of albumin by use of an anti-HSA immunoaffinity column. The albumin-depleted fraction was resolved with SDS-PAGE (see Figure 5.6, lane 6). The gel lanes were sliced and processed for identification by tryptic digestion and proteins were identified by analysis of peptides using 2D LC with online ion trap MS/MS.

This approach allowed identification of 52 unique proteins that were nonspecifically bound to the Cibacron Blue column (Table 5.3). Analysis of the Cibacron Blue flow-through fraction indicates that some serum proteins such as hemopexin, Complement C3, C4, ITIH 1, ITIH 2, and ITIH 4 were bound quantitatively to the Cibacron Blue column, as these proteins were not detected at all in the flow-through fraction and were thus retained on the column.

5.4 CONCLUSIONS

The described immunoaffinity protein depletion column is based on affinity-purified polyclonal antibodies, prepared using whole proteins as immunogen. The use of polyclonal antibodies offers a significant advantage, since purification will rely on multiple epitopes and not a single one, as in the case of the monoclonal antibodies. We have observed that this allows the removal of proteolytic fragments and multiple variants of targeted proteins attributed to posttranslational modifications and sequence variations.

The excellent selectivity and specificity of immunoaffinity depletion is likely a result of the stringent affinity purification procedure used to prepare the captured antibodies. The column is highly specific for human bodily fluid samples—serum, plasma,[16] and CSF.[17] Additional studies have shown that efficient depletion of targeted proteins from monkey serum can be achieved,[18] whereas other tested samples of serum from various species, including swine, canine, mouse, and rat, have shown incomplete removal of targeted proteins. Immunoaffinity depletion columns specific for mouse albumin, transferrin, and IgG have also been prepared, with results highly similar to those obtained for humans, in terms of efficiency of immunodepletion, as well as species specificity. This indicates that species-specific polyclonal antibody removal devices offer a generally applicable approach to high-abundant protein removal for a broad range of samples, particularly when compared to competitive technologies currently available.

The chromatographic separation of proteins can be influenced by mobile-phase conditions such as pH, ionic strength, and added constituents of the mobile phases. The two proprietary column buffers were optimized to avoid run-to-run variability and promote column specificity, reproducibility, and lifetime. The immunoaffinity column performed reproducibly over 200 run cycles, as judged by retention characteristics, as well as by electrophoretic analysis of collected fractions. Highly specific depletion of high-abundant proteins results in the removal of up to 98 to 99% of the targeted proteins and their fragments, while minimizing nonspecific base material and protein-protein interactions. This specificity is determined by the characteristics of the binding buffer.

Reproducible and stable operation of the immunodepletion column for many sample injections requires that the material must be chemically and physically stable. An underrated consideration is the requirement that the affinity binding capacity remains stable during extended use of the material. A simple calculation shows that a 1% loss of binding capacity per use of the column will result in less than 14% remaining capacity after 200 consecutive sample injections. Using the conditions and materials described herein, we have observed negligible loss of column capacity (measured with purified proteins) after 200 consecutive injections, implying that the affinity selector remains completely available over this regime. In addition, no cross contamination is noted between runs, as indicated by the absence of protein elution when blank injections follow serum samples. This ability to have a clean and stable affinity surface is determined by the chemistry of surface construction, as well as the choice of the elution Buffer B, which functions to both desorb the antibody-bound protein, as well as maintain its solubility in the mobile phase. This latter consideration is critical in the choice of elution buffer, as illustrated by our experience with human alpha-2-macroglobulin, a very large serum protein with challenging solubility characteristics. An acceptable elution regime has yet to be developed to support stable capacity for this serum protein.

Reproducibility and specificity of the depletion are major concerns when choosing the proteomic sample preparation technique. The four nontargeted proteins detected by MS/MS in the column bound fraction appear to represent a small fraction of those detected in the flow-through and are most likely due to a relatively stable but nonspecific association of these proteins with the targeted proteins or the base material. We have observed that small amounts of these same proteins tend to appear in almost all immunoaffinity separations, regardless of the antibody immobilized on the chromatographic surface. To date, no systematic analysis has been conducted to quantify the removal of these proteins during the procedure, but the intensities of MS signals generated in downstream analyses are small, in comparison to those apparent in the flow-through fraction, indicating that at worst these are small relative losses. In contrast, we have demonstrated that a variety of serum proteins, in addition to serum albumin, bind to the Cibacron Blue resin (52 proteins identified), while many of these proteins are quantitatively removed by the described Cibacron Blue selection procedure. Although the specificity of Cibacron Blue can be somewhat manipulated by use of alternative binding buffer conditions, increasing contamination of the flow-through sample by albumin or albumin fragments results.

The immunoaffinity column depletion technology described offers rapid and simultaneous removal of six high-abundant proteins in 20 to 25 minutes per sample. This methodology is robust, scalable, and easily automated for multiple sample processing, and it is compatible with downstream 1D and 2D SDS-PAGE, LC, LC-MS, and enzymatic or chemical fragmentation methods, or both.

The approach presented here enables an expanded dynamic range for the detection of low-abundant proteins in these complex proteomic samples and thereby assists in the search for novel biomarkers of disease states and intervention strategies.

REFERENCES

1. Anderson, N.L. and Anderson, N.G., The human plasma proteome: History, character, and diagnostic prospects, *Mol. Cell. Proteomics*, 1, 845–867, 2002.
2. Brzeski, H., Katenhusen, R.A., Sullivan, A., Russell, S., George, A., Somiari, R.I., and Shriver, G., Albumin depletion method for improved plasma glycoprotein analysis

by two-dimensional difference gel electrophoresis, *BioTechniques*, 35, 1128–1132, 2003.
3. Georgiou, H.M., Rice, G.E., and Baker, M.S., Proteomic analysis of human plasma: Failure of centrifugal ultrafiltration to remove albumin and other high molecular weight proteins, *Proteomics*, 1, 1503–1506, 2001.
4. Ahmed, N., Barker, G., Oliva, K., Garfin, D., Talmadge, K., Georgiou, H., Quinn, M., and Rice, G., An approach to remove albumin for the proteomic analysis of low abundance biomarkers in human serum, *Proteomics*, 3, 1980–1987, 2003.
5. Burgess-Cassler, A., Johansen, J.J., and Kendrick, N., Immunodepletion of albumin from human serum samples, *Clin. Chim. Acta*, 183, 359–366, 1989.
6. Govorukhina, N.I., Keizer-Gunnink, A., van der Zee, A.G.J., de Jong, S., Bruijn, H W A., and Bischoff, R., Sample preparation of human serum for the analysis of tumor markers: Comparison of different approaches for albumin and γ-globulin depletion, *J. Chromatogr. A*, 1009, 171–178, 2003.
7. Björhall, K., Miliotis, T., and Davidsson, P., Comparison of different depletion strategies for improved resolution in proteomic analysis of human serum samples, *Proteomics* 5, 307–317, 2005.
8. Adkins, J.N., Varnum, S.M., Auberry, K.J., Moore, R.J., Angell, N.H., Smith, R.D., Springer, D.L., and Pounds, J.G., Toward a human blood serum proteome: Analysis by multidimensional separation coupled with mass spectrometry, *Mol. Cell. Proteomics*, 1, 947–955, 2002.
9. Huse, K., Böhme, H.-J., and Scholz, G.H., Purification of antibodies by affinity chromatography, *J. Biochem. Biophys. Methods*, 51, 217–231, 2002.
10. Wang, Y.Y., Cheng, P., and Chan, D.W., A simple affinity spin filter method for removing high-abundant common proteins or enriching low-abundant biomarkers for serum proteomic analysis, *Proteomics*, 3, 243–248, 2003.
11. Pieper, R., Su, Q., Gatlin, C.L., Huang, S.-T., Anderson, N.L., and Steiner, S., Multi-component immunoaffinity subtraction chromatography: An innovative step towards a comprehensive survey of the human plasma proteome, *Proteomics*, **2003**, 3, 422–432.
12. Lee, W.-C. and Lee, K.H., Applications of affinity chromatography in proteomics, *Anal. Biochem.*, 324, 1–10, 2004.
13. Jack, G.W., Immunoaffinity chromatography, *Mol. Biotech.*, 1, 59–86, 1994.
14. Harlow, E. and Lane, D., *Antibodies: A Laboratory Manual*, Cold Spring Harbor Laboratory, New York, 1988, 726 pp.
15. Leatherbarrow, R.J., *GraFit Version 5*, Erithacus Software Ltd., Horley, U.K., 2001, 262 pp.
16. Agilent Technolopgies, http://www.chem.agilent.com/temp/rad6D493/00045136.pdf (last accessed August 2005).
17. Maccarrone, G., Milfay, D., Birg, I., Rosenhagen, M., Grimm, R., Bailey, J., Zolotarjova, N., Miller, C., and Turck, C., Mining the human CSF proteome by immunodepletion and shotgun mass spectrometry, *Electrophoresis*, 25, 2402–2412, 2004.
18. Agilent Technolopgies, http://www.chem.agilent.com/temp/radCF34D/00047172.pdf (last accessed August 2005).

6 Isolation of Plasma Membrane Proteins for Proteomic Analysis

Catherine Fenselau and Amir Rahbar

CONTENTS

6.1 INTRODUCTION

It is estimated that about 30% of the many kinds of proteins in the cell are embedded in cellular membranes[1] and that more that 50% of current drug targets are membrane proteins.[2,3] Consequently, isolation of membranes and analysis of membrane proteins is of interest across a variety of disciplines. Isolation of the total cellular membrane fraction is straightforward; however, it is more difficult to separate membranes from individual organelles. The isolation of the plasma membrane is a particular challenge, due to its low abundance in relation to mitochondria, endoplasmic reticulum, and other membranes in the cell, its similarity to these other membrane components, and its propensity to exist in a variety of structures during isolation, including both open sheets and vesicles.[4,5] The procedure reported here is a modification of Jacobson's cationic colloidal silica technique,[6,7] which leads to 20-fold enrichment of the plasma membrane. Proteins are solubilized from the isolated plasma membrane into Laemmli buffer suitable for sodium dodecyl sulfate–polyacrylamide gel electrophoresis (SDS PAGE) and proteomic analysis. At least 50% of the proteins identified are traditionally assigned to the plasma membrane, and some theoretical membrane proteins are characterized.

6.2 PREPARATION OF THE PLASMA MEMBRANE FRACTION

The method is described here as it is applied to the isolation of plasma membranes from human multiple myeloma RPMI 8226 cells grown in suspension and human breast cancer MCF-7 cells cultured in monocellular layers. We thank Dr. R. Fenton of the Greenebaum Cancer Center, University of Maryland School of Medicine, for the multiple myeloma cells, and Dr. K.H. Cowan of the Eppley Institute, University of Nebraska Medical Center, for the MCF-7 cell line.

6.2.1 Materials

Cationic colloidal silica (Ludox CL) and polyacrylic acid (100,000 average molecular weight fraction) were obtained from Sigma Aldrich (St. Louis, Mo.). Nycodenz, Laemmli buffer, growth media, phosphate-buffered saline (PBS), and other chemicals were obtained from either Bio-Rad (Hercules, Calif.) or Sigma Aldrich.

Membrane coating buffer A: 20 mM MES, 150 mM NaCl, 800 mM sorbitol, pH 5.3
Membrane coating buffer B: 20 mM MES, 135 mM NaCl, 0.5 mM $CaCl_2$, 1 mM $MgCl_2$, pH 5.3
Polyacrylic acid solution: 10 mg/ml polyacrylic acid in membrane coating buffer A or B, pH 6.5.
Lysis buffer: 2.5 mM imidazole pH 7

6.2.2 Cell Suspensions

The RPMI 8226 cells are grown in RPMI 1640 media and heat-inactivated fetal calf serum[8] in 150 cm^2 cell culture flasks at 37°C under 5% CO_2. The RPMI cells are recovered by centrifugation at 900*g*. About 1.5 g wet weight cells are washed in 50 ml plasma membrane coating buffer A (see above) and added dropwise into 15 ml of a suspension of 10% (by volume) cationic colloidal silica in the same membrane coating buffer. The suspension is rocked gently on ice for 15 minutes. The silica-coated cells are then sedimented at 900*g* for 5 minutes, and the supernatant suspension is removed by decantation.

The silica-coated cells are resuspended in 50 ml buffer A and sedimented once more at 900*g* for 5 minutes to remove any excess silica. The cells are resuspended in 7 ml buffer solution A, added dropwise to 15 ml of polyacrylic acid solution (see above), and placed on ice for 15 minutes with gentle rocking. The silica and polyacrylic acid coated cells are then sedimented at 900*g* for 5 minutes, and the supernatant is removed by decantation. The cells are washed with 50 ml plasma membrane coating buffer A, recovered by centrifugation at 900*g* for 5 min, and placed in 50 ml lysis buffer with 0.5 ml of Sigma protease inhibitor cocktail and left on ice for 30 minutes.

The swollen cells are then lysed using nitrogen cavitation at 1500 psi. The cell lysate is centrifuged at 900*g* for 30 minutes to sediment the silica-coated plasma membrane pieces, along with nuclei. The pellet is resuspended in 10 ml lysis buffer and diluted with an equal amount of 100% Nycodenz in lysis buffer. This solution is layered over a 70% Nycodenz lysis buffer solution in six 4 ml centrifuge tubes. Lysis buffer is then layered onto each tube to fill it to the top, and the tubes are spun at 60,000*g* in an SW60Ti rotor for 23 minutes. The silica-coated plasma membranes are collected in the pellet, leaving the nuclei at the 50%:70% Nycodenz interface. The supernatant is drawn off, and the silica-coated plasma membrane pellets are resuspended in lysis buffer and centrifuged at top speed on a bench top microfuge to remove the excess Nycodenz. The pellet is washed in this way two more times with lysis buffer and then three times with 100 mM Na_2CO_3 (pH 11.4) for a total of six washes.

The purified plasma membrane proteins are then recovered from their silica coating by suspension directly in Laemmli loading buffer, incubation in a 60ºC water bath for 30 minutes, sonication 5 times for 10 seconds each at maximum setting, and incubation in the 60ºC water bath for an additional 30 minutes. Finally the suspension is spun at maximum speed in a microfuge for 15 minutes to pellet the silica. The supernatant, which now contains the solubilized plasma membrane proteins, is drawn off and stored at 80ºC, and the pellet is discarded.

6.2.3 Cells Growing in Monolayers

MCF-7 cells were grown to confluence in 150 cm^2 flasks with Improved Minimal Essential Medium containing 5% fetal calf serum and antibiotics at 37ºC and 5% CO_2.[9] Typically twelve flasks are processed in parallel.

The medium is drawn off and the adherent cells are washed twice with PBS containing 1 mM $MgCl_2$ and 1 mM $CaCl_2$ and then with plasma membrane coating buffer B (see above). The cells are then coated with 15 ml of a 5% suspension of cationic colloidal silica in the membrane coating buffer B and left on ice for 1 minute. The silica suspension is removed, followed by a wash with membrane buffer solution B to remove excess silica.

The coated cells are treated with 15 ml of a 10 mg/ml solution of polyacrylic acid in membrane buffer B, pH 6.0 to 6.5, and left on ice for 1 minute. The polyacrylic acid solution is removed, and the cells are washed with 15 ml membrane buffer B.

The adherent cells are washed quickly with 15 ml lysis buffer (see above), and then 15 ml lysis buffer containing 0.5 ml Sigma protease inhibitor cocktail is added to the flasks, which are left on ice for 30 minutes. Once the cells have swollen, the flasks are placed on a bench top and allowed to reach room temperature.

The apical part of the plasma membranes, which has been coated by silica and polyacrylic acid, is sheared from the rest of the cell by pipetting the lysis buffer in each flask up and down over the cells. The sheared coated apical membrane sheets are drawn off with a pipette and treated as described in the previous section, finally solubilizing the proteins in Laemmli loading buffer.

The basolateral portions of the plasma membranes are still attached to the cell culture surface. Following lysis and removal of the apical fragments, any remaining lysis buffer is poured off and 15 ml of 5 M NaCl solution is added to each flask.

After the flask is rocked for 5 minutes at room temperature, the NaCl solution is poured off and 15 ml PBS containing 10 mM EDTA is added to each flask and rocked for 5 minutes at room temperature. The PBS-EDTA solution is poured off, and the flask is washed quickly with 15 ml of 100 mM Na_2CO_3, pH 11.4, to remove excess PBS-EDTA. Then 15 ml of 100 mM Na_2CO_3, pH 11.4, is again added to each flask and incubated for 5 minutes at room temperature with rocking. This is poured off and a third 15 ml aliquot of 100 mM Na_2CO_3 is added to each flask.

The basolateral membranes are scraped from the bottom of the flask with a cell scraper and spun at 14,000*g* in an SW28 rotor for 20 minutes. The purified basolateral fragments are recovered in the pellet, resuspended in a minimal amount of 100 mM Na_2CO_3, and transferred to a 1.5 ml microfuge tube and spun at maximum speed in a microfuge for 20 minutes to pellet the plasma membranes. The basolateral plasma membrane pellet can then be solubilized in 2% SDS or directly in Laemmli loading buffer, augmented by sonication and a 60ºC water bath as described above.

6.3 PROTEOMICS ANALYSES

The solubilized plasma membrane proteins can be subjected to various proteomic strategies. Dissolution in Laemmli buffer facilitates loading the protein fraction directly onto a ID SDS-PAGE gel for electrophoretic separation. (Alternatively, the protein mixture can be digested in solution with appropriate proteases for peptide analysis by HPLC and tandem mass spectrometry (shotgun strategy[1,3]). Thus, 80 μg of protein is loaded onto four 15% gels (13.3 × 8.7 cm W × L)) and separated according to the specifications of the Bio-Rad Criterion precast gel system. For Western blotting, the protein bands separated by electrophoresis are transferred to a nitrocellulose membrane using the Bio-Rad Mini trans-blot cell. A Sigma ProteoQwesta Colorimetric Western Blotting Kit and an antibody against the human Na/K ATPase can be used.[10] The developed nitrocellulose membranes are scanned using the Bio-Rad GS-800 densitometer. Comparison is made to the abundance of ATPase in whole cell lysate. Quantitation of the images is carried out with the ImageQuant image analysis software by Amersham Biosciences (Piscataway, N.J.). For example, enrichments were determined as 18-fold for plasma membranes from RPMI 8226 suspension cells, 8-fold for apical membrane fragments from surface bound MSR MCF-7 cells, and 20-fold for the basolateral domains.

The SDS-PAGE gel is cut into 48 bands, and tryptic digestion is carried out on each slice.[11] After extraction from the gel bands, the tryptic peptides are desalted using Zip Tip C18 pipette tips (Millipore, Billerica, Mass.). The acetonitrile/trifluoroacetic acid is removed using a SpeedVac (ThermoElectron Corp., Woburn, Mass.), and the peptides are resuspended in a solution of methanol/water/acetic acid (50/50/2) in preparation for static infusion nanospray tandem MS (MS/MS) analysis, or placed in 0.1% formic acid in preparation for analysis by nanoflow liquid chromatography-tandem mass spectrometry. Reverse-phase conditions are as follows. Solvent A, 97.5% H_2O/2.5% acetonitrile/0.1% formic acid; B, 97.5% acetonitrile/2.5% H_2O/0.1% formic acid, with a 70-minute gradient from 5 to 35% B on a PepMap 75 μM I.D., 15 cm, 3 μm, 100 column (LC Packings, Sunnyvale CA). Each protein is identified based on

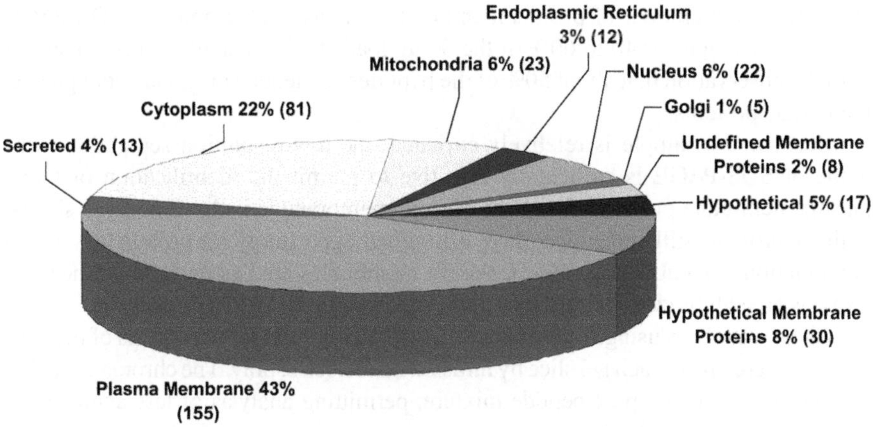

FIGURE 6.1 Distribution of the proteins identified, based on the SwissProt database.

sequences from two or more peptides, using the integrated QStar software Analyst QS, with Bioanalyst and ProID for the LC-MS experiments, and the MASCOT search engine from Matrix Science for the infusion nanospray experiments.[12]

In a typical experiment with the RMPI plasma membrane fraction analyzed by infusion nanospray, 47 proteins were identified, of which 20 were located in the plasma membrane according to SwissProt annotation.[13] Although plasma membrane localization of some of the other proteins cannot be excluded, at a minimum 43% of the identified proteins were plasma membrane proteins. Analysis of the basolateral fraction of the MCF-7 line by infusion provided identification of 42 proteins, of which 21 had previously been reported as localized in the plasma membrane.

Analysis of peptides produced from proteins associated with the basolateral plasma membrane of the mitoxantrone resistant MCF-7 breast cancer cell line by nanoLC MS provided partial sequences and positive identification of a total of 1695 peptides. Of these, 918, or 54%, were assigned to plasma membrane proteins. A total of 365 proteins was identified, among which 155, or 43%, had been previously reported as being located in the plasma membrane. The distribution of proteins, based on SwissProt annotation,[13] is shown in Figure 6.1.

6.4 CONCLUSIONS

Although the pellicular coating method described here was first reported more than two decades ago,[6] recent advancements in mass spectrometry and the introduction of bioinformatics allow plasma membrane proteins to be analyzed in sufficient detail both to characterize the method and to exploit it for proteomics applications. The additional washing steps introduced in the present procedure are credited with increasing the enrichment of the plasma membranes. It is likely, of course, that achievable enrichment levels will vary from one cell line to another. Estimating that the plasma membrane makes up 2 to 3% of the cell mass, and assuming a 20-fold enrichment, the amount of plasma membrane proteins loaded onto the SDS-PAGE gel would account for 40 to 60% of the 80 μg loaded. This calculation is consistent with the observation that about 50% of the peptides sequenced originate from plasma membrane proteins.

Because the sample is relatively enriched, the low-resolution separation technique of SDS-PAGE is sufficiently effective to permit the identification of many plasma membrane proteins. Since the plasma membrane is isolated as large sheets, with the proteins still embedded, there is minimal opportunity for protein loss due to precipitation in solubilizing steps. Once the membranes are isolated and washed, the proteins are solubilized directly into SDS, followed by SDS-PAGE analysis.

Clearly, analysis using coupled nanoLC-MS/MS is superior to analysis of the peptides recovered from each gel slice by infusion nanoelectrospray. The chromatographic step separates the complex peptide mixture, permitting analysis of less abundant or less basic peptides, or both.

Although not addressed here, the opportunities are obvious for applying the pellicular method to study polarization of plasma membrane proteins in cells growing on surfaces. Similarly the application to cells isolated from fresh tissue samples would appear to have good potential.

ACKNOWLEDGMENTS

The method discussed here was developed as part of Amir Rahbar's Ph.D. thesis, supported by a grant from the National Institutes of Health (GM 21248).

REFERENCES

1. Wu, C.C. and Yates, J.R., The application of mass spectrometry to membrane proteomics, *Nat. Biotechnol.*, 21, 262–267, 2003.
2. Drews, J., Drug discovery: A historical perspective, *Science,* 287, 1960–1964, 2000.
3. Wu, C.C., MacCoss, M.J., Howell, K.E., and Yates, J.R., A method for the comprehensive proteomic analysis of membrane proteins, *Nat. Biotechnol.*, 21, 532–538, 2003.
4. Kearney, P. and Thibalt, P., Bioinformatics meets proteomics—bridging the gap between mass spectrometry data analysis and cell biology, *J. Bioinformatics Comput. Biol.*, 1, 183–200, 2003.
5. Schirmer, E.C., and Gerace, L., Organellar proteomics: The prizes and pitfalls of opening the nuclear envelope, *Genome Biol.*, 3, 1008.1–1008.4, 2002.
6. Chaney, L.K. and Jacobson, B.S., Coating cells with colloidal silica for high yield isolation of plasma membrane sheets and identification of transmembrane proteins, *J. Biol. Chem.*, 258, 62–67, 1983.
7. Stolz, D.B. and Jacobson, B.S., Examination of transcellular membrane protein polarity of bovine aortic endothelial cells in vitro using the cationic colloidal silica microbead membrane-isolation procedure, *J. Cell Sci.*, 103, 39–51, 1992.
8. Dvorakova, K., Waltmire C.N., Payne C.M., Tome M.E., Briehl M.M., and Dorr R.T., Induction of mitochondrial changes in myeloma cells by imexon, *Blood*, 97, 3544–3551, 2001.

9. Hathout, Y., Riordan, K., Gehrmann, M., and Fenselau, C., Differential protein expression in the cytosol fraction of an MCF-7 breast cancer cell line selected for resistance toward melphalan., *J. Proteome Res.,* 1, 435–442, 2002.
10. Spector, D.L., Goldman, R.D., and Leinwald, L.A., Culture and biochemical analysis of cells, in *Cells: A Laboratory Manual,* Spector, D.L. and Leinwald, R.D., Eds., Cold Spring Harbor Laboratory Press, Cold Spring Harbor, NY, 1998, vol. 1, pp. 34.1–34.94.
11. Rahbar, A.M., and Fenselau, C., Integration of Jacobson's pellicle method into proteomic strategies for plasma membrane proteins, *J. Proteome Res.,* 3, 1267–1277, 2004.
12. Perkins, D.N., Pappin, D.J., Creasy, D.M., and Cottrell, J.S., Probability-based protein identification by searching sequence databases using mass spectrometry data, *Electrophoresis*, 20, 3551–3567, 1999.
13. Swiss-Prot Protein Knowledgebase, http://us.expasy.org/sprot/ (last accessed August 2005).

7 New Ultrafiltration and Solid Phase Extraction Techniques Improve Serum Peptide Detection

Elena Chernokalskaya, Sara Gutierrez, Aldo M. Pitt, Alexander V. Lazarev, and Jack T. Leonard

CONTENTS

7.1 INTRODUCTION

A biomarker can be defined as a molecule that indicates an alteration in physiology. Biomarkers play an essential role in the drug discovery and development process. They provide powerful clues to genetic susceptibility, disease progression, and predisposition, and they offer information on physiological and metabolic profiling of diseases and drug response. Biomarkers can also provide valuable diagnostic and prognostic information that can facilitate personalized medicine. Peptide and protein patterns have been linked to ovarian cancer, breast cancer, prostate cancer, and astrocytoma.[1–6]

Most diagnostic tests are based on blood or urine analysis.[5] Serum is a key source of putative protein biomarkers, and, by its nature, can elucidate organ-confined events. The use of mass spectroscopy coupled with bioinformatics is capable of distinguishing serum protein pattern signatures of ovarian cancer in patients with early- and late-stage disease.[7]

One of the major impediments to the discovery of new biomarkers is the fact that plasma or serum contains a significant number of salts, proteins, and lipids that make it difficult to detect and analyze peptides by mass spectrometry (MS). When untreated serum is spotted onto a MALDI plate, it does not produce any useable signal in MS. Multiple protocols have been developed to extract and enrich peptides from tissues and body fluids, such as extraction with 0.1% trifluoroacetic acid (TFA) or 50% acetonitrile to selectively precipitate large proteins while enhancing the solubility of smaller proteins and peptides and batch reversed-phase chromatography over C18 resin.

Ultrafiltration (UF) has been a valuable separations tool for protein research since the early 1970s (extensively reviewed in[8,9]). Protein and other biomolecule concentration, desalting, or buffer exchange have been the major applications of UF. UF has been well accepted for the recovery of larger (>10 kDa) biomolecules. Until recently, the UF technique was rarely utilized to recover low molecular weight biomarkers because the high protein and lipid concentration in serum interfered with the separation. However, improved techniques and design of UF devices allows better separation of serum components.[10,11] A similar application of UF is the separation of free drugs from plasma protein-bound drugs in plasma.[12,13] In this chapter, we briefly review UF technology and demonstrate its value as a bioseparations tool to assist in the investigation of the proteome.

UF membranes can be composed of a wide selection of polymeric materials. However, UF membranes are primarily made from either polyethersulfone (PES) or regenerated cellulose (RC). A cross-sectional scanning electron microscopy image of an RC membrane is shown in Figure 7.1, where the huge internal surface area of the filters is easily observed. Each material has characteristic physical-chemical properties making it better suited for different applications. The PES membranes have higher nonspecific binding (NSB) properties for proteins (10 to 75 μg albumin/cm^2), while the RC membranes generally exhibit very low protein binding (1 μg albumin/cm^2).[14] The PES membranes, however, have higher flow, offering much higher volumetric flux rates (permeate volume/surface area/time or mL/cm^2/sec). These properties make PES membranes suitable for manufacturing processes where rapid protein concentration is a significant selection criterion. RC UF membranes, however, generally have lower fluxes, but their low NSB make them ideal candidates for analytical protein separations such as those required for proteomic sample preparation. Bioprocessing UF systems typically utilize PES filters and large-scale positive pressure forces to effect the separation, while lab-scale separations are conveniently achieved by devices optimized and designed for centrifugation. The low NSB of RC membranes also makes them ideal for the preparation of low molecular weight fractions by the collection of the filtrate.

State of the filtration technology not only requires advanced membrane production capabilities but also device design expertise. Millipore's Amicon® Ultra-4 centrifugal filter devices (as shown in Figure 7.2) are configured in a vertical design, which minimizes protein polarization and subsequent fouling of the membrane. The vertical design and large relative surface area provide for fast sample processing, high retentate recovery (typically greater than 90% of dilute protein

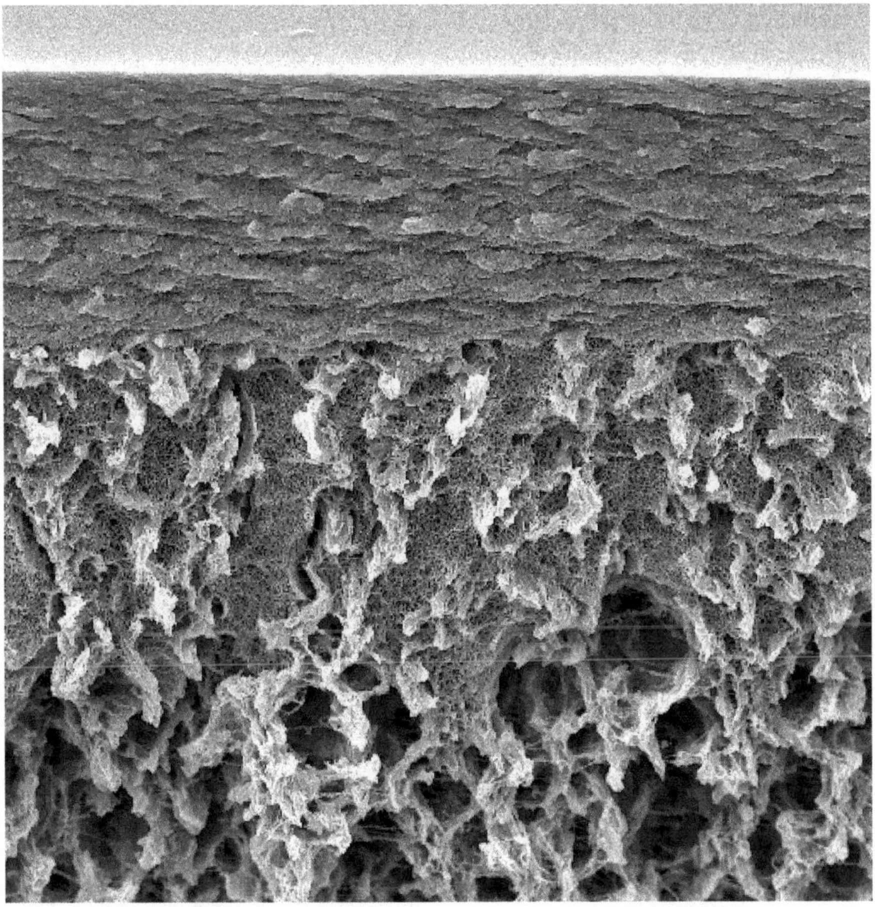

FIGURE 7.1 Cross-sectional SEM of regenerated cellulose UF membrane.

concentrate), and the capability for very high concentration factors (>80-fold). The Amicon Ultra devices are suitable for fast separation and concentration of viscous solutions, such as serum or plasma and cell lysate samples, thus making it ideal for biomarkers studies where large molecular weight species should be removed from small peptides.

7.2 METHODS

7.2.1 Preparation of Serum Peptides by UF and SPE

One milliliter of rat serum, with or without acetonitrile, was filtered in Amicon Ultra-4 10,000 MWCO centrifugal device. The UF devices were centrifuged in a swinging bucket rotor for 15 to 30 minutes at 3000*g*. Ten microliters of the filtrate was acidified

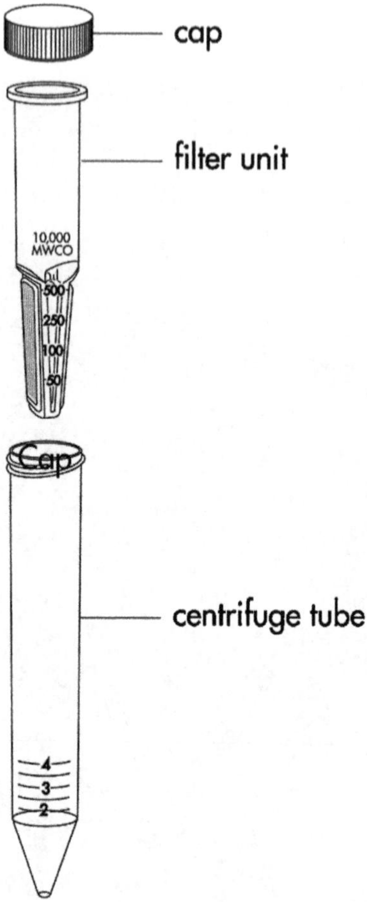

FIGURE 7.2 Amicon Ultra-4 design.

with 5 µl of 1% TFA, desalted, and cleaned with ZipTip$_{\mu C18}$ pipette tips. Coelution was performed directly onto the MALDI target with 2 µl of alpha-cyano-4-hydroxy cinnamic acid matrix (5 mg/ml in 50% acetonitrile, 0.1% TFA). If acetonitrile was added to the serum prior to the UF, the samples were briefly evaporated in a speed-vac centrifuge before ZipTip purification.

7.2.2 Preparation of Serum Peptides by Acetonitrile Precipitation

Acetonitrile was added to rat serum in a 1:1 (v/v) ratio, and samples were centrifuged to precipitate the larger proteins. The supernatant fluid was then dried in a speed-vac centrifuge, resuspended in 0.1% TFA, and subsequently desalted and concentrated with ZipTip$_{\mu C18}$ pipette tips. Coelution was performed directly onto the MALDI target with 2 µl of alpha-cyano-4-hydroxy cinnamic acid matrix (5 mg/ml prepared in 50% acetonitrile, 0.1% TFA).

7.2.3 Peptide Analysis by Mass Spectrometry

Peptide containing ultrafiltrates from serum were acidified with 1% TFA and concentrated on ZipTip$_{C18}$ or ZipTip$_{SCX}$ following manufacturer's directions. All samples were overlaid with 1 µl of alpha-cyano-4-hydroxy cinnamic acid matrix (5 mg/ml in 50% acetonitrile, 0.1%TFA) and analyzed on Voyager-DE™ Workstation (Applied Biosystems) in linear mode. High-resolution MALDI-TOF spectra and post-source decay (PSD) sequencing of peptides was performed on Axima CFR+ mass spectrometer equipped with the curved field reflectron (Kratos Analytical, Ltd., Manchester, U.K.). The amino acid sequence of unknown serum peptides was determined by Mascot tandem MS search (Matrix Science, Ltd., Manchester, U.K.) against NCBI nonredundant sequence database.

7.2.4 Two-Dimensional (2D) Gel Electrophoresis

Protein samples were resuspended and solubilized by brief sonication in a denaturing ProteomIQ sample buffer and separated on ElectrophoretIQ3 multifunctional electrophoresis instrument (Proteome Systems, Woburn, Mass.). The 11-cm-long immobilized pH gradient (IPG) strips pH 4–7 and 6–15% polyacrylamide gradient gels (GelChips) were also from Proteome Systems. The IPG strips were rehydrated overnight and focused for 8 hours at 10,000 V following an 8-hour concave voltage gradient from 100 V to 10,000 V to accumulate approximately 75,000 volt-hours of isoelectric focusing. The second-dimension run was performed at 30 mA per gel for approximately 2.5 hours.

7.3 RESULTS

Complexity of blood serum polypeptide composition has been attracting attention of numerous research efforts. It has been reported that human and mammalian serum contains peptides and protein at concentrations ranging over at least nine orders of magnitude. Considerable efforts have been made to characterize low-abundance protein biomarkers in serum via sample prefractionation, concentration, and removal of highly abundant protein species;[15] however, few efforts to investigate small peptide profiles in serum or plasma have been reported.[10,16] Nevertheless, these limited reports clearly illustrate that small proteolytic or endogenous peptides found in the circulatory system can serve as clear biomarkers of physiological state and could be more abundant than their intact protein precursors.

While serum contains numerous peptides and small proteins, they are not accessible by direct MS analysis. Even after reversed-phase concentration and desalting, only a few peptides are detectable in the mass spectrum (data not shown). This can be explained by the high concentration of proteins and lipids competing with the peptides for the binding to the reverse-phase resin.

One of the common methods for serum peptide enrichment prior to MS is acetonitrile fractionation, where addition of 50 to 70% acetonitrile precipitates larger proteins, while soluble peptides remain in the supernatant. Another way to produce relatively protein-free filtrates is UF. Figure 7.3 presents the MALDI TOF spectra of untreated rat serum and rat serum supernatant after 50% acetonitrile precipitation. The results show that neither of these two methods is sufficient for serum peptides analysis.

UF is an alternative method for preparation of low molecular weight serum proteome. UF of serum was reportedly used to remove the large abundant proteins

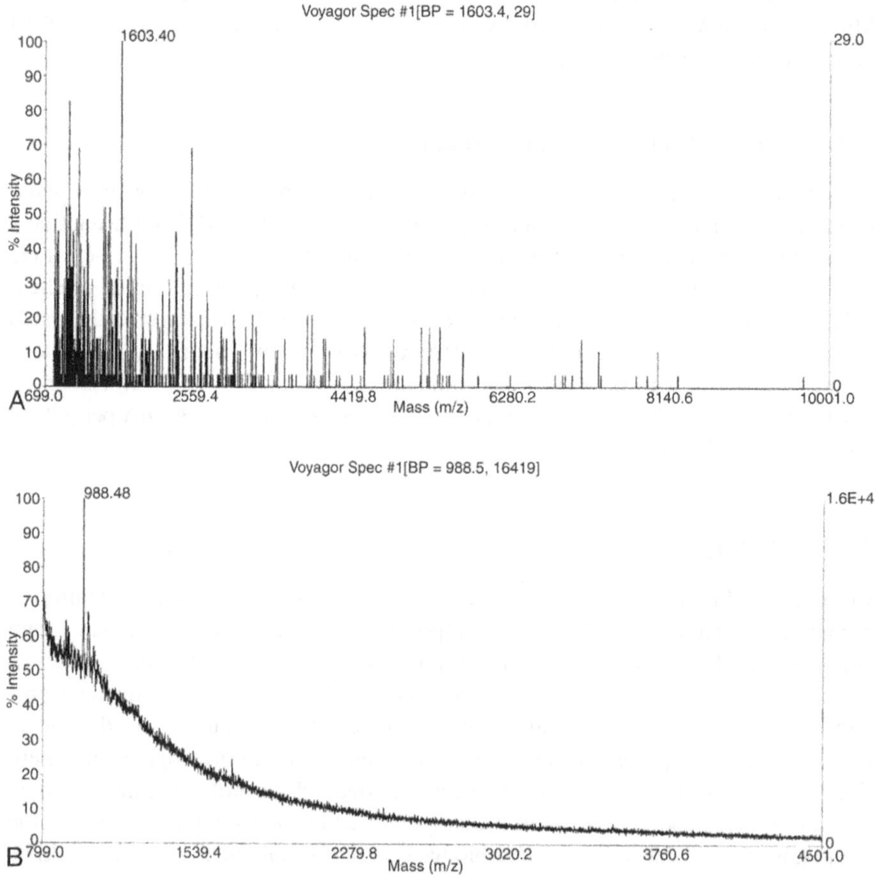

FIGURE 7.3 MALDI TOF spectra of (A) untreated rat serum; and (B) serum peptides in 50% acetonitrile supernatant.

from serum resulting in the enrichment of the low molecular weight proteins or peptides. This <30 kDa fraction contained several classes of physiologically important proteins such as cytokines, chemokines, and peptide hormones, as well as proteolytic fragments of larger proteins.[10,17] We applied UF through three different nominal molecular weight cutoff (NMWCO) membranes to evaluate suitability of the filtrates for peptide analysis. Because of the low peptide concentration and high salt content, it is impractical to use the filtrates directly for MS (data not shown). Instead, the filtrates were acidified and concentrated and desalted by reversed-phase chromatography on $ZipTip_{C18}$ pipette tips prior to MS analysis. The resulting mass spectra are shown in Figure 7.4. Regardless of UF membrane cut off, the protocol significantly improved MALDI TOF resolution. Similar peptide profiles were obtained in all three experiments. However, more peptides were observed after 10,000 NMWCO filtration, and some of the observed peptides were unique for each NMWCO.

We further investigated the efficacy of serum peptide preparation by applying $ZipTip_{C18}$ purification to straight serum, 50% acetonitrile supernatant, and serum processed by UF (Figure 7.5). Although it required one more step, the last method provided a stronger MALDI-TOF signal and a higher signal-to-noise ratio, and it detected approximately twice as many peptides. Similar results were obtained with plasma peptides (Figure 7.6). It was also shown that peptide spectrum is different depending on the NMWCO used in the UF step (Figure 7.6).

We have also evaluated whether addition of acetonitrile to serum prior to UF improves the detection of serum peptides (MALDI spectrum shown in Figure 7.5D). Further improvement of the spectrum was attained especially on the *m/z* 1000 area of the spectrum. The solvent may help to release some peptides from protein complexes and thus result in higher overall number of detected peptide species.

UF offers another valuable sample fraction that should not be overlooked. In the UF process, proteins with molecular weights larger than the value of MWCO are concentrated in the retentate, while low molecular weight polypeptides are found in the ultrafiltrate and are essentially free of larger species. Analysis of both retentate and filtrate presents a unique opportunity to look for biomarkers within the entire range of molecular weights.

To prove this concept, we separated plasma proteins from the 10,000 MWCO retentate on a 2D electrophoresis gel. Since addition of 20% acetonitrile to the serum and plasma had a positive effect on the peptide detection, we have tested the retentate fraction from that experiment (Figure 7.7). Evidently the sample has sufficient protein purity and low conductivity to obtain a high-quality 2D gel.

Small peptide profiling using UF can be a valuable source of biomarkers, as well as of diagnostic and prognostic information. Profiling of small polypeptides has not been popular, presumably due to apparent complications in peptide identification by MS. Unlike large protein species, peptides cannot be identified reliably by enzymatic digestion followed by peptide mass fingerprinting. Thus, successful identification of small peptides is now realized by *de novo* sequencing using automated high-sensitivity

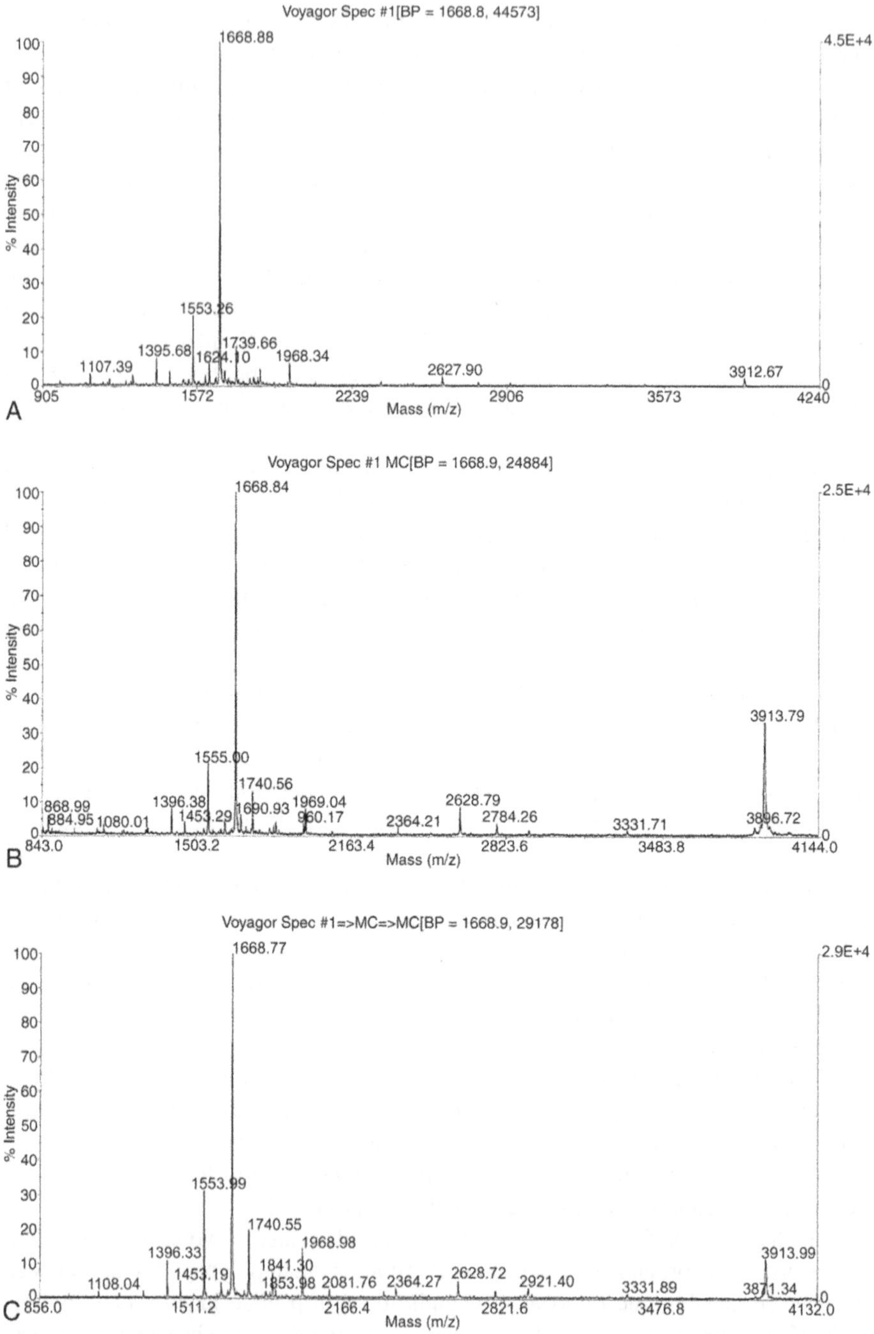

FIGURE 7.4 MALDI TOF spectra of rat serum peptides after UF on Amicon Ultra (A) 10,000; (B) 30,000; and (C) 50,000 devices, followed by concentration and desalting on ZipTip$_{C18}$ pipette tip.

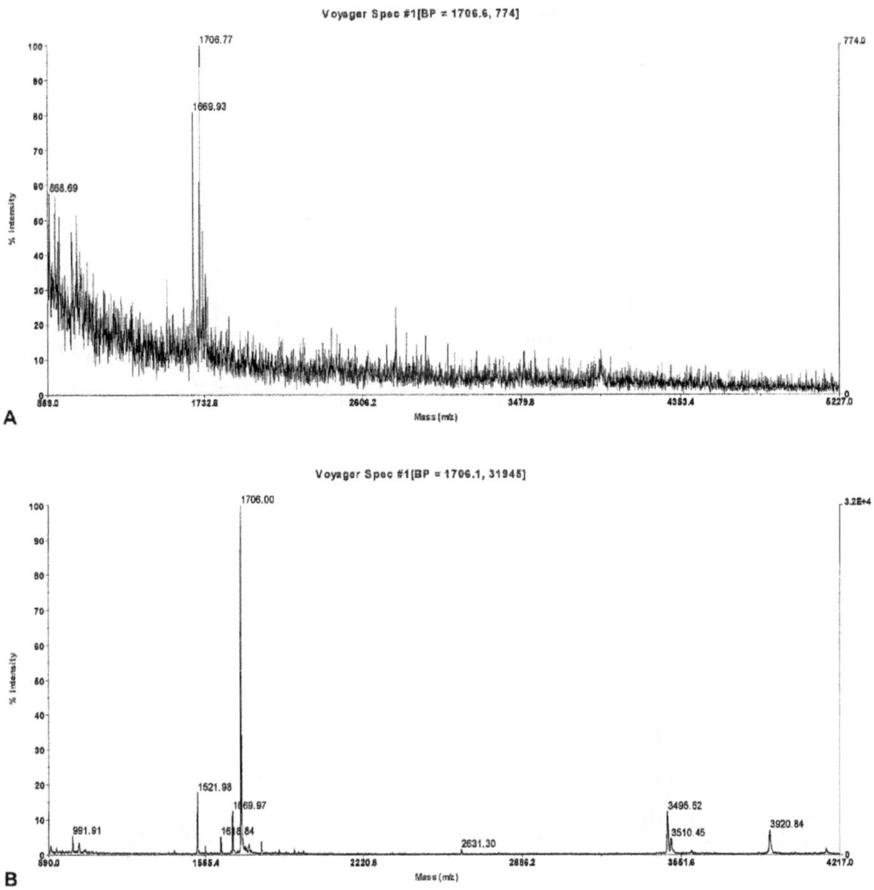

FIGURE 7.5 (Continued)

postsource decay fragmentation and tandem MS. A submilliliter amount of serum sample was shown to be sufficient to identify many peptide peaks. Several serum peptides (Table 7.1) were identified with a significant ion score using automated PSD sequencing on an Axima CFR+ MALDI-TOF mass spectrometer. Figure 7.8 provides an example of PSD sequencing data quality. The modern high-resolution tandem mass spectrometers provide MS/MS peptide sequencing capability at higher sensitivity, which will enable identification of an even greater number of potential peptide biomarkers.

7.4 CONCLUSION

In the analysis of serum peptides, complexity reduction by eliminating higher molecular weight proteins is critical for high-resolution MS. Efficient separation of peptides from

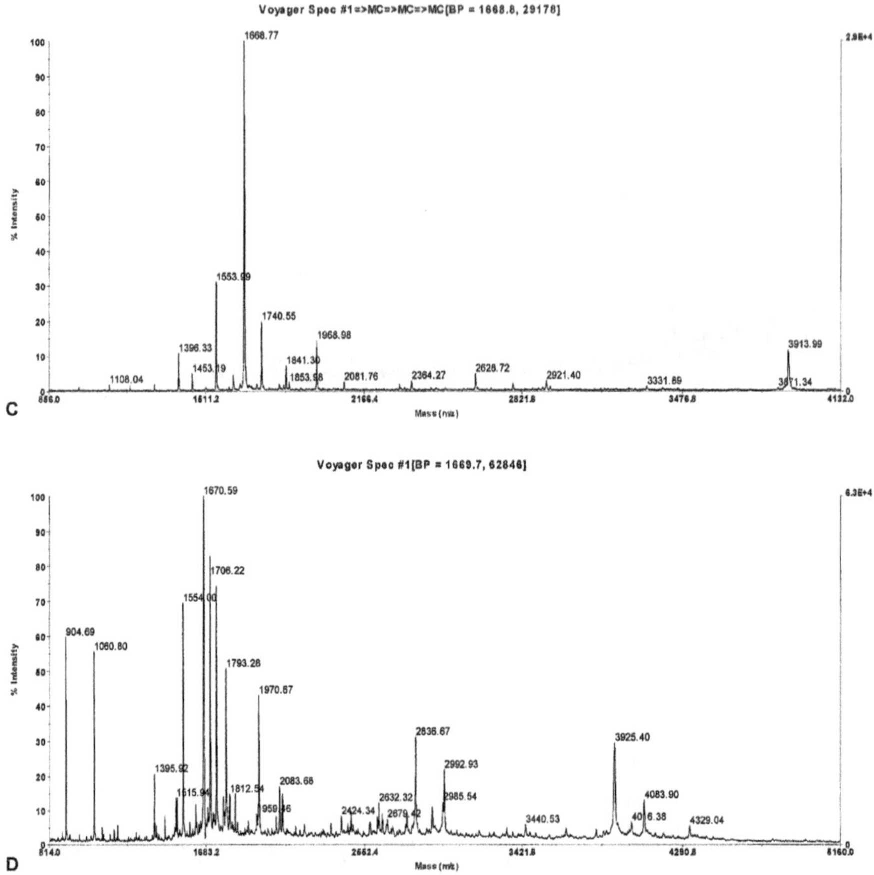

FIGURE 7.5 MALDI TOF spectra of rat serum peptides after concentration and desalting by reversed-phase chromatography: (A) rat serum processed with ZipTip$_{C18}$ pipette tip; (B) serum supernatant after 50% acetonitrile precipitation processed with ZipTip$_{C18}$ pipette tip; (C) 10 K serum ultrafiltrate processed with ZipTip$_{C18}$ pipette tip; and (D) the same as (C) but 20% acetonitrile was added to the serum prior to UF.

majority of proteins and salts can be achieved by sample UF. We have shown use of Amicon Ultra 4 30,000 centrifugal device for the peptides preparation. The method can be used in combination with ZipTip$_{C18}$ pipette tip directly for peptide identification by MS/MS or as a first step prior to further surface-mediated enrichment used in SELDI-TOF methods. In addition, the retentate from those experiments can be used for 2D electrophoresis, thus greatly improving the comprehensive proteome analysis.

REFERENCES

1. Ardekani, A.M., Liotta L.A., and Petricoin, E.F., III, Clinical potential of proteomics in the diagnosis of ovarian cancer, *Expert Rev. Mol. Diagn.*, 2, 312–320, 2002.

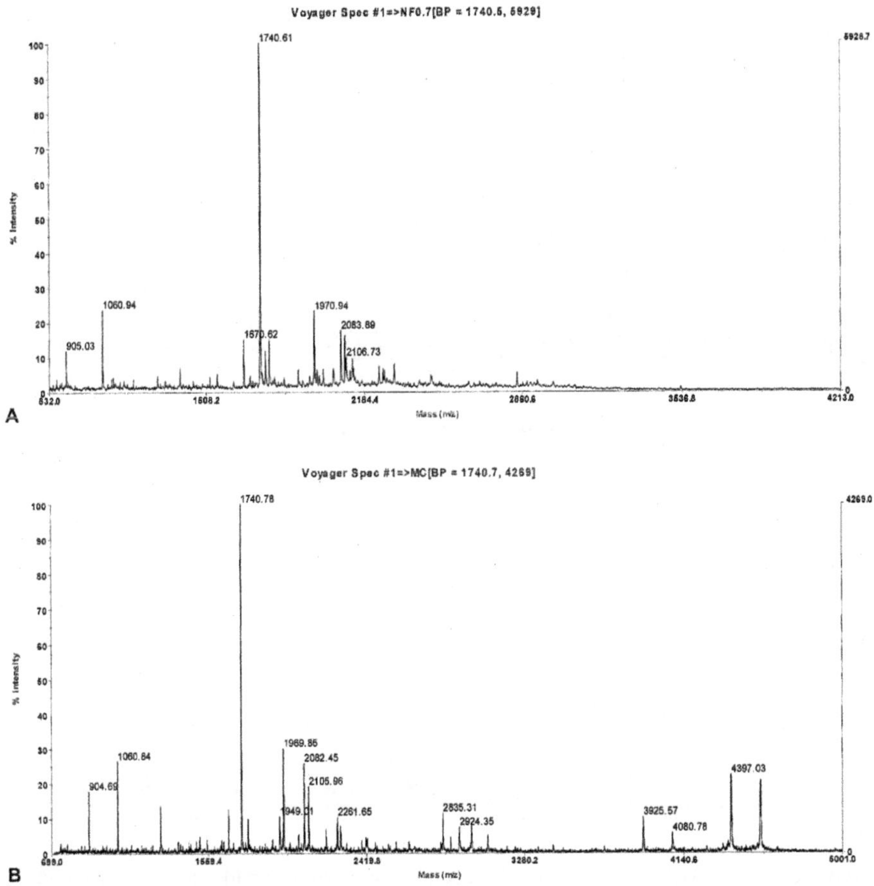

FIGURE 7.6 (Continued)

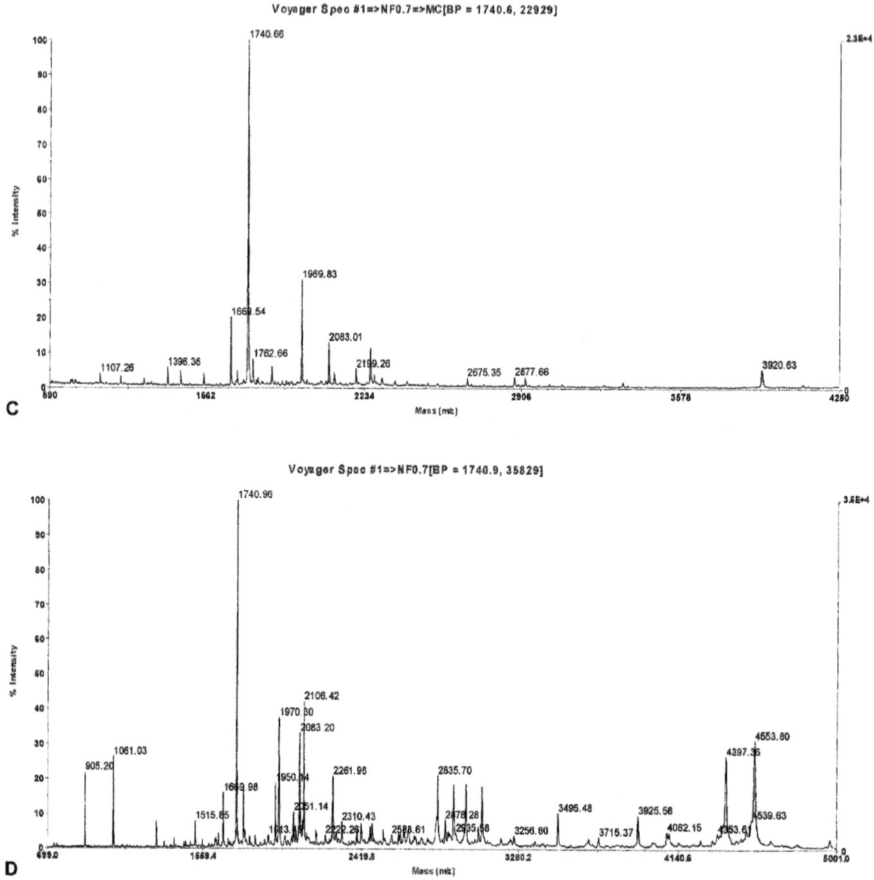

FIGURE 7.6 MALDI TOF spectra of rat serum and plasma peptides after (A) acetonitrile precipitation; (B) UF through 10,000; (C) 30,000; and (D) 50,000 cut-off membrane, all followed by concentration and desalting by reversed-phase chromatography on $ZipTip_{C18}$ pipette tip.

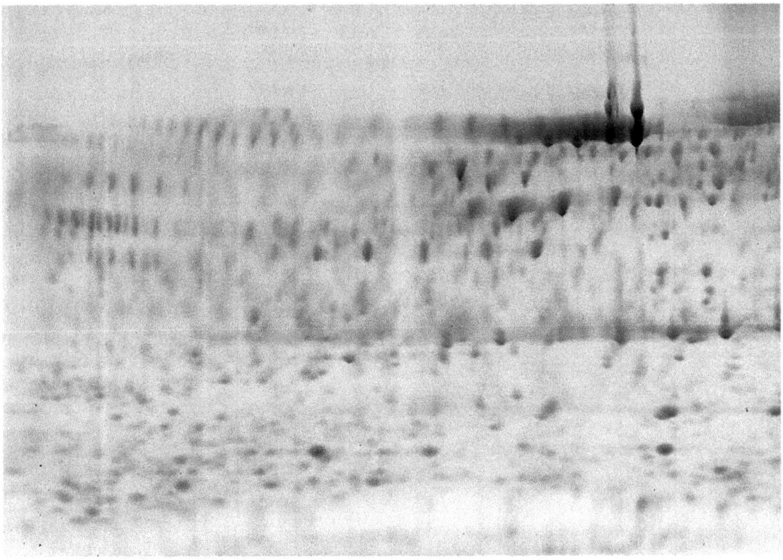

FIGURE 7.7 2D gel of plasma proteins concentrated in 10,000 MWCO Amicon Ultra device.

TABLE 7.1 Summary of MALDI-TOF-PSD Sequencing of 15 Nontryptic Rat Serum Peptides

Polypeptide Name	NCBI Accession	Parent Peptide Mass	Mascot MS/MS Ion Score
Fibrinopeptide A	650771P	1739.81	68
α_1-acid glycoprotein	OMRT1	1816.01	61
Albumin	NP_599153	2407.23	59
Keratin 9	NP_775151	1249.62	42
Angiotensin I	P01015	1296.68	31
α_1-antytripsin	P17475	1494.58	24
Corticotropin precursor	P01194	1623.77	23
Haptoglobin	P06866	1721.72	21
Prothrombin precursor	P18292	2075.84	19
Plasminogen precursor	Q01177	1084.48	19
Complement factor C6	Q811M5	1604.74	17
Atrial natriuretic factor precursor	AWRT	2084.96	16
Apolipoprotein B	NP_062160	1970.01	15
Transferrin	AAH87021	2171.95	13
Hemoglobin α-2 chain	NP_001007723	1571.88	11

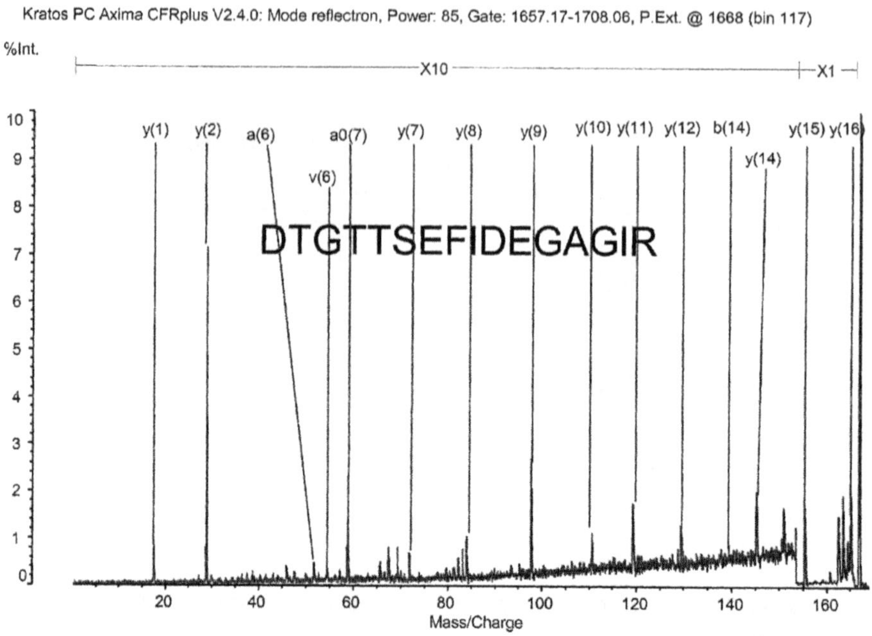

FIGURE 7.8 MALDI postsource decay sequencing of one of the rat serum peptides, fibrinopeptide A, found in all rat serum peptide samples. Sequence of the peptide has been identified using Mascot MS/MS search.

2. Carter, D., Douglass, J.F., Cornellison, C.D., Retter, M.W., Johnson, J.C., Bennington, A.A., Fleming T.P., Reed, S.G., Houghton, R.L., Diamond, T.S., and Vedvick, T.S., Purification and characterization of the mammaglobin/lipophilin B complex, a promising diagnostic marker for breast cancer, *Biochemistry*, 41, 6714–6722, 2002.
3. Krieg, R.C., Fogt, F., Braunschweig, T., Herrmann, P.C., Wollscheidt, V., and Wellmann, A., ProteinChip Array analysis of microdissected colorectal carcinoma and associated tumor stroma shows specific protein bands in the 3.4 to 3.6 kDa range, *Anticancer Res.*, 24, 1791–1796, 2004.
4. Petricoin, E.F., Ardekani, A.M., Hitt, B.A., Levine, P.J., Fusaro, V.A., Steiberg, S.M., Mills, G.B., Simone, C., Fishman, D.A., Kohn, E.C., and Liotta, L.A., Use of proteomic patterns in serum to identify ovarian cancer, *Lancet*, 359, 572–577, 2002.
5. Bischoff, R. and Luider, T.M., Methodological advances in the discovery of protein and peptide disease markers, *J.Chrom. B*, 803, 27–40, 2004.
6. Basso, D., Valerio, A., Seraglia, R., Mazzza, S., Piva, M.G., Greco, E., Fogar, P., Gall, N., Pedrazzoli, S., Tiengo, A., and Plebani, M., Putative pancreatic cancer-associated diabetogenic factor: 2030 MW peptide, *Pancreas*, 24, 8–14, 2002.
7. Stevens, E.V., Liotta, L.A., and Kohn, E.C., Proteomic analysis for early detection of ovarian cancer: A realistic approach? *J. Gynecol. Cancer*, 13,133–139, 2003.
8. Zeman, L.J. and Zydney, A.L., *Microfiltration and Ultrafiltration: Principles and Applications*, Marcel Dekker, New York, 1996.
9. Ho, W.S.W. and Sirkar, K.K., Eds., *Membrane Handbook,* Van Nostrand Reinhold, New York, 1992.

10. Tirumalai, R.S., Chan, K.C., Prieto, D.A., Issaq, H.J., Conrads, T.P., and Veenstra, T.D., Characterization of the low molecular weight human serum proteome, *Mol. Cell. Proteomics*, 2, 1096–1103, 2003.
11. Chernokalskaya, E., Gutierrez, S., Pitt, A.M., and Leonard, J.T., Ultrafiltration for proteomic sample preparation, *Electrophoresis*, 25, 2461–2468, 2004.
12. Barre, J., Chamouard, J.M., Houin, G., and Tillement, J.P., Equilibrium dialysis, ultrafiltration, and ultracentrifugation compared for determining the plasma-protein-binding characteristics of valproic acid, *Clin. Chem.*, 31, 60–64, 1985.
13. Perucca, E., Plasma protein binding of phenytoin in health and disease-relevance to therapeutic drug monitoring, *Ther. Drug Monit.*, 2, 331–334, 1980.
14. Pitt, A.M., The nonspecific binding of polymeric microporous membranes, *J. Parenteral Sci. Tech.*, 41, 110–113, 1987.
15. Adkins, J.N., Varnum, S.M., Auberry, K.J., Moore, R.J., Angell, N.H., Smith, R.D., Springer, D.L., and Pounds, J.G., Toward a human blood serum proteome: Analysis by multidimensional separation coupled with mass spectrometry, *Mol. Cell. Proteomics*,1, 947–955, 2002.
16. Ornstein, D.K., Rayford, W., Fusaro, V.A., Conrads, T.P., Ross, S.J., Hitt, B.A., Wiggins, W.W., Veenstra, T.D., Liotta, L.A., and Petricoin, E.F., III, Serum proteomic profiling can discriminate prostate cancer from benign prostates in men with total prostate specific antigen levels between 2.5 and 15.0 ng/ml, *J. Urol.*, 172, 1302–1305, 2004.
17. Prazeres, S., Santos, M.A., Ferreira, H.G., and Sobrinho, L.G., A practical method for the detection of macroprolactinaemia using ultrafiltration, *Clin. Endocrinol. (Oxf)*, 58, 686–690, 2003.

Part II

Sample Prefractionation and Analysis

8 Tools for Sample Preparation and Prefractionation in Two-Dimensional Gel (2D) Electrophoresis

Anton Posch, Aran Paulus, and Mary Grace Brubacher

CONTENTS

8.1 INTRODUCTION

Two-dimensional gel electrophoresis (2D), with its unique capacity to resolve thousands of proteins in a single run, is a fundamental research tool for nearly all protein-related scientific projects.[1] Over the last two decades, numerous researchers in academia and in industry have improved the technology to the point where a novice user is capable of achieving respectable gel separations on the first try. In addition, 2D technology as practiced today has seen enormous gains in reproducibility, resolution, and automation,[2,3] all of which contribute to its widespread use. Nevertheless, 2D is still a technically demanding method.

The quest to map and characterize each and every protein in a given cell type, tissue, or organism has given 2D an additional boost as the separation methodology of choice for many proteomics laboratories. However, the task list for a proteomics researcher is daunting: the number of proteins in a biological sample, although unknown at this time, is believed to be in the hundreds of thousands, covering a concentration range of seven or more orders of magnitude. In addition, the proteome is extremely dynamic, with protein expression depending on the cell state and further complicated by posttranslational modifications such as phosphorylation or glycosylation, to name just two possible changes to proteins in a functional biological system.

As more and more laboratories start up their own proteomic effort or ramp up existing programs, they realize that meticulous attention to 2D methodology is only one critical aspect when identifying differentially expressed proteins or investigating a particular biological pathway.

The information content of 2D is heavily influenced by a proper sample preparation strategy. Interestingly, not much attention was paid to this area during 2D methodology development.

This chapter will provide a broad overview of the principles and recent developments of sample preparation tools prior to the first step of 2D. Examples from three strategies for sample preparation, based on solution chemistry, chromatography, and electrophoresis, are discussed in detail and are used to illustrate how these key areas can be applied to general-purpose sample cleanup and sample fractionation for enrichment of low-abundance proteins.

8.2 SAMPLE PREPARATION BASICS

Due to the great diversity of protein sample types and origins, no universal sample preparation method applicable for all proteins exists.[4] Unfortunately, an optimal procedure must be determined empirically in most cases and tailored for each sample type. This can be a daunting task. However, some general sample preparation guidelines should be kept in mind to avoid a number of pitfalls during sample preparation for proteomic studies. Sample preparation is not only a prerequisite for a successful 2D experiment but also the key factor to meaningful data evaluation.

The major application of 2D is the differential display of two or more samples, one being a meaningful control sample. Since only proteins in solution can be analyzed in 2D and the objective is to catalogue and compare as many differentially expressed proteins as possible, solubilization is an essential step. The first step of 2D, isoelectric focusing (IEF), is usually performed under denaturing conditions. However, the charge and molecular weight characteristics of the polypeptide chains need to be preserved. An effective sample preparation strategy for 2D will provide the following:[5]

1. Reproducibly solubilize proteins of all classes, including hydrophobic proteins
2. Enhance spot resolution
3. Prevent protein aggregation and loss of solubility during focusing
4. Prevent postextraction chemical modification, including enzymatic or chemical degradation of the protein sample
5. Remove or thoroughly digest nucleic acids and other interfering molecules
6. Yield proteins of interest at detectable levels, which may require the removal of interfering abundant proteins or non-relevant classes of proteins

With these criteria in mind, proteomic sample preparation can be categorized into two major classes: (*a*) general-purpose cleanup and (*b*) fractionation to reduce complex protein samples into smaller subsamples that are easier to manipulate and separate. In some sample preparation strategies, a fractionation protocol can be followed by a general-purpose cleanup protocol. The overall strategy is determined by fractionation approach and the need to remove contaminants introduced by the fractionation method prior to the 2D experiment. Again, customization may be necessary and will depend on the nature of the proteins being studied.

8.2.1 Cell Disruption Methods

The effectiveness of a particular cell disruption method determines the accessibility of all intracellular proteins to extraction. Different biological materials require individual cell disruption strategies. In general, these methods can be divided into two main categories: gentle and harsher methods.

Gentle cell disruption protocols are generally employed when the sample of interest consists of cells that lyse easily, such as red blood cells and tissue culture cells. The most popular and widespread methods are as follows:

- *Osmotic lysis.* Cells swell and burst when they are suspended in hypotonic solution, releasing all cellular contents.[6]
- *Freeze-thaw lysis.* Cells can be lysed by subjecting them to one or more cycles of quick freezing using liquid nitrogen and subsequent thawing.
- *Detergent lysis.* Cellular membranes can be solubilized just by suspending the cells in many detergent-containing solutions.[7,8]
- *Enzymatic lysis.* Cells can be lysed in isoosmotic solutions by enzymes that specifically digest the cell wall (e.g., cellulase and pectinase for plant cells, lyticase for yeast cells, and lysozyme for bacterial cells).

Biological material with tough cell walls and many tissue types require more rigorous methods, which are mainly based on mechanical rupture.

- *Sonication.* A cell suspension, cooled on ice to avoid heating, can be disrupted by shear forces using short bursts of ultrasonic waves.[9,10]
- *French press.* Cells are lysed by shear forces by forcing a cell suspension through a small orifice at high pressure.[11]
- *Grinding.* Cells of solid tissues and microorganisms can be broken with a mortar and pestle. Usually, the mortar is filled with liquid nitrogen and the tissue or cells are ground to a fine powder.[12,13]
- *Mechanical homogenization.* This approach is ideally suited for soft, solid tissues.[14] Handheld devices are known under the brand names Dounce and Potter-Elvehjem homogenizers. When larger samples have to be processed, the use of blenders or other motorized devices is advisable.
- *Glass-bead homogenization.* Vortexed glass beads generate abrasion that breaks cell walls.[15] Usually 1 to 3 g of glass beads is added per gram of wet cells.

During cell lysis, the compartmentalization of a cell will be partly or fully destroyed, depending on the lysis method applied. As a consequence, hydrolases (phosphatases, glycosidases, and proteases) will be present in a homogeneous protein solution and possibly alter the protein composition of the lysed cells. Differential expression is only meaningful when the protein composition of the cellular state to be examined is preserved. Therefore, enzymatic degradation must be avoided by placing the freshly disrupted sample in solutions with strong denaturing agents such as 8 M urea, 2% sodium dodecyl sulfate (SDS), or 10 to 20% trichloroacetic acid (TCA). In this environment, enzymatic activity is often negligible. In addition, cell disruption is often performed at low temperatures to diminish enzymatic activity. Furthermore, because enzymatic activity is pH dependent, unwanted hydrolysis can often be inhibited by lysing the protein samples at a pH higher than 9 using either sodium carbonate or Tris as a buffering agent in the lysis solution.[16]

Contrary to relatively labile phosphatases and glycosidases, some proteases are fairly resistant to denaturation, pH, and temperature shift. In these cases it is advisable to consider the addition of a chemical protease inhibitor. Since individual protease inhibitors are specific to different classes of proteases, broad-range protease inhibitor cocktails should be used. Examples of broad-range inhibitors are phenylmethylsulfonyl fluoride (PMSF), aminoethyl benzylsulphonyl fluoride (AEBSF), tosyl lysine chloromethyl ketone (TLCK), tosyl phenyl chloromethyl ketone (TPCK), ethylenediaminetetraacetic acid (EDTA), benzamidine, and peptide protease inhibitors (e.g., leupeptin, pepstatin, aprotinin, bestatin). Figure 8.1 demonstrates a dramatic example of two-dimensional (2D) maps of yeast proteins processed in the presence or absence of proper inhibitor cocktails.

Phosphorylation is one of the most important posttranslational modifications determining cellular function. In functional proteomics studies, phosphatase inhibitors such as okadaic acid, calyculin A, and vanadate are necessary to block the phosphatases involved in phosphorylation pathways.[17]

8.2.2 Sample and Lysis Buffer Constituents

Solubilization, disaggregation, denaturation, and reduction of proteins during or right after cell disruption is achieved by the use of chaotropic agents, detergents, reducing agents, buffers, and carrier ampholytes. These buffer constituents must be compatible with IEF, both electrically and chemically.

8.2.2.1 Chaotropic Agents

Since the introduction of nondenaturing 2D by Klose[18] and O'Farrell,[19] urea, a neutral chaotropic agent, has been used as a denaturing agent in sample solutions at concentrations as high as 9.8 M, but more typically at 8 M. It effectively disrupts noncovalent and ionic bonds between amino acid residues. When working with urea solutions, the user must be aware that some spontaneous degradation of urea to cyanate is occurring even at room temperature. Cyanate ions are reacting with the amine

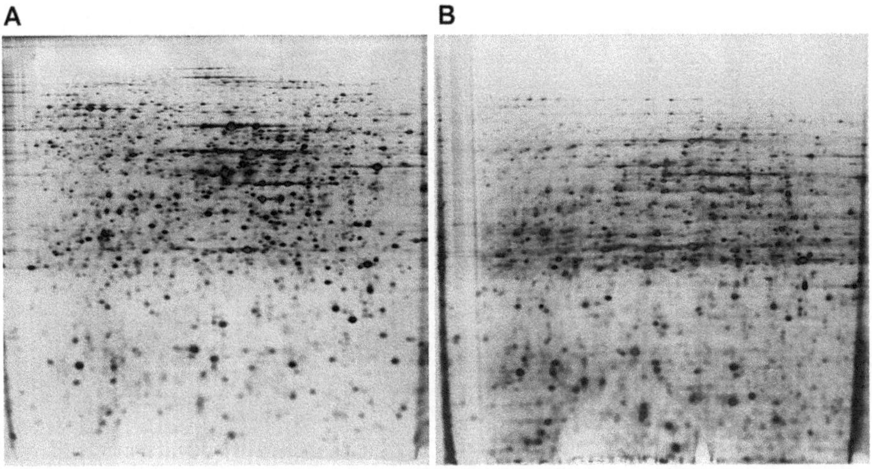

FIGURE 8.1 The effect of protease inhibitors on 2D pattern quality. Yeast (*Saccharomyces cerevisiae*) cells were lysed in 9 M urea, 4% CHAPS, 1% DTT, and 0.5% (w/v) carrier ampholytes. (A) Sample buffer with protease inhibitor cocktail; (B) sample buffer without protease inhibitors leads to massive protein degradation.

groups of proteins (causing carbamylation) and removing the positive charge of the amine, thereby affecting the pI of the respective proteins. Since the generation of cyanate ions is temperature dependent, urea containing solutions may not be heated above 37°C to avoid increasing the rate of these reactions.[20]

The solubilizing power of urea containing sample solutions can be dramatically increased by the addition of thiourea.[21] This effect was first reported by Thierry Rabilloud; his 2D gels of nuclear proteins, integral membrane proteins, and tubulin showed very high recovery rates compared to classical solubilization approaches without thiourea. The improved solubilizing power of thiourea containing sample solutions, especially for membrane proteins, is shown in Figure 8.2, where *Escherichia coli* outer membrane proteins have been successfully extracted on the addition of thiourea to the protein extraction cocktail.

Typically, thiourea is used at 2 M concentration in conjunction with 5 to 7 M urea. Highly concentrated urea solutions are essential for solubilizing thiourea, which by itself exhibits poor solubility in water.

8.2.2.2 Detergents

Hydrophobic interactions within a polypeptide chain or between proteins in protein complexes can be disrupted in the presence of detergents. An additional benefit to the use of detergents is increased solubility, especially of membrane proteins. Detergents operate synergistically with chaotropic agents during the solubilization process, as they prevent hydrophobic interactions between hydrophobic protein stretches exposed by the chaotropic agents.

Detergents consist of a hydrophobic tail and a hydrophilic head moiety, which may be anionic, cationic, zwitterionic, or nonionic. To allow proteins to migrate according

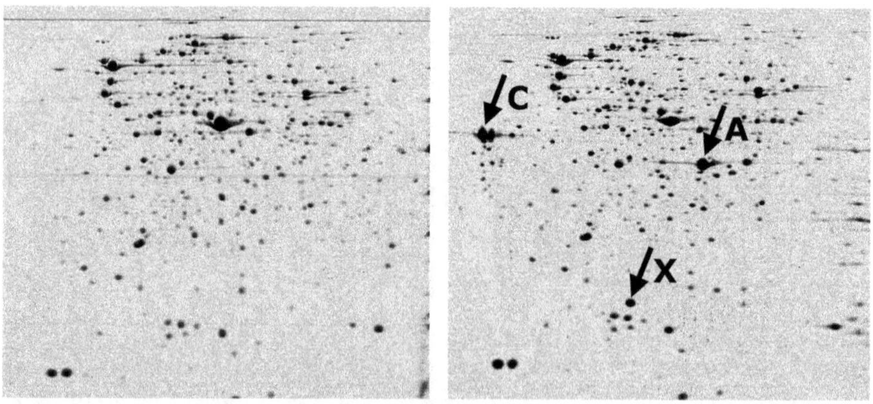

FIGURE 8.2 Increasing the elution power of 2D sample buffer by adding thiourea. *E. coli* cells were lysed in 9 M urea, 4% CHAPS, 1% DTT, 0.5% (w/v) carrier ampholytes (left) and 7 M urea, 2 M thiourea, 4% CHAPS, 1% DTT, 0.5% (w/v) carrier ampholytes (right). The thiourea-containing sample buffer enables the extraction of outer membrane proteins (OMP) C, A, and X, as indicated by arrows.

to their own charge during IEF, zwitterionic or nonionic detergents are preferred.[22] Early 2D work described the use of Nonidet P-40 or Triton X-100, both nonionic detergents, in concentrations ranging from 0.5 to 4%. More recently, nonionic detergents have frequently been replaced in the solubilization cocktail by the zwitterionic detergent 3[(cholamidopropyl)dimethylammonio]-1-propane sulfonate (CHAPS), a linear sulphobetaine surfactant. Several studies have indicated the superior solubilization power of CHAPS,[23–25] particularly for membrane proteins. As a result, linear sulphobetaines, especially those with a carbon tail at least 10 carbon units long, are emerging as probably the most versatile surfactants. Prominent members of this class of surfactants are SB3–10 and ASB-14. The selection of the optimal detergent in a solubilization cocktail should also take into account how the detergent interacts with high concentrations of urea. For example, when using SB3-10, the urea concentration is limited to 5 M, whereas ASB-14 can be used with 9 M urea. Chevallet et al.[26] published a comprehensive review about the structural requirements for the optimal sulphobetaine for an appropriate solubilization solution.

SDS is a detergent with very wide use in a number of biochemical solution-based protein assays. Although SDS is an ionic detergent, it can be used during the IEF step and does not interfere with the focusing when used in combination with an excess of a nonionic or zwitterionic detergent (NP-40, CHAPS, Triton X-100). According to Ames and Nikaido,[27] the NP-40:SDS concentration ratio should be at least 8:1 to avoid detrimental effects of SDS in IEF. In addition, its final concentration in the sample solution buffer should be below 0.25%.[10]

8.2.2.3 Reducing Agents

Thiol reducing agents such as mercaptoethanol, dithioerythritol, and dithiothreitol (DTT) are commonly used to disrupt intramolecular and intermolecular disulfide bonds. A large excess of reducing agent will shift the equilibrium of the oxidation-reduction reaction toward the fully reduced protein state. Therefore, reducing agents are typically used in large excess. For example, DTT is present in immobilized pH gradient (IPG) cocktails at concentrations ranging from 20 to 100 mM to assure reduction of the protein disulfide bonds. Thus, it is necessary that DTT be evenly distributed over the whole length of the IPG strip during IEF in order to maintain the reduced state of all cystines present in the protein. However, because DTT is a weak acid, it will not remain at a constant concentration in the basic regions of an IPG strip. It will migrate toward the anode.[28] As a consequence, some disulfide bonds will reform, leading to precipitation of some disulfide rich proteins (e.g., wool keratins). Artifactual spots will also be generated due to the formation of so-called scrambled disulfide bridges.[29] To avoid this behavior, an alkylation step after the reduction step and as a sample preparation treatment is recommended, especially for basic proteins.

Görg and coworkers[30] tried to circumvent the DTT migration problem by placing extra filter paper soaked with DTT close to the cathodic end of the IPG strip to serve as a reservoir to avoid depletion of DTT in the basic range.

Phosphines such as tributylphosphine (TBP) offer an alternative to thiols as reducing agents because they are uncharged and reduce cystines stoichiometrically at concentrations as low as 2 mM. Tributylphosphine is a more effective reducing agent than DTT because it facilitates improved protein solubility. This has been demonstrated by Herbert et al.[31] for highly cross-linked wool keratins. In contrast to DTT, short chain phosphines are chemically more difficult to handle, because they are volatile, toxic, and highly flammable in concentrated stock solutions. However, TBP formulations safe for shipping and laboratory use have been developed by commercial suppliers.

8.2.2.4 Carrier Ampholytes

Carrier ampholytes are usually included at a concentration of ≤0.2% (w/v) in sample solutions for IPG strip focusing. They appear to reduce protein-matrix hydrophobic interactions, which tend to occur at the basic end of the IPG strip, and lead to streaking caused by precipitation.[32] Furthermore, the addition of carrier ampholytes overcomes detrimental effects from salt and helps to compensate for insufficient salt in a sample. Even in the presence of detergents, certain samples may have stringent salt requirements to maintain the solubility of some proteins. Although IPG strips are quite tolerant to salts, salt should be in the sample only if they are absolutely required, and then only at a total concentration less than 40 mM. During IEF, any salt will be removed and proteins that need salt for solubility are then subject to precipitation. As a general rule, salts forming strong acids and bases (e.g., NaCl, Na_2HPO_4) should be avoided, because strongly alkaline cationic and strongly acidic anionic boundaries are formed by the migration of the salt's ionic constituents. Salts formed from weak acids and bases such as Tris-acetate and Tris-glycine should be used instead.[33]

8.2.2.5 General Solubilization Cocktails and Sample Preparation Guidelines

The sample and lysis buffer constituents can be combined together in a variety of formulations to address the solubility requirements of the protein sample. Some of these formulations have led to standardized solubilization cocktails and are summarized in Table 8.1. When choosing a suitable solubilization cocktail from Table 8.1, the following general sample preparation guidelines should always be kept in mind:

- Each additional step in sample preparation bears the risk of protein loss. The best strategy is the simplest one and takes the compatibility of each step into consideration.
- Solutions for sample preparation should be freshly prepared with high-quality chemicals or stored as frozen aliquots.
- Protein containing samples should be stored in reasonably sized aliquots at −80°C or readily used immediately in the IEF step. Repeated freezing and thawing should be avoided.
- Urea containing solutions should always be kept below 37°C to avoid the formation of isocyanate and the possible carbamylation of proteins.
- Prior to the IEF step, the sample should be subjected to high-speed centrifugation to remove all solid particles that can clog the pores of the gels used for separation.

TABLE 8.1 Protein Solubilization Cocktails for 2D Electrophoresis

Solubilization Buffer	Organism	Comment	Reference	Year
7 M urea, 2 M thiourea, 4% CHAPS, 65 mM DTT, 2% (v/v) ampholytes pH 3–10	Yeast	Samples boiled in 1% SDS and then diluted with buffer	[10]	1999
7 M urea, 2 M thiourea, 2% C8phi, 0.5% TX-100, 1.2% ampholytes	*Arabidopsis thaliana*	Isolation of plasma membrane. Alkaline buffer treatment before extraction was essential	[71]	1999
7 M urea, 2 M thiourea, 2 mM TBP, 1% ASB 14, 0.5% ampholytes pH 3–10	*E. coli*	Isolation of outer membrane proteins (after carbonate washing)	[72]	2000
5 M urea, 2 M thiourea, 2% CHAPS, 2% SB3-10, 20 mM dithiothreitol, 5 mM TCEP, 0.5% carrier ampholytes, pH 4–6.5, 0.25% carrier ampholytes, pH 3–10	Maize (endosperm)	The efficiency of several solubilization buffers for plant proteins were compared	[73]	2003
7 M urea, 2 M thiourea, 0.9% DTT, 4% CHAPS 14, 4% (v/v) ampholytes pH 3–10	Rat kidney	Isolation of intrarenal structures such as glomeruli, vessels, and tubules	[4]	2004

8.2.3 Technologies for Protein Sample Preparation

Sample preparation can be viewed in two different ways, based on either the technology used or the sample preparation problem. Both views are expressed in this chapter: the technology-centered view and the sample-preparation view. Table 8.2 summarizes both approaches and may serve as a guide to which problem is best addressed by which technology.

Many useful technologies in protein chemistry can be applied to sample preparation. These technologies can be categorized into three major groups: solution chemistry, chromatographic methods, and electrophoretic methods. All these approaches can be used to clean up proteomic samples, reduce their complexity, or both.

8.2.3.1 Solution Chemistry

The first and arguably simplest technology for contaminant removal as well as for fractionating protein mixtures is the use of differential solubility of proteins or protein classes. Combining detergents and other chemicals at appropriate temperatures provides solubility conditions that precipitate the proteins of interest or alternatively selectively precipitate unwanted proteins or contaminants. Examples of this type of

TABLE 8.2 Proteomic Sample Prep by Technology and Application

	Application								
	General-Purpose Cleanup					Fractionation			
Technology	Salt Removal	Detergent, Polysaccharide, Phenolic Compound, Lipid Removal	Albumin Removal	Disulfide Bond Removal	Total Protein Extraction	Cellular Location	Solubility Hydrophobicity	Size (MW)	Charge/pI
Solution Chemistry	X	X	—	X	X	X	X	—	—
Chromatography	X	—	X	—	—	—	—	X	X
Electrophoresis	—	—	—	—	—	—	—	X	X

approach include precipitation methods for sample cleanup using TCA, acetone, or ammonium sulfate. In addition, complex protein mixtures can be simplified by precipitation of proteins in a biological sample followed by resolubilization in a series of solutions with increasing solubilization strength. This serial resolubilization effectively enriches for increasingly hydrophobic proteins with each step. This is an effective means of reducing protein complexity.

8.2.3.2 Chromatographic Methods

Chromatographic methods offer a broad range of strategies for protein purification and thus for prefractionation in proteomic studies. Both denaturing and nondenaturing chromatographic methods for proteins have been described. For the discussion here, we concentrate on ion-exchange and affinity approaches and briefly discuss multidimensional chromatography. Ion-exchange chromatography (IEX) is widely used for proteins and separates according to the protein's intrinsic charge. A variety of stationary phases with either anionic or cationic properties and the pH of the elution buffer dictate which proteins are captured by the stationary phase and which are eluted. For complex samples, both salt and pH gradients are possible. An experimental advantage of chromatography is that it is used almost exclusively in a column format. This makes fraction collection straightforward.

Fountoulakis and coworkers extensively used chromatography to prefractionate the proteomes of *Haemophilus influenza* and *E. coli* by using heparin and hydroxyapatite matrices, chromatofocusing, and hydrophobic interaction chromatography.[34–36] Although this successful and innovative work expanded the possibilities for proteome analysis of low-abundance proteins, the interfacing of chromatography and 2DE might bear some inherent problems. Usually, huge amounts of salts (~1 M NaCl) are needed for elution from chromatographic columns filled with ion-exchange material, and then subsequent 2D analysis of chromatographically derived fractions is therefore not directly possible. These samples have to be desalted and often concentrated prior to2D analysis. This poses the risk of protein loss during the various manipulations. Additionally, the fractions obtained by chromatography usually do not represent narrow or discrete pI cuts (see Chapter 4, this volume) and the fractions have to be analyzed on wide-range (pH 3–10) instead of narrow-range IPG strips of one or two pH units.

Affinity chromatography methods can be used either to isolate a particular set of proteins[37] or to deplete a sample of high-abundance components that ordinarily mask other proteins.[38] Affinity chromatography occupies a unique place among all chromatographic methods since it enables the enrichment of proteins on the basis of their biological function or individual chemical structure.

Serum samples are unique because of their high content of albumin and IgGs. Albumin constitutes 50 to 60% and IgGs constitute 10 to 20% of proteins present in serum. These abundant proteins in a sample present a problem for 2D electrophoresis because they limit the amount of sample that can be loaded on first dimension IPG strips to display nonabundant proteins. However, since the major components of serum are already identified, they can safely be removed to look more closely at the remaining 20% of serum proteins, believed to contain thousands of unknown proteins. An illustration of the use of affinity chromatography to address the selective removal of albumin and IgG with Affi-Gel Blue resins is shown in Figure 8.3. Recently, further selectivity has been achieved by immobilizing antibodies specific to human albumin and IgGs on a stationary phase.[39]

Multidimensional chromatography is increasingly popular as a method to increase the theoretical maximum capacity of a system. The success of 2D gels with two independent modes of separation initiated the search for orthogonal chromatographic methods that can be combined and automated. Experimentally, coupling two or more chromatographic methods is simple, since the effluent of one column can be directly injected onto the second column. In practice, some buffer incompatibilities are observed. For example, a high salt buffer used to elute from an antibody column requires desalting to be compatible with an ion-exchange column. A combination of ion exchange and reversed phase has been used very successfully for the separation of peptide mixtures[40] but would be less efficient for proteins. Lubman et al. described a protein chromatographic separation scheme to mimic 2D gels using chromatofocusing as the first dimension and reversed phase as the second dimension.[41,42]

8.2.3.3 Electrophoretic Methods

Electrophoresis separates proteins in an electric field according to differences in their charge, mass, or mass-to-charge ratio. In IEF, all proteins are separated according to their charge as they move in response to an electrical field until their pI matches the pH of an external pH gradient. Sodium dodecyl sulfate–polyacrylamide gel electrophoresis (SDS-PAGE) separates proteins according to their molecular mass. In this technique, SDS is added to a protein mixture to both denature the proteins and coat them with a net negative charge. This SDS treatment essentially creates a uniform mass-to-charge ratio for all proteins in the sample. The proteins then migrate and are separated in a porous medium according to their mass.

Electrophoresis may be performed in either matrix-based (e.g., gel-based) or matrix-free (liquid-based) systems. Matrix-free, liquid-based electrophoretic fractionation

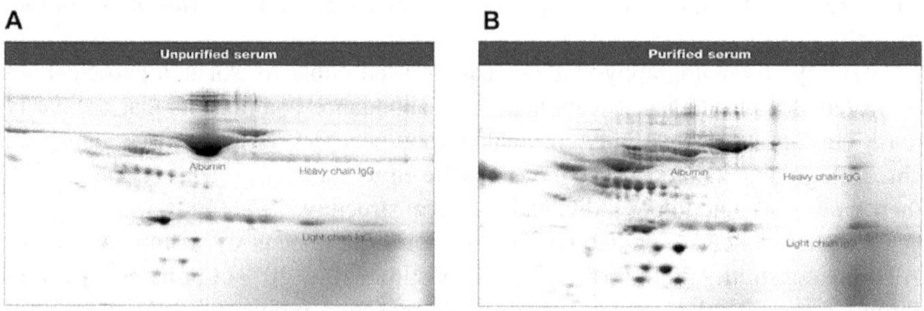

FIGURE 8.3 Removal of abundant proteins in serum. Total protein (1.32 mg) was loaded onto an Aurum serum protein mini column. Then 200 μg total protein from each fraction was loaded onto IPG strips (pH 3–10, 11 cm). (A) Unpurified serum; (B) purified serum.

protocols provide an interesting alterative to chromatography since they are also compatible with subsequent 2D separation. All matrix-free electrophoretic techniques exploit the same separation mechanism as IEF. A variety of liquid-based instruments for the pI-dependent separation of proteins (see Table 8.3) have been described. Some of them favor compartmentalization utilizing a membrane, grid, or matrix (batchwise processing of protein samples), while others rely on a continuous separation mode, such as free-flow electrophoresis instrumentation. An excellent review by Righetti[43] provides more detail.

8.3 GENERAL-PURPOSE CLEANUP FOR IMPROVED RESOLUTION AND REPRODUCIBILITY

Success or failure of any protein analysis depends on sample purity. Contaminants, such as salts, detergents, and ionic compounds, as well as high-abundance proteins, can compromise a 2D experiment by interfering with the protein separation or masking proteins of interest. It is crucial to eliminate these contaminants prior to analysis.

8.3.1 Removal of Interfering Contaminants

Interfering substances are any contaminants that negatively impact IEF, SDS-PAGE, or both. Contaminants of this nature include salts, detergents, small compounds, nucleic acids, lipids, phenolic compounds, and polysaccharides. Therefore, sample preparation has to include procedures to eliminate these substances from the original biological sample. An excellent review by Rabilloud[44] discusses approaches in detail for the removal of interfering contaminants, and we refer here to his recommendations. The most versatile method to selectively separate proteins from other contaminants consists

TABLE 8.3 Liquid-Based Instruments for the pI-Dependent Separation of Proteins

Instrument Name	Manufacturer	Recent Publications
Rotofor® System	Bio-Rad Labs, Hercules, California	[60,74–78]
Gradiflow™ BF 400 System	Gradipore, French's Forest, Australia	[79–84]
IsoelectrIQ² System	Proteome Systems, North Ryde, Australia	[62,85]
Off-Gel™ electrophoresis	DiagnoSwiss, Monthey, Switzerland	[64,65]
ZOOM® IEF fractionator	Invitrogen, Carlsbad, California	[63,86,87]
Free-Flow-Electrophoresis System	FFEWeber GmbH, Planegg/Munich, Germany	[88–92]

of protein precipitation followed by resolubilization in an IEF sample solution. Recently, Jiang et al.[45] compared four widely applied precipitation methods for sample cleanup of human plasma, which used TCA, acetone, chloroform/methanol, and ammonium sulfate, as well as ultrafiltration. They found that precipitation with TCA and acetone as well as ultrafiltration resulted in efficient sample concentration and desalting. However, independent of precipitation and resolubilization method, protein recoveries in the liquid phase are rarely 100%. All protocols require careful optimization to assure a true protein representation in the final 2D image.

A variety of commercial vendors provide kits that can simplify and standardize laboratory procedures for protein isolation from biological samples. For example, the ReadyPrep 2D cleanup kit (Figure 8.4) uses a modification of the traditional TCA protocol, claiming quantitative protein recovery as well as ensuring easy and reproducible removal of interfering substances.

8.3.1.1 Removal of Salt

Salt concentration above 40 mM should be avoided to assure a high quality IEF step. Ionic compounds such as salts limit the applied voltage in the IEF step and therefore increase the focusing time and risk burning the IPG strip. High salt concentrations may also trigger electroendoosmotic flow, resulting in water loss in parts of the IPG strip. Salts in samples can be removed by dialysis, gel filtration, protein concentration devices, or precipitation followed by resolubilization. Figure 8.5 compares the focusing quality of *E. coli* protein extracts containing 1 M NaCl before and after removal of salts with the ReadyPrep 2D cleanup kit.

8.3.1.2 Nucleic Acids

Quickly dividing cells and cultured cells are usually nucleic-acid-rich samples. The presence of nucleic acids tends to increase sample viscosity and clog the pores of the polyacrylamide matrix. Consequently, nucleic acids can inhibit protein entry and

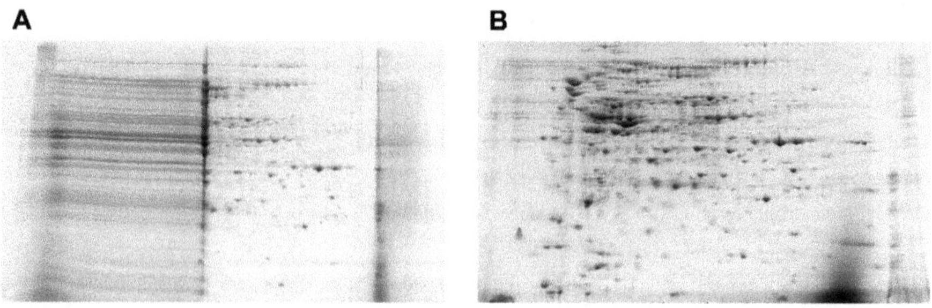

FIGURE 8.4 Detergent removal using the ReadyPrep 2D Cleanup Kit. *E. coli* extracts containing 1% SDS were electrophoresed (A) before; and (B) after treatment with the 2D Cleanup kit. The samples were focused using 11 cm ReadyStrip pH 3–10 IPG strips and run on Criterion 8–16% Tris-HCl precast gels for the second dimension.

slow migration in the IPG strip. In addition, DNA potentially binds to proteins and causes artifactual migration and streaking on the 2D gel. Enzymatic digestion with endonucleases is the most straightforward approach to DNA removal. The enzyme can be added prior to loading the sample onto the first-dimension IPG strip. Addition of carrier ampholytes with subsequent ultracentrifugation or precipitation with a basic polyamine at high pH have also been reported.[3]

8.3.1.3 Polysaccharides

Similarly to nucleic acids, high molecular weight polysaccharides can interfere in IEF by obstructing gel pores. Additional problems may be caused by negatively charged polysaccharides such as mucins and dextrans. They may bind proteins through ionic interactions, leading again to streaking on the 2D gels. Removal of polysaccharides is recommended by precipitation in TCA, ammonium sulfate, or phenol and ammonium acetate.

8.3.1.4 Lipids

Proteins bind to lipids by hydrophobic interactions, giving rise to artifactual heterogeneity on 2D gels. Lipid-protein complexes may be completely insoluble in aqueous solution and consequently may not move into the polyacrylamide matrix at all. Addition of excess detergent is one way to break lipid-protein interactions. In the case of lipid-rich tissues such as brain, chemical removal of lipid with organic solvents prior to sample resolubilization may be necessary.

8.3.1.5 Insoluble Material

As a general rule, samples should always be clarified from insoluble material by high-speed centrifugation (~20,000*g*, 30 minutes) prior to first-dimension IEF. Insoluble material not removed in this manner can clog gel pores and will result in poor focusing,

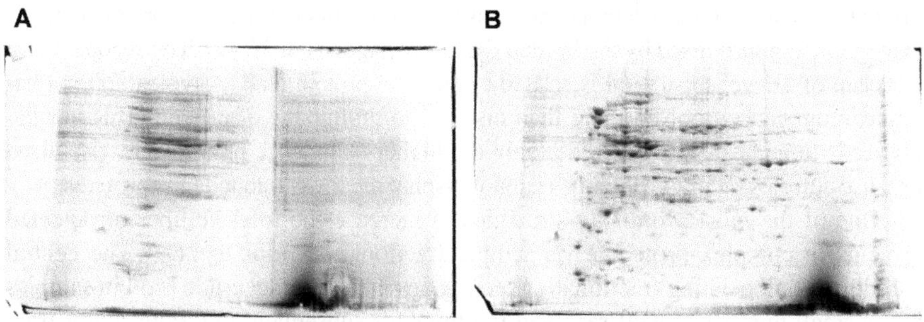

FIGURE 8.5 Salt removal using the ReadyPrep 2D cleanup kit. *E. coli* extracts containing 1 M NaCl were electrophoresed (A) before; and (B) after treatment with the 2D cleanup kit. The samples were focused using 11 cm IPG strips pH 3–10 and run on Criterion 8–16% Tris-HCl precast gels for the second dimension.

especially when the sample is applied using sample cups. During high-speed centrifugation, the sample temperature must be carefully controlled and should be around 20°C, which corresponds to the conditions for IEF in IPG gels. Centrifugation below 10°C may pose the risk of urea precipitation. Centrifugation at room temperature (e.g., 25 to 30°C) will help to keep proteins in solution that would otherwise precipitate out under the conditions of IEF.

8.3.2 Removal of Disulfide Bonds

As already discussed, DTT and TBP are used as reducing agents to break intramolecular and intermolecular disulfide bonds to improve protein solubilization. DTT is ionized at high pH and in an electrical field migrates to the anode. Reoxidation of reduced S–S bonds will occur in these DTT-depleted gel areas resulting in the formation of scrambled disulfide bridges and spurious proteins spots.

The use of TBP is favored over DTT because it is used in a stoichiometric reaction at a 2 mM concentration. Moreover, TBP is uncharged and can therefore avoid the electroendoosmotic flow-related problems of DTT. To overcome general problems with preserving the reduced state of a protein sample, Herbert et al.[29] suggested protein reduction and alkylation of the resulting free –SH groups. Alkylation is an irreversible process and will prevent the phenomena outlined above. For more details, refer to Chapter 16 (this volume).

8.4 FRACTIONATION: THE QUEST FOR LOW-ABUNDANCE PROTEINS

The classical approach for comparing and characterizing complex protein mixtures relies on separation by 2D and protein identification by mass spectrometry. Although this approach has been successfully demonstrated in numerous studies, there are two major criticisms of the 2D gel separation. First, the 2D gels have a limited dynamic range of less than 10^4 when fluorescently stained. Biological samples have a protein concentration range of at least seven orders of magnitude. The study of low-abundance proteins such as cellular receptors and transcription factors in crude cell protein extracts is therefore compromised by the limited dynamic range of the 2D gel. The second major criticism of 2D gel separation is related to the total protein load. A typical 2D gel has a maximum capacity of no more than one to two milligrams of protein. This inevitably leads to a cut of the proteome: only the higher-abundance proteins are visualized because current staining techniques cannot display the lowest-abundance proteins.

One of the most promising strategies to increase the total number of detected proteins in complex proteomes is sample fractionation prior to 2D.[46] The central paradigm of all existing fractionation techniques is to separate crude protein samples into a small number of well-resolved fractions for subsequent 2D. Four major fractionation approaches are discussed:

1. Fractionation by subcellular location
2. Fractionation by differential solubility

3. Fractionation by protein size
4. Fractionation by protein charge or isoelectric point (pI)

8.4.1 Fractionation by Subcellular Location

A well-established approach to dividing cellular protein components into subgroups is to isolate subcellular fractions. Purification of subcellular fractions offers great potential to obtain highly purified and enriched protein preparations. The benefits of this approach are obvious. The protein diversity and complexity are reduced because only a subset of proteins of the entire proteome is selected, and the analysis of biologically associated proteins is made easier.[11,47,48] For example, the key to recovering enriched fractions of membrane proteins is dependent on solubilization strategies for hydrophobic proteins. Traditional approaches include temperature-dependent solubilization in various detergents such as Triton X-100 or Triton X-114.[49] Another approach uses sodium carbonate to extract highly hydrophobic proteins.[50]

Subcellular fractions can also be obtained through centrifugation of subcellular organelles. A centrifuge-based cell fractionation scheme is often used for this purpose. By applying different forces between 1,000*g* and 150,000*g* in a series of centrifugations, subcellular organelles such as nuclei, mitochondria, peroxisomes, lysosomes, and small vesicles are isolated. Although relatively pure preparations can be obtained, subsequent detailed analysis by electron microscopy or organelle-specific enzyme identification is necessary to assure sample quality. Additionally, these methods are rather slow and low throughput, and endogenous proteolysis, which would alter the protein composition of the sample prior to proteomic analysis, is a danger.

8.4.2 Fractionation by Differential Solubility

Fractionation by differential solubility takes advantage of solubility as an independent means of protein separation for reducing sample complexity.[51] Proteins are sequentially extracted with increasingly hydrophobic solubilization solutions. More protein spots are resolved by applying each solubility class to a separate gel, thereby enriching for particular proteins while simplifying the 2D patterns in each gel. Molloy et al.[52] used this approach of sequential solubilization to extract *E. coli* membrane proteins in a three-step procedure. In a first step, 40 mM Tris base was used to solubilize many cytosolic proteins. The pellet was then subjected to extraction with a solution of intermediate solubilization power containing 8 M urea, 4% CHAPS, 2 mM TBP, 40 mM Tris base, and 0.2% carrier ampholytes (pH 3–10). Finally, the membrane-rich protein pellet was partially solubilized using a potent combination of 5 M urea, 2 M thiourea, 2% CHAPS, 2% SB3–10, 2 mM TBP, 40 mM Tris, and 0.2% carrier ampholytes (pH 3–10). Many of the extracted membrane proteins had not been previously identified on a 2D gel, as verified by N-terminal sequencing and peptide mass fingerprint analysis. The key to this approach was the increasing solubilization strength of each subsequent extraction solution. Figure 8.6 shows this approach using a commercial kit (ReadyPrep Sequential Extraction Kit). A similar kit (ReadyPrep Protein Extraction Kit [soluble/insoluble]) follows a similar theme and

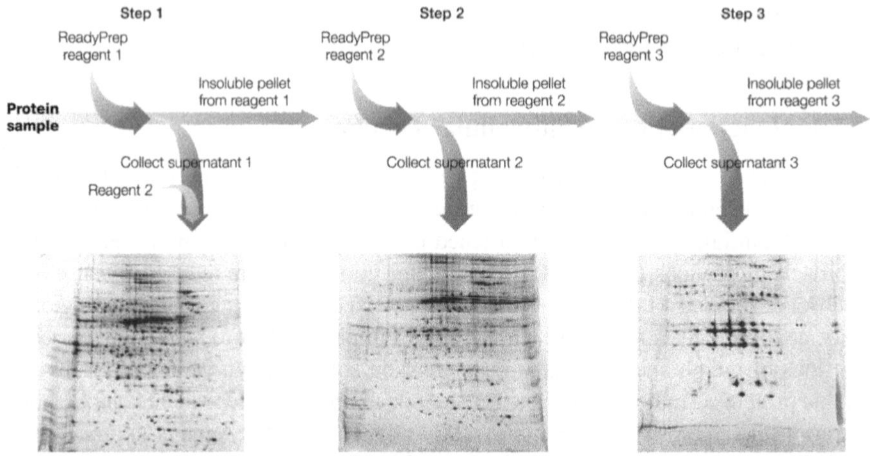

FIGURE 8.6 The ReadyPrep Sequential Extraction Kit distributes proteins into three different fractions based on differential solubility, providing increased resolution of proteins on 2D gels.

generates only two fractions, with the second solubilizing solution containing the detergent ASB-14. ASB-14 is thought to be a stronger detergent than SB3–10.

8.4.3 Fractionation by Protein Size

Two methods may be used to separate proteins by their size, or molecular weight: size exclusion chromatography and gel electrophoresis, particularly continuous-elution electrophoresis. Whereas size exclusion chromatography provides a convenient method for crude fractionation and removal of contaminants, continuous-elution electrophoresis provides superior resolution.

In continuous-gel electrophoresis, protein separation may be conducted via native PAGE or denaturing SDS-PAGE. Separated bands migrate off the bottom of a gel and are collected as liquid fractions. Commercially available instruments for this approach are the Model 491 prep cell and the mini prep cell from Bio-Rad Laboratories (Figure 8.7). These devices allow proteins to be eluted directly and continuously from the gel into an elution chamber. The flow of buffer then passes the isolated bands to an elution tube and finally on to a fraction collector. Purified polypeptides are recovered in test tubes ready for analysis in as little as 5 hours.[53–55] The method is particularly beneficial when studying protein families or posttranslational modifications.[56,57]

8.4.4 Fractionation by Charge and pI

The pI is an important characteristic of every protein. A number of instrumental approaches have been developed to take advantage of the pI as a discriminating factor in preparative fractionation schemes. All of them are liquid based, meaning that the proteins of interest stay in solution. Contrary to chromatographic methods, where proteins are adsorbed onto and desorbed from a stationary phase, which risk

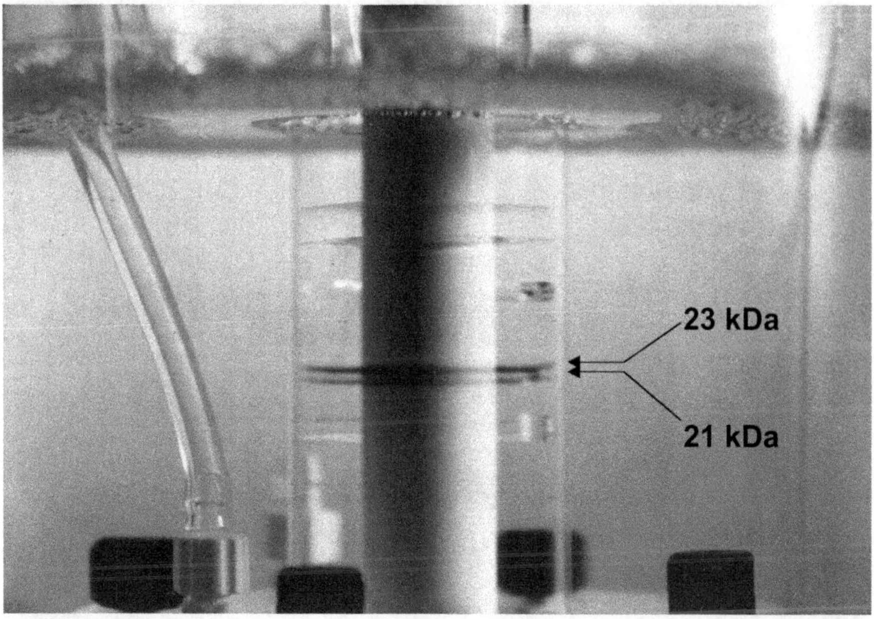

FIGURE 8.7 Separation of phycocyanin subunits of 21 kDa and 23 kDa by SDS-PAGE on the Model 491 Prep Cell.

irreversible structural changes or permanent adsorption, liquid-based methods are gentler and keep a protein-friendly environment. Below, we discuss five examples of pI-based fractionation instrumentation.

8.4.4.1 Rotofor System

The Rotofor system (Figure 8.8), developed by Bier's group[58] and commercially available from Bio-Rad, has a long history in preparative free-solution IEF. The Rotofor cell consists of a cylindrical focusing chamber that holds a plastic core dividing the chamber into 20 compartments separated by polyester screens. The screens offer resistance to fluid convection but do not hinder current flow or the transport of proteins. The protein sample is diluted into a mixture of water and ampholytes and loaded into the assembled chamber. For proteomic applications, high concentrations of urea, nonionic detergents, et cetera may also be included if required to enhance the solubility of specific proteins. Proteins, which are initially dispersed uniformly throughout the chamber, migrate in response to an electrical field to the compartments that are at pH values nearest to their pI values; often, this is a single compartment. At the recommended power of 12 to 15 W, a standard focusing run takes only about 4 hours. Fractions are recovered directly into test tubes using a vacuum manifold. Two focusing chambers are available, capable of holding up to 18 or 55 ml of sample, respectively. The utility of this fractionation method is enhanced when it is followed by the analysis of resulting fractions to narrow and micro range IPG strips, as all proteins outside the pH range of interest are effectively removed.[59,60]

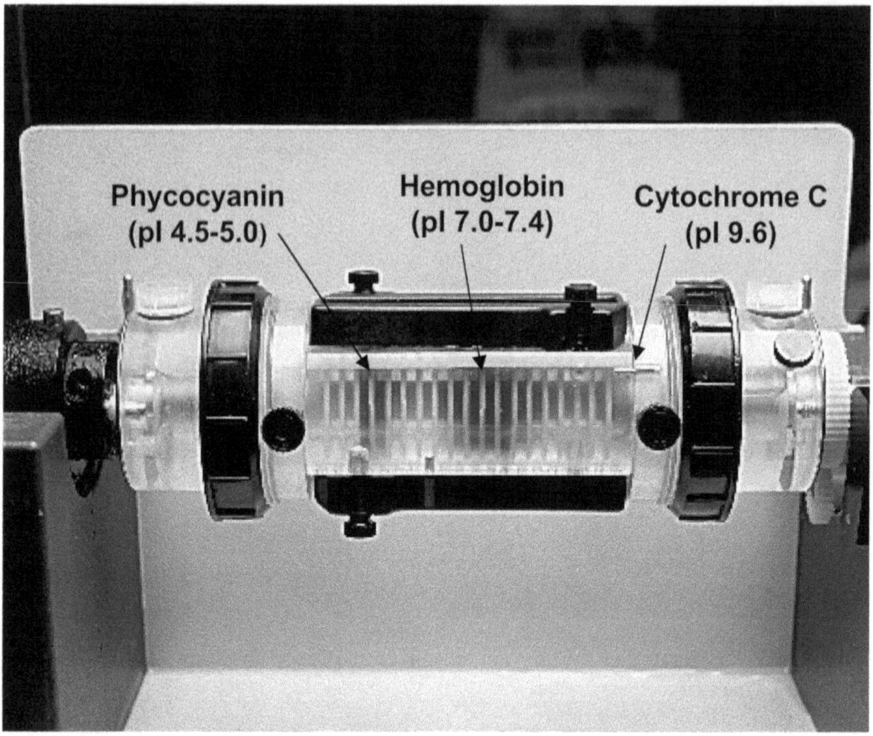

FIGURE 8.8 Protein separation by charge and pI. A pooled sample of three naturally colored proteins was separated by pI on the Rotofor. The three proteins are shown in the Rotofor focusing chamber: blue phycocyanin (pI 4.5–5.0), red hemoglobin (pI 7.0–7.4), and orange cytochrome C (pI 9.6).

8.4.4.2 Multicompartment Electrolyzers

Immobilized pH gradient technology has also been incorporated into a preparative IEF apparatus. Righetti[61] described a multicompartment electrolyzer where each compartment is separated by a polyacrylamide gel membrane with a specific pH. The membranes are custom-made before each separation and act as pH barriers in the system. However, this instrument, commercialized under the name IsoPrime, has not been used widely as a prefractionation tool in proteomic research because the total volume of each of the chambers is 38 ml. To cope with the requirements of modern proteomics, the apparatus has been miniaturized by Herbert and Righetti.[62]

Based on the same separation principle originally published by Righetti, two miniaturized devices are commercially available. The IsoelectrIQ MCE is offered by Proteome Systems Limited and the ZOOM IEF fractionator, which is based on the microscale solution IEF (μsol-IEF) device devised by Zuo and Speicher,[63] is offered by Invitrogen Corporation. Both devices use a series of precast immobilized disks of predetermined pH values to separate smaller volumes (as little as 3.5 ml) of sample into five fractions.

8.4.4.3 Off-Gel IEF in Multicompartment Devices

Off-gel electrophoresis is the latest development in preparative techniques based on IPG technology.[64] Protonation of an ampholyte occurs in the thin layer of solution close to an immobilized gel solution interface. In off-gel electrophoresis, a protein sample is applied to a liquid chamber that is positioned on top of an IPG gel. The gel buffers the thin layer of the solution in the liquid chamber, and the proteins become charged according to their pI and to the pH imposed by the gel. Upon application of a voltage gradient perpendicular to the liquid chamber, the electrical field penetrates into the channel and causes the movement of all charged species (those that have pI values above and below the pH of the IPG gel) from the sample cup. After separation, only the neutral species (those having a pI equal to the pH of the IPG gel) remain in solution and are subsequently recovered. This technique offers high separation efficiency and allows easy recovery of the purified compounds directly in the liquid phase.[65]

8.4.4.4 Continuous Free-Flow Electrophoresis (FFE)

This liquid-based electrophoresis technique was first described by Hannig.[66] Free-flow electrophoresis utilizes a continuous laminar flow of a thin film of separation buffer between two narrowly spaced plates and an electrical field applied perpendicular to the flow. Free-flow electrophoresis results in a differential deflection of charged samples as they move toward the collection ports.[67] FFE supports all modes of electrophoresis, such as zone electrophoresis (ZE), field-step electrophoresis (FSE), isotachophoresis (ITP), and IEF. However, IEF is the method of choice for prefractionation applications since it offers the highest resolution for proteins. Recently, Zischka et al.[68] used FFE for the purification and characterization of yeast mitochondria isolated by differential centrifugation.

8.4.4.5 The Gradiflow System

This instrument belongs to the membrane-based instruments and has been commercialized by Gradipore (Sydney, Australia). The Gradiflow system is a multifunctional electrokinetic membrane apparatus that can process and purify a protein solution in a number of different ways, for example by pI, mobility, size, and affinity.[69] While the Gradiflow system does fractionate samples by a variety of physical properties, it is primarily used as a tool for pI separations. The use of the Gradiflow as a separation step prior to 2D was demonstrated by Corthals et al.,[70] who adopted this instrument for the prefractionation of native human serum and enrichment of protein fractions.

8.5 CONCLUDING REMARKS

Expression proteomics aims to separate, quantitate, characterize, and identify every protein present in a given biological sample. The number of possible different proteins in a cell or tissue sample is believed to be in the several hundreds of thousands, spanning concentration ranges from the level of a single molecule to micromolar amounts. No single analytical method developed today is capable of resolving and detecting such a diverse sample. Mapping an entire proteome by a single analytical method is an elusive goal.

The 2D gel approach offers the highest available resolving power available today. However, it still falls short of separating more than 5,000 proteins in a single run and has a limited dynamic range. Therefore, a strategy to clean up, concentrate, and fractionate the protein content of the sample of interest is necessary. We discussed three methodologies—solution chemistry, chromatography and electrophoresis—that can be used prior to 2D gel electrophoresis to simplify the biological sample. Typically, all steps prior to 2D are referred to as sample preparation. However, they effectively add a third or fourth separation dimension to the separation strategy.

Each biological sample is unique, and the proteomic questions of interest vary from laboratory to laboratory. Mapping and characterizing the complete proteome, biomarker discovery, differential expression analysis, and pathway elucidation are just a few of the typical applications. The sample preparation strategy has to consider the type of sample as well as the type of biological question being addressed. The complex nature of proteins requires a multitude of sample preparation options. This chapter attempts to rationalize the available choices to prepare a protein sample for successful 2D separation. The methods described here are not limited to 2D and can also be applied to other protein separation techniques. As the proteomics effort gains momentum, it will be interesting to see what future developments are made in this area. Adding methods and procedures to simplify the protein content of a biological sample without losing relevant biological information is a key element to these future developments.

REFERENCES

1. Fey, S.J. and Larsen, P.M., 2D or not 2D, *Curr. Opin. Chem. Biol.*, 5, 26–33, 2001.
2. Gorg, A. et al., The current state of two-dimensional electrophoresis with immobilized pH gradients, *Electrophoresis*, 21, 1037–1053, 2000.
3. Rabilloud, T., Valette, C., and Lawrence, J.J., Sample application by in-gel rehydration improves the resolution of two-dimensional electrophoresis with immobilized pH gradients in the first dimension, *Electrophoresis*, 15, 1552–1558,1994.
4. Thongboonkerd, V., Klein, E., and Klein, J.B., Sample preparation for 2-D proteomic analysis, *Contrib. Nephrol.* 141, 11–24, 2004.
5. Garfin, D.E., Two-dimensional gel electrophoresis: An overview, *Trends Anal. Chem.*, 22, 263–272, 2003.
6. Bohring, C. and Krause, W., The characterization of human spermatozoa membrane proteins: Surface antigens and immunological infertility, *Electrophoresis*, 20, 971–976, 1999.
7. Sanchez, J.C. et al., Inside SWISS-2DPAGE database, *Electrophoresis*, 16, 1131–1151, 1995.
8. Portig, I. et al., Identification of stress proteins in endothelial cells, *Electrophoresis*, 17, 803–808, 1996.
9. Wheeler, T.T., Jordan, T.W., and Ford, H.C., A double-label two-dimensional procedure for the analysis of membrane proteins, *Electrophoresis*, 9, 270–287, 1988.
10. Harder, A. et al., Comparison of yeast cell protein solubilization procedures for two-dimensional electrophoresis, *Electrophoresis*, 20, 826–829, 1999.

11. Cull, M., and McHenry, C.S., Preparation of extracts from prokaryotes, *Methods Enzymol.,* 182, 147–153, 1990.
12. Davidsen, N.B., Two-dimensional electrophoresis of acidic proteins isolated from ozone-stressed Norway spruce needles (*Picea abies* L. Karst): Separation method and image processing, *Electrophoresis,* 16, 1305–1311, 1995.
13. Posch, A. et al., Sequence analysis of wheat grain allergens separated by two-dimensional electrophoresis with immobilized pH gradients, *Electrophoresis*, 16, 1115–1119, 1995.
14. Watarai, H. et al., Proteomic approach to the identification of cell membrane proteins, *Electrophoresis,* 21, 460–464, 2000.
15. Nandakumar, M.P. and Marten, M.R., Comparison of lysis methods and preparation protocols for one- and two-dimensional electrophoresis of *Aspergillus oryzae* intracellular proteins, *Electrophoresis,* 23, 2216–2222, 2002.
16. Castellanos-Serra, L. and Paz-Lago, D., Inhibition of unwanted proteolysis during sample preparation: Evaluation of its efficiency in challenge experiments, *Electrophoresis,* 23, 1745–1753, 2002.
17. Robaye, B. et al., Apoptotic cell death analyzed at the molecular level by two-dimensional gel electrophoresis, *Electrophoresis,* 15, 503–510, 1994.
18. Klose, J., Protein mapping by combined isoelectric focusing and electrophoresis of mouse tissues: A novel approach to testing for induced point mutations in mammals, *Humangenetik,* 26, 231–243, 1975.
19. O'Farrell, P., High resolution two-dimensional electrophoresis of proteins, *J. Biol. Chem.,* 250, 4007–4021, 1975.
20. McCarthy, J. et al., Carbamylation of proteins in 2-D electrophoresis—myth or reality? *J. Proteome Res.,* 2, 239–242, 2003.
21. Rabilloud, T. et al., Improvement of the solubilization of proteins in two-dimensional electrophoresis with immobilized pH gradients, *Electrophoresis,* 18, 307–316, 1997.
22. Satta, D. et al., Solubilization of plasma membranes in anionic, non-ionic and zwitterionic surfactants for iso-dalt analysis: A critical evaluation, *J. Chromatogr.*, 299, 57–72, 1984.
23. Persson, H. and Overholm, T., Two-dimensional electrophoresis of membrane proteins: Separation of myelin proteins, *Electrophoresis,* 11, 642–648, 1990.
24. Pasquali, C., Fialka, I., and Huber, L.A., Preparative two-dimensional gel electrophoresis of membrane proteins, *Electrophoresis,* 18, 2573–2581, 1997.
25. Fountoulakis, M. and Takacs, B., Effect of strong detergents and chaotropes on the detection of proteins in two-dimensional gels, *Electrophoresis,* 22, 1593–1602, 2001.
26. Chevallet, M. et al., New zwitterionic detergents improve the analysis of membrane proteins by two-dimensional electrophoresis, *Electrophoresis,* 19, 1901–1909, 1998.
27. Ames, G.F. and Nikaido, K., Two-dimensional gel electrophoresis of membrane proteins, *Biochemistry*, 15, 616–623, 1976.
28. Hoving, S. et al., Preparative two-dimensional gel electrophoresis at alkaline pH using narrow range immobilized pH gradients, *Proteomics,* 2, 127–134, 2002.
29. Herbert, B. et al., Reduction and alkylation of proteins in preparation of two-dimensional map analysis: Why, when, and how? *Electrophoresis,* 22, 2046–2057, 2001.
30. Gorg, A. et al., Recent developments in two-dimensional gel electrophoresis with immobilized pH gradients: Wide pH gradients up to pH 12, longer separation distances and simplified procedures, *Electrophoresis,* 20, 712–717, 1999.

31. Herbert, B.R. et al., Improved protein solubility in two-dimensional electrophoresis using tributyl phosphine as reducing agent, *Electrophoresis,* 19, 845–851, 1998.
32. Righetti, P.G. and Gianazza, E., Isoelectric focusing in immobilized pH gradients: Theory and newer methodology, *Methods Biochem. Anal.*, 32, 215–278, 1987.
33. Righetti, P.G., Chiari, M., and Gelfi, C., Immobilized pH gradients: Effect of salts, added carrier ampholytes and voltage gradients on protein patterns, *Electrophoresis,* 9, 65–73, 1988.
34. Karlsson, K. et al., Enrichment of human brain proteins by heparin chromatography, *Electrophoresis,* 20, 2970–2976, 1999.
35. Fountoulakis, M. et al., Enrichment and purification of proteins of *Haemophilus influenzae* by chromatofocusing, *J. Chromatogr. A*, 806, 279–291, 1998.
36. Fountoulakis, M. et al., Enrichment of low abundance proteins of *Escherichia coli* by hydroxyapatite chromatography, *Electrophoresis,* 20, 2181–2195, 1999.
37. Lopez, M.F. et al., High-throughput profiling of the mitochondrial proteome using affinity fractionation and automation, *Electrophoresis,* 21, 3427–3440, 2000.
38. Wang, Y.Y., Cheng, P., and Chan, D.W., A simple affinity spin tube filter method for removing high-abundant common proteins or enriching low-abundant biomarkers for serum proteomic analysis, *Proteomics*, 3, 243–248, 2003.
39. Nicol, G., Zolotarjova, N., Martosella, J., Zhang, K., and Boyes, B.E., Specific Removal of Multiple High Abundance Proteins from Human Sera, presented at American Society of Mass Spec, 2003.
40. Washburn, M.P., Wolters, D., and Yates, J.R., III, Large-scale analysis of the yeast proteome by multidimensional protein identification technology, *Nat. Biotechnol.,* 19, 242–247, 2001.
41. Zheng, S. et al., Two-dimensional liquid chromatography protein expression mapping for differential proteomic analysis of normal and O157:H7 *Escherichia coli*, *Biotechniques*, 35, 1202–1212, 2003.
42. Chong, B.E. et al., Chromatofocusing nonporous reversed-phase high-performance liquid chromatography/electrospray ionization time-of-flight mass spectrometry of proteins from human breast cancer whole cell lysates: A novel two-dimensional liquid chromatography/mass spectrometry method, *Rapid Commn. Mass Spectrom.*, 15, 291–196, 2001.
43. Righetti, P.G. et al., Prefractionation techniques in proteome analysis, *Proteomics*, 3, 1397–1407, 2003.
44. Rabilloud, T., Solubilization of proteins for electrophoretic analyses, *Electrophoresis,* 17, 813–829, 1996.
45. Jiang, L., He, L., and Fountoulakis, M., Comparison of protein precipitation methods for sample preparation prior to proteomic analysis, *J. Chromatogr. A,* 1023, 317–320, 2004.
46. Righetti, P.G., Castagna, A., and Herbert, B., Prefractionation techniques in proteome analysis, *Anal. Chem.,* 73, 320A–326A, 2001.
47. Reichert, G.H., Two-dimensional gel analysis of proteins from human trisomy 21 fetal liver tissue after DEAE-Sepharose chromatography, *Hum. Genet.,* 73, 250–253,1986.
48. Hanson, B.J. et al., A novel subfractionation approach for mitochondrial proteins: A three-dimensional mitochondrial proteome map, *Electrophoresis,* 22, 950–959, 2001.
49. Prime, T.A. et al., A proteomic analysis of organelles from *Arabidopsis thaliana, Electrophoresis,* 21, 3488–3499, 2000.

50. Burkhardt, J. et al., Gaining insight into a complex organelle, the phagosome, using two-dimensional gel electrophoresis, *Electrophoresis,* 16, 2249–2257, 1995.
51. Weiss, W., Postel, W., and Gorg, A., Application of sequential extraction procedures and glycoprotein blotting for the characterization of the 2-D polypeptide patterns of barley seed proteins, *Electrophoresis,* 13, 770–773, 1992.
52. Molloy, M.P. et al., Extraction of membrane proteins by differential solubilization for separation using two-dimensional gel electrophoresis, *Electrophoresis,* 19, 837–844, 1998.
53. Viard, M., Blumenthal, R., and Raviv, Y., Improved separation of integral membrane proteins by continuous elution electrophoresis with simultaneous detergent exchange: Application to the purification of the fusion protein of the human immunodeficiency virus type 1, *Electrophoresis,* 23, 1659–1666, 2002.
54. Treuheit, M.J. et al., Purification of the alpha and beta subunits of (Na,K)-ATPase by continuous elution electrophoresis, *Prep. Biochem.,* 23, 375–387, 1993.
55. Mandal, M. et al., Identification, purification and partial characterization of tissue inhibitor of matrix metalloproteinase-2 in bovine pulmonary artery smooth muscle, *Mol. Cell Biochem.,* 254, 275–287, 2003.
56. Fountoulakis, M. and Juranville, J.F., Enrichment of low-abundance brain proteins by preparative electrophoresis, *Anal. Biochem.,* 313, 267–282, 2003.
57. Zugaro, L.M. et al., Characterization of rat brain stathmin isoforms by two-dimensional gel electrophoresis-matrix assisted laser desorption/ionization and electrospray ionization–ion trap mass spectrometry, *Electrophoresis,* 19, 867–876, 1998.
58. Bier, F.F. and Egen, N.B., Large-scale recycling electrofocusing, in *Electrofocus/78,* Haglund, H. and Westerfield, J.G., Eds., Elsevier, New York, 1979.
59. Puchades, M. et al., Proteomic studies of potential cerebrospinal fluid protein markers for Alzheimer's disease, *Brain Res. Mol. Brain Res.,* 118, 140–146, 2003.
60. Davidsson, P. et al., Identification of proteins in human cerebrospinal fluid using liquid-phase isoelectric focusing as a prefractionation step followed by two-dimensional gel electrophoresis and matrix-assisted laser desorption/ionisation mass spectrometry, *Rapid Commn. Mass Spectrom.,* 16, 2083–2088, 2002.
61. Righetti, P.G. et al., A horizontal apparatus for isoelectric protein purification in a segmented immobilized pH gradient, *J. Biochem. Biophys. Methods,* 15, 189–198, 1987.
62. Herbert, B. and Righetti, P.G., A turning point in proteome analysis: Sample prefractionation via multicompartment electrolyzers with isoelectric membranes, *Electrophoresis,* 21, 3639–3648, 2000.
63. Zuo, X. and Speicher, D.W., Comprehensive analysis of complex proteomes using micro scale solution isoelectric focusing prior to narrow pH range two-dimensional electrophoresis, *Proteomics,* 2, 58–68, 2002.
64. Ros, A. et al., Protein purification by Off-Gel electrophoresis, *Proteomics,* 2, 151–156, 2002.
65. Michel, P.E. et al., Protein fractionation in a multicompartment device using Off-Gel isoelectric focusing, *Electrophoresis,* 24, 3–11, 2003.
66. Hannig, K., Die trägerfreie kontinuierliche Elektrophorese und ihre Anwendung, *Z. Anal. Chem.,* 181, 244–254, 1961 (in German).
67. Hannig, K. and Heidrich, H.G., *Free-Flow Electrophoresis,* GIT Verlag, Darmstadt, Germany, 1990.
68. Zischka, H. et al., Improved proteome analysis of *Saccharomyces cerevisiae* mitochondria by free-flow electrophoresis, *Proteomics*, 3, 906–916, 2003.

69. Horvath, Z.S. et al., Multifunctional apparatus for electrokinetic processing of proteins, *Electrophoresis,* 15, 968–971, 1994.
70. Corthals, G.L. et al., Prefractionation of protein samples prior to two-dimensional electrophoresis, *Electrophoresis,* 18, 317–323, 1997.
71. Santoni, V. et al., Towards the recovery of hydrophobic proteins on two-dimensional electrophoresis gels, *Electrophoresis,* 20, 705–711, 1999.
72. Molloy, M.P. et al., Proteomic analysis of the *Escherichia coli* outer membrane, *Eur. J. Biochem.,* 267, 2871–2881, 2000.
73. Mechin, V. et al., An efficient solubilization buffer for plant proteins focused in immobilized pH gradients, *Proteomics,* 3, 1299–1302, 2003.
74. Davidsson, P. et al., Proteome studies of human cerebrospinal fluid and brain tissue using a preparative two-dimensional electrophoresis approach prior to mass spectrometry, *Proteomics,* 1, 444–452, 2001.
75. Groleau, P.E. et al., Fractionation of beta-lactoglobulin tryptic peptides by ampholyte-free isoelectric focusing, *J. Agric. Food Chem.,* 50, 578–583, 2002.
76. Wall, D.B., Parus, S.J., and Lubman, D.M., Three-dimensional protein map according to pI, hydrophobicity and molecular mass, *J. Chromatogr. B Analyt. Technol. Biomed. Life Sci.,* 774, 53–58, 2002.
77. Wang, H. et al., A protein molecular weight map of ES2 clear cell ovarian carcinoma cells using a two-dimensional liquid separations/mass mapping technique, *Electrophoresis,* 23, 3168–3181, 2002.
78. Gagne, J.P. et al., A proteomic approach to the identification of heterogeneous nuclear ribonucleoproteins as a new family of poly(ADP-ribose)-binding proteins, *Biochem. J.,* 371, 331–340, 2003.
79. Rylatt, D.B. et al., Electrophoretic transfer of proteins across polyacrylamide membranes, *J. Chromatogr. A,* 865, 145–153, 1999.
80. Gilbert, A., Evtushenko, M., and Nair, H., Purification of fibrinogen and virus removal using preparative electrophoresis, *Ann N.Y. Acad. Sci.,* 936, 625–629, 2001.
81. Locke, V.L. et al., Gradiflow as a prefractionation tool for two-dimensional electrophoresis, *Proteomics,* 2, 1254–1260, 2002.
82. Rothemund, D.L., Thomas, T.M., and Rylatt, D.B., Purification of the basic protein avidin using Gradiflow technology, *Protein Expr. Purif.,* 26, 149–152, 2002.
83. Thomas, T.M. et al., Preparative electrophoresis: A general method for the purification of polyclonal antibodies, *J. Chromatogr. A,* 944, 161–168, 2002.
84. Rothemund, D.L. et al., Depletion of the highly abundant protein albumin from human plasma using the Gradiflow, *Proteomics,* 3, 279–287, 2003.
85. Pedersen, S.K. et al., Unseen proteome: Mining below the tip of the iceberg to find low abundance and membrane proteins, *J. Proteome Res.,* 2, 303–311, 2003.
86. Zuo, X. and Speicher, D.W., A method for global analysis of complex proteomes using sample prefractionation by solution isoelectrofocusing prior to two-dimensional electrophoresis, *Anal. Biochem.,* 284, 266–278, 2000.
87. Zuo, X. et al., Towards global analysis of mammalian proteomes using sample prefractionation prior to narrow pH range two-dimensional gels and using one-dimensional gels for insoluble and large proteins, *Electrophoresis,* 22, 1603–1615, 2001.
88. Kobayashi, H. et al., Free-flow electrophoresis in a micro fabricated chamber with a micro module fraction separator: Continuous separation of proteins, *J. Chromatogr. A,* 990, 169–178, 2003.
89. Mohr, H. and Völkl, A., Isolation of peroxisomal subpopulations from mouse liver by immune free-flow electrophoresis, *Electrophoresis,* 23, 2130–2137, 2002.

90. Hoffmann, P. et al., Continuous free-flow electrophoresis separation of cytosolic proteins from the human colon carcinoma cell line LIM 1215: A non two-dimensional gel electrophoresis-based proteome analysis strategy, *Proteomics,* 1, 807–818, 2001.
91. Mazereeuw, M. et al., Free flow electrophoresis device for continuous on-line separation in analytical systems: An application in biochemical detection, *Anal. Chem.,* 72, 3881–3886, 2000.
92. Sengelov, H. and Borregaard, N., Free-flow electrophoresis in subcellular fractionation of human neutrophils, *J. Immunol. Methods,* 232, 145–152, 1999.

9 Optic Nerve Fractionation for Proteomics

Sanjoy K. Bhattacharya, John S. Crabb, Suresh P. Annangudi, Karen A. West, Xiaorong Gu, Jian Sun, Vera L. Bonilha, Gary B. Smejkal, Karen Shadrach, Joe G. Hollyfield, and John W. Crabb

CONTENTS

9.1 INTRODUCTION

Progressive degeneration of the optic nerve is a common feature of ocular diseases such as glaucoma, optic neuritis, and ischemic optic neuropathy.[1] Optic nerve damage can also result from accidental eye trauma or exposure to a variety of chemicals.[2,3] Over 70 million people worldwide suffer from a variety of glaucomas, a group of irreversible blinding diseases. In primary open angle glaucoma, most patients exhibit increased intraocular pressure that leads to optic nerve degeneration.[4,5] In response to elevated pressure, protein expression patterns change *in vitro*[6–10] and similar modulation appears likely *in vivo*. Indeed, *in vivo* expression of optic nerve myocilin is reduced in glaucoma and under conditions of elevated intraocular pressure.[11–13] Proteomic analysis of diseased and damaged optic nerve tissue offers a promising approach to better understanding pathogenic mechanisms in glaucoma and other optic neuropathies.

Optic nerve is a myelinated tissue from which protein fractionation remains challenging. Myelin is a unique membrane composed of about 70% lipid and 30% protein.[14] The myelin basic proteins (MBPs) and proteolipid protein isoforms account for up to 80% of myelin protein, with the remainder being a large number of lower abundance proteins.[14] Myelin has both active and passive functions in the central nervous system, including regulation of axon diameter and formation of axon microtubular networks.[15,16] Historically, biochemical fractionation of myelinated tissues has involved differential gradient centrifugation and yielded low but significant levels of axolemmal proteins from highly enriched myelin preparations.[17] More recently, two-dimensional (2D) gel electrophoresis has been used to study protein synthesis during myelination in rat optic nerve[18] and to seek functional insights into myelin biology.[19] However, relatively few protein identifications have been obtained. Proteomic analyses of optic nerve offer the potential to identify protein differences between normal and glaucomatous optic nerve. In this chapter, we describe the fractionation of human optic nerve by solution-phase isoelectric focusing (IEF) and 2D polyacrylamide gel electrophoresis (2D PAGE) and report the identification of 149 proteins. The multi-compartment electrolyzer (MCE)[20,21] was used for the solution-phase IEF, and although this technology is still evolving, the present results demonstrate the potential utility of this fractionation methodology for differential proteomic analyses of optic nerve from diseased and normal eye donors.

9.2 MATERIALS AND METHODS

9.2.1 Human Tissue Procurement

Thirty-five eyes from normal human donors (15 female and 20 male, ages 43–89) were used in this study. Eyes were obtained through the Cleveland Eye Bank (Cleveland, Ohio) and the Eye Donor Program of the Foundation Fighting Blindness, Inc. (Owens Mills, Md.). Eyes were enucleated within 8 hours of death and either placed in Optisol GS medium or flash frozen in liquid nitrogen.

9.2.2 Optic Nerve Sample Preparation

The stalk region of the optic nerve was carefully dissected (about 1–2 mm in thickness) and weighed using a Mettler AE163 calibrated balance. Tissues were stored at −80°C until analyzed in 20 mM Tris buffer pH 7.5 containing protease inhibitors (Product No. P2714, Sigma Chemical Co., St. Louis), 5 μM dithiothrietol and 5 μM butylated hydroxytoluene to protect against *in vitro* oxidation and degradation. To develop protein extraction methods, small portions of optic nerve explants (20 ± 1.2 mg wet weight) were subjected to homogenization and protein extraction using a battery powered, handheld tissue homogenizer (Pellet Pestle Kontes, South Jersey Precision Tool and Mold Inc, N.J.) and 100 μl of detergent per extraction. The protein extraction efficacy was compared for four different detergent solutions containing 7 M urea and 2 M thiourea, namely 6% C7BzO, 3% amidosulfobetaine-14 (ASB-14), 3% dodecylmaltoside, and a proprietary detergent formulation from Proteome Systems (Woburn, Mass.). Following extraction, protein was precipitated with acetone, resuspended in Laemmeli sample buffer[22] and subjected to one-dimensional (1D) sodium dodecyl sulfate–polyacrylamide gel electrophoresis (SDS-PAGE).

Human optic nerve explants from 34 eyes (~2.7 g total tissue wet weight) were used for preparative fractionation and protein identification. Protein extraction was performed batchwise with about four optic nerve explants per batch (~320 mg wet weight). The tissue was first minced with a sharp scalpel then homogenized in 3% dodecylmaltoside containing 7 M urea and 2 M thiourea (2 ml). Following 6 extractions of 2 ml each, protein was precipitated with acetone and resuspended in the same detergent solution. Reduction and alkylation was then performed under argon with tributylphosphine (5 mM) and acrylamide (10 mM) for 90 minutes at room temperature. All extraction buffers and the MCE kit were obtained from Proteome Systems, Inc. Protein recovery was quantified by a modified Bradford assay.[23]

9.2.3 Acetone Precipitation

Acetone precipitation was performed according to Proteome Systems (Proteome Systems, Woburn, Mass.). Briefly, four volumes of acetone were added to one volume (typically 12 ml) of the detergent solubilized protein preparation and incubated at room temperature for 20 minutes with intermittent, gentle mixing. The suspension was then centrifuged at 2500*g* for 15 minutes at room temperature, and the clear supernatant fluid was carefully removed with a pipette. The precipitate was recentrifuged at 2500*g* for 5 minutes, and the residual supernatant was removed. The pellet was air dried for 15 minutes at room temperature and then resuspended in a detergent solution for further investigation.

9.2.4 Solution-Phase IEF

IEF in the MCE was performed according to Pedersen et al.[21] except that 1–7.5 mg of acrylamide alkyated optic nerve protein was prepared in 3% dodecylmaltoside, 7 M urea, 2 M thiourea (3.5 ml). Immobilized pH membranes were used to construct IEF trapping chambers of pH 3–5, 5–6.5, 6.5–8, and 6–11. The original protein solution was diluted with the same detergent solution to obtain the desired sample amount in 3.5 ml and loaded into the central pH 5–6.5 MCE chamber. Detergent solution was placed in the other trapping chambers of the MCE assembly. Electrode solution supplied by Proteome Systems was placed in the terminal chambers. IEF fractionation was carried out initially with a 100–1500 V ramp over 13 hours and subsequently at a constant 1500 V for an additional 5 hours. Throughout the fractionation, current was limited at 0.8 A. After IEF, chamber solutions were transferred to tubes and protein was concentrated by acetone precipitation. Protein was resuspended in the aforementioned detergent solution (100–200 μl) quantified by the modified Bradford procedure[23] and subjected to 2D PAGE.

9.2.5 2D Gel Electrophoresis

The resuspended fractions from MCE separation (in 3% dodecylmaltoside, 7 M urea and 2 M thiourea) were centrifuged briefly to remove particulate material (14,000*g*, 10 minutes) and 0.001% bromophenol blue was added. Protein samples were transferred to 11 cm IPGphore coffins (Amersham Pharmacia Biosciences, Piscataway, N.J.), and the volume was adjusted to 300 μl with additional detergent solution. An 11 cm precast ProteomIQ™ immobilized linear pH gradient strip gel (IPG, pH 3–10) was placed in the coffin and overlaid with 750 μl of mineral oil. First dimension IEF was performed with the Pharmacia IPGphor and the following programmed voltage gradient.[24,25] The IPG strips were rehydrated with sample at 30 V for 6 hours and then 300 V for 1 minute, followed by a linear increase to 3500 V for 8 hours, held at 3500 V for 1 hour, then at 8000 V for 21 hours to reach a total of approximately 90 kVh. For the second dimension, IPG strips were equilibrated in 50 mM Tris acetate pH 7.0, 3 M urea, 2% SDS, bromophenol blue (24 minutes) and then embedded in 0.7% w/v agarose on the top of a 6–15% acrylamide linear gradient gel (ProteomeIQ™ Gelchip, Proteome Systems). Second dimension SDS-PAGE (5 hours at 300 mA constant current) was performed with Proteome Systems electrophoresis equipment. The gels were stained and destained using ProteomeIQ™ Blue gel stain kit (Proteome Systems) and scanned with a GS-710 Imaging Densitometer (BioRad).

9.2.6 Protein Identification

Select 2D gel spots were excised, destained, and digested *in situ* with trypsin as documented elsewhere.[24–27] Tryptic peptides were analyzed by liquid chromatography electrospray tandem mass spectrometry (LC MS/MS) using a CapLC system and a quadrupole time-of-flight mass spectrometer (QTOF2, Waters Corporation, Milford, Mass.).[25–27] Protein identifications from MS/MS data utilized ProteinLynx™ Global Server (Waters Corporation) and Mascot (Matrix Science) search engines and the Swiss-Protein and NCBI protein sequence databases.

9.3 RESULTS

9.3.1 Extraction of Protein from Optic Nerve

Comparison of the optic nerve protein extraction efficacy of four different detergent solutions is presented in Table 9.1 and Figure 9.1. Optic nerve extracts from all four detergents produced similar 1D SDS-PAGE profiles and protein recoveries (relative yield, 0.7–0.9%). However, ASB-14 and dodecylmaltoside yielded greater Coomassie blue staining intensity for select gel bands (Figure 9.1) and slightly more overall soluble

TABLE 9.1 Solubilization of Optic Nerve Protein

Detergent Solution	Protein (μg) Recovered per mg Tissue	Total Protein Recovered (μg)	Relative Yield (%)
6% C7BzO, 7 M urea, 2 M thiourea	8.0 ± 0.8	160.4 ± 2.5	0.8
3% ASB-14, 7 M urea, 2 M thiourea	8.7 ± 0.6	174.0 ± 1.8	0.9
3% Dodecylmaltoside, 7 M urea, 2 M thiourea	8.9 ± 0.7	176.3 ± 1.5	0.9
Proprietary formulation, Proteome Systems	6.7 ± 0.4	134.4 ± 2.8	0.7

Note: Optic nerve explants (20 ± 1.25 mg wet weight) were extracted (1×) with the indicated detergent solution (100 μl) and soluble protein quantified by a modified Bradford assay as described in Methods. Average results (± standard deviation) are shown from four independent experiments.

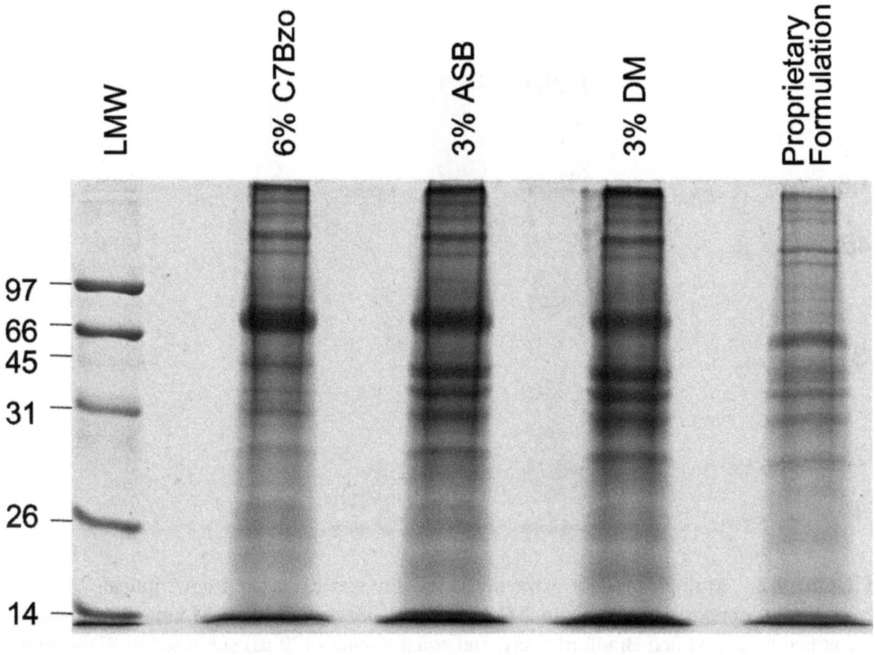

FIGURE 9.1 SDS-PAGE of optic nerve proteins extracted in different detergents. Protein was extracted from equal amounts of optic nerve (~20 mg wet weight) using the indicated detergents (100 μl per extraction) and 15 μg protein per lane subjected to SDS-PAGE on a 10% gel.

protein (Table 9.1). These results are consistent with reports for other myelin-containing membrane-rich tissues.[28] Dodecylmaltoside was selected for further studies because the critical micelle concentration (0.1–0.6 mM) is higher than for ASB-14 (<0.1 mM). As shown in Figure 9.2, multiple extractions with 200 μl yielded significantly more soluble optic nerve protein than a single extraction (Figure 9.2) with the combined yield from six repetitive extractions approaching 1.3% (w/w) relative to the tissue wet weight. For comparison, tissue samples from mouse liver (115 mg wet weight) and mouse skeletal muscle (90 mg wet weight) were extracted under identical conditions with 3% dodecylmaltoside, 7 M urea, 2 M thiourea, quantified as for the optic nerve preparations, and found to yield 4.9% and 2.7% total protein, respectively, relative to the tissue wet weights.

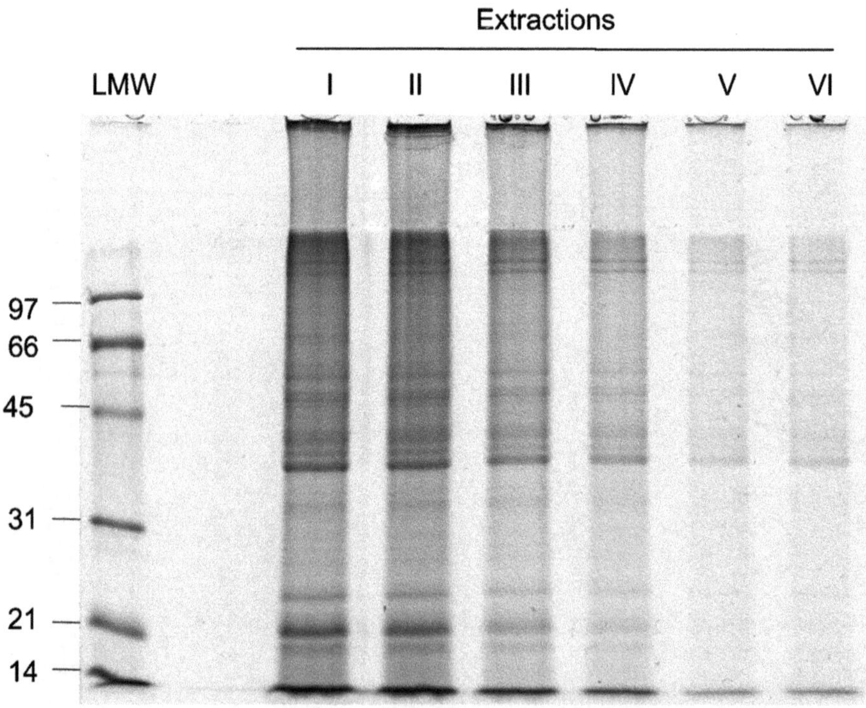

FIGURE 9.2 Multiple protein extractions. Protein was extracted from triplicate 200 mg (wet weight) optic nerve explants in 200 μl of 3% dodecylmaltoside, 7 M urea, 2 M thiourea, quantified by a modified Bradford assay, and equal volumes (30 μl) subjected to SDS-PAGE. The average recovery (± standard deviation) for six successive extractions was as follows: 793 ± 22 μg, 664 ± 12 μg, 567 ± 25 μg, 327 ± 14 μg, 101 ± 8 μg, 77 ± 11 μg. The total combined recovery from the six extractions was ~2.53 mg (1.3% relative to the tissue wet weight).

9.3.2 Optic Nerve IEF Using the Multicompartment Electrolyzer

Soluble protein for IEF and 2D PAGE analyses were prepared from normal human optic nerve explants by repetitive extractions as described in the Methods section. Following acetone precipitation, ~35 mg soluble protein was recovered from 34 optic nerve explants and redissolved in 3% dodecylmaltoside, 7 M urea, 2 M thiourea for IEF (16.6 ml final volume). Solution state IEF was performed with the MCE using sample amounts of about 3 mg, 5 mg, and 7.5 mg optic nerve protein. Recovery from the MCE trapping chambers following IEF was determined to be about 31–39% relative to the amounts applied (Table 9.2). Following IEF, protein precipitate was evident in the pH 5–6.5 trapping chamber where the sample was applied and where the protein concentration remained the highest. Recovery of optic nerve protein was found to be lower (8–22%) following MCE fractionation of smaller sample amounts (~1 mg) under the same conditions.

While recovery from the current procedure was relatively low, 2D PAGE analyses of the MCE trapping chamber solutions demonstrated that significant separation of optic nerve proteins was achieved. 2D electrophoretic analyses of the MCE fractions are shown in Figure 9.3. The relatively wide isoelectric charge separation of optic nerve proteins achieved by MCE fractionation can be readily appreciated by comparison of the 2D gel profiles from the MCE fractions and the starting sample with no solution state IEF. Consistent with the manufacturer's claims, greater 2D PAGE separation of the optic nerve MCE fractions was observed with appropriate narrow range IPG strips. For example, proteins enriched in trapping chamber pH 3–5 were resolved to a greater extent on IPG strips of pH 3–5 than on pH 3–10 (not shown).

TABLE 9.2 Recovery of Optic Nerve Protein from MCE Fractionation

MCE Chamber	3.0 mg Protein	5.0 mg Protein *protein recovery (μg)*	7.5 mg Protein
Anode	—	—	—
pH 3–5	107	223	290
pH 5–6.5	560	1242	1100
pH 6.5–8	120	231	875
pH 8–11	105	192	590
Cathode	45	47	45
Total Protein (μg)	937	1935	2900
Percent Recovery	31	39	39

Note: Optic nerve protein in 3% dodecylmaltoside, 7 M urea, 2 M thiourea was fractionated by solution state IEF with the MCE and recovered protein quantified by a modified Bradford assay (see Methods).

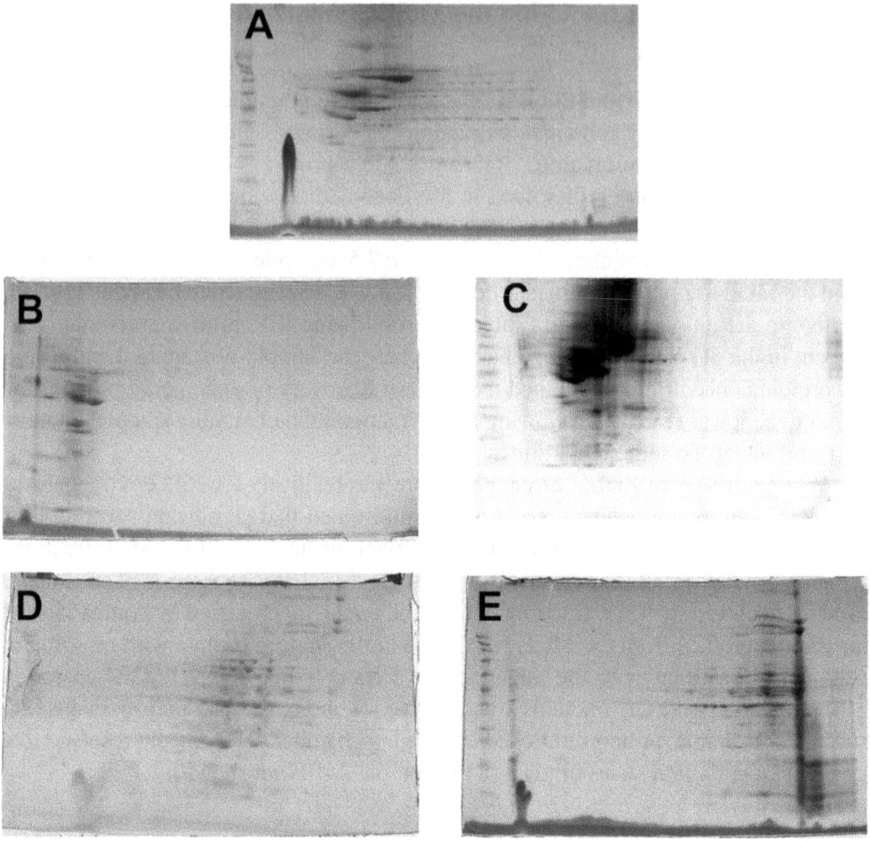

FIGURE 9.3 2D PAGE analyses of MCE fractions. Following solution phase IEF of optic nerve protein (5 mg) using the MCE, 2D PAGE analysis of the MCE fractions was performed using ProteomIQ immobilized linear pH gradient strips (pH 3–10, 11 cm), 6–15% gradient ProteomIQ Gels, and the ProteomeIQ Blue gel staining kit (Proteome Systems). (A) Optic nerve protein without MCE fractionation, 100 μg; (B) MCE chamber pH 3–5, 100 μg; (C) MCE chamber pH 5–6.5, 100 μg; (D) MCE chamber pH 6.5–8, 100 μg; (E) MCE chamber pH 8–11, 100 μg.

9.3.3 Identification of Optic Nerve Proteins following MCE Fractionation

For protein identification, 2D gel spots were excised (Figure 9.3) and subjected to in-gel tryptic digestion, and peptides were analyzed by LC MS/MS and bioinformatics methods. One hundred forty-nine proteins identified from analyses of the gel spots derived from the optic nerve MCE fractions are listed in Table 9.3. As expected, a number of proteins previously identified in brain tissue[19] were also found in optic nerve (Table 9.3), for example, myelin basic protein, mimecan, and neuron

TABLE 9.3 Human Optic Nerve Proteins Identified after MCE and 2D Gel Fractionation[a]

Protein ID	Accession No.[b]	Peptide Matches	Molecular Mass (kDa)		pI	
			Calculated[c]	Observed[d]	Calculated[c]	Observed[d]
MCE Chamber pH 3–5						
Alpha-1-acid glycoprotein 1	P02763	6	23	25	5.0	3.5
Annexin A2	P07355	4	38	55	7.6	4.5
Apolipoprotein D	P05090	1	21	30	5.1	4.7
Cullin homolog 5	Q93034	1	91	24	8.3	3.5
Desmin	P17661	4	53	55	5.2	4.5
Desmocollin	Q02487	1	99	22	5.2	4.5
Elongation factor 1-beta	P24534	3	25	30	4.5	4.7
F-box only protein 2	Q9UK22	4	33	65	4.3	3.8
Glial fibrillary acidic protein[e]	P14136	24	50	45	5.5	4.8
Glutamine synthetase	P32288 (S)	2	42	22	6.3	4.5
Glyceraldehyde 3-phosphate dehydrogenase	P04406	8	36	55	8.8	4.5
Long-chain-fatty-acid—CoA ligase 2	P33121	1	78	45	7.2	4.8
Mimecan[e]	P20774	3	34	45	5.5	4.8
Nasopharyngeal epithelium specific protein 1	Q9UL16	1	46	30	10.3	4.7
Paired box protein Pax-3	P23760	1	53	55	9.0	4.5
PITSLRE serine-threonine kinase CDC2L1	P21127	1	93	30	5.6	4.7
Placental thrombin inhibitor	P35237	2	43	45	5.2	4.7
Plasticin	P31393 (C)	4	53	55	5.5	4.5

(Continued)

TABLE 9.3 Human Optic Nerve Proteins Identified after MCE and 2D Gel Fractionation[a] (Continued)

Protein ID	Accession No.[b]	Peptide Matches	Molecular Mass (kDa) Calculated[c]	Molecular Mass (kDa) Observed[d]	pI Calculated[c]	pI Observed[d]
Proteasome subunit alpha type 5	P28066	2	26	25	4.8	4.7
Protein 14-3-3, Epsilon	P42655	26	29	30	4.7	4.7
Protein 14-3-3, eta	Q04917	8	28	25	4.8	4.7
Protein 14-3-3, tau	P27348	8	28	25	4.7	4.7
Protein 14-3-3, zeta/delta	P29312	11	28	25	4.8	4.7
Protein FAN	Q92636	1	104	30	6.1	4.7
Protein phosphatase 1 regulatory subunit 12B	O60237	1	110	45	5.6	4.8
Pyruvate kinase, M1 isozyme	P14618	3	58	20	8.2	4.5
Rho GDP-dissociation inhibitor 1	P52565	2	23	25	5.0	4.7
Vimentin	P08670	30	54	55	5.1	4.5
WD-repeat protein 9	Q9NSI6	1	257	22	8.8	3.1
MCE chamber pH 5–6.5						
Actin, Beta	P60709	22	42	40	5.3	5.5
Annexin A5	P08758	19	36	36	5.0	5.1
Apolipoprotein A-I	P02647	22	31	30	5.7	5.5
Apolipoprotein D	P05090	1	21	30	5.1	5.5
Creatine kinase	P12277	18	43	45	5.5	5.5
Cytoplasmic antiproteinase	P35237	1	43	36	5.2	5.1

Dihydropyrimidinase related protein-2[e]	Q16555	2	62	35	6.0	4.9
Enolase, Alpha	P06733	20	47	50	7.5	6.5
Extracellular superoxide dismutase	P08294	1	26	30	6.5	5.5
Ferritin heavy	P02794	2	21	30	5.5	5.5
Glial fibrillary acidic protein, astrocyte[e]	P14136	1	50	30	5.4	5.5
Glutathione S-transferase Mu 3	P21266	2	26	30	5.4	5.5
Glutathione S-transferase P	P09211	9	23	24	5.5	5.5
Heat shock 27 kDa protein	P04792	1	23	30	6.3	5.5
Histone-lysine N-methyltransferase	Q15047	1	143	36	5.9	5.5
Hypothetical protein DJ845O24.1	O60809	1	53	24	8.7	4.9
Ig gamma-1 chain C region	P01857	4	36	36	8.6	6.5
Ig kappa chain V-I region A	P01593	1	12	30	5.9	5.5
Ig kappa chain V-I region N	P01613	1	12	30	5.4	5.5
Ig lambda chain C regions	P01842	2	11	30	7.4	5.5
Junction plakoglobin	P14923	1	81	35	6.3	4.7
Mimecan[e]	P20774	4	34	36	5.5	5.1
NDRG2 protein[e]	Q9UN36	1	40	36	5.2	5.5

(Continued)

TABLE 9.3 Human Optic Nerve Proteins Identified after MCE and 2D Gel Fractionation[a] (Continued)

Protein ID	Accession No.[b]	Peptide Matches	Molecular Mass (kDa) Calculated[c]	Molecular Mass (kDa) Observed[d]	pI Calculated[c]	pI Observed[d]
Neurofilament triplet M protein[e]	P07197	1	102	24	4.9	4.9
Neuron cytoplasmic protein[e]	P09936	5	25	24	5.4	5.5
Peroxiredoxin 2	P32119	1	22	30	5.9	5.5
Protein 14-3-3, Beta/alpha	P31946	7	28	24	4.8	4.9
Protein 14-3-3, Epsilon	P42655	14	29	30	4.7	4.7
Protein 14-3-3, Gamma	P35214	6	28	29	4.8	4.9
Proto-oncogene C-crk	P46108	2	34	30	5.5	5.5
Rab GDP dissociation inhibitor, Beta	P50395	2	51	50	6.1	6.5
Rho GDP-dissociation inhibitor 1	P52565	1	23	24	5	4.9
Septin 7 (CDC10 protein homolog)	Q16181	5	49	50	9.1	6.5
Serum albumin	P02768	2	69	30	6.2	6.2
TRF1-interacting ankyrin-related ADP-ribose polymerase	O95271	1	141	50	7	6.5
Triosephosphate isomerase	P60174	1	26	30	7.2	5.5
Tropomyosin 1, Alpha	P09493	4	33	35	4.7	4.7
Tubulin, Beta-5	P04350	4	50	36,50	4.8	5.1
UMP-CMP kinase	P30085	2	22	30	5.5	5.5
Vimentin	P08670	1	53	24	5	4.9

MCE chamber pH 6.5–8						
Acyl-CoA dehydrogenase, very-long-chain specific	P49748	2	70	65	9.1	7.2
Aldo-keto reductase family 1 member C2	P52895	2	37	36	7.6	6.7
Aldo-keto reductase family 1 member C3	P42330	2	37	36	8.3	6.7
Annexin A2	P07355	17	38	36	8.0	6.7
Aspartyl-tRNA synthetase	Q9KDG1 (H)	2	67	65	5.1	6.7
Basic membrane protein A	O31280 (B)	2	37	36	5.2	7.0
Beta-1-syntrophin	Q13884	1	58	50	9.0	7.3
Biglycan precursor	P21810	1	41	36	7.0	6.7
cAMP-dependent chloride channele	P13569	1	168	36	9.0	6.7
Collagen alpha 1(I)	P02452	2	139	120	5.8	7.8
Collagen alpha 2(I)	P08123	2	129	116	9.2	7.8
Crystallin Alpha, B	P02511	9	20	20	7.3	6.7
Cytoskeleton-like bicaudal D protein homolog 1	Q96G01	1	110	65	5.8	7.2
G2/mitotic-specific cyclin B1	P14635	1	48	36	7.6	6.7
Glial fibrillary acidic protein, astrocytee	P14136	1	49	65	5.4	7.2
Glyceraldehyde 3-phosphate dehydrogenase	P04406	17	36	36	8.8	6.7
Hermansky-Pudlak syndrome 3 protein	Q969F9	1	113	55	6.4	7

(Continued)

TABLE 9.3 Human Optic Nerve Proteins Identified after MCE and 2D Gel Fractionation[a] (Continued)

Protein ID	Accession No.[b]	Peptide Matches	Molecular Mass (kDa) Calculated[c]	Molecular Mass (kDa) Observed[d]	pI Calculated[c]	pI Observed[d]
Ig gamma-1 chain C region	P01857	1	36	50	8.6	7.3
Kinesin-like protein KIF12	Q96FN5	1	60	55	9.4	7
M-phase inducer phosphatase 2	P30305	1	65	36	6.1	6.7
Neurexin 3-alpha precursor (Neurexin III-alpha)e	Q9Y4C0	1	169	116	6.1	7.8
Nuclear pore complex protein Nup88	Q99567	1	84	36	5.6	6.7
Palmitoyl-protein thioesterase 1	P50897	1	34	36	6.7	6.7
Peroxiredoxin 1	Q06830	3	22	20	8.5	6.7
Peroxisome assembly factor-2	Q13608	1	104	50	6.3	6.8
Phosphatidylethanolamine-binding protein	P30086	10	21	20	7.6	6.7
Prolargin	P51888	2	44	50	9.6	7.3
Protein-glutamine glutamyltransferase 4	P49221	1	77	65	6.7	7.2
Pyruvate kinase, M1 isozyme	P14618	24	58	55	8.2	7.0
Restricted expression proliferation associated protein	Q9ULW0	1	86	65	9.6	7.2
Rho-related GTP-binding protein RhoG	P35238	1	21	20	8.5	6.7
Ribose 5-phosphate isomerase	P49247	1	26	116	7.6	7.8
T-complex protein 1, eta subunit	Q99832	1	59	55	7.9	7
Transcription factor BF-1e	P55315	1	65	50	9.9	7.3
Transketolase	P29401	19	68	65	7.9	7.2

MCE chamber pH 8–11						
2',3'-cyclic nucleotide 3'-phosphodiesterase[e]	P09543	18	48	50	9.4	8.6
Acylphosphatase, muscle type isozyme	P14621	1	11	14	9.8	9.5
ATP synthase, Alpha	P25705	5	60	60	9.4	8.6
Caspase recruitment domain protein 4	Q9Y239	1	107	65	7.1	7.9
Collagen, Alpha 1(I)	P02452	3	139	150	5.8	9.0
Collagen, Alpha 1(II)	P02458	2	134	250	8.7	9.0
Collagen, Alpha 2(I)	P08123	4	129	116	9.2	9.1
Disco-interacting protein 2 homolog[e]	Q14689	1	165	250	8	9
DNA polymerase gamma subunit 1	P54098	1	139	60	6.8	8.6
DNA-directed RNA polymerase	O00411	2	139	50	9.2	8.6
Dystrobrevin alpha (Dystrobrevin-alpha)[e]	Q9Y4J8	1	84	30	6.8	9.5
Elongation factor 1-alpha	P18624 (D)	3	50	250	9.1	9.0
Fibrinogen alpha/alpha-E	P02671	2	95	20	5.9	9.5
FK506-binding protein 1A	P20071	1	12	14	8.5	8.6
Fructose-bisphosphate aldolase A	P04075	6	39	50	8.5	8.6
Glyceraldehyde 3-phosphate dehydrogenase	P80534	11	39	40	9.0	7.1
Guanine nucleotide-binding protein[e]	Q9UBI6	1	8	14	9.4	9.5
Hemoglobin, Alpha	P01922	7	15	15	9.0	8.6
Hemoglobin, Delta	P02042	3	16	15	8.4	8.6

(*Continued*)

TABLE 9.3 Human Optic Nerve Proteins Identified after MCE and 2D Gel Fractionation[a] (Continued)

Protein ID	Accession No.[b]	Peptide Matches	Molecular Mass (kDa)		pI	
			Calculated[c]	Observed[d]	Calculated[c]	Observed[d]
Heterogeneous nuclear ribonucleoprotein C	P07910	2	34	15	5.0	9.5
Histone deacetylase 5	Q9UQL6	1	122	250	6.2	9
Histone H2A2	P28001	2	14	14	11.2	9.5
Histone H2B.d (H2B/d)	Q99877	4	14	14	10.6	9.5
Histone H3	P06351	3	15	14	11.3	9.5
Histone H4	P02304	3	11	14	11.4	9.5
Hypothetical protein JHP1324	Q9ZJI5 (P)	2	36	116	8.8	9.1
Ig gamma-1 chain C region	P01857	3	36	60	8.6	8.6
Mast/stem cell growth factor receptor	P10721	1	110	60	6.8	8.6
Myelin basic protein[e]	P02686	6	33	30	10.1	9.5
Na(+)-translocating NADH-quinone reductase	P00387	2	34	15	7.3	8.6
NADH-ubiquinone oxidoreductase, 15 kDa	O43920	1	12	11	9.5	9.5
NADH-ubiquinone oxidoreductase, B8	O43678	1	10	14	9.9	9.5
Nuclear pore complex protein Nup88	Q99567	1	83	65	5.6	7.9
Phosphoglycerate kinase	P07205	2	45	50	9.0	8.6
Phosphoglycerate kinase 1	P00558	7	45	50	8.5	8.6
Potential phospholipid-transporting ATPase	P98198	1	137	60	7	8.6
Prolargin	P51888	4	44	116	9.6	9.1
Protein rdxB	P54932 (R)	2	54	45	8.5	7.1

Putative HTH-type transcriptional regulator HI0570	P44757 (E)	2	23	250	9.6	9.0
Putative transmembrane protein NMB	Q14956	1	63	11	6.6	9.5
Pyruvate kinase, M1 isozyme	P14618	3	58	60,36	8.2	9.5
Ribosomal protein L12, 60S	P30050	2	18	20	9.8	9.5
S-arrestin	P10523	1	45	14	6.5	9.5
Septin 7 (CDC10 protein homolog)	Q16181	1	49	60	9.1	8.9
Serum albumin	P02768	2	69	15	6.2	9.5
Shwachman-Bodian-Diamond syndrome protein	Q9Y3A5	1	28	14	9.1	9.5
Signal recognition particle 14 kDa protein	P37108	1	14	14	10.4	9.5
Trifunctional enzyme, Alpha	P40939	2	83	150	9.4	9.0
Ubiquinol-cytochrome C reductase complex	O14949	1	9	11	10.2	9.5

[a] Spots were excised from the 2D gel shown in Figure 9.3 and proteins identified by LC MS/MS as described in Methods.

[b] Swiss-Protein database accession numbers are shown. Links to Swiss-Protein accession numbers use the EXPASY server at http://us.expasy.org/sprot/. Parenthesies following accession numbers designate identifications based on homology with other species: B = Borrelia afzelii, C = Carassius auratus, D = Dictyostelium discoideum, E = Haemophilus influenzae, H = Bacillus halodurans, P = Helicobacter pylori J99, R = Rhodobacter sphaeroides, S = Saccharomyces cerevisiae.

[c] Based on protein sequence.

[d] Approximate mass and pI observed on 2D gels.

[e] Also found in brain tissue.[19]

cytoplasmic protein. For most of the optic nerve proteins in Table 9.3, the sequence calculated pI was in good agreement with the pH range of the trapping chamber from which it was found. For the proteins exhibiting inconsistent observed and calculated pIs, the apparently altered pIs may be due to posttranslational modifications, covalent aggregation, or proteolytic processing.

9.4 DISCUSSION

Optic nerve contains significant lipid-rich myelin membranes and constitutes one of the more difficult tissues from which to extract soluble protein. Detergents and thiourea are essential for solubilization of many membrane proteins,[28–30] and here we found that ASB-14 and dodecylmaltoside are relatively effective in extracting optic nerve proteins. Our results are consistent with other efforts to solubilize central nervous system myelin.[28,31,32] However in our hands, even after multiple extractions, only ~1.3% soluble protein was recovered relative to the wet weight of the optic nerve tissue, significantly less than that obtained from detergent extraction of mouse skeletal muscle (2.7%) or liver (4.9%). Reported efforts to extract protein from glaucomatous optic nerve tissues have not been quantitative and more efficient extraction methods have yet to be established.[33,34] Of the detergents we tested, ASB-14 and dodecylmaltoside offered the greatest potential for solubilization of optic nerve proteins for subsequent fractionation and identification. As expected, multiple extractions were more efficient than a single extraction.

Fractionation of detergent solubilized optic nerve proteins by solution-state IEF-enhanced subsequent 2D gel separations and facilitated protein identification by mass spectrometry and bioinformatics. Accordingly, the MCE instrument from Proteome Systems appears to provide another useful technique in the arsenal of methods available for separating complex mixtures of proteins.[35] With the MCE instrument, optic nerve proteins were clearly resolved into fractions based on isoelectric pH. Protein precipitation was observed in the chamber where the optic nerve preparation was applied to the apparatus (pH 5–6.5), however this potential problem appears to have occurred late in the IEF process based on the 2D gel analyses which demonstrated charge separations consistent with the isoelectric trapping membranes. In support of the MCE instrument, isoelectric protein precipitation is a common occurrence in IEF procedures, and in the present application, such precipitation had little apparent effect on the overall resolution of the proteins.

Despite the above benefits, the current MCE technology could be improved. The foremost problem with the present instrument design is the relatively large sample volume and amount of protein required as starting material. The original MCE system required chamber volumes of ~5 ml. The manufacturer subsequently provided silicon wafer inserts, which reduced the chamber volumes to 3.5 ml; however, these chambers are still exceedingly large for small sample amounts. Many biological samples of interest can only be obtained in microgram amounts and for such samples, the present instrument does not appear suitable due to poor protein recovery. In our hands, larger amounts of optic nerve protein (5–7.5 mg) provided the highest protein recovery (39%) while 1 mg samples yielded ≤22% recovery. Miniaturization and optimization of the

MCE chambers for small sample amounts would constitute a significant improvement in instrument design. Another significant issue of concern is the acetone precipitation step used to concentrate protein in the trapping chambers for subsequent preparative 1D or 2D gel analyses. Substantial protein losses can occur during acetone precipitation, and such losses likely contributed to the low recovery of optic nerve protein from the MCE. A more efficient concentration method would enhance the present technology. Another minor concern is possible leakage, which was occasionally encountered. This issue might be minimized by the reduction of the number of components required for assembly and disassembly of the MCE chambers, perhaps through the use of solid blocks with insertable membranes.

In summary, solution-phase IEF with the MCE followed by 2D gel analyses has provided a productive approach to the fractionation of milligram amounts of optic nerve proteins for proteomic analyses. The detergents ASB-14 and dodecylmaltoside were useful for the solubilization and extraction of optic nerve proteins. The proteins identified in this study represent the most extensive catalogue of optic nerve proteins to date and contribute to the identification of the human optic nerve proteome. Optic nerve degeneration is associated with a number of visual disorders. We anticipate that differential proteomic analyses of diseased and normal optic nerve tissues will provide insights into the pathogenic mechanisms of glaucoma and other optic neuropathies of unknown etiology.

ACKNOWLEDGMENTS

This study was supported in part by a grant from the National Glaucoma Research Program of American Health Assistance Foundation (SKB), NIH grants EY6603, EY14239, and EY015638, a Research Center Grant from The Foundation Fighting Blindness, and funds from the Cleveland Clinic Foundation. We thank Professors M. Rosario Hernandez and Steven Pfeiffer for their comments on the manuscript.

REFERENCES

1. Margalit, E. and Sadda, S.R., Retinal and optic nerve diseases, *Artif. Organs.*, 27, 963–974, 2003.
2. Gupta, B.N., Stefanski, S.A., Bucher, J.R., and Hall, L.B., Effect of methyl isocyanate (MIC) gas on the eyes of Fischer 344 rats, *Environ. Health Perspect.*, 72, 105–108, 1987.
3. McKellar, M.J., Hidajat, R.R., and Elder, M.J., Acute ocular methanol toxicity: Clinical and electrophysiological features, *Aust. N. Z. J. Ophthalmol.*, 25, 225–230, 1997.
4. Flammer, J., Orgul, S., Costa, V.P., Orzalesi, N., Krieglstein, G.K., Serra, L.M., Renard, J.P., and Stefansson, E., The impact of ocular blood flow in glaucoma, *Prog. Retin. Eye Res.*, 21, 359–393, 2002.
5. van Buskirk, E.M. and Cioffi, G.A., Glaucomatous optic neuropathy, *Am. J. Ophthalmol.*, 113, 447–452, 1992.
6. Neufeld, A.H. and Liu, B., Comparison of the signal transduction pathways for the induction of gene expression of nitric oxide synthase-2 in response to two different stimuli, *Nitric Oxide*, 8, 95–102, 2003.

7. Neufeld, A.H., Sawada, A., and Becker, B., Inhibition of nitric-oxide synthase 2 by aminoguanidine provides neuroprotection of retinal ganglion cells in a rat model of chronic glaucoma, *Proc. Natl. Acad. Sci. USA*, 96, 9944–9948, 1999.
8. Hernandez, M.R., Pena, J.D., Selvidge, J.A., Salvador-Silva, M., and Yang, P., Hydrostatic pressure stimulates synthesis of elastin in cultured optic nerve head astrocytes, *Glia*, 32, 122–136, 2000.
9. Ricard, C.S., Kobayashi, S., Pena, J.D., Salvador-Silva, M., Agapova, O., and Hernandez, M.R., Selective expression of neural cell adhesion molecule (NCAM)-180 in optic nerve head astrocytes exposed to elevated hydrostatic pressure in vitro, *Brain Res. Mol. Brain Res.*, 81, 62–79, 2000.
10. Salvador-Silva, M., Ricard, C.S., Agapova, O.A., Yang, P., and Hernandez, M.R., Expression of small heat shock proteins and intermediate filaments in the human optic nerve head astrocytes exposed to elevated hydrostatic pressure in vitro, *J. Neurosci. Res.*, 66, 59–73, 2001.
11. Ricard, C.S., Agapova, O.A., Salvador-Silva, M., Kaufman, P.L., and Hernandez, M.R., Expression of myocilin/TIGR in normal and glaucomatous primate optic nerves, *Exp. Eye Res.*, 73, 433–447, 2001.
12. Ahmed, F., Torrado, M., Johnson, E., Morrison, J., and Tomarev, S.I., Changes in mRNA levels of the Myoc/Tigr gene in the rat eye after experimental elevation of intraocular pressure or optic nerve transection, *Invest Ophthalmol. Vis. Sci.*, 42, 3165–3172, 2001.
13. Clark, A.F., Kawase, K., English-Wright, S., Lane, D., Steely, H.T., Yamamoto, T., Kitazawa, Y., Kwon, Y.H., Fingert, J.H., Swiderski, R.E., Mullins, R.F., Hageman, G.S., Alward, W.L., Sheffield, V.C., and Stone, E.M., Expression of the glaucoma gene myocilin (MYOC) in the human optic nerve head, *FASEB J.*, 15, 1251–1253, 2001.
14. Pfeiffer, S.E., Warrington, A.E., and Bansal, R., The oligodendrocyte and its many cellular processes, *Trends Cell Biol.*, 3, 191–197, 1993.
15. Peles, E. and Salzer, J.L., Molecular domains of myelinated axons, *Curr. Opin. Neurobiol.*, 10, 558–565, 2000.
16. Boiko, T., Rasband, M.N., Levinson, S.R., Caldwell, J.H., Mandel, G., Trimmer, J.S., and Matthews, G., Compact myelin dictates the differential targeting of two sodium channel isoforms in the same axon, *Neuron*, 30, 91–104, 2001.
17. Menon, K., Rasband, M.N., Taylor, C.M., Brophy, P., Bansal, R., and Pfeiffer, S.E., The myelin-axolemmal complex: Biochemical dissection and the role of galactosphingolipids, *J Neurochem.*, 87, 995–1009, 2003.
18. Colello, R.J., Fuss, B., Fox, M.A, and Alberti, J., A proteomic approach to rapidly elucidate oligodendrocyte-associated proteins expressed in the myelinating rat optic nerve, *Electrophoresis*, 23, 144–151, 2002.
19. Taylor, C.M., Marta, C.B., Claycomb, R.J., Han, D.K., Rasband, M.N., Coetzee, T., and Pfeiffer, S.E., Proteomic mapping provides powerful insights into functional myelin biology, *Proc. Natl. Acad. Sci. USA*, 101, 4643–4648, 2004.
20. Herbert, B. and Righetti, P.G., A turning point in proteome analysis: Sample prefractionation via multicompartment electrolyzers with isoelectric membranes, *Electrophoresis*, 21, 3639–3648, 2000.
21. Pedersen, S.K., Harry, J.L., Sebastian, L., Baker, J., Traini, M.D., McCarthy, J.T., Manoharan, A., Wilkins, M.R., Gooley, A.A., Righetti, P.G., Packer, N.H., Williams, K.L., and Herbert, B.R., *J. Proteome Res.*, 2, 303–311, 2003.
22. Laemmli, U.K., Cleavage of structural proteins during the assembly of the head of bacteriophage T4, *Nature*, 227, 680–685, 1970.

23. Ramagli, L.S., Quantifying protein in 2-D PAGE solubilization buffers, in *Methods in Molecular Biology: 2-D Proteome Analysis Protocols*, vol. 112, Link, A.J., Ed., Totowa, NJ Human Press, 1999, pp. 99–103.
24. West, K.A., Yan, L., Miyagi, M., Crabb, J.S., Marmorstein, A.D., Marmorstein, L., and Crabb, J.W., Proteome survey of proliferating and differentiating rat RPE-J cells, *Exp. Eye Res.*, 73, 479–491, 2001.
25. West, K.A., Yan, L., Shadrach, K., Sun, J., Hasan, A., Miyagi, M., Crabb, J.S., Hollyfield, J.G., Marmorstein, A.D., and Crabb, J.W., Protein database, human retinal pigment epithelium, *Mol. Cell. Proteomics*, 2, 37–49, 2003.
26. Crabb, J.W., Miyagi, M., Gu, X., Shadrach, K., West, K.A., Sakaguchi, H., Kamei, M., Hasan, A., Yan, L., Rayborn, M.E., Salomon, R.G., and Hollyfield, J.G., Drusen proteome analysis: An approach to the etiology of age-related macular degeneration, *Proc. Natl. Acad. Sci. USA*, 99, 14,682–14,687, 2002.
27. Miyagi, M., Sakaguchi, H., Darrow, R.M., Yan, L., West, K.A., Aulak, K.S., Stuehr, D.J., Hollyfield, J.G., Organisciak, D.T., and Crabb, J.W., Evidence that light modulates protein nitration in rat retina, *Mol. Cell. Proteomics*, 1, 293–303, 2002.
28. Taylor, C.M. and Pfeiffer, S.E., Enhanced resolution of glycosylphosphatidylinositol-anchored and transmembrane proteins from the lipid-rich myelin membrane by two-dimensional gel electrophoresis, *Proteomics*, 3, 1303–1312, 2003.
29. Rosenow, M.A., Magee, C.L., Williams, J.C., and Allen, J.P., The influence of detergents on the solubility of membrane proteins, *Acta Crystallogr. D Biol. Crystallogr.*, 58, 2076–2081, 2002.
30. Rabilloud, T., Use of thiourea to increase the solubility of membrane proteins in two-dimensional electrophoresis, *Electrophoresis*, 19, 758–760, 1998.
31. Meller, K., Cryo-electron microscopy of myelin treated with detergents, *Cell Tissue Res.*, 276, 551–558, 1994.
32. Yamamoto, Y., Yoshikawa, H., Nagano, S., Kondoh, G., Sadahiro, S., Gotow, T., Yanagihara, T., and Sakoda, S., Myelin-associated oligodendrocytic basic protein is essential for normal arrangement of the radial component in central nervous system myelin, *Eur. J. Neurosci.*, 11, 847–855, 1999.
33. Dahl, D., The vimentin-GFA protein transition in rat neuroglia cytoskeleton occurs at the time of myelination, *J. Neurosci. Res.*, 6, 741–748, 1981.
34. Brooks, D.E., Samuelson, D.A., Gelatt, K.N., and Smith, P.J., Morphologic changes in the lamina cribrosa of beagles with primary open-angle glaucoma, *Am. J. Vet. Res.*, 50, 936–941, 1989.
35. Righetti, P.G., Castagna, A., Herbert, B., Reymond, F., and Rossier, J.S., Prefractionation techniques in proteome analysis, *Proteomics*, 3, 1397–1407, 2003.

10 Fractionation of Retina for Proteomic Analyses

Sanjoy K. Bhattacharya, Karen A. West, Xiaorong Gu, John S. Crabb, Kutralanathan Renganathan, Zhiping Wu, Jian Sun, and John W. Crabb

CONTENTS

10.1 INTRODUCTION

The neural retina is the complex, multicell type tissue at the back of the eye responsible for converting light stimuli into neural impulses directed to the brain. Visual pigment proteins such as rhodopsin in rod photoreceptor cells and blue, green, and red opsins in cone photoreceptor cells absorb light via photoisomerization of protein-bound 11-*cis*-retinal, triggering visual signal transduction, that is, phototransduction. These light-absorbing proteins are integral membrane proteins located in the outer segments of photoreceptor cells, within densely packed membranous discs containing other proteins involved in phototransduction. Additional cells in the retinal circuitry, including bipolar and ganglion cells, process and relay the signal from the photoreceptors through the nerve fiber layer to the optic nerve, modulated in part by horizontal and amacrine cells and supported by retinal glial cells such as Müller cells, astrocytes, and microglia. Although the neural retina also has an internal vasculature, the photoreceptor cells are separated from their principal blood supply in the choroid by the retinal pigment epithelium (RPE), a polarized single-cell layer that plays key roles in retinal physiology. Worldwide efforts continue toward a better understanding of the complex processes ongoing in retina, including regulation of visual sensitivity and dark adaptation, regeneration of 11-*cis*-retinal for rod and cone visual pigments (the visual cycle), and neuronal transmission.[1] Proteomic approaches offer significant potential for deciphering molecular mechanisms in healthy and diseased retina. Nevertheless, such studies remain challenged by the many different and difficult-to-purify cell types in the retina, a correspondingly complex retinal proteome, and a relatively high proportion of membrane proteins.

Our recent proteomic studies have focused in large part on the RPE,[2,3] the visual cycle,[4–8] the role of oxidative protein modifications in age-related macular degeneration,[9,10] and mechanisms involved in retinal light damage.[11–15] Two-dimensional polyacrylamide gel electrophoresis (2D PAGE) has been a major fractionation tool in several of these studies, and as reported by others,[16] we have experienced difficulty with the detection of membrane proteins. More specifically, we have not detected rhodopsin, the most abundant protein in rod outer segments (ROS), in our 2D PAGE[11–15] because the protein typically remains embedded in the first-dimension immobilized pH gradient. In this chapter, we describe the fractionation of bovine retina by solution-state isoelectric-focusing (IEF) and one-dimensional polyacrylamide gel electrophoresis (1D PAGE). The present results demonstrate the utility of this approach for the identification of retinal proteins, including integral membrane proteins such as rhodopsin.

10.2 MATERIALS AND METHODS

10.2.1 Bovine Retina Sample Preparation

Bovine eyes were obtained from Tucker Packing Company (Orrville, Ohio). Retinas were dissected from the globe without the RPE within 7 hours of death[3] and stored until analyzed at −80°C in 20 mM Tris buffer pH 7.5 containing protease inhibitors (product P2714, Sigma Chemical Co., St. Louis), 5 μM dithiothrietol, 10 μM butylated hydroxytoluene, and 2 mM ethylenediaminetetraacetic acid (EDTA). To probe membrane protein extraction efficiency, small portions of retinal tissue (~23 mg wet weight) were homogenized with a small, battery powered tissue homogenizer[17] in 7 M urea and 2 M thiourea containing one of eight different detergent solutions (~116 μl) obtained from Proteome Systems (Woburn, Mass.). These detergents and the concentrations used were as follows: 3% amidosulfobetaine-14 (ASB-14), 4% 3-[(3-cholamidopropyl)dimethylammonio]-1-propane sulfonate (CHAPS), 2% digitonin, 3% dodecyl maltoside, 1% phosphatidyl choline, 1% SDS, 1.5% Triton X100, and 1% C7BzO. Western blot analysis was used to evaluate the efficacy of these detergents to extract rhodopsin from retina. Protein extracts prepared from 272 to 280 mg bovine retina (wet weight) with 3% ASB-14, 4% CHAPS, or 3% dodecyl maltoside solutions were quantified by a modified Bradford procedure[18] and fractionated by solution-state IEF.

10.2.2 Western Blot Analysis and 1D Electrophoresis

For Western blot analyses, retinal extracts in IEF solvent were diluted 1:1 with Laemmli SDS sample buffer,[19] and following 1D electrophoresis, protein was blotted to PVDF membrane (Millipore, Bedford, Mass.)[20,21] and probed with anti-rhodopsin monoclonal antibody B6–30N.[22] Purified bovine rod outer segments (ROS) (10 μg) were used as a positive control for rhodopsin.[23] Immunoreactivity was detected by chemiluminescence (Amersham Pharmacia Biotech, Inc., San Francisco). 1D SDS-PAGE was performed on 12% acrylamide gels using either a Bio-Rad Mini-Protean II electrophoresis system with 8 × 10 cm gels or a Bio-Rad Protein II xi system with 16 × 20 cm gels.[3] For proteomic analyses, gels were stained with colloidal Coomassie blue (Pierce Code Blue, detection limit ~10 ng protein).

10.2.3 Solution-State IEF

Solution-state IEF was performed using the Multicompartment Electrolyzer (MCE) according to Proteome Systems[24] with bovine retinal protein in 7 M urea, 2 M thiourea containing either 3% ASB-14, 3% dodecyl maltoside, or 4% CHAPS. IEF trapping chambers of pH 3–5, 5–6.5, 6.5–8, and 8–11 were prepared with immobilized pH membranes supplied by the vendor. Retinal protein extracts were loaded into the central pH 5–6.5 chamber and diluted with the same detergent to 3.5 ml. Detergent solution was placed in the other trapping chambers, and electrode solution supplied by the vendor was placed in the terminal MCE chambers. IEF fractionation was carried out initially with a 100–1500 V ramp over 13 hours and subsequently at a constant 1500 V for an additional 5 hours. Throughout the fractionation, maximum current was set at 0.8 mA. After IEF, chamber solutions were transferred to tubes, and protein was concentrated by acetone precipitation, resuspended in Laemmli SDS sample buffer (100–200 μl), and subjected to 1D PAGE.[3,19]

10.2.4 Acetone Precipitation

Acetone precipitation was performed according to Proteome Systems. Four volumes of acetone were added to one volume of the detergent solubilized protein solution and incubated at room temperature for 20 minutes with intermittent, gentle mixing. The suspension was then centrifuged at 2500*g* for 15 minutes at room temperature and the clear supernatant fluid carefully removed with a pipette. The precipitate was recentrifuged at 2500*g* for 5 minutes, and the residual supernatant was removed. The pellet was resuspended in Laemmli sample buffer, quantified by the bicinchoninic (BCA) protein assay (Pierce Biotechnology, Inc., Rockford, Ill.) and subjected to1D PAGE.

10.2.5 Protein Identification

Select 1D gel bands were excised and digested *in situ* with trypsin using methods documented elsewhere.[2,3,9,11] Band selection was arbitrary except effort to identify rhodopsin by mass spectrometry, which was guided by Western blot results. Tryptic peptides were analyzed by liquid chromatography electrospray tandem mass spectrometry (LC MS/MS) using a CapLC system and a quadrupole time-of-flight mass spectrometer (QTOF2, Waters Corporation, Milford, Mass.).[3,9,11] Protein identifications from MS/MS data utilized ProteinLynx™ Global Server (Waters Corporation) and Mascot (Matrix Science) search engines and the Swiss-Protein and NCBI protein sequence databases. The MS/MS spectra were examined manually to verify determined sequences.

10.3 RESULTS

10.3.1 Comparison of Detergent Extracts of Bovine Retina

Toward the development of improved proteomic methods for analyzing retinal membrane proteins, bovine retinal extracts were prepared using several different detergent solutions. The overall protein profiles of these detergent extracts as revealed by 1D PAGE and colloidal Coomassie blue detection were similar in pattern and staining intensity (Figure 10.1). Western blot analysis using anti-rhodopsin antibody was used to monitor recovery of the major integral membrane protein in rod photoreceptor cells (Figure 10.2). All the detergent extracts exhibited rhodopsin immunoreactivity; however, differences were apparent in the aggregation state of the protein. Retinal extracts prepared in 1% SDS revealed intense immunoreactivity for monomeric rhodopsin (mass of 39,007 Da calculated from the amino acid sequence) and little high mass aggregates (>100 kDa). Extracts in 3% ASB-14 contained apparent monomeric and dimeric forms of rhodopsin but less high mass aggregates than any other detergent except SDS. Retinal extracts prepared in CHAPS exhibited intense rhodopsin immunoreactivity in the high mass range but variable and sometimes no apparent monomeric protein. Western blot analysis of extracts prepared in 2% digitonin (not shown) provided similar rhodopsin profiles to that obtained with 3% dodecyl maltoside, 1% phosphatidyl choline, and 1% C7BzO. Soluble protein recovery (w/w) relative to the tissue wet weights for the retinal extracts in ASB-14, dodecyl maltoside, and CHAPS were 1.7% (4.8 mg), 1.4% (3.9 mg), and 1.3% (3.6 mg), respectively.

10.3.2 Retina Fractionation by Solution-State IEF and 1D PAGE

Bovine retinal protein extracted in 7 M urea and 2 M thiourea containing either 3% dodecyl maltoside, 3% ASB-14, or 4% CHAPS was separated by solution-state IEF with the MCE apparatus into pH fractions 3–5, 5–6.5, 6.5–8, and 8–11. Following IEF, no retinal protein precipitation was apparent in the trapping chambers with any of these detergent solutions. Total protein recovery in the MCE trapping chambers relative to the amount applied (3 mg) was in the 25–34% range for the three detergents tested (after acetone precipitation), with ASB-14 providing the greatest estimated yield (Table 10.1). 1D SDS-PAGE analyses after IEF

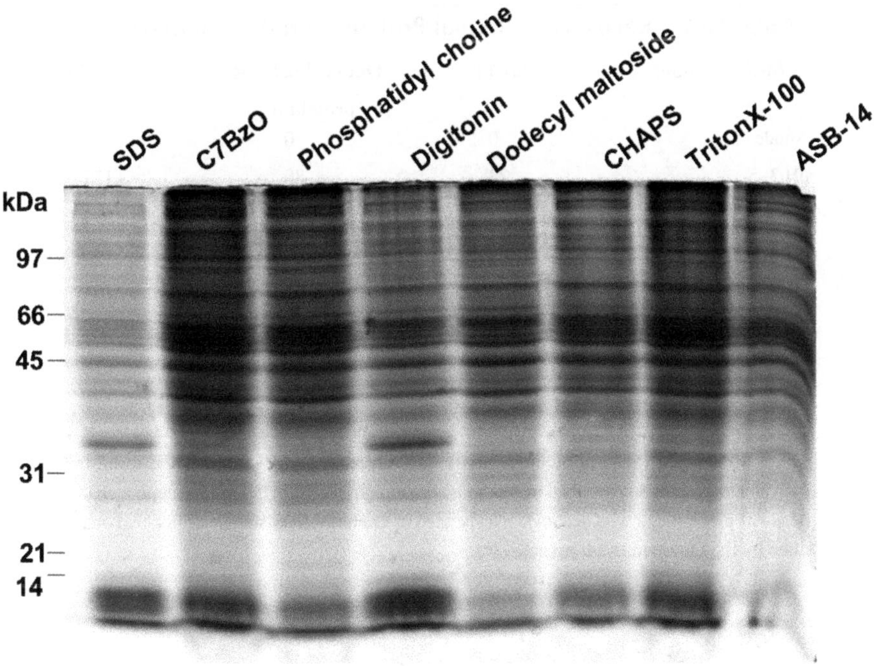

FIGURE 10.1 Detergent extracts of bovine retina. Small portions of bovine retina (~23 mg wet weight) were extracted using 7 M urea, 2 M thiourea containing the indicated detergent and soluble protein (20 μg) subjected to 1D PAGE on 12% acrylamide gels (8 × 10 cm) and detected with colloidal Coomassie blue.

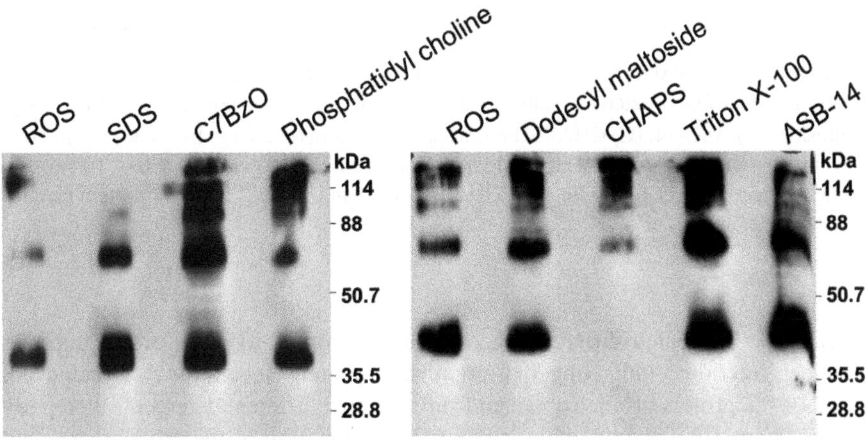

FIGURE 10.2 Western analyses of retinal extracts for rhodopsin. Bovine retinal extracts were prepared with the indicated detergents, subjected to 1D PAGE, blotted to PVDF membrane, and probed with anti-rhodopsin antibody.

TABLE 10.1 Recovery of Retinal Protein from IEF Fractions

MCE Chamber	ASB-14	Dodecylmaltoside	CHAPS
		protein (μg)	
Anode	0	0	0
pH 3–5	78	80	152
pH 5–6.5	675	388	354
pH 6.5–8	209	226	146
pH 8–11	51	60	164
Cathode	0	0	0
Total protein (μg)	1013	754	816
Protein recovery (%)	34	25	27

Note: Solubilized retinal protein (3.0 mg) was subjected to IEF in the MCE using the indicated detergent as described in Methods. Protein recovered in the MCE trapping chambers was precipitated with acetone, resuspended in Laemmli SDS sample buffer, and quantified by the modified Bradford assay.[18]

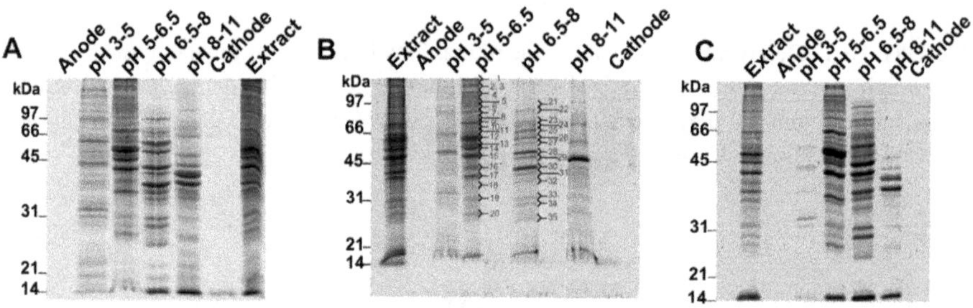

FIGURE 10.3 1D PAGE of IEF fractions. Solution-state IEF of bovine retinal extracts was performed in different detergent solutions as described in the Methods section. (A) Dodecyl maltoside; (B) ASB-14; (C) CHAPS. Following IEF, protein (~20 μg) from each pH fraction was further separated by 1D PAGE on 12% polyacrylamide minigels (0.1 × 8 × 10 cm) and stained with colloidal Coomassie blue. The numbered bands were excised for protein identification (Table 10.2).

(Figure 10.3) revealed different electrophoretic profiles for each of the trapping chamber fractions, indicating that significant protein separation was achieved. Notably, the protein profiles obtained from the three different detergents were very similar, supporting reproducible isoelectric fractionation of the proteins detected by colloidal Coomassie blue staining. Western blot analysis (Figure 10.4) following IEF in dodecyl maltoside revealed intense rhodopsin immunoreactivity in the

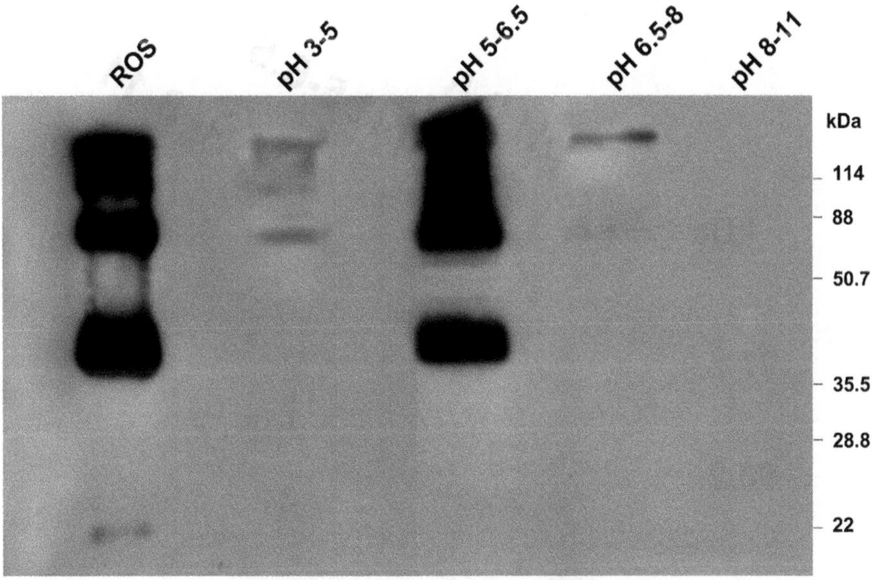

FIGURE 10.4 Western blot analyses of IEF fractions for rhodopsin. Following solution-state IEF of bovine retinal extract in dodecyl maltoside, protein (20 μg) in the pH fractions was subjected to 1D PAGE, blotted to PVDF membrane, and probed with anti-rhodopsin antibody.

pH 5–6.5 trapping chamber and weak immunoreactivity in the pH 3–5 and pH 6.5–8 fractions, with monomeric rhodopsin apparent only in the pH 5–6.5 fraction.

10.3.3 Protein Identification

Following IEF in 3% ASB-14, protein fractions from the MCE trapping chambers were subjected to 1D PAGE, select bands were excised as shown in Figure 10.3B and Figure 10.5 and processed for protein identification by mass spectrometric and bioinformatic analyses. About two thirds of the recovered protein was trapped in the pH 5–6.5 chamber from which LC MS/MS analysis of 57 gel bands yielded the identity of 201 proteins (Table 10.2). The pH 3–5 fraction yielded 59 protein identifications from 13 gel bands, the pH 6.5–8 fraction yielded 33 protein identifications from 15 gel bands, and the pH 8–11 fraction provided 64 identifications from analysis of 11 gel bands (Table 10.2). Rhodopsin was identified from the pH 5–6.5 fraction in gel bands ranging in apparent mass from ~40 kDa to over 100 kDa. In addition, 31 other integral membrane proteins and 33 membrane associated proteins were identified and are highlighted within Table 10.2. Similar mass spectrometric analyses following IEF in dodecyl maltoside provided detection of rhodopsin but only in high mass bands near the top of the 1D gels (results not shown).

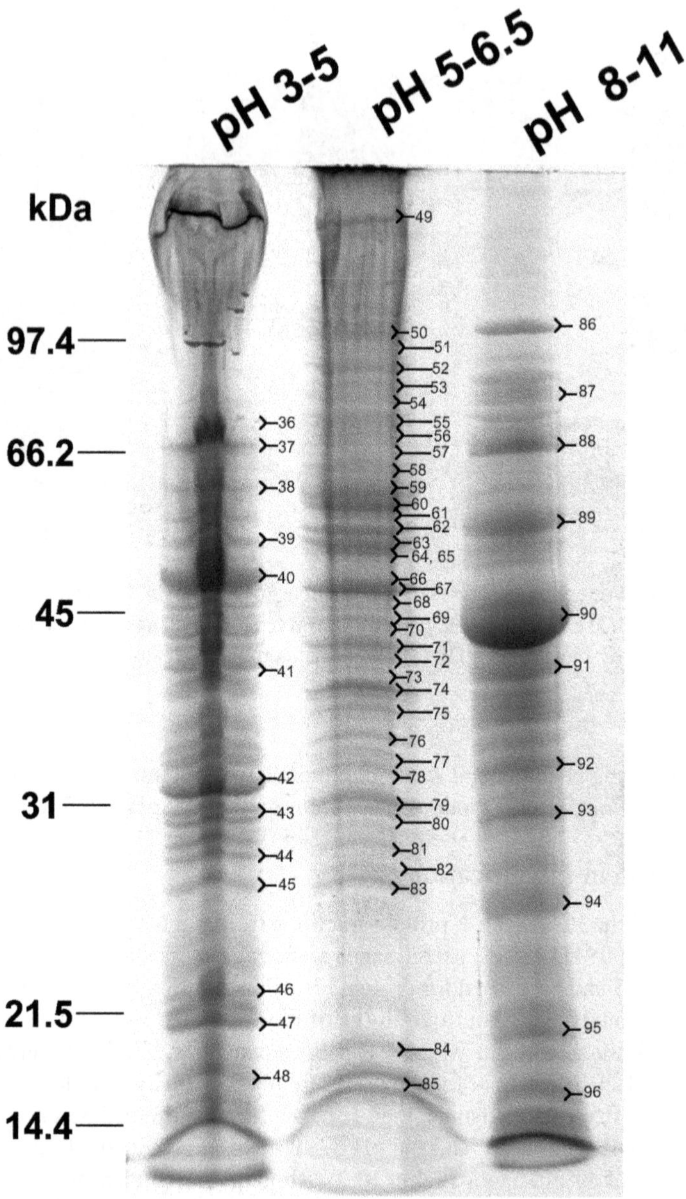

FIGURE 10.5 IEF fractions separated on large 1D gels. Bovine retinal extracts were fractionated by solution-state IEF in ASB-14 then protein (50 μg) from the indicated pH fractions was subjected to 1D PAGE on 12% polyacrylamide gels (0.1 × 16 cm × 20 cm). The numbered bands were excised for protein identification (Table 10.2).

TABLE 10.2 Proteins Identified from Bovine Retina after Solution-State IEF and 1D PAGE[a]

Protein	Accession Number[b]	Peptide Matches	pI Calc[c]	Molecular Mass (kDa) Calc[c]	Molecular Mass (kDa) Obs[d]	SDS-PAGE Band Number
MCE Chamber pH 3–5						
Acidic leucine-rich nuclear phosphoprotein	P51122	1	4.95	20	31	42
Acidic ribosomal protein P2, 60S	P42899	1	4.49	12	17	48
Bcl-2-like protein 13	Q9BXK5	1	4.45	53	41	41
Calbindin	P04467	13	4.66	30	25, 26	44, 45
Calmodulin	P62157	5	4.08	17	17, 19	47, 48
Calreticulin 1	P52193	11	4.31	48	66, 71	37, 36
Calreticulin 3	Q96L12	2	6.56	45	71	36
Calumenin	043852	4	4.47	37	56	39
cAMP-Dependent protein kinase	P00515	7	4.82	45	60	38
Chromobox protein homolog 1	P23197	3	4.88	21	25	45
Clathrin light chain A	P04973	3	4.44	27	41	41
Cytochrome b5	P00171	2	5.00	15	17	48
Desmin	062654	3	5.56	53	48	40
Dickkopf related protein-3 precursor	Q9UBP4	1	4.58	38	66	37
DNA-binding protein A	P16989	2	9.95	40	71	36
DNA-directed RNA polymerase	Q923G2 (M)	3	4.49	17	17	48
Eukaryotic translation initiation factor 6	Q9TU47	3	4.58	27	29	43
GA binding protein beta chain	Q06547	1	4.82	42	56	39
Glial fibrillary acidic protein	Q28115	9	5.38	49	17, 41	41, 48
Guanylyl cyclase activating protein 1	P46065	4	4.41	23	20	46
Hepatoma-derived growth factor	Q9XSK7	3	4.74	27	41	41
Heterogeneous nuclear ribonucleoprotein K	Q07244	2	5.46	51	56	39

(Continued)

TABLE 10.2 Proteins Identified from Bovine Retina after Solution-State IEF and 1D PAGE[a] (Continued)

Protein	Accession Number[b]	Peptide Matches	pI Calc[c]	Molecular Mass (kDa) Calc[c]	Molecular Mass (kDa) Obs[d]	SDS-PAGE Band Number
HIRA-interacting protein 5	Q9UMS0	1	4.21	22	26	44
Limbic system-associated membrane protein	Q13449	4	6.63	37	66, 71	37, 36
Membrane associated progesterone receptor	Q8MIL9	6	4.54	21	25	45
Metal-response element-binding transcription factor 2	Q9Y483	1	9.14	67	41	41
Methylosome subunit	P35521 (D)	2	4.03	26	41	41
Myosin regulatory light chain	P19105	5	4.66	20	20	46
Myristoylated alanine-rich C-kinase substrate	P12624	1	4.57	31	48	40
Neural cell adhesion molecule 1	P31836	1	4.91	94	66	37
Neurofilament triplet L protein	P02548	7	4.62	62	66	37
Neuronal axonal membrane protein	P80724	8	4.56	23	56, 60, 66, 71	36–39
Neurotrimin	Q9P121	2	8.20	38	66	37
Nuclease sensitive element binding protein 1	P16991	5	10.06	36	56, 71	39, 36
Nucleophosmin	P06748	1	4.65	33	19, 20, 25	45–47
Nucleosome assembly protein	P55209	6	4.37	45	60, 66	37, 38
Parathymosin	P08814	1	4.11	11	20	46
Peroxisomal farnesylated protein	Q60415 (H)	2	4.24	33	41	41
Prefoldin subunit 4	Q9NQP4	1	4.43	15	17	48
Proteasome subunit alpha 5	P28066	2	4.78	26	29	43
Protein CGI-100	Q9Y3A6	1	4.73	26	25	45
Protein kinase C inhibitor 14-3-3 epsilon	P62261	25	4.65	29	31, 41	42, 41
Protein kinase C inhibitor 14-3-3 zeta/delta	P29312	9	4.75	28	31, 41	42, 41

TABLE 10.2 Proteins Identified from Bovine Retina after Solution-State IEF and 1D PAGE[a] (Continued)

Protein	Accession Number[b]	Peptide Matches	pI Calc[c]	Molecular Mass (kDa)		SDS-PAGE Band Number
				Calc[c]	Obs[d]	
Protein kinase C inhibitor 14-3-3 beta/alpha	P29358	9	4.83	28	29	43
Prothymosin alpha	P01252	2	3.67	12	17	48
Ras-related protein Rab-3A	P11023	2	4.86	25	26	44
Reticulocalbin 2	Q14257	2	4.26	37	48	40
Ribonuclease inhibitor	P10775 (P)	2	4.79	49	48	40
S-phase kinase-associated protein 1A	P34991	5	4.40	19	19, 20	46, 47
Synaptosomal-associated protein	P13795	4	4.67	23	26, 29	43, 44
Synaptotagmin I	P48018	13	8.50	48	66, 71	37, 36
Synuclein beta	P33567	4	4.47	14	20	46
TATA-binding protein-associated phosphoprotein	Q01658	2	4.72	19	20	46
Trehalase	Q9W2M2 (F)	2	4.90	68	17	48
Tropomyosin 1 alpha	P09493	2	4.72	33	31	42
Tropomyosin beta	P07951	4	4.69	33	41	41
Tubulin beta	Q7M372	9	4.81	50	17, 19, 20, 26, 60	38, 44, 46–48
Vimentin	P48616	23	5.25	54	26, 48, 56, 60, 71	36, 38–40, 44
Visinin-like protein 2	P28677	1	4.73	22	20	46
MCE Chamber pH 5–6.5						
Acidic leucine-rich nuclear phosphoprotein	P51122	4	4.95	20	32	78
Acidic ribosomal protein P0, 60S	Q95140	3	5.67	33	42, 43	17, 72
Actin	P60712	13	5.31	42	48, 54, >97	1, 15, 66, 67
Actinin 4, alpha	043707	4	5.38	105	68	51

(Continued)

TABLE 10.2 Proteins Identified from Bovine Retina after Solution-State IEF and 1D PAGE[a] (Continued)

Protein	Accession Number[b]	Peptide Matches	pI Calc[c]	Molecular Mass (kDa)		SDS-PAGE Band Number
				Calc[c]	Obs[d]	
Adapter-related protein complex 2 alpha 2	P17427 (M)	2	6.89	104	60, 68	50, 51
Adapter-related protein complex 2 beta 1	P21851	2	5.29	105	60	50
Adenosylhomocysteinase	P50247 (M)	5	6.45	48	29, 31	79, 80
ADP, ATP carrier protein	P32007	11	10.07	33	26, 31, 52, 55, >97	1, 61, 64, 79, 83
Annexin A4	P13214	2	5.60	36	34, 39	18, 77
Annexin A5	P81287	3	4.96	36	34, 36	76, 77
Annexin A6	P79134	7	5.52	70	70, 79	9, 55
Arginine/serine-rich splicing factor 10	Q15815	2	11.27	34	42, 46	69, 72
Aspartate aminotransferase	P12344	14	9.37	47	44, 46	69, 70
ATP synthase beta chain	P00829	17	5.21	56	25, 44, 55, 62, >97	3, 13, 61, 62, 70, 82
ATP synthase gamma	P05631	6	9.59	33	34, 39	18, 77
ATP synthase, mitochondrial	P19483	6	9.43	60	55	61
Calcium/calmodulin-dependent protein kinase type II	Q01061	1	7.96	59	55	61
Calnexin	Q8HYW3	3	4.48	68	60	50
Calretinin	P22676	3	5.11	32	29	80
cAMP-dependent protein kinase	P05131	3	8.99	40	47	68
Cellular retinaldehyde-binding protein	P10123	5	5.05	36	36, 38, 39, 43	17, 18, 75, 76
Centractin, alpha	P42024	3	6.56	43	48, 50	65, 66
cGMP-specific 3', 5'-cyclic phosphodiesterase	P23439	7	5.31	98	68, 78	52, 51
Clathrin heavy chain 1	P49951	9	5.63	192	>97	4, 5
Core histone macro-H2A.1	075367	2	10.14	39	46	69
Creatine kinase	P05124 (D)	9	5.68	43	47, 48, 54	15, 68, 66
Crystallin Mu	Q14894	2	6.63	34	42, 43	17, 72, 73
Cytochrome c1	P00125	2	6.94	27	32	78

TABLE 10.2 Proteins Identified from Bovine Retina after Solution-State IEF and 1D PAGE[a] (Continued)

Protein	Accession Number[b]	Peptide Matches	pI Calc[c]	Molecular Mass (kDa)		SDS-PAGE Band Number
				Calc[c]	Obs[d]	
DB83 protein	P57088	2	9.87	28	26	83
Dihydropyrimidinase related protein-1	P97427 (M)	6	7.06	62	65	57
Dihydropyrimidinase related protein-2	O02675	14	6.29	62	62, 67, 68, 75	10, 12, 56, 58
Dynamin 1	Q05193	5	7.37	97	68, 97	7, 51
Dynamin 3	Q9UQ16	1	8.58	97	76	53
Dynamin-like 120 kDa protein	O60313	1	8.16	112	76	53
Elongation factor 1-gamma	*BAC56510*	1	6.68	50	60	14
Elongation factor Tu	P49410	5	7.16	49	50	65
Endoplasmin	Q95M18	3	4.94	83	60, 68, 97, >97	6, 7, 50, 51
Enolase alpha	Q9XSJ4	12	6.83	47	52, 53, 55, 60, 62	13, 14, 63–64
Enolase beta	P13929	6	8.02	47	53	63
Enolase gamma	P09104	10	4.94	47	50, 52, 60	14, 64, 65
Eukaryotic initiation factor 4A-II	Q14240	5	5.41	46	50	65
Eukaryotic translation initiation factor 2C 1	*NP_001001133*	2	9.42	97	68	51
Eukaryotic translation initiation factor 5A	P10159	3	5.13	17	18	84
Excitatory amino acid transporter 1	P46411	2	8.50	60	76, >97	1, 49
F-actin capping protein alpha-2	AAR16258	2	5.75	33	42	73
Fatty aldehyde dehydrogenase	*NP_787005*	2	8.23	55	55	61
Fructose-bisphosphate aldolase C	P09972	2	6.82	39	47	68
Ganglioside-induced differentiation-associated protein	Q8TB36	1	9.33	41	43	71
Gelsolin	Q28372	2	5.76	81	68, 97	7, 51

(Continued)

TABLE 10.2 Proteins Identified from Bovine Retina after Solution-State IEF and 1D PAGE[a] (Continued)

Protein	Accession Number[b]	Peptide Matches	pI Calc[c]	Molecular Mass (kDa)		SDS-PAGE Band Number
				Calc[c]	Obs[d]	
Glial fibrillary acidic protein	Q28115	10	5.38	49	42–50, 52–60	14–16, 62, 64, 65, 68, 71, 72
Glucose-regulated protein, 78 kDa	P19120	8	5.10	72	73, 97	8, 54
Glutamate dehydrogenase	P00366	1	8.52	56	55	61
Glutamine synthetase	P15104	3	6.83	42	48, 50, 54	15, 65, 66
Glutathione S-transferase Mu 5	P48774	4	7.38	27	27	81
Glyceraldehyde 3-phosphate dehydrogenase	P10096	4	8.71	36	42	72
Glycogen phosphorylase	P79334	2	6.78	97	68	51
GTPase-activating protein 1 Rho	Q07960	1	6.18	50	55	61
Guanine nucleotide-binding protein G	P08239	16	5.54	40	43, 44, 47	16, 71, 70
Heat shock 70 kDa protein 1	Q27975	2	5.66	70	70	55
Heat shock 70 kDa protein 3	P34933	5	5.60	70	68	56
Heat shock cognate 71 kDa protein	P19120	13	5.58	71	70, 73, 79	9, 55, 54
Heat shock protein 60 kDa	P18687 (H)	6	5.97	61	65, 69	11, 57
Heat shock protein 75 kDa	Q12931	1	8.32	80	76, 97	7, 53
Heat shock protein 83	002192 (F)	6	4.91	82	78	52
Heat shock protein HSP 90-alpha	Q9GKX7 (E)	31	5.04	83	68, 76, 78, 97	7, 51–53
Heat shock protein HSP 90-beta	Q76LV1	17	5	83	78	52
Heterogeneous nuclear ribonucleoprotein H	P31943	3	6.21	49	55	61
Heterogeneous nuclear ribonucleoprotein H3	P31942	1	6.78	37	43	71
Heterogeneous nuclear ribonucleoprotein K	P61978	4	5.46	51	62, 65, 68, 69, 75	10, 11, 56–58
Heterogeneous nuclear ribonucleoproteins A2/B1	P22626	3	9.12	37	42	72, 73

TABLE 10.2 Proteins Identified from Bovine Retina after Solution-State IEF and 1D PAGE[a] (Continued)

Protein	Accession Number[b]	Peptide Matches	pI Calc[c]	Molecular Mass (kDa)		SDS-PAGE Band Number
				Calc[c]	Obs[d]	
Heterogeneous nuclear ribonucleoproteins C1/C2	P07910	5	4.97	34	44	70
Heterogenous nuclear ribonucleoprotein U	Q00839	1	5.95	90	76	53
Hexokinase, type I	P19367	4	6.82	102	>97	2, 5, 6
Histone H1x	Q92522	1	11.08	22	32	78
Histone H2B	Q8WNR4	5	10.73	14	16, 18, 26	83–85
Importin beta-1	Q14974	2	4.7	97	68	51
Importin, alpha	P55060	1	5.81	110	68	51
Inorganic pyrophosphatase	P37980	1	5.35	33	42	73
Internexin, alpha	Q16352	7	5.42	55	55, 62, 65	57, 58, 61
Interphotoreceptor retinoid-binding protein	P12661	12	5.23	140	>97	2, 5
Isocitrate dehydrogenase	O77784	4	9.01	42	44, 46	69, 70
Lactoglobulin, beta	P02754	1	4.97	20	46	69
Lamin B1	P20700	4	5.15	66	70	55
Lon protease homolog	P36776	2	6.32	106	60	50
Lupus La protein	P10881	1	7.1	47	50	65
Malate dehydrogenase, cytoplasmic	P11708 (P)	5	6.5	36	38, 42	73, 75
Malate dehydrogenase, mitochondrial	P08249 (M)	1	8.97	36	39	74
Matrin 3	P43243	2	6.17	95	>97	2, 5
Microtubule-associated protein 1B	O41515	1	4.77	271	34	77
Mitochondrial inner membrane protein	Q16891	3	6.40	84	78	52
Myosin heavy chain, nonmuscle type B	Q27991	2	5.53	229	76	49
Myosin X	*NP_776819*	1	8.98	239	46	69
NADH-ubiquinone oxidoreductase 30 kDa	P23709	4	7	30	29, 31, 33	19, 79, 80
NADH-ubiquinone oxidoreductase 39 kDa	P34943	5	9.81	43	38	75

(Continued)

TABLE 10.2 Proteins Identified from Bovine Retina after Solution-State IEF and 1D PAGE[a] (Continued)

Protein	Accession Number[b]	Peptide Matches	pI Calc[c]	Molecular Mass (kDa) Calc[c]	Molecular Mass (kDa) Obs[d]	SDS-PAGE Band Number
NADH-ubiquinone oxidoreductase 42 kDa	P34942	2	6.94	39	44	70
NADH-ubiquinone oxidoreductase 49 kDa	P17694	6	6.28	49	48, 50, 54	15, 65, 66
NADH-ubiquinone oxidoreductase 75 kDa	P15690	4	6.07	79	73, 97	8, 54
Neural cell adhesion molecule 1	P31836	3	4.91	94	64, >97	3, 88
Neurofilament triplet L protein	P02548	5	4.62	62	50	65
Neurofilament triplet M protein	O77788	2	4.9	91	44	70
NG, NG-dimethylarginine dimethylaminohydrolase 1	P56965	2	5.92	31	47	16
Nucleolin	P19338	5	4.6	76	60, 73	50, 54
Nucleophosmin	P06748	1	4.65	33	43	71
Osmotic stress protein 94	P48722 (M)	2	5.68	94	60, >97	6, 50
Oxoglutarate/malate carrier protein	P22292	13	10.10	34	31	92
Oxoglutarate dehydrogenase	Q02218	3	7.03	113	60	50
Oxoisovalerate dehydrogenase beta	P21839	2	5.97	43	43	71
Peroxiredoxin 2	Q9BGI3	4	5.5	22	26, 27	83, 20
Peroxiredoxin 4	Q9BGI2	2	6.35	31	27	81
Peroxiredoxin 6	O77834	3	6.3	25	27, 29	80, 81
Phosducin	P19632	2	4.94	28	32, 34	77, 78
Phosphofructokinase	P52784 (D)	6	8.5	85	73, 76	53, 54
Phosphogluconate dehydrogenase	*BAC56480*	1	9.39	28	50	65
Phosphoglycerate mutase 1	*BAC56419*	2	7.15	29	29	80
Progesterone receptor 1	O00264	2	4.54	22	29	80
Progesterone receptor 2	O15173	1	4.75	24	25	82
Prohibitin	P35232	4	5.7	30	26, 33	19, 83

TABLE 10.2 Proteins Identified from Bovine Retina after Solution-State IEF and 1D PAGE[a] (Continued)

Protein	Accession Number[b]	Peptide Matches	pI Calc[c]	Molecular Mass (kDa) Calc[c]	Molecular Mass (kDa) Obs[d]	SDS-PAGE Band Number
Proteasome subunit alpha type 3	P25788	1	5.26	28	31	79
Protein disulfide isomerase A3	P38657	9	6.58	57	62, 69	58, 11
Protein kinase C inhibitor 14-3-3 tau	P27348	2	4.7	28	31	79
Protein kinase C inhibitor 14-3-3 beta/alpha	P29358	3	4.85	28	27, 29	80, 81
Protein kinase C inhibitor 14-3-3 epsilon	P42655	4	4.65	29	32	78
Protein kinase C inhibitor 14-3-3 gamma	P29359	3	4.83	28	31, 33	19, 79
Protein kinase C inhibitor 14-3-3 zeta/delta	P29312	3	4.75	28	29, 33	19, 80
Protein kinase C, alpha	P04409	4	7.01	77	76, 97	8, 53
Purine nucleoside phosphorylase	P55859	2	6.27	32	32	78
Pyrroline-carboxylate synthetase, delta	P54886	3	7.07	87	76	53
Pyruvate dehydrogenase E1 alpha	P08559	2	8.46	43	50	65
Pyruvate dehydrogenase E1 beta	P11177	1	6.58	39	43	17
Pyruvate kinase, M2 isozyme	P52480 (M)	7	7.63	58	62	58
Rab GDP dissociation inhibitor alpha	P21856	5	5.03	51	53, 65, 69	63, 57, 11
Ras-related protein Rab-1A	P11476	5	6.14	23	26	83
Ras-related protein Rab-3A	P11023	3	4.86	25	27	81
Recoverin	P21457	10	5.39	23	25, 26, 27	82, 83, 20
Reticulon 4	Q9NQC3	1	4.44	130	47	68
Reticulon protein 3	O95197	1	8.86	26	25, 27	81, 82
Retinol dehydrogenase 11	Q8TC12 (H)	1	9.91	35	36, 38	75, 76
Rhodopsin	P02699	4	6.20	39	40, 41, 80, 96, >100	1, 2, 51, 49, 54, 73, 74

(Continued)

TABLE 10.2 Proteins Identified from Bovine Retina after Solution-State IEF and 1D PAGE[a] (Continued)

Protein	Accession Number[b]	Peptide Matches	pI Calc[c]	Molecular Mass (kDa)		SDS-PAGE Band Number
				Calc[c]	Obs[d]	
Rhodopsin kinase	P28327	3	6.18	63	65	57
Ribosomal protein L10, 60S	Q9XSI3	2	10.43	24	25	82
Ribosomal protein L10a, 60S	P53025	1	10.30	25	29	80
Ribosomal protein L19, 60S	P14118	1	11.55	23	27	81
Ribosomal protein L27, 60S	P61356	2	10.85	16	18	84
Ribosomal protein L3, 60S	P39872	2	10.56	46	52	64
Ribosomal protein L5A, 60S	P15125 (X)	2	9.90	34	36	76
Ribosomal protein L6, 60S	Q02878	1	10.90	33	34, 36	76, 77
Ribosomal protein L7, 60S	P18124	5	10.92	29	31	79
Ribosomal protein L7a, 60S	P11518	3	10.92	30	32	78
Ribosomal protein L9, 60S	Q862F0	4	10.27	22	26	83
Ribosomal protein S13, 40S	Q02546	2	10.81	17	18	84
Ribosomal protein S17, 40S	P08636 (G)	4	9.92	15	18	84
Ribosomal protein S18, 40S	Q861U5	3	11.16	18	18	84
Ribosomal protein S2, 40S	O18789	1	10.60	31	34	77
Ribosomal protein S24, 40S	P16632	2	11.07	15	18	84
Ribosomal protein S3a, 40S	Q862H3	3	10.09	30	34	77
ROD outer segment membrane protein 1	P52205	6	6.16	37	38, 39	74, 75
S-arrestin	P08168	19	6.39	45	55, 62, 78	13, 24, 62, 61
Septin 2	Q15019	3	6.52	41	47	68
Serine/threonine protein phosphatase 2A	Q61151 (M)	1	5.50	45	55	61
Serine/threonine protein phosphatase alpha-1	P48452	2	6.62	37	42	72
Serum albumin	P02769	1	6.10	69	31	79
Sideroflexin 1	Q9H9B4	1	9.38	36	34	77
Sodium/potassium-transporting ATPase beta-2	Q28030	1	8.52	33	53, >97	1, 63
Soluble NSF attachment protein, alpha	P81125	2	5.52	33	34, 36	76, 77

TABLE 10.2 Proteins Identified from Bovine Retina after Solution-State IEF and 1D PAGE[a] (Continued)

Protein	Accession Number[b]	Peptide Matches	pI Calc[c]	Molecular Mass (kDa)		SDS-PAGE Band Number
				Calc[c]	Obs[d]	
Solute carrier family 2, member 1	P27674	2	9.05	54	>97	1
Spectrin alpha	P16086 (R)	7	5.28	285	76	49
Spectrin beta	Q01082	19	5.54	275	76	49
Succinate dehydrogenase	P31039	2	7.68	73	68, 70	55, 56
Synapsin I	P17599	5	10.19	75	73	54
Synaptophysin	P20488	1	5.03	34	43	71
Synaptosomal-associated protein 25	P60878 (G)	2	4.67	23	27, 29	80, 81
Synaptotagmin I	P48018	5	8.50	48	65	57
Syntaxin 1B	P41414	2	5.33	33	36, 38	75, 76
Syntaxin 3	Q13277	3	5.40	33	38, 39	74, 75
Syntaxin binding protein 1	P61763	15	6.89	68	68, 75	10, 56
T-complex protein 1, alpha	P11984 (M)	5	5.99	60	65	57
T-complex protein 1, beta	P80314 (M)	4	6.32	57	55, 65, 67, 69	11, 12, 57, 61
T-complex protein 1, epsilon	P48643	1	5.57	60	69	11
Transcriptional activator protein PUR-alpha	Q00577	1	6.37	35	46	69
Transforming protein RhoA	P61585	2	6.53	22	25	82
Transitional endoplasmic reticulum ATPase	P55072	5	5.18	89	68, >97	6, 51
Triosephosphate isomerase	P17751 (M)	3	7.47	27	25, 27	81, 82
Tubulin alpha	P81947	12	5.01	50	38, 42, 53, 60–69	11, 12, 63, 59, 58, 72, 75
Tubulin beta	Q7M372	23	4.81	50	55–67, >97	2, 12, 13, 59–61
Tumor necrosis factor receptor	O19131	1	6.60	50	48	66

(Continued)

TABLE 10.2 Proteins Identified from Bovine Retina after Solution-State IEF and 1D PAGE[a] (Continued)

Protein	Accession Number[b]	Peptide Matches	pI Calc[c]	Molecular Mass (kDa) Calc[c]	Molecular Mass (kDa) Obs[d]	SDS-PAGE Band Number
Ubiquinol-cytochrome C reductase	P23004	2	8.99	48	48	66
Ubiquitin-activating enzyme E1	AAC63374	1	5.66	118	>97	6
Unc-119 protein homolog	Q13432	3	6.33	27	36	76
Vacuolar ATP synthase A	P31404	10	5.51	68	70, 79	9, 55
Vacuolar ATP synthase B	P31408	3	5.86	57	55, 67	61, 12
Vacuolar ATP synthase C	P21282	2	7.55	44	46	69
Vacuolar ATP synthase H	046563	2	6.68	56	55	61
Vacuolar protein sorting 35	Q96QK1	3	5.43	92	76	53
Vesicle-associated membrane protein 2	*NP_776908*	3	8.50	12	18	84
Vesicle-fusing ATPase	P46459	8	6.75	83	73, 97	8, 54
Vimentin	P48616	3	5.25	54	50, 67	12, 65
Voltage-dependent anion-selective channel 1	P45879	1	8.76	32	34	77
Voltage-dependent anion-selective channel 2	P45880	5	6.72	38	36	76
Voltage-dependent anion-selective channel 3	Q9MZ13	2	9.12	31	32	78
Wee1-like protein kinase	P47817 (X)	1	8.87	62	67	26
XPA-binding protein 1	Q9HCN4	1	4.83	42	57	28
MCE Chamber pH 6.5–8						
Aconitate hydratase	P20004	7	8.27	85	97	22
Aminobutyrate aminotransferase	Q9BGI0	2	8.55	57	64	27
Annexin A2	P04272	2	7.46	38	44	31
Aspartate aminotransferase, cytoplasmic	P33097	5	7.63	46	52	29
Aspartate aminotransferase, mitochondrial	P12344	2	9.37	47	52	29
ATP synthase, mitochondrial	P19483	13	9.43	60	64, 67	26, 27
Citrate synthase, mitochondrial	075390	1	8.37	52	57	28

TABLE 10.2 Proteins Identified from Bovine Retina after Solution-State IEF and 1D PAGE[a] (Continued)

Protein	Accession Number[b]	Peptide Matches	pI Calc[c]	Molecular Mass (kDa)		SDS-PAGE Band Number
				Calc[c]	Obs[d]	
Creatine kinase, ubiquitous mitochondrial	Q9TTK8	6	8.72	47	57	28
Enolase alpha	Q9XSJ4	4	6.83	47	64	27
Enolase gamma	057391 (G)	2	4.87	47	64	27
Fructose-bisphosphate aldolase A	P00883 (O)	3	8.47	39	52	29
Fructose-bisphosphate aldolase C	P05063 (M)	2	7.12	39	52	29
Fumarate hydratase, mitochondrial	P07954	2	9.08	55	64	27
Glutathione S-transferase Mu	P48774 (M)	2	7.38	27	31	34
Glyceraldehyde 3-phosphate dehydrogenase	Q28554 (A)	8	8.16	35	48	30
Heterogeneous nuclear ribonucleoproteins A2/B1	P22626	1	9.12	37	44, 48	30, 31
Hydroxyacyl-CoA dehydrogenase type II	002691	1	8.67	27	31	34
Lactate dehydrogenase A chain	P19858	8	8.37	36	44	31
Malate dehydrogenase, mitochondrial	P08249 (M)	6	8.97	36	44	31
Mitochondrial aspartate glutamate carrier 1	075746	2	8.72	75	78	24
Phosphatidylethanolamine-binding protein	P13696	2	7.60	21	26	35
Phosphoglycerate kinase 1	P00558	6	8.52	45	57	28
Phosphoglycerate mutase 1	P18669	5	7.15	29	34	33
Putative phosphoglycerate mutase 3	Q8N0Y7	4	6.57	29	34	33
Pyruvate kinase, M1 isozyme	P14618	8	8.20	58	67, 74	25, 26
S-arrestin	P08168	1	6.39	45	57	28
Septin 7	Q16181	3	9.08	49	64	27

(Continued)

TABLE 10.2 Proteins Identified from Bovine Retina after Solution-State IEF and 1D PAGE[a] (Continued)

Protein	Accession Number[b]	Peptide Matches	pI Calc[c]	Molecular Mass (kDa)		SDS-PAGE Band Number
				Calc[c]	Obs[d]	
Transketolase	P29401	3	7.65	68	78	24
Triosephosphate isomerase	P60174	1	6.88	27	31	34
Ubiquinol-cytochrome C reductase protein 2	P23004	3	8.99	48	57	28
Voltage-dependent anion-selective channel 2	P45880	1	6.72	38	44	31
Wee1-like protein kinase	P47817 (X)	1	7.87	62	67	26
XPA-binding protein 1	Q9HCN4	1	4.83	42	57	28
MCE Chamber pH 8–11						
Actin dependent regulator of chromatin	Q96GM5	2	9.39	55	64	88
Adaptor related protein complex (AP)-3 delta	*BAA36591*	1	6.58	136	18	96
Adenine nucleotide translocator 1	P02722	1	9.84	33	26	94
ADP-ribosylation factor 2	A45422	2	6.16	21	21	95
Aspartate aminotransferase	P12344	21	9.37	47	26–50, 64, 80, 97	86–92, 94
ATP synthase, mitochondrial	P13621	8	10.21	23	21, 26, 64	88, 94, 95
Beta-adrenergic receptor kinase 1	P21146	1	7.30	80	50	89
Brain-specific protein, 25 kDa	Q27957	4	9.80	23	30	93
Carnitine/acylcarnitine carrier protein	Q9Z2Z6 (M)	1	9.41	33	31	92
Cold-inducible RNA-binding protein	Q14011	4	9.64	19	21	95
Dicarboxylate carrier	Q9UBX3	4	9.78	31	30	93
Dihydropyrimidinase related protein-2	O02675	1	6.29	62	18	96
DNA-directed RNA polymerase beta chain	P57146	1	8.15	151	26	94
Elongation factor 1-alpha 1	*AAB65435.1*	13	9.22	34	50, 97	86, 89
Endonuclease G	P38447	1	9.34	32	30	93

TABLE 10.2 Proteins Identified from Bovine Retina after Solution-State IEF and 1D PAGE[a] (Continued)

Protein	Accession Number[b]	Peptide Matches	pI Calc[c]	Molecular Mass (kDa) Calc[c]	Molecular Mass (kDa) Obs[d]	SDS-PAGE Band Number
Glyceraldehyde 3-phosphate dehydrogenase	P10096	5	8.90	34	30, 31	92, 93
GTP:AMP phosphotransferase	P08760	1	9.27	26	30	93
Heterogeneous nuclear ribonucleoprotein G	P38159	1	10.18	42	50, 97	86, 89
Heterogeneous nuclear ribonucleoprotein R	043390	4	8.42	71	80	87
Histone H2B	Q8WNR4	3	10.32	14	18	96
LIM and SH3 domain protein 1	Q14847	1	7.05	30	21	95
NADH-ubiquinone oxidoreductase 18 kDa	Q02375	5	10.47	20	21	95
NADH-ubiquinone oxidoreductase B16.6	Q95KV7	3	9.40	17	18	96
NADH-ubiquinone oxidoreductase B17	Q02367	1	9.83	15	21	95
NADH-ubiquinone oxidoreductase PDSW	Q02373	6	8.87	21	26	94
NADH-ubiquinone oxidoreductase B17.2	097725	6	9.89	17	18	96
Neuronal kinesin heavy chain	P33175 (M)	1	5.89	117	30	93
NipSnap1 protein	055125 (M)	4	9.67	33	30	93
Nuclear receptor coactivator 5	Q9HCD5	1	9.76	66	64	88
Nucleoporin p58/p45	Q8R332 (M)	1	9.61	59	64	88
Oxoglutarate/malate carrier protein	P22292	13	10.10	34	31	92
Peroxiredoxin 1	*NP_776856.1*	3	8.81	22	26	94
Peroxisomal multifunctional enzyme type 2	P51659	1	9.16	80	80	87
Polyadenylate-binding protein 1	CAB96752	8	9.52	71	80	87
Ran GTPase-activating protein 1	P46060	1	4.66	64	50	89

(Continued)

TABLE 10.2 Proteins Identified from Bovine Retina after Solution-State IEF and 1D PAGE[a] (Continued)

Protein	Accession Number[b]	Peptide Matches	pI Calc[c]	Molecular Mass (kDa)		SDS-PAGE Band Number
				Calc[c]	Obs[d]	
Rho guanine nucleotide exchange factor 5	Q12774	1	7.99	60	18	96
Rho-related GTP-binding protein RhoG	Q9TU25	3	7.52	21	26	94
Ribosomal protein L12, 60S	P61284	1	9.48	18	21	95
Ribosomal protein L22, 60S	P35268	2	9.49	15	18	96
Ribosomal protein S11, 28S	P82911	1	8.59	5	21	95
Ribosomal protein S16, 40S	P17008	2	10.47	16	18	96
Ribosomal protein S20, 40S	P17075	1	10.32	13	18	96
Ribosomal protein S3, 40S	*BAC56417*	8	9.86	23	31	92
Ribosomal protein S5, 40S	P46782	2	10.01	23	26	94
RNA and export factor binding protein 2	Q9JJW6 (M)	2	10.30	24	31	92
RNA-binding protein FUS	Q28009	9	9.48	52	50, 64	88, 89
Septin 7	Q16181	2	9.08	49	50	89
Shwachman-Bodian-Diamond syndrome protein	Q9Y3A5	2	9.14	29	31	92
Sideroflexin 1	Q9H9B4	1	9.38	36	31	92
Signal recognition particle receptor beta	Q9Y5M8	1	9.39	30	31	92
similar to ribosomal protein S10	*BAC56342*	5	9.90	17	21	95
Sin3 associated polypeptide p18	O00422	2	9.62	18	21	95
Splicing factor, arginine/serine-rich 1	Q07955	1	10.49	28	31	92
Splicing factor, proline- and glutamine-rich	P23246	11	9.70	76	64, 97	86, 88
Succinate dehydrogenase	*AAC72370*	4	5.87	16	30	93
Synapsin I	P17599	10	10.19	75	50, 64, 80	87–89
Tetratricopeptide repeat protein 11	Q9Y3D6	3	9.08	17	18	96
Tetratricopeptide repeat protein 9	Q92623	1	10.03	35	18	93

TABLE 10.2 Proteins Identified from Bovine Retina after Solution-State IEF and 1D PAGE[a] (Continued)

Protein	Accession Number[b]	Peptide Matches	pI Calc[c]	Molecular Mass (kDa)		SDS-PAGE Band Number
				Calc[c]	Obs[d]	
Thioesterase superfamily member 2	Q9NPJ3	1	9.51	15	18	96
Trafficking protein particle complex subunit 6B	Q86SZ2	2	9.06	18	18	96
U2 small nuclear ribonucleoprotein B"	P08579	2	10.02	25	21	93
Ubiquinone biosynthesis monooxgenase COQ6	Q9Y2Z9	1	7.26	51	30	96
Ubiquitin-conjugating enzyme	P51966	1	8.89	18	30	95
Voltage-dependent anion-selective channel 3	Q9MZ13	9	9.12	31	31	92

[a] Bands were excised from the gels shown in Figure 10.3B and Figure 10.5, and proteins were identified by LC MS/MS as described in Methods. Membrane-associated proteins are highlighted in bold; integral membrance proteins are shown in bold italics.

[b] Swiss-Protein (in plain font) and NCBI (in italics) accession numbers are shown. Links to Swiss-Protein accession numbers use the EXPASY server at http://us.expasy.org/sprot/. Links to NCBI accession numbers use the Entrez server at http://www.ncbi.nlm.nih. Parenthesis following accession numbers designate identifications based on homology with species other than bovine and human. A, *Ovis aries* (sheep); D, *Cannis familiaris* (dog); E, *Equus caballus* (horse); F, *Drosophila melanogaster* (fruit fly); G, *Gallus gallus* (chicken); H, *Cricetulus griseus* (Chinese hamster); L, *Loigo forbesi* (northern European squid); M, *Mus musculus* (house mouse); O, *Oryctolagus cuniculus* (rabbit); P, *Sus scrofa* (pig); R, *Rattus norvegicus* (Rat); T, *Clostridium tetani*; X, *Xenopus laevis* (African clawed frog).

[c] Based on protein sequence.

[d] Approximate mass observed by SDS-PAGE.

10.4 DISCUSSION

The usual lack of identification of rhodopsin in 2D PAGE analyses in our laboratory[11–15] and elsewhere[25,26] prompted the present study to evaluate solution-state IEF followed by 1D PAGE as a method for facilitating the identification of membrane proteins from retina by mass spectrometry. Historically, rhodopsin has been purified chromatographically in the presence of detergents such as Ammonyx LO, octyl glucoside, and dodecyl maltoside,[27,28] and differences have been reported in the solubilization and extractability of rhodopsin using emulphogene, CHAPS, taurocholate, octyl glucoside, and digitonin.[29] Here we compared the ability of eight different detergents to extract bovine retinal proteins and observed little overall difference by1D PAGE. However, Western blot analysis for rhodopsin showed that except for SDS, ASB-14 yielded the least apparent amount of high mass aggregates and strong immunoreactivity for the apparent monomeric protein. This result suggested that ASB-14 was possibly a useful detergent for fractionating retinal membrane proteins by solution-state IEF.

Retinal protein extractability and IEF fractionation were compared using the zwitterionic detergent ASB-14, dodecyl maltoside, a nonionic detergent commonly used in studies of retina, and CHAPS, perhaps the most common zwitterionic detergent used in IEF. Bovine retinas were extracted in 7 M urea and 2 M thiourea containing either ASB-14, dodecyl maltoside, or CHAPS, and more soluble protein was recovered with ASB-14. The protein recovery from retina (1.3–1.7% relative to the tissue wet weight) was comparable to that previously reported from optic nerve (1.3%) using the dodecyl maltoside solvent but less than that extracted from skeletal muscle (2.7%) or liver (4.9%).[17] These differences in yield appear to be related to membrane protein content of the tissues. Following IEF, total protein recovery was also greater with ASB-14 compared with dodecyl maltoside and CHAPS. Overall protein recovery in the MCE trapping chambers from 3 mg retinal extract (25–34%) was in the range obtained from 3 mg optic nerve extract (31%) but less than that obtained from 5 to 7.5 mg optic nerve (39%).[17] As noted elsewhere, optimization of the MCE instrument for smaller sample amounts and a more efficient protein concentration method than acetone precipitation could improve recovery.[17]

Following IEF, Coomassie blue 1D PAGE profiles for protein in the MCE trapping chambers were distinct from each other but very similar among the three detergents tested. Mass spectrometric analyses of 96 1D gel bands were pursued following IEF in ASB-14. On average about 4 proteins per gel band were identified, providing 329 unique protein identifications from these select proteomic analyses of retina. Generally the sequence-calculated isoelectric points (pIs) of the retinal proteins agreed well with the pH range of the trapping chamber from which they were identified. For proteins trapped in chambers inconsistent with their calculated pIs, many exhibited altered masses, suggesting the presence of posttranslational modifications, covalent aggregation, or proteolytic processing. However, some proteins trapped in pH fractions inconsistent with their calculated pIs did not exhibit significant mass changes and therefore may contain small, charge-altering modifications, for example due to oxidative damage.[9–15]

Rhodopsin was clearly identified by mass spectrometric analyses from the pH 5–6.5 fraction, consistent with Western blot results, and in gel bands supporting the presence of monomeric, dimeric, and high mass aggregate forms of the protein. Approximately 20% of the other identified proteins were membrane-associated or integral membrane proteins such as vacuolar ATP synthase, neuronal axonal membrane protein, and rod outer segment membrane protein 1. A number of proteins associated with phototransduction and the retinoid visual cycle were also detected, including membrane-associated rhodopsin kinase, rod cGMP-specific 3',5'-cyclic phosphodiesterase, and retinol dehydrogenase 11, as well as guanylyl cylcase activating protein, cellular retinaldehyde binding protein, interphotoreceptor retinoid binding protein, recoverin, and arrestin.

In summary, solution-state IEF followed by 1D PAGE provides an attractive alternative to 2D gel analyses with immobilized pH gradients for proteomic studies of retina. In this study, ASB-14 was the most effective detergent for the solubilization and extraction of retina and identification of rhodopsin. We anticipate that multiple, smaller pH range trapping chambers will further enhance the resolving power of this protein fractionation methodology.

ACKNOWLEDGMENTS

This study was supported in part by NIH grants EY6603, EY14239, and EY015638, a Research Center Grant from The Foundation Fighting Blindness, and funds from Alcon Laboratories, Inc. and the Cleveland Clinic Foundation. Rhodopsin antibody and bovine ROS were generous gifts from Dr. Paul Hargrave, University of Florida, and Dr. George Hoppe, Cleveland Clinic Foundation, respectively.

REFERENCES

1. Lamb, T.D., and Pugh, E.N., Jr., Dark adaptation and the retinoid cycle of vision, *Prog. Retin. Eye Res.*, 23, 307–380, 2004.
2. West, K.A., Yan, L., Miyagi, M., Crabb, J.S., Marmorstein, A.D., Marmorstein, L., and Crabb, J.W., Proteome survey of proliferating and differentiating rat RPE-J cells. *Exp. Eye Res.* 73, 479–491, 2001.
3. West, K.A., Yan, L., Shadrach, K., Sun, J., Hasan, A., Miyagi, M., Crabb, J.S., Hollyfield, J.G., Marmorstein, A.D., and Crabb, J.W., Protein database, human retinal pigment epithelium, *Mol. Cell. Proteomics*, 2, 37–49, 2003.
4. Golovleva, I., Bhattacharya, S., Wu, Z., Shaw, N., Yang, Y., Andrabi, K., West, K.A., Burstedt, M.S., Forsman, K., Holmgren, G., et al., Disease-causing mutations in the cellular retinaldehyde binding protein tighten and abolish ligand interactions, *J. Biol. Chem.*, 278, 12,397–12,402, 2003.
5. Wu, Z., Yang, Y., Shaw, N., Bhattacharya, S., Yan, L., West, K., Roth, K., Noy, N., Qin, J., and Crabb, J.W., Mapping the ligand binding pocket in the cellular retinaldehyde binding protein, *J. Biol. Chem.*, 278, 12,390–12,396, 2003.
6. Nawrot, M., West, K., Huang, J., Possin, D.E., Bretscher, A., Crabb, J.W., and Saari, J.C., Cellular retinaldehyde-binding protein interacts with ERM-binding phosphoprotein 50 in retinal pigment epithelium, *Invest. Ophthalmol. Vis. Sci.*, 45, 393–401, 2004.
7. Wu, Z., Hasan, A., Liu, T., Teller, D., Saari, J.C., and Crabb, J.W., Identification of CRALBP ligand interactions by photoaffinity labeling, hydrogen/deuterium exchange and structural modeling, *J. Biol. Chem.*, 279, 27,357–27,364, 2004.
8. Bonilha, V.L., Bhattacharya, S.K., West, K.A., Crabb, J.S., Sun, J., Rayborn, M.E., Nawrot, M., Saari, J.C., and Crabb, J.W., Support for a proposed retinoid-processing protein complex in apical retinal pigment epithelium, *Exp. Eye Res.*, 79, 419–422, 2004.

9. Crabb, J.W., Miyagi, M., Gu, X., Shadrach, K., West, K.A., Sakaguchi, H., Kamei, M., Hasan, A., Yan, L., Rayborn, M.E., et al., Drusen proteome analysis: An approach to the etiology of age-related macular degeneration, *Proc. Natl. Acad. Sci. USA*, 99, 14,682–14,687, 2002.
10. Gu, X., Meer, S.G., Miyagi, M., Rayborn, M.E., Hollyfield, J.G., Crabb, J.W., and Salomon, R.G., Carboxyethylpyrrole protein adducts and autoantibodies: Biomarkers for age-related macular degeneration, *J. Biol. Chem.*, 278, 42,027–42,035, 2003.
11. Miyagi, M., Sakaguchi, H., Darrow, R.M., Yan, L., West, K.A., Aulak, K.S., Stuehr, D.J., Hollyfield, J.G., Organisciak, D.T., and Crabb, J.W., Evidence that light modulates protein nitration in rat retina, *Mol. Cell. Proteomics* 1, 293–303, 2002.
12. Sakaguchi, H., Miyagi, M., Darrow, R.M., Crabb, J.S., Hollyfield, J.G., Organisciak, D.T., and Crabb, J.W., Intense light exposure changes the crystallin content in retina, *Exp. Eye Res.*, 76, 131–133, 2003.
13. Renganathan, K., Sun, M., Darrow, R., Shan, L., Gu, X., Salomon, R.G., Hazen, S., Organisciak, D., and Crabb, J.W., Light induced protein modifications and lipid oxidation products in rat retina, *Invest. Opthalmol. Vis. Sci*, 44, E-abstract 5129, 2003.
14. Renganathan, K., Collier, R., Gu, X., Salomon, R.G., Kapin, M., Hollyfield, J.G., and Crabb, J.W., Similarities in oxidative damage from AMD and retinal light damage, *Invest. Opthalmol. Vis. Sci*, 45, E-abstract 1795, 2004.
15. Gu, X., Renganathan, K., Grimm, C., Wenzel, A., Salomon, R.G., Reme, C.E., and Crabb, J.W., Rapid changes in retinal oxidative protein modifications induced by blue light, *Invest. Opthalmol. Vis. Sci*, 45, E-abstract 3473, 2004.
16. Santoni, V., Kieffer, S., Desclaux, D., Masson, F., and Rabilloud, T., Membrane proteomics: Use of additive main effects with multiplicative interaction model to classify plasma membrane proteins according to their solubility and electrophoretic properties, *Electrophoresis*, 21, 3329–3344, 2000.
17. Bhattacharya, S.K., Crabb, J.S., Annangudi, S.P., West, K.A., Gu, X., Sun, J., Bonilha, V.L., Smejkal, G., Shadrach, K., Hollyfield, J.G., et al., chap. 9, this volume.
18. Ramagli, L., Quantifying protein in 2D PAGE solubilization buffers, in *Methods in Molecular Biology: 2-D Proteome Analysis Protocols*, vol. 112, Link, A., Ed., Humane Press, Totowa, NJ, 1999, pp. 99–103.
19. Laemmli, U.K., Cleavage of structural proteins during the assembly of the head of bacteriophage T4, *Nature*, 227, 680–685, 1970.
20. Crabb, J.W., Gaur, V.P., Garwin, G.G., Marx, S.V., Chapline, C., Johnson, C.M., and Saari, J.C., Topological and epitope mapping of the cellular retinaldehyde-binding protein from retina, *J. Biol. Chem.*, 266, 16,674–16,683, 1991.
21. Kennedy, B.N., Goldflam, S., Chang, M.A., Campochiaro, P., Davis, A.A., Zack, D.J., and Crabb, J.W., Transcriptional regulation of cellular retinaldehyde-binding protein in the retinal pigment epithelium: A role for the photoreceptor consensus element, *J. Biol. Chem.*, 273, 5591–5598, 1998.
22. Adamus, G., Zam, Z.S., Arendt, A., Palczewski, K., McDowell, J.H., and Hargrave, P.A., Anti-rhodopsin monoclonal antibodies of defined specificity, characterization, and application, *Vision Res.*, 31, 17–31, 1991.
23. Papermaster, D.S., Preparation of retinal rod outer segments, *Methods Enzymol.*, 81, 48–52, 1982.
24. Pedersen, S.K., Harry, J.L., Sebastian, L., Baker, J., Traini, M.D., McCarthy, J.T., Manoharan, A., Wilkins, M.R., Gooley, A.A., Righetti, P.G., et al., Unseen proteome, mining below the tip of the iceberg to find low abundance and membrane proteins, *J. Proteome Res.*, 2, 303–311, 2003.

25. Nishizawa, Y., Komori, N., Usukura, J., Jackson, K.W., Tobin, S.L., and Matsumoto, H., Initiating ocular proteomics for cataloging bovine retinal proteins: Microanalytical techniques permit the identification of proteins derived from a novel photoreceptor preparation, *Exp. Eye Res.*, 69, 195–212, 1999.
26. Organisciak, D.T., Henkels, K.M., West, K., Sun, J., Crabb, J.W., and Darrow, R.M., Intense light- and time-dependent changes in rat rod outer segment crystallins, *Invest. Opthalmol. Vis. Sci*, 45, E-abstract 733, 2004.
27. Hargrave, P.A., The amino-terminal tryptic peptide of bovine rhodopsin: A glycopeptide containing two sites of oligosaccharide attachment, *Biochim. Biophys. Acta.*, 492, 83–94, 1977.
28. McDowell, J.H., Nawrocki, J.P., and Hargrave, P.A., Phosphorylation sites in bovine rhodopsin, *Biochemistry*, 32, 4968–4974, 1993.
29. Aveldano, M.I., Phospholipid solubilization during detergent extraction of rhodopsin from photoreceptor disk membranes, *Arch. Biochem. Biophys.*, 324, 331–343, 1995.

11 Reducing Protein Sample Complexity with Free Flow Electrophoresis (FFE)

Askar Kuchumov, Gerhard Weber, and Christoph Eckerskorn

CONTENTS

11.1 INTRODUCTION

Progress in proteome analysis relies heavily on development of technologies to reduce the complexity of the sample prior to systematic application of such analytical techniques as gel electrophoresis, liquid chromatography, and mass spectrometry. Free flow electrophoresis (FFE), available from Becton, Dickinson and company (BD™), is reemerging as one of the most versatile semipreparative to preparative fractionation and separation techniques.[1–4] It offers a unique and highly powerful approach to the separation of charged species including proteins, peptides, cellular organelles, and whole cells.[5] Such a variety of FFE applications is achieved via different separation techniques: isoelectric focusing (IEF), zone electrophoresis (ZE), and isotachophoresis (ITP).[6–9] This chapter focuses on the IEF mode of the FFE as it applies to analysis of proteins. Due to its matrix-free fractionation principle, FFE offers considerable advantages over traditional chromatographic and gel-based techniques: exceptional sample recovery, fast fractionation times, and high throughput.[10] These features allow FFE to offer a unique approach to reducing complexity of any proteome, often as a complementary technique to one-dimensional (1D) or two-dimensional (2D) gel electrophoresis.[11] At the same time, continuous flow principle allows for virtually unlimited preparative fractionations.[12] This combination of effective fractionation with high loading capacity provides for a practical methodology to enrich for low-abundant proteins and to segregate highly abundant proteins in biological samples such as cell lysates or body fluids. This chapter is meant to give an overview of practical aspects of the free flow isoelectric focusing (FFE–IEF) and to demonstrate its capabilities using two challenging applications: display of low-abundant proteins in rat liver lysate and fractionation of human plasma. A detailed description of FFE protocols and a comprehensive review of literature and theoretical considerations have been recently published.[13,14]

11.2 PRINCIPLE OF FREE FLOW ELECTROPHORESIS

The main functional part of an FFE device is a precisely manufactured chamber formed by two narrowly spaced plates (the distance between the plates is 0.4 or 0.5 mm), with the length of 500 mm and a width of 100 mm. Along the sides of the chamber, there are two electrodes that provide high voltage during separation. A sample is continuously infused into a thin, laminar layer of one or more separation buffers flowing through the chamber. A voltage is applied perpendicularly to the flow direction. As separation buffer and samples are moving through the chamber, the electric field leads to a deflection of different sample components according to their charge. As a result, sample components that entered the chamber as a mixture at one end will leave the chamber at the other end as separated components that can be collected (Figure 11.1).

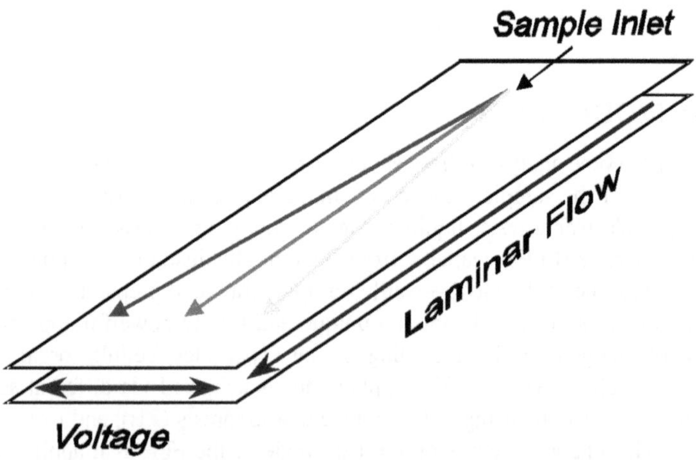

FIGURE 11.1 Principle of FFE: A high voltage between electrodes generates an electric field perpendicular to a laminar flow causing charged species to migrate toward electrodes as they are carried by the flow.

In the BD™ Free Flow Electrophoresis system (Figure 11.2), up to 13 inlets can be used to introduce buffers under positive pressure. Variation in buffer composition and number of inlets used allow for a variety of applications and pH gradients to be developed. Sample can be infused via one of four inlets located on the same end of chamber. The choice usually depends on the nature of the sample and separation conditions (native or denaturing). At the opposite end, the chamber has 96 outlets that split the flowing liquid into 96 fractions collected into a standard microtiter plate. Close to fractionation outlets, there are seven additional inlets that deliver so-called counterflow medium in the direction opposite to main media stream (Figure 11.3). The counterflow concept has helped to solve many technical challenges faced by early developers of free flow systems. First, by merging with the main separation media stream at the point of 96 outlets, counterflow helps avoid turbulence and thus preserve fractionation pattern upon transition of the flow from the chamber into the 96 tubes. Second, counterflow ensures uniform flow of all 96 tubes independent of the separation media flow rate. And third, it allows to introduce an on-line pH shift or boost salt concentration immediately after fractionation for the applications where pH or ionic strength are critical.[15]

Another noteworthy feature of the BD Free Flow Electrophoresis system is the presence of stabilization solutions that run along the electrodes and effectively

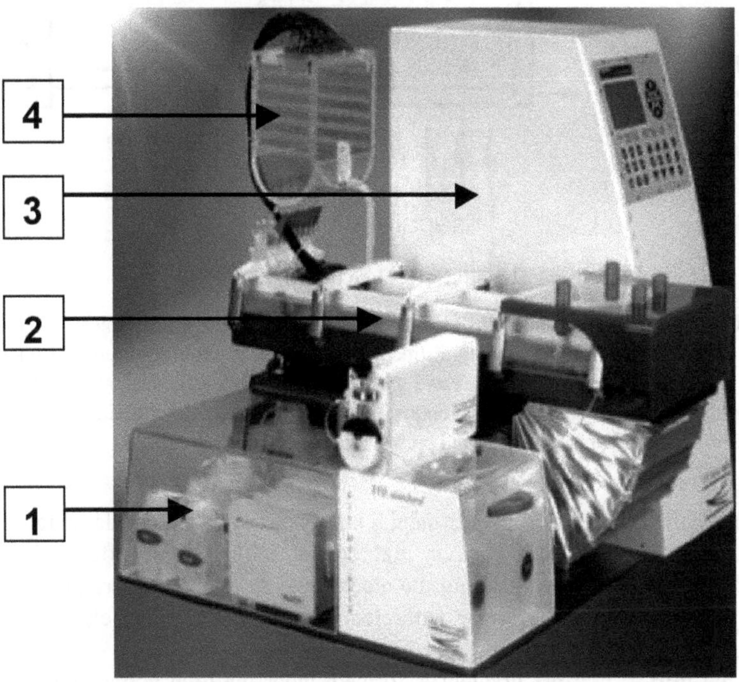

FIGURE 11.2 BD Free Flow Electrophoresis system. (1) Housing for buffers, media, and sample pumps; (2) separation chamber; (3) electrical operating unit including high voltage power supply; (4) collection block.

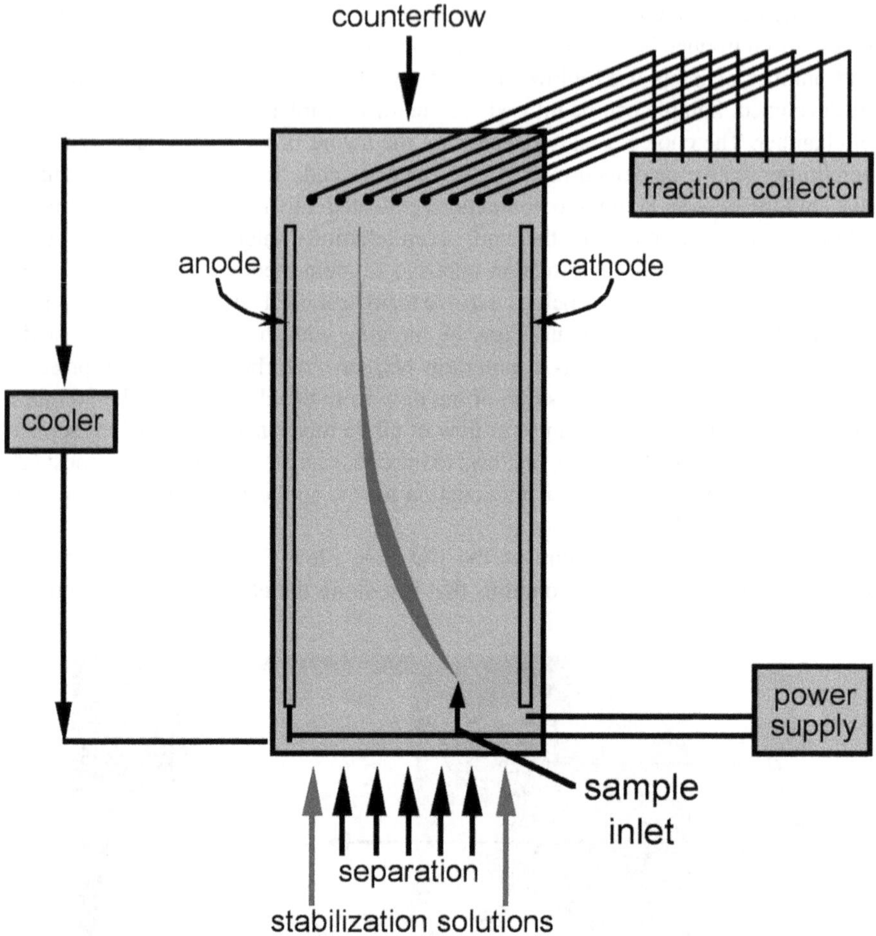

FIGURE 11.3 Operating principle of IEF by FFE. As proteins are carried toward the collection tubes, they focus into sharp lanes with distinct pI values. Stabilization solutions prevent proteins from a direct contact with electrodes and ensure full recovery. Counterflow merges with the laminar flow of separation buffers as it enters the collection tubes. This provides turbulence-free transfer of the buffers into the tubes and preserves a separation pattern.

protect separation media from detrimental influences of the electrodes (Figure 11.3). The stabilization solutions have very high conductivity and thus effectively prevent proteins or peptides from reaching the electrodes. As a result, there are no protein losses associated with interaction with electrodes, which in turn helps elevate total sample recovery of the method.

In the FFE–IEF, a pH gradient is formed under the high-voltage conditions by multiple separation buffers positioned between stabilization media (Figure 11.3). Compared to other modes of FFE separation, IEF yields the highest resolution of amphoteric compounds such as proteins and peptides. Depending on their isoelectric point (pI), proteins or peptides will be shifting toward anode or cathode until a region with pH identical to their pI is found. At this point, a sample component will no longer carry a net charge, will stop its migration to an electrode, and will only continue moving with the flow of buffers forming a narrow lane. As a true focusing technique, IEF leads to very sharp bands, or lanes, resulting in high-resolution fractionation, because any diffusion will place a compound in a pH region away from its pI and the compound will acquire a charge and be forced to migrate back into its lane.

Although linear pH gradients can be formed by polymer-based ampholytes or low molecular-weight buffer pairs,[16] the best resolution is achieved when using BD™ FFE Reagents—proprietary mixtures of low-molecular-weight organic acids, bases, and zwitterions (MW <300 Da) that are well defined and can be reproducibly manufactured.[12] Due to their physicochemical properties, BD FFE Reagents allow for highly reproducible runs and can be easily removed from the final FFE fractions. The latter fact renders FFE fractions amenable to practically any downstream protein analysis techniques such as gel electrophoresis, chromatography, or mass spectrometry applications.

A wide-range, linear pH 2.5–12 gradient is often useful for the initial analysis of a new type of sample. However, for many samples, even better resolution can be achieved in the region pH 4–7, in which the pI values of many proteins occur. In addition, the most abundant proteins in a sample (e.g., albumin in plasma) often can precipitate when the sample's loading rate reaches certain threshold. This can happen due to two factors: first, minimal solubility of a protein at its pI, and, second, very high local concentration of the protein in the tightly focused lane. The presence of 8 M urea, or 7 M urea/2 M thiourea, can raise the threshold but still might not be sufficient to keep proteins with high concentration in solution. This problem can be overcome to some extent with the use of a nonlinear, or stepwise, gradients, in which the region of interest (pH 4–7, or pH 5.5–6.5) contains a much flatter gradient than in the acidic and basic sides. Each nonlinear gradient requires adjacent introduction of different buffers and is usually developed separately for each type of sample. A case with albumin separation from plasma is a good example of stepwise gradient application and is reviewed later in the chapter.

11.3 PRACTICAL CONSIDERATIONS

11.3.1 Separation Buffers

As mentioned earlier in the chapter, BD FFE Reagents represent the running buffer elements that establish pH gradient in solution when placed in an electrical field. In addition to these reagents, all commercially available chemicals common to typical IEF applications can be used, but their concentration needs to be optimized for each application.

Similar to 2D electrophoresis (2DE) analysis, sample solubility is one of the most critical factors for successful protein separation by FFE–IEF. Ideally, denaturing separation conditions should lead to the disruption of all noncovalently bound protein complexes and aggregates into a solution of individual polypeptides. The zwitterionic detergent, 3[(cholamidopropyl)dimethylammonio]-1-propane sulfonate (CHAPS), has been found to be effective for solubilization of membrane proteins,[17] particularly when used at a concentration of 4% w/v in combination with a mixture of 2 M thiourea and 8 M urea.[18] Thiourea is a much stronger denaturant than urea but cannot be used alone because it is weakly soluble in water. However, it is more soluble in a concentrated solution of urea, so urea–thiourea mixtures exhibit improved solubilizing power. The combination of 2 M thiourea and 7 M urea was found to provide optimal conditions for a denaturing FFE–IEF separation of complex eukaryotic protein samples. Up to 1% CHAPS can also be added to FFE separation buffers. Triton X-100, NP-40, ASB-14, and Zwittergent 3–10 are some other examples of possible nonionic detergents that can be used in FFE–IEF. To maintain cysteine-containing proteins in reduced state, 10 mM DTT is commonly added to the FFE separation buffers.

Other additives may be used in the FFE–IEF separation buffers depending on the sample requirements and desired post-FFE application. They include glycerol for increased viscosity and protein stability, hydroxypropylmethylcellulose (HPMC) for coating of chamber walls in order to minimize electroosmotic flow, and mannitol, a monomeric sugar, which replaces HPMC in the applications that require a high degree of concentration of fractions.

11.3.2 Making an FFE Run

The notion of run is more suited to traditional separation methods such as gel electrophoresis and chromatography that work in a batchwise mode where sample injection or loading is followed by a separation step (a run). On the other hand, FFE is performed in a continuous fashion when sample infusion and separation happen simultaneously. In this context, the term "FFE run" is not an accurate description of the process, although it is used for convenience.

The current FFE system is designed for semipreparative to preparative scale separations. For standard IEF protocols, 1–5 mg/ml of total protein concentration can be viewed as a reasonable working range. However, special protocols are being developed for an analytical mode that can manage sample concentrations as low as 50 μg/ml. As far as the throughput, the most common sample application rate is between 0.5 ml/hr and 5 ml/hr. Depending on the sample concentration, this leads to a total protein loading rate of up 50 mg/hr for common methods. Higher throughput can be achieved under special protocols.

11.3.3 Sample Preparation

The chemical and physical properties of the sample and the separation buffers (media) should be as close as possible. To match the density, viscosity, and conductivity of the sample solution and separation media, original samples are prepared in a concentrated form and then diluted at least 1:5 with an appropriate separation buffer. Turbid protein samples should be cleared by filtration or centrifugation before infusion into the FFE chamber.

The total salt concentration of the sample should be no higher than 25 or 50 mM for classical FFE-IEF protocols. However, custom protocols can be now developed for samples that must maintain stronger ionic force.

11.3.4 Anticipated Results

When a narrow focused lane of proteins at certain pI is ready to enter one of the 96 collection tubes, it is possible that these proteins can get into a single tube. In this case, proteins with a certain pI will be collected in one well. In practice, one, two or three wells per protein lane, or band, should be expected, with urea-containing media leading to slightly worse resolution compared to native conditions. A few highly abundant proteins would form wider bands and therefore would give broader well distribution.

The sample recovery after FFE-IEF depends to a large extent on the amount of the sample. As noted earlier, the matrix-free separation features much higher recoveries as compared to other methods. Also, both the internal chamber and the fractionation outlets have a relatively low surface area; they are made of low-binding materials and can be treated with appropriate additives that reduce surface adsorption. Thus recoveries higher than 95% are common for milligram amounts of sample if no precipitation occurs. Due to the semipreparative dimensions of the instrument, the losses can be more substantial (up to 50%) when micrograms of protein in microliters of volume are separated.

To evaluate the dilution factor after the FFE separation, it is important to remember that the separation process distributes all the proteins from a complex mix into different wells. Thus it is useful to consider what happens to an individual protein rather than to the complex sample (1 ml), which is diluted into 96 wells (1:96). The total flow rate (media plus counterflow) contributing to a collection plate is approximately 100 ml/hr, which translates into ~1 ml/hr/well for a 96-well plate. Assuming that a sample is applied at a rate of 1 ml/hr and that each particular protein is collected into three wells, the dilution factor for each protein should be close to 3.

11.3.5 Post-FFE Treatment of Fractions

The choice of post-FFE treatment of the fractions depends on the composition of the separation buffers used in the FFE-IEF and solution requirements for a downstream application. FFE fractionation of proteins has been recently demonstrated to couple successfully with LC/MS/MS analysis.[19] In cases when FFE fractions can not be used directly, ultrafiltration with low-binding molecular-weight cutoff membranes or solid phase extractions (SPE) procedures can be performed. Ultrafiltration is used in majority of cases because it allows simultaneous concentration of the sample and a buffer exchange. The SPE techniques vary depending on protein retention mechanisms. The most common resins include reversed-phase, ion-exchange, hydrophilic, and hydrophobic interactions, and mixed resins. Usually, the SPE resin type is chosen based on elution conditions that should match a downstream analytical method, nature of the sample, and protein recovery values.

Hydrophilic interaction chromatography (HILIC) has been found to yield the best results in terms of protein recovery and degree of purification. The method and its basic principles have been extensively described.[20]

11.4 APPLICATIONS

11.4.1 Liver Proteome Fractionation

With complex samples such as eukaryotic cell extracts, 2DE on a single wide-range pH gradient reveals only a small percentage of the whole proteome because of insufficient spatial resolution and the difficulty of visualizing low copy number proteins in the presence of the more abundant species. One approach to overcoming the problem of high dynamic range and diversity of the proteins expressed in eukaryotic tissues is sample prefractionation. This can be achieved by methods such as subcellular fractionation, electrophoresis in the liquid phase, adsorption chromatography, and selective precipitation.[21] An alternative approach is the sequential extraction of proteins from cells or tissues on the basis of their solubility in a series of buffers with increasingly powerful solubilizing properties.[18,22]

In the experiment described here, the advantages of the FFE prefractionation are combined with zoom gels—multiple, overlapping narrow-range IPGs spanning 1 to 1.5 pH units.[23] The overall workflow is outlined in Figure 11.4. The technical details and specific conditions are listed in Table 11.1. The preparation of a rat liver homogenate resulted in a solution with 30 mg/ml total protein concentration. After dilution with

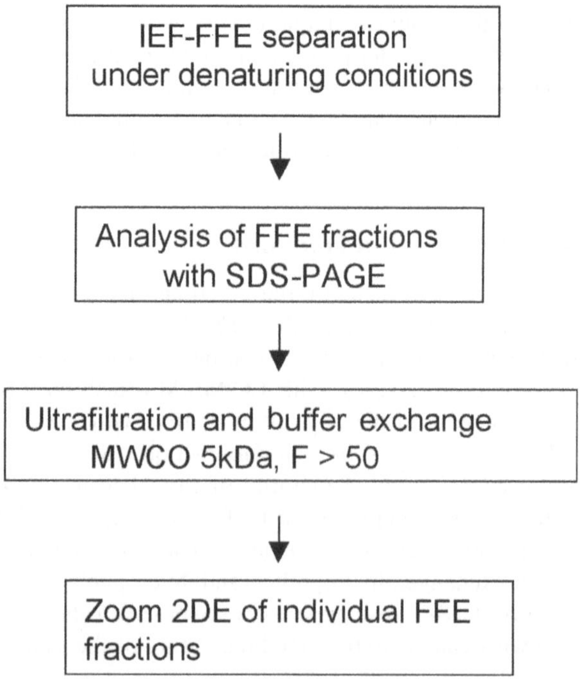

FIGURE 11.4 A diagram outlining a rat liver proteome fractionation with IEF-FFE and subsequent zoom 2DE of individual FFE fractions (see text).

appropriate FFE separation buffer, 30 mg of protein was subjected to the FFE to ensure sufficient amounts of fractionated proteins for subsequent 2DE analysis. Continuous nature of FFE separation allowed for large amounts of protein in relatively large volume (10 ml) to be separated.

TABLE 11.1 Experimental Conditions for a Rat Liver Proteome Fractionation

	Sample Preparation	
Total protein concentration in original lysate		30 mg/ml
Concentration after dilution with FFE buffer		3 mg/ml
Centrifugation to remove insoluble components		30 min. @ 25,000 rpm @ 18°C
Total volume infused into FFE for fractionation		10 ml
Rate of sample infusion		6 mg/hr
	IEF-FFE conditions	
	Inlets	*Composition*
Anode stabilization buffer	1, 2	100 mM H_2SO_4, 7 M urea, 2 M thiourea, 250 mM mannitol
BD FFE Reagent 1 buffer	3, 4	BD FFE Reagent 1 (30% w/w), 7 M urea, 2 M thiourea, 250 mM mannitol, pH 4.41
BD FFE Reagent 2 buffer	5, 6, 7	BD FFE Reagent 2 (50% w/w), 7 M urea, 2 M thiourea, 250 mM mannitol, pH 7.38
BD FFE Reagent 3 buffer	8, 9	BD FFE Reagent 3 (37% w/w), 7 M urea, 2 M thiourea, 250 mM mannitol, pH 9.73
Cathode stabilization buffer	10–13	100 mM NaOH, 7 M urea, 2 M thiourea, 250 mM mannitol
Counterflow		7 M urea, 2 M thiourea, 250 mM mannitol
Flow rate of separation buffers		62 g/hr
Voltage applied		1200 V
Current under equilibrium		22 mA
Temperature		10°C
	Post-FFE Treatment of Fractions	
Average volume of each fraction		10 ml
Ultrafiltration device MW cutoff		5 kDa
Concentration factor		$F > 50$
FFE buffer exchanged to		IPG rehydration buffer
Final volume and protein concentration before loading on IPG		1 ml, 2 mg/ml

In this experiment, increased solubility provided by a combination of thiourea and urea and CHAPS allowed for high recovery of proteins from FFE across the entire pH range and a high loading capacity of 6 mg / hr without significant precipitation.

In total, 10 ml of each fraction was collected. Every other fraction from a collection 96-well plate was analyzed by sodium dodecyl sulfate–polyacrylamide gel electrophoresis (SDS-PAGE) and visualized with Silver Stain. As evident from the gel images in Figure 11.5, FFE-IEF resulted in efficient separation of the original complex protein sample into a number of fractions with unique protein content.

For further analysis of FFE fractions by 2DE, it is necessary not only to concentrate the fractions, but also to replace BD EEF Reagents because they have high buffering capacity and prevent efficient focusing by IPG. Ultrafiltration devices are ideally suited for achieving both of these goals at the same time. In this experiment, 6 ml ultrafiltration devices with a molecular weight cutoff of 5 kDa (Vivascience, Hannover, Germany) were used. FFE fractions were concentrated more than 50-fold and exchanged with the typical IPG loading buffer. The final volume of FFE fractions prior to 2DE was approximately 0.2 ml and contained 0.1 to 1mg of total protein.

Figure 11.6 shows a panel of zoom 2DE images obtained from 20 FFE fractions. As expected, each fraction generated a unique pattern slightly overlapping with adjacent fractions. Reduced complexity of individual FFE fractions, and sufficient amounts of protein fractionated during a continuous flow separation, allowed for detection of significantly increased number of spots compared to crude extract 2DE analysis. When these multiple overlapping zoom IPG gels were used to generate an electronic 2DE protein map of rat liver proteome, the total number of unique features quantified in this master gel between pH 3.5 and pH 7.5 was 15,457 by Sypro™ Ruby staining, whereas a 2DE image (pH 3–10) of crude liver extract obtained with the same staining resolved only 2,033 features.[24]

In summary, the combination of FFE prefractionation with zoom 2DE analysis of FFE fractions results in dramatically increased resolution of detectable features and offers a practical approach to accessing lower-abundance proteins in complex samples.

11.4.2 Plasma Proteome Fractionation

The dynamic range of proteins present in human plasma is extremely high ($>10^{10}$), preventing differential analysis of whole plasma preparations.[25] Approximately 50% (30–50 mg/ml) of the total protein content of human plasma is albumin. Many proteins that have important functions in serum, such as immunoglobulins, complement components, and macroglobulins, are also present at milligrams per milliliter levels. In fact, 7 to 12 proteins make up more than 90% of the protein bulk present in human serum.[26] In contrast, many tissue leakage proteins and proteins secreted by tumors (i.e., potential biomarkers) are most likely present at low nanograms per milliliter to picograms per milliliter levels.[27] Cytokines, the rarest proteins measured clinically, are present in low- to sub-picograms per milliliter quantities.[28] Identification of protein biomarkers that are likely to be in low abundance is of primary importance to the study and diagnosis of many human disease states.

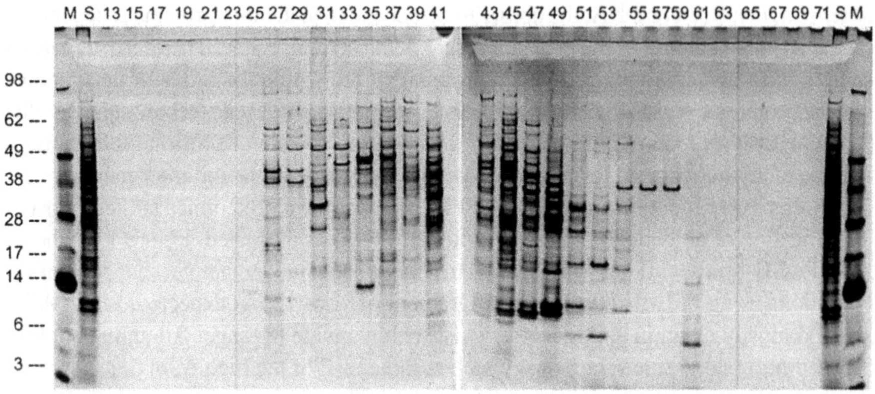

FIGURE 11.5 A rat liver proteome separation using FFE in IEF mode. Silver-stained SDS-PAGE of every other FFE fraction is shown. M, protein molecular weight markers; S, original liver sample before fractionation. Numbers above the gel indicate FFE fractions. Numbers on the left side of the gel indicate molecular weight of marker proteins. Seven microliters of an FFE fraction is loaded per lane.

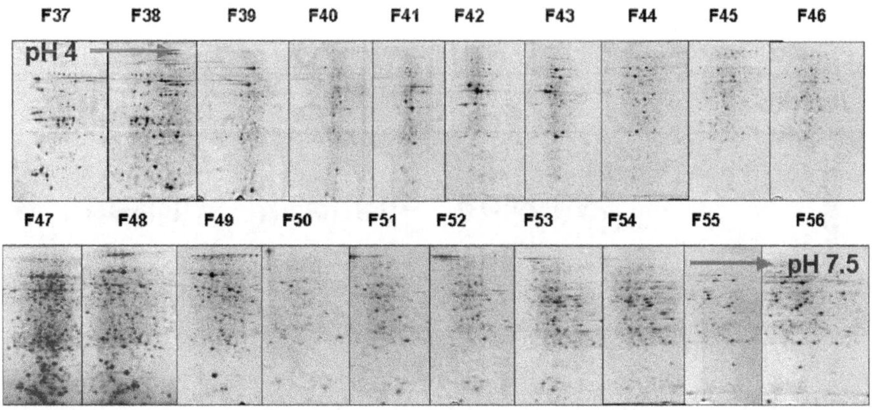

FIGURE 11.6 A panel of narrow pH range zoom 2DE images obtained from 20 FFE fractions (silver stain) from a rat liver fractionation (see text).

Current approaches to reduce the dynamic range combine orthogonal fractionation methods such as chromatography, immunoaffinity subtraction, preparative IEF, or precipitation with 2DE and MS. FFE distinguishes itself from other fractionation technologies such as chromatography or gel-based electrophoresis in several ways. The most significant differences are the matrix-free fractionation principle that allows high sample recoveries and the continuous flow system that allows sample loads virtually unlimited by volume.

One of the primary obstacles to proteomic analysis of plasma or serum is the low concentration of new potential biomarker proteins. Because proteins cannot be amplified *in vitro,* as can RNA or DNA, the ability to detect and then identify proteins is dependent on the absolute amount of the protein present in the serum sample. Therefore, the initial sample must contain amounts of each protein that are within the limits of detection of current proteomic technologies. Since the most sensitive proteomic technologies for differential analysis require high picogram to nanogram protein quantities, the minimum volume of serum required for biomarker discovery will be in the milliliter range.

In this experiment, the utility of the FFE as a part of a proteomics strategy for the identification of potential biomarkers in human serum is evaluated. Figure 11.7 shows SDS-PAGE profiles of the human serum fractionated under native and denaturing conditions using FFE-IEF with a linear gradient pH 2.5–12. As expected, SDS-PAGE showed mostly albumin and IgGs that exist within a wide pI range. Albumin is tightly focused into two to three fractions, which is indicative of the high resolving power of this technology. But, at the same time, the risk of albumin precipitation with increased serum loads is augmented due to the high local concentration of albumin created within narrow lane formed under the linear pH gradient separation. The presence of 7 M urea, 2 M thiourea, and 1% CHAPS (denaturing conditions) elevate the threshold of albumin precipitation and allow safe sample loading rates of up to 30 mg/hr. However, this semipreparative scale will not be sufficient to effectively process large volumes of serum for the discovery of low-abundant proteins.

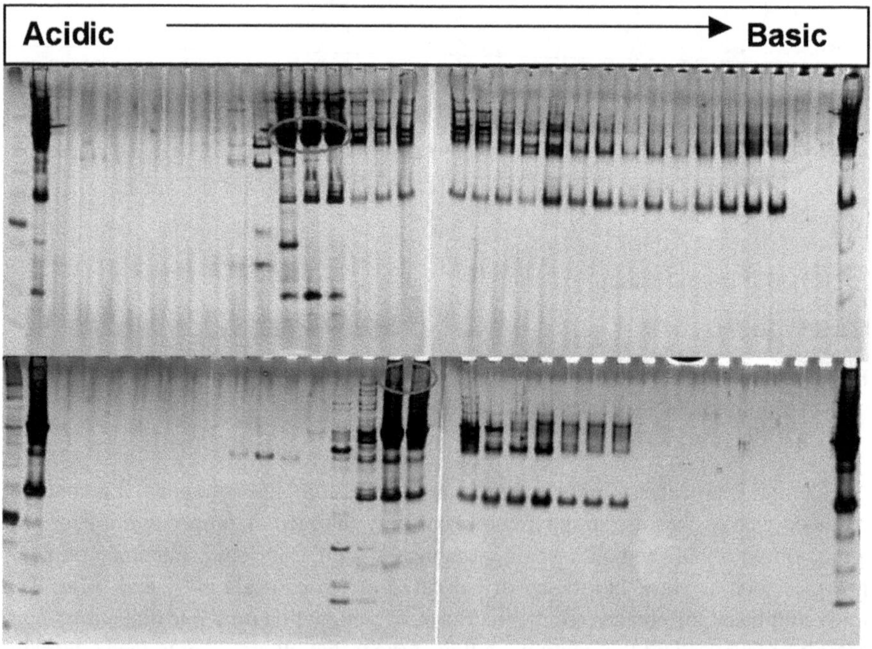

FIGURE 11.7 SDS-PAGE profiles of the human serum fractionated under native (upper panel) and denaturing (lower panel) conditions using FFE–IEF with a linear gradient pH 2.5–12. Oval indicates albumin-containing fractions.

To address the problem, a new, stepwise gradient has been designed and optimized that minimizes albumin precipitation by flattening the pH range around albumin pI (Figure 11.8) and eluting the protein into 12 to 15 fractions instead of 2 or 3. As a result, plasma or serum is effectively fractionated into at least three main pools: acidic pool (all the proteins that focus on the acidic side of human serum albumin [HSA]), an HSA pool, and a basic pool. When performed under native conditions and in a matrix-free environment characteristic of the FFE, this method can be considered a novel albumin segregation technique, with the advantages of near-full recovery of protein material and virtually unlimited loading capacity.

While these segregation strategies do not increase the concentration of low-abundance proteins, they effectively increase their ratio in a fraction, thereby facilitating the analysis of greater numbers of proteins in a single experiment. In order to demonstrate the full advantage of an all-liquid HSA segregation and enrichment for lower abundant proteins in plasma, the experiment schematically outlined in Figure 11.9 was designed. In essence, three plasma protein pools are generated after the first fractionation under native conditions. Once sufficient amounts of protein in each pool are collected under continuous flow fractionation, these pools are concentrated and separately reapplied to the FFE separation under denaturing conditions. The resulting multiple fractions can now be analyzed with gel electrophoresis followed by MS analysis.

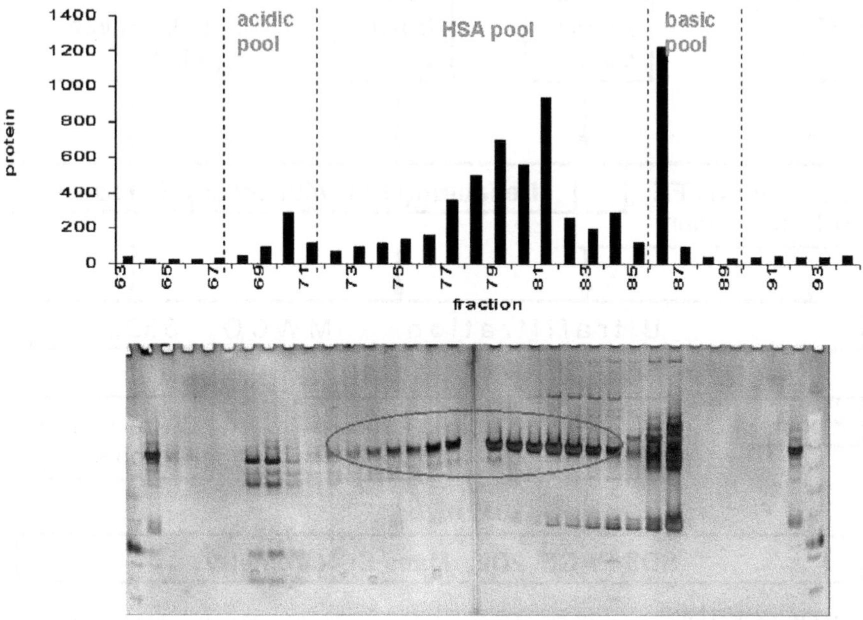

FIGURE 11.8 Lower panel: SDS-PAGE profile of the human serum fractionated under native conditions using FFE-IEF with a nonlinear, step gradient. Oval indicates albumin-containing fractions. Upper panel: protein assay results by Bradford corresponding to the fractions shown on the gel. Dashed lines indicate boundaries of pooled fractions.

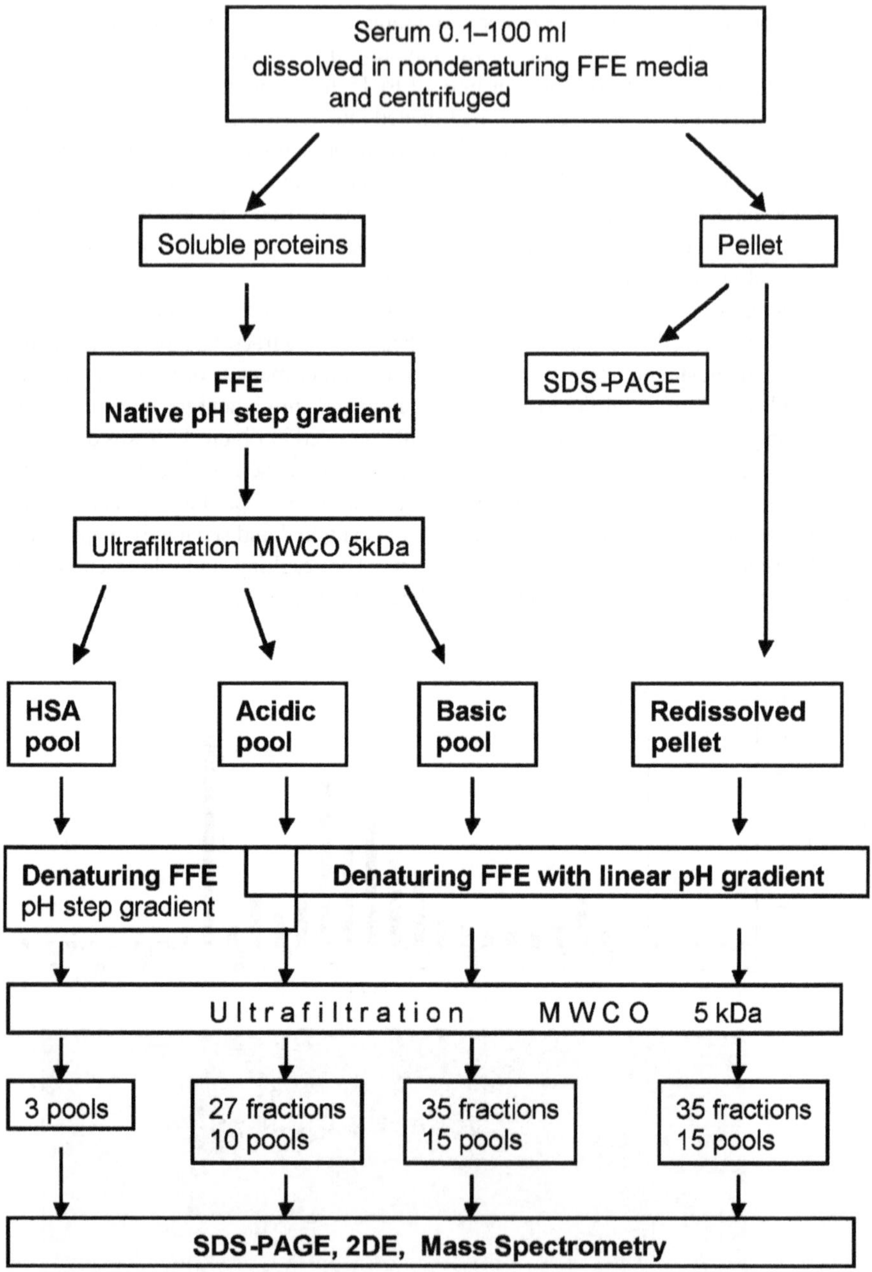

FIGURE 11.9 A diagram outlining the experiment on albumin separation and low-abundant proteins enrichment in human serum. Nonlinear pH gradient under native conditions is applied to a soluble serum to achieve fractionation into three main fractions (acidic pool, albumin pool, and basic pool). These pools can be analyzed by 2DE and mass spectrometry or subjected to further FFE-IEF under denaturing conditions to achieve higher resolution.

Figure 11.10 offers an excellent illustration of the power of native fractionation of plasma using a new step gradient of pH. Three pools (acidic, HSA, and basic) obtained after native FFE fractionation of soluble plasma proteins display clearly unique band patterns. Fraction 55 is on the borderline between HSA and basic pools and thus has been processed separately from both pools. The last lane on the SDS-PAGE represents insoluble plasma protein fraction pelleted by centrifugation of plasma after dilution with the FFE separation buffer.

The native plasma fractionation using the stepwise pH gradient offers a unique practical approach to albumin separation due to high resolving power of the FFE-IEF. In addition to clean separation of acidic and basic protein pools from albumin, a possibility for considerable enrichment of lower-abundant proteins can be observed by comparing the original serum lane with those of the five fractions including the insoluble one. The high quality of albumin segregation (>95%) and significant enrichment of lower-abundant proteins is even better illustrated with higher resolution 2DE analysis of respective pools (Figure 11.11). It should be noted that the above mentioned results were obtained only after the first FFE-IEF plasma fractionation under native conditions (Figure 11.9) and that even better resolution and enrichment can be achieved by subjecting isolated pools to a second round of FFE, this time under denaturing conditions, as indicated in Figure 11.9. In case of albumin pool, this type of double treatment with FFE-IEF can potentially yield access to proteins and peptides that form complexes with albumin under native conditions and thus focus in the same pH range.

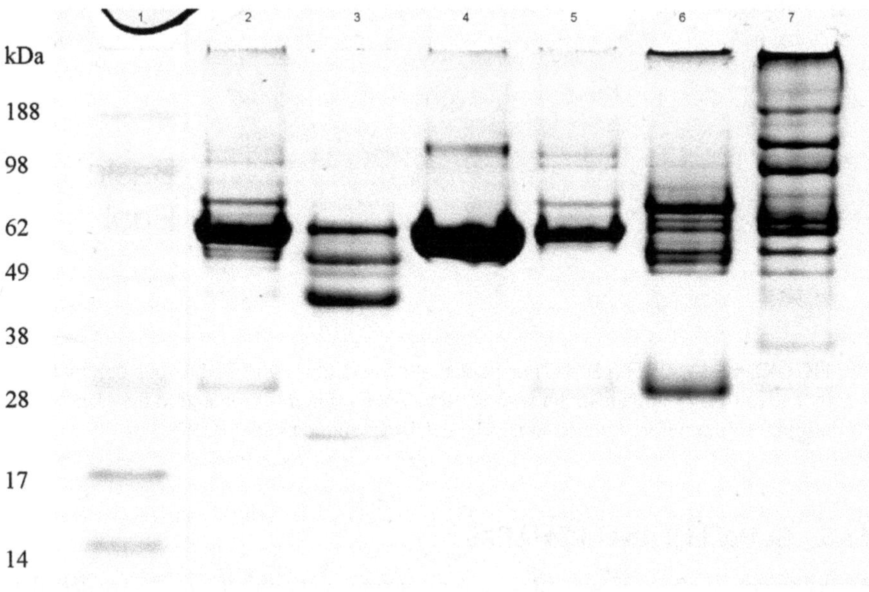

FIGURE 11.10 SDS-PAGE (Coomassie stain) of pooled and concentrated FFE fractions obtained by native FFE of human serum with nonlinear pH gradient. (1) Molecular weight standards; (2) original serum before fractionation; (3) acidic pool; (4) albumin pool; (5) FFE fraction 55; (6) basic pool; (7) insoluble serum fraction.

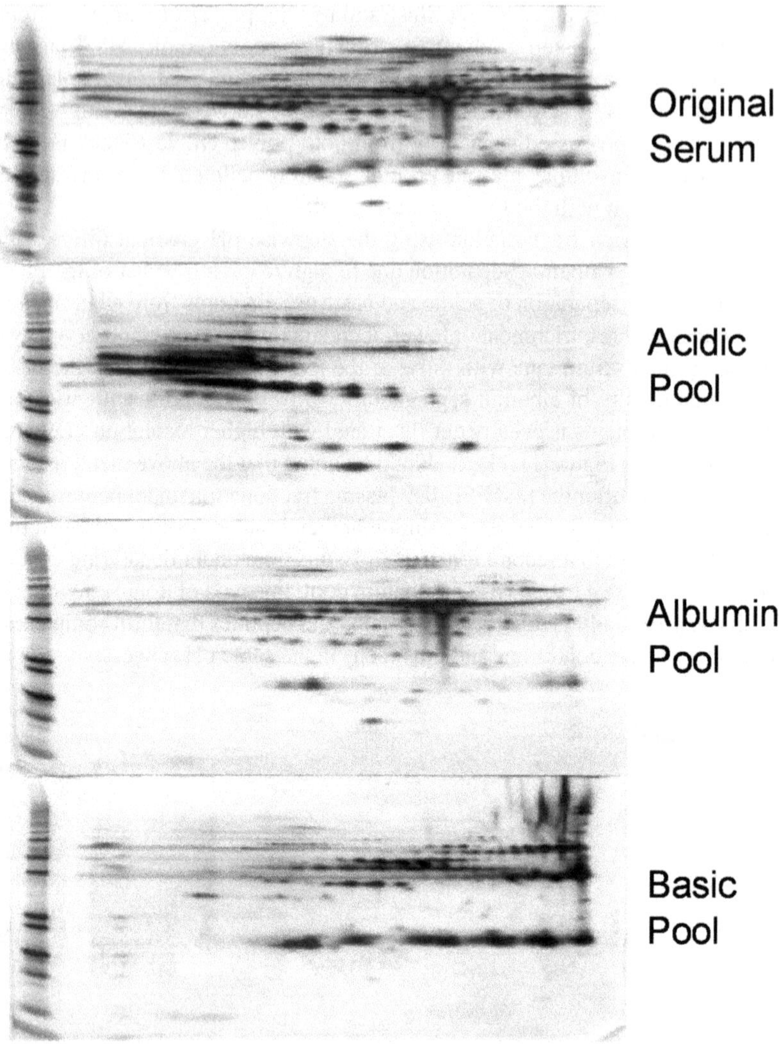

FIGURE 11.11 2DE (pH 4–7) of pooled and concentrated FFE fractions obtained by native FFE of human serum with nonlinear pH gradient, compared to the 2DE of original serum before fractionation. The gels were stained with Silver.

11.5 CONCLUDING REMARKS

A recent compilation of several serum proteome mapping studies[26] revealed that approximately 1000 different proteins can be identified in serum using a combination of three proteomic strategies. Less than 5% of the proteins identified were found in all three methods, suggesting that many complementary methods will be needed to construct a comprehensive map of the human serum proteome. This conclusion reflects the growing understanding in the proteomics community that due to the complexity of protein analysis, no single technology platform is capable of providing comprehensive answers to the task of identifying all proteins. Every approach has its own specific advantages, and a complementary combination of many is needed to build a full catalog of proteins within a particular proteome.

BD Free Flow Electrophoresis system is quickly becoming a recognized tool for proteomics research because of the unique capabilities and advantages it offers. Unlimited loading capacity combined with matrix-free high-resolution separation allows the most challenging limitations in proteomics such as access to low-abundant proteins and depletion of highly abundant proteins to be addressed. Furthermore, FFE is fully compatible with other analytical methods in proteomics and thus is well positioned to both serve as a separation technique on its own or to add another dimension to any other approach.

REFERENCES

1. Roman, M.C. and Brown, P.R., Free-flow electrophoresis, *Anal. Chem.*, 66, 86–94, 1994.
2. Krivankova, L. and Bocek, P., Continuous free-flow electrophoresis, *Electrophoresis*, 19, 1064–1074, 1998.
3. Bauer, J., Advances in cell separation: Recent developments in counterflow centrifugal elutriation and continuous flow cell separation, *J. Chromatogr. B*, 722, 55–69, 1999.
4. Pasquali, C., Fialka, I., and Huber, L.A., Subcellular fractionation, electromigration analysis and mapping of organelles, *J. Chromatogr. B*, 722, 89–102, 1999.
5. Hannig, K. and Heidrich, H.G., *FreeFlow Electrophoresis*, GIT Verlag, Darmstadt, Germany, 1990.
6. Kuhn, R. and Wagner, H., Free flow electrophoresis as a method for the purification of enzymes from *E. coli* cell extract, *Electrophoresis*, 10, 165–172, 1989.
7. Burggraf, D., Weber, G., and Lottspeich, F., Free flow–isoelectric focusing of human cellular lysates as sample preparation for protein analysis, *Electrophoresis*, 16, 1010–1015, 1995.
8. Hoffstetter-Kuhn, S., Kuhn, R., and Wagner, H., Free flow electrophoresis for the purification of proteins. I. Zone electrophoresis and isotachophoresis, *Electrophoresis*, 11, 304–309, 1990.
9. Wagner, H. and Mang, V., Preparative free flow isotachophoresis, in *Analytical and Preparative Isotachophoresis*, Holloway, C.J., Ed., Walter de Gruyter, Berlin, 1984, pp. 357–363.
10. Wagner, H., Free-flow electrophoresis, *Nature*, 341, 669–670, 1989.
11. Hoffmann, P., Ji, H., Moritz, R.L., Connolly, L.M., Frecklington, D.F., Layton, M.J., Eddes, J.S., and Simpson, R.J., Continuous free-flow electrophoresis separation of cytosolic proteins from the human colon carcinoma cell line LIM 1215: A non two-dimensional gel electrophoresis-based proteome analysis strategy, *Proteomics*, 1, 807–818, 2001.
12. Weber, G. and Bocek, P., Recent developments in preparative free flow isoelectric focusing, *Electrophoresis*, 19, 1649–1653, 1998.
13. Weber, P.J.A., Weber, G., and Eckerskorn, C., Isolation of organelles and prefractionation of protein extracts using free-flow electrophoresis, in *Short Protocols in Protein*

Science, Coligan, J.E., Dunn, B.M., Speicher, D.W., Wingfield, P.T., Eds., Wiley & Sons, New York, 2003, pp. 22.5.1–22.5.21.

14. Weber, P.J.A., Weber, G., and Eckerskorn, C., Protein purification using free flow electrophoresis, in *Purifying Proteins for Proteomics: A Laboratory Manual*, Simpson, R.J., Ed., Cold Spring Harbor Laboratory Press, New York, 2004.
15. Weber, G. and Bocek, P., Optimized continuous flow electrophoresis, *Electrophoresis*, 17, 1906–1910, 1996.
16. Bier, M., Ostrem, J., and Marquez, R.B., A new buffering system and its use in electrophoresis and isoelectric focusing, *Electrophoresis*, 14, 1011–1018, 1993.
17. Perdew, G.H., Schaup, H.W., and Selivonchick, D.P., The use of zwitterionic detergent in two-dimensional gel electrophoresis of trout liver microsomes, *Anal. Biochem.*, 135, 453–455, 1983.
18. Molloy, M.P., Herbert, B., Walsh, B.J., Tyler, M.I., Traini, M., Sanchez, J.C., Hochstrasser, D.F., Williams, K.L., and Gooley, A.A., Extraction of membrane proteins by differential solubilization for separation using two-dimensional electrophoresis, *Electrophoresis*, 19, 837–844, 1998.
19. Crawford, F., Kuchumov, A., Gillece-Castro, B., and Mullan, M., Mouse Brain Proteins Were Identified by Coupling A Free Flow Electrophoresis (FFE) Fractionation with Q-TOF LC/MS/MS Analyses, paper presented at the 52nd American Society for Mass Spectrometry Conference, Nashville, May 23–27, 2004.
20. Alpert, A.J., Hydrophilic-interaction chromatography for the separation of peptides, nucleic acids and other polar compounds, *J. Chromatogr.*, 499, 177–196, 1990.
21. Corthals, G.L., Wasinger, V.C., Hochstrasser, D.F., and Sanchez, J.C., The dynamic range of protein expression: A challenge for proteomic research, *Electrophoresis*, 21, 1104–1115, 2000.
22. Cordwell, S.J., Nouwens, A.S., Verrils, N.M., Basseal, D.J., and Walsh, B.J., Subproteomics based upon protein cellular location and relative solubilities in conjunction with composite two-dimensional gels, *Electrophoresis*, 21, 1094–1103, 2000.
23. Wildgruber, R., Harder, A., Obermaier, C., Boguth, G., Weiss, W., Fey, S.J., Larsen, P.M., and Gorg, A., Towards higher resolution: 2D-Electrophoresis of *S. cerevisiae* proteins using overlapping narrow IPGs, *Electrophoresis*, 21, 2610–2616, 2000.
24. Eckerskorn, C., Weber, G., Weber, P.J.A., Kussian, R., Sukop, U., Korner, S., Shadid, R., Schneider, U., Nissum, M., Kuhfuss, S., Mast, M., Obermaier, C., Hauptmann, M., Wildgruber, R., Rein, K., and Posch, A., Pushing the Limits of Protein Analytic through 3-D Electrophoresis, paper presented at the 51st American Society for Mass Spectrometry Conference, Montreal, May 23–27, 2003.
25. Anderson, N.L. and Anderson, N.G., The human plasma proteome: History, character, and diagnostic prospects, *Mol. Cell. Proteomics*, 1, 845–867, 2002.
26. Anderson, N.L., Polanski, M., Pieper, R., Gatlin, T., Tirumala, R.S., Conrads, T.P., Veenstra, T.D., Adkins, J.N., Pounds, J.G., Fagan, R., and Lobley, A., The human plasma proteome: A non-redundant list developed by combination of four separate sources, *Mol. Cell. Proteomics*, 3, 311–326, 2004.
27. Perkins, G., Slater, E., Sanders, G., and Prichard, J., Serum tumor markers, *Am. Fam. Physician*, 68, 1075–1082, 2003.
28. Keller, E., Wanagat, J., and Ershler, W.B., Molecular and cellular biology of Interleukin-6 and its receptor, *Frontiers Biosci.*, 1, 340–357, 1996.

Part III

Applications of Electrophoresis in Proteomics

12 Destreaking Strategies for Two-Dimensional Electrophoresis

Fengju Bai, Sheng Liu, and Frank A. Witzmann

CONTENTS

12.1 TWO-DIMENSIONAL GEL ELECTROPHORESIS (2DE) AND BASIC END STREAKS

Two-dimensional gel electrophoresis (2DE) is a powerful technique that resolves complex protein mixtures in the first dimension by isoelectric point (pI) and in the second dimension by molecular weight. The introduction of immobilized pH gradient (IPG) strips has significantly improved 2DE separation in a reproducible manner.[1] However the number of the proteins that can be resolved and visualized by 2DE is limited.[2] A large format gel (20 × 25 cm) can only portray approximately 1500 to 2000 protein spots, while a mammalian cell probably contains more than 20,000 protein species, and a single mammalian tissue may represent a mixture of more than 50,000 protein species.[3] The limitations in protein detection and quantification are mainly due to the huge dynamic range of cellular protein expression and the presence of poorly resolved protein smears or streaks instead of distinct protein spots in the gel. While strategies have been employed to address the issue of dynamic range, for example, the application of various sample complexity reduction techniques (subcellular fractionation, solution-phase isoelectric focusing [IEF], affinity enrichment or depletion, and chromatographic separation) and the application of more sensitive gel staining techniques,[4] efforts have also been made to reduce or eliminate protein streaks.

12.2 CHEMICAL NATURE OF PROTEIN STREAKING

These smears and streaks are proteins cross-linked through the formation of intramolecular and intermolecular disulfide bridges (R_1–S–S–R_2, when R_1 = R_2 represents intramolecular disulfides) following the oxidation of the cysteinyl thiol groups (–SH).[5,6] Therefore utilizing a reductant in sample solubilization buffer, such as the most commonly used dithiothreitol (DTT) or its isomer dithioerythritol (DTE), is a common practice to reduce disulfide bond and to help unfolding and disaggregating the protein prior to the first dimension IEF. However DTT and DTE are weak acids that can migrate toward the anode during IEF, leading to the loss of the reducing agent from the basic portion of the IPG strip.[7,8] In an environment lacking a reducing agent, the highly reactive cysteinyl thiol groups tend to cross-link again, leading to the spontaneous restoration of disulfide bridges and therefore the presence of a highly streaky basic end (pH $>$ 7). The electrochemical oxidation at cathode may also facilitate the oxidation event.

12.3 DESTREAKING TECHNIQUES

Various attempts have been made to reduce the basic pH range streaks, yet achieving an optimal IEF in the alkaline region remains a challenge. For example, decreasing the protein sample concentration, anodic cup-loading, shortening IEF duration, addition of a DTT reservoir at the cathode to replenish DTT, or using an alternative reducing agent such as hydroxyethyldisulphide (HED) to form mixed disulfides with cysteinyl thiols have all been attempted.[7–9] Strategies such as decreasing protein loading compromise the detection sensitivity because low-abundant protein signals are lost and is not efficient since the primary reason for the generation of streaks (disulfide bridge formation) has not been addressed. Continuously supply of DTT to the basic end is rather variable due to the dynamic influx of DTT during IEF from sample to sample. The combination of several of these aforementioned techniques can achieve a better result, yet it is still not 100% effective.[7,8] Figure 12.1 shows the silver stained 2D images of human brain basic proteins with various combinations of the destreak efforts. The streaks are still apparent in these gels.

An alternate approach is to prepare samples in a way that prevents the regeneration of the cross-links between cysteinyl thiols by a covalent modification of the free thiol groups through alkylation.[6,10] This modification is commonly accomplished through a reduction and alkylation process using DTT to reduce the disulfide bonds to free thiol groups (Figure 12.2, I) and iodoacetamide or acrylamide derivatives (such as *N,N*-dimethylacrylamide [DMA]) to alkylate free thiols (Figure 12.2, II).[11,12] Unfortunately both iodoacetamide and acrylamide can change the gel patterns or generate extra artifactual spots in 2DE maps.[7] The problem concerns the spurious alkylation of peptides. Both iodoacetamide and acrylamide can react with other amino acid residues (e.g., lysine), leading to the formation of multiple alkylating products with various pI or molecular weight. In a study by Luche et al., matrix-assisted laser-desorption ionization mass spectrometry (MALDI MS)

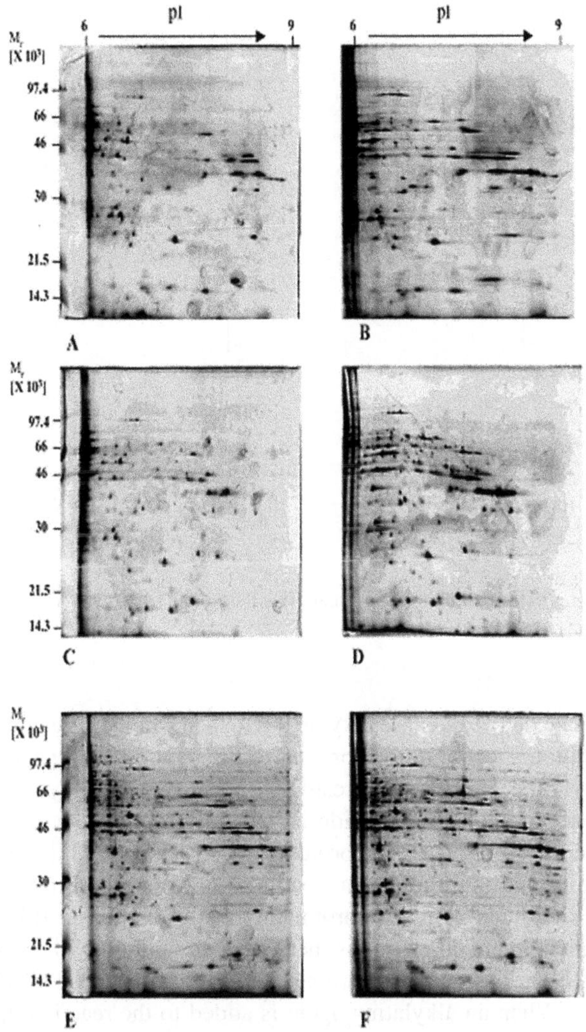

FIGURE 12.1 2D images (silver stain, 18 × 20 cm) of human brain basic proteins with various combinations of the destreaking efforts. IEF was performed on IPG 6-9 IPG strips. Samples were applied to IPG using either in-gel rehydration (A, C, E) or anodic cup-loading (B, D, F). IPG strips were rehydrated in either 8 M urea, 0.5% CHAPS, 0.2% DTT, and 0.2% Pharmalyte pH 3-10 (A, B, C, D) or Destreak Rehydration Solution containing HED (E, F) overnight. (C, D) A wick soaked in 15mM DTT was applied at the cathodic end. Adapted from Pennington et al.[8] with permission from *Proteomics*.

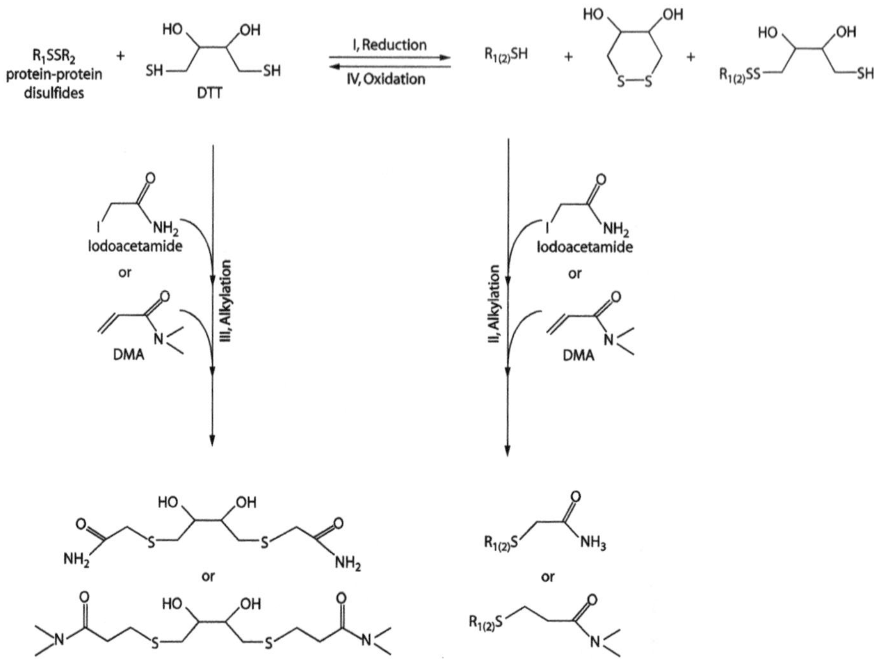

FIGURE 12.2 Reduction and alkylation reactions using DTT and DMA.

analysis of malate dehydrogenase alkylated with acrylamide detected five apparent spurious alkylation spectra, corresponding to the peptides indigestible by trypsin due to the alkylation on the lysine residue.[13] Moreover the alkylation of the free thiols by iodoacetamide or acrylamide is generally incomplete, as no more than 80% of all the thiol groups in the protein can be derivatized.[11,14] The formation of intramolecular and intermolecular disulfides between these residual free thiol groups is still significant, thus compromising resolution during IEF. The primary reason for an incomplete alkylation is the excess amount of DTT (~20-fold molar excess) in the 2DE protein solubilization buffer, which is needed to achieve a complete reduction. When an alkylating agent is added to the reaction, the remaining DTT in the sample competes with the cysteinyl thiols to react with the alkylating agent (Figure 12.2, III), driving the reaction toward the regeneration of disulfide cross-links (Figure 12.2, IV).

Studies have been carried out to compare the destreaking efficiency and sufficiency of different types of cysteine-blocking agents, when the reactions were carried out at various pH ranges or for different length of reaction time (see Figure 12.3). Organic disulfides, such as dithiodiethanol (DTDE) and dithiodiglycerol, were shown to be the most effective among tested compounds in improving protein resolution on 2D gels, but the effect cannot be extended to basic proteins due to a combination of the incomplete blocking of cysteinyl thiols (~85% derivatization) and the consumption

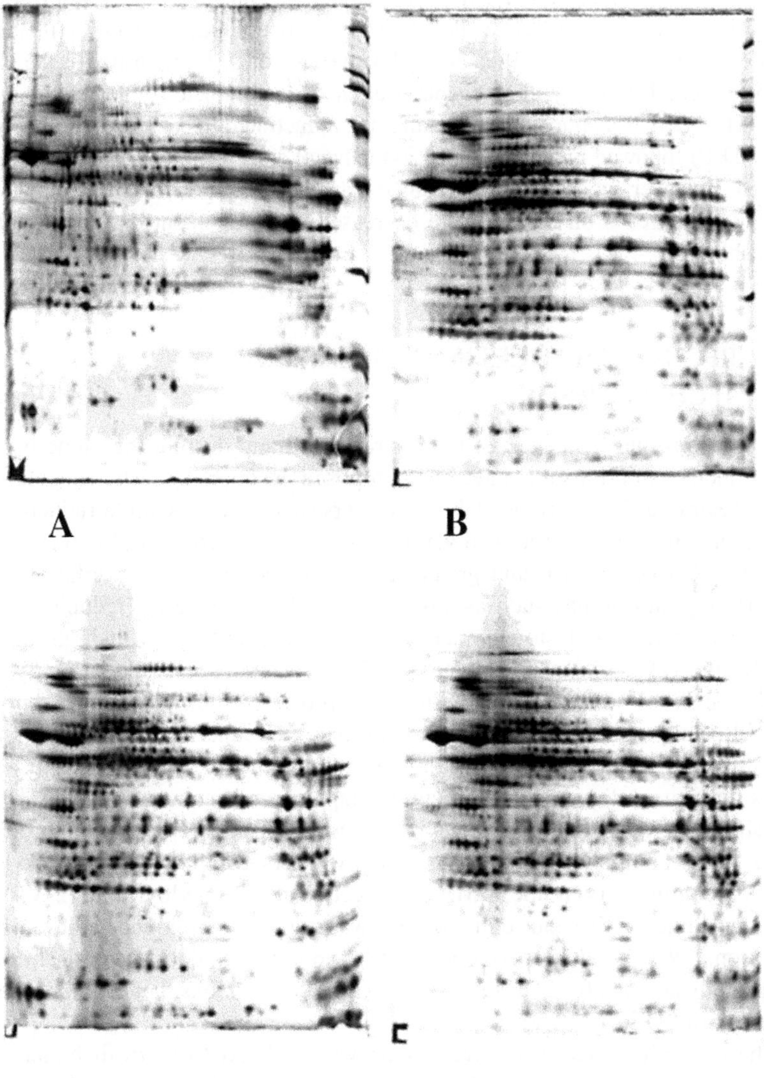

FIGURE 12.3 Cysteine blocking with DTET and maleimide derivatives. Bovine mitochondrial proteins were separated by 2DE. First dimension: linear pH 3.75–10.5 IPG. Equilibration after IPG was performed using the DTT-iodoacetamide two-step method. Second dimension: 10% T gel. Detection was performed with silver staining. Proteins were reduced (A) in 5 mM TBP separated run in a gel containing 100 mM DTDE; (B) with 5 mM tributylphosphine and alkylated prior to IPG with 20 mM methylmaleimide for 6 hours at pH 6; (C) as described in B, but alkylation was performed with ethyl maleimide; (D) as described in B, but alkylation was performed with maleic hydrazide. Adapted from Luche et al.[13] with the permission from *Proteomics*.

of organic disulfides at cathodic end.[13] A compatible enhancement in spot resolution was observed when less-reactive alkylating agents, maleimide derivatives (methylmaleimide, ethyl maleimide, and maleic hydrazide), were used. Interestingly, the alkylation of cysteinyl thiols by these agents was far from complete (~30%).[13] In another study, alkylation was almost complete after the reduction of lysozyme by tributylphosphine (TBP) followed by alkylation by iodoethanol at 37°C, pH 10, after a 60-minute reaction. However, when using this reduction and alkylation method to treat rat serum for 2DE, the streaks on the gel were decreased to some extent but not completely diminished.[15]

12.4 RECENT ADVANCES IN DESTREAKING TECHNIQUES

As mysterious and persistent as these streaks are and despite the numerous partially effective solutions that have been proposed, a simple but efficient solution is highly desirable. Recently our laboratory experimented on sample reduction and alkylation using tris(2-carboxyethyl)-phosphine hydrochloride (TCEP) and vinylpyridine (VP) before IEF and greatly improved the basic end resolution in 2D gels. Trialkylphosphines such as TBP and TCEP are powerful reducing agents, which can readily and stoichiometrically reduce disulfides with high specificity.[16] Unlike DTT, trialkylphosphines do not react with some alkylating agents.[16] It has been shown that TBP greatly improves protein solubility when used prior to 2DE.[17] Because of its solubility, odor, and toxicity, reduction with TCEP is more favorable than with TBP. VP is one of several commonly used protein-alkylating agents for protein digests prior to peptide mapping and sequence analysis.[10] It has been shown to react with the cysteinyl thiols with 100% specificity, and the alkylation is 100% complete.[18] Sebastiano and coworkers compared several alkylating agents, including acrylamide, DMA, iodoacetic acid, 2-vinylpyridine (2-VP), and 4-vinylpuridine (4-VP) and found that VP was the only compound achieved 100% alkylation.[18] We therefore developed a reduction and alkylation approach to prepare protein samples for 2DE to reduce basic pH end streaks, hence improving the resolution of 2DE.

The cytosolic fraction of liver tissue was isolated from fresh human liver (a generous gift from Dr. C. Max Schmidt, Departments of Surgery and Biochemistry and Molecular Biology, Indiana University School of Medicine, Cancer Research Institute) by differential centrifugation. Liver was homogenized in a buffer containing 0.25 M sucrose and 10 mM Tris-HCl, pH 7.4, and centrifuged at 100,000*g* for 45 minutes using a Beckman Type 45 Ti Rotor. Protein denaturation, solubilization, and reduction were performed in a portion of the supernate (cytosol) by the addition of urea, 3-[(3-cholamidopropyl)dimethylammonio]-1-propanesulfonate (CHAPS), and DTT. Carrier ampholyte (pH 3–10) was also added. The final concentrations for these reagents were 9 M urea, 4% CHAPS, 65 mM DTT, and 0.5% pH 3–10 ampholyte. Another portion of the supernate was subjected to the same denaturation, solubilization, and reduction process except TCEP was used instead of DTT. The final TCEP concentration was 10 mM. The sample that had been reduced by TCEP was alkylated with 1/20 volumes of 4-VP (400 mM) for 1 hour while vortexing. The reaction was quenched by the addition of the same volume of DTT (400 mM), which destroys excess VP. Frozen mouse brain (Harlan Sprague-Dawley, Indianapolis) was minced and homogenized in solubilization buffer containing 9 M urea, 4% CHAPS, 65 mM DTT, and 0.5% pH 3–10 ampholyte. Samples were centrifuged at 100,000*g* for 20 minutes using a Beckman TL-100 ultracentrifuge (Beckman Coulter, Fullerton, Calif.) to remove nucleic acid and insoluble materials, and the supernate was collected. Similar to the liver cytosolic proteins, reduction and alkylation with TCEP and VP were carried out for a portion of the brain protein lysate using the same solubilization buffer, but TCEP (10 mM) was used instead of DTT. Some of the alkylated and unalkylated sample was prefractionated by microscale solution IEF using the Zoom® IEF fractionator (Invitrogen, Carlsbad, Calif.). The protein fractions (alkylated and unalkylated) that were enriched in basic proteins (pI 7–10) were subjected to subsequent IEF on IPGs. In some experimental stages, a protein assay was performed using the RC DC Protein Assay kit (Bio-Rad, Richmond, Calif.) according to the manufacturer's protocol to determine protein concentration.

First-dimensional IEF was performed on IPG strips (pH 3–10, 24 cm, Bio-Rad, Hercules, Calif.). The protein lysate was diluted with rehydration buffer (8 M urea, 2% CHAPS, 15 mM DTT, 0.2% ampholytes pH 3–10). Rehydration of IPG strips was carried out overnight at room temperature. The proteins were focused at $\leq$50 μA/strip at 20°C, using progressively increasing voltage up to 10,000 V for a total of 100,000 Vh. Two 10-minute equilibration steps were carried out in equilibration buffer I (6 M urea, 0.375 M Tris, pH 8.8, 2% SDS, 20% glycerol, 2% DTT) and equilibration buffer II (6 M urea, 0.375 M Tris, pH 8.8, 2% SDS, 20% glycerol, 2.5% iodoacetamide), respectively, for the samples that went through the conventional DTT reduction right before the IPG strips were loaded on to the slab gels. A 20-minute equilibration step was carried out after IEF for those samples that had been treated with TCEP and VP using equilibration buffer III containing 6 M urea, 0.375 M Tris, pH 8.8, 2% SDS, and 20% glycerol. Second-dimension separation was accomplished on linear 11–19% acrylamide gradient slab gels (20 cm $\times$ 25 cm $\times$ 1.5 mm). Gels were run simultaneously for approximately 18 hours at 160 V and 8°C. Slab gels were stained using a colloidal Coomassie Brilliant Blue G-250 procedure.[19]

Alkylation with VP effectively eliminated the streaks in the basic pH range in 2DE maps. This destreaking method was efficient in various sample types, including total tissue lysate (Figure 12.4A, B) and protein lysates that had been prefractionated by subcellular fractionation (Figure 12.4C, D) or Zoom IEF fractionation (Figure 12.4E, F). The absence of streaks in the cytosolic protein fraction was not surprising (Figure 12.4C, D) because streaky proteins, often membrane proteins, are associated with such cellular organelles as endoplasmic reticulum (microsomes) and mitochondria. The remarkable efficiency of VP alkylation in streak reduction was

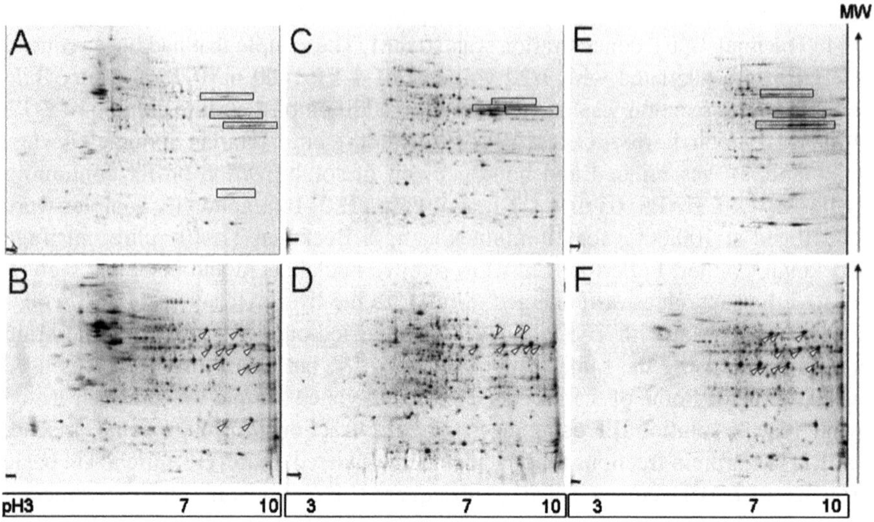

FIGURE 12.4 2DE patterns for various tissue protein lysates. (A) Total mouse brain lysate; (B) total mouse brain lysate alkylated with VP; (C) human liver cytosolic proteins; (D) human liver cytosolic proteins alkylated with VP; (E) mouse brain protein lysated-enriched in basic proteins obtained by Zoom IEF fractionation; (F) mouse brain protein lysated-enriched in basic proteins alkylated with VP.

further demonstrated in more complex samples, which either contained membrane protein or was enriched in basic proteins. The streaks shown in alkaline region of the 2DE map for the mouse brain protein lysate (Figure 12.4A) were fully eliminated in the corresponding sample that had undergone VP alkylation (Figure 12.4B). Furthermore, the basic protein-enriched fraction (pI 7–10) from Zoom IEF, which was shown as the protein streaks in the unalkylated sample (Figure 12.4E), was completely resolved in the 2DE map of the same sample that had been alkylated with VP (Figure 12.4F).

There are several advantages of using TCEP/VP reduction and alkylation for sample preparation prior to 2DE separation. This approach features a complete alkylation of the thiol groups in the protein (Figure 12.5). The reducing agent TCEP does not react with VP, therefore driving the alkylation reaction to 100% completion (Figure 12.5, II). VP is neither inhibited nor destroyed by neutral or zwitterionic surfactants and urea/thiourea, common components of the solubilization cocktails for 2DE.[19] Therefore, it is 2DE compatible. Unlike the charged alkylating agents, such as iodoacetamide, VP alkylation does not change the pI of a protein. The elimination of the streaks in the 2D gel will greatly increase the possibility of the identification and the characterization of those proteins undergoing quantitative (up–down regulation, presence, absence, etc.) or qualitative (posttranslational modification) changes in the basic pH range. The improvement in resolution also allows loading of more concentrated samples, therefore enhancing the intensity of

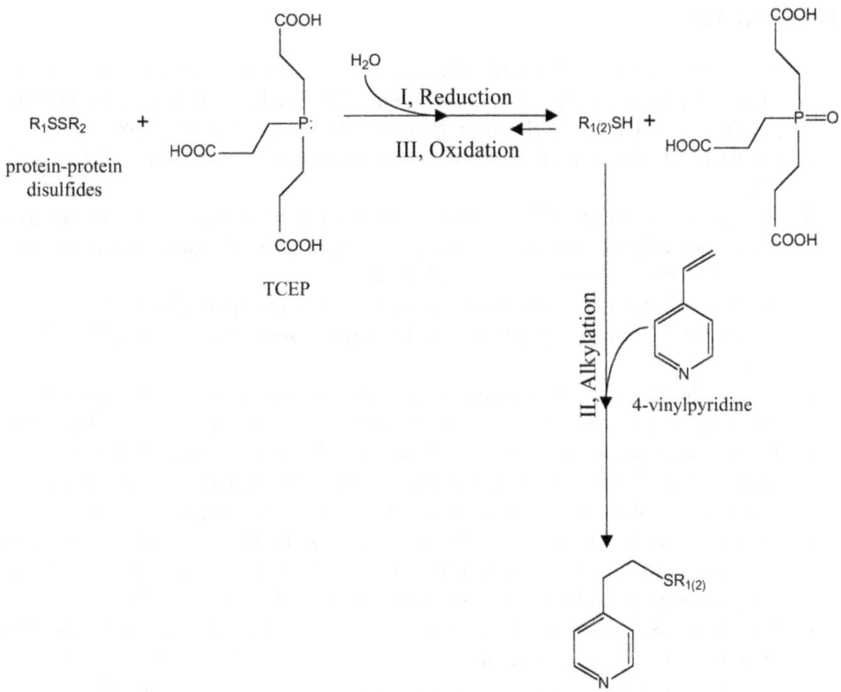

FIGURE 12.5 Chemical reactions of reduction and alkylation using TCEP and VP.

the faint protein spots and increasing the number of proteins that can be identified by peptide mass fingerprinting. For example, Figure 12.4B reflects a loading of approximately twice the amount of protein that was used in Figure 12.4A, yet the resolution is perfect and still with the potential to load more. In addition, since the alkylation has been accomplished with 100% efficiency in the sample preparation step before 2DE analysis, the reduction and alkylation steps before tryptic digestion and MS analysis can be skipped. This sample preparation method is also applicable to samples subjected to microscale solution IEF (Zoom IEF) prior to narrow pH range 2DE, in which streaking is a major problem for proteins in the basic pH fraction (Figure 12.4E, F). More importantly, it is a very simple approach, needing no special instrumentation.

Handling of VP in a fume hood is strongly recommended, due to its odor and air sensitivity. However, according to a recent summary of the toxicology data on pyridine derivatives that had been submitted to EPA by Reilly Industries, Inc.,[20] the weight of the evidence from *in vitro* studies indicates that 2-VP is negative for mutagenic activity. Repeated dose testing in rats found no systemic effects through histological analysis. Although a slight but statistically significant increase (~20%) in relative testes and ovary to body weight ratios were observed in rats following high dose repeated oral gavages, no gross pathology and histopathological effects were observed.

REFERENCES

1. Bjellqvist, B., Ek, K., Righetti, P.G., Gianazza, E., Gorg, A., Westermeier, R., and Postel, W., Isoelectric focusing in immobilized pH gradients: Principle, methodology and some applications, *J. Biochem. Biophys. Methods*, 6, 317–339, 1982.
2. Quadroni, M. and James, P., Proteomics and automation, *Electrophoresis*, 20, 664–677, 1999.
3. Zuo, X. and Speicher, D.W., Comprehensive analysis of complex proteomes using microscale solution isoelectrofocusing prior to narrow pH range two-dimensional electrophoresis, *Proteomics*, 2, 58–68, 2002.
4. Beranova-Giorgianni, S., Proteome analysis by two dimensional gel electrophoresis and mass spectrometry: Strengths and limitations, *Trends Anal. Chem.*, 22, 273–281, 2003.
5. Altland, K., Becher, P., Rossmann, U., and Bjellqvist, B., Isoelectric focusing of basic proteins: The problem of oxidation of cysteines, *Electrophoresis*, 9, 474–485, 1988.
6. Herbert, B., Galvani, M., Hamdan, M., Olivieri, E., MacCarthy, J., Pedersen, S., and Righetti, P.G., Reduction and alkylation of proteins in preparation of two-dimensional map analysis: Why, when, and how? *Electrophoresis*, 22, 2046–2057, 2002.
7. Olsson, I., Larsson, K., Palmgren, R., and Bjellqvist, B., Organic disulfides as a means to generate streak-free two-dimensional maps with narrow range basic immobilized pH gradient strips as first dimension, *Proteomics*, 2, 1630–1632, 2002.
8. Pennington, K., McGregor, E., Beasley, C.L., Everall, I., Cotter, D., and Dunn, M.J., Optimization of the first dimension for separation by two-dimensional gel electrophoresis of basic proteins from human brain tissue, *Proteomics*, 4, 27–30, 2004.
9. Hoving, S., Gerrits, B., Voshol, H., Muller, D., Roberts, R.C., and van Oostrum, J., Preparative two-dimensional gel electrophoresis at alkaline pH using narrow range immobilized pH gradients, *Proteomics*, 2, 127–134, 2002.
10. Simpson, R.J., *Proteins and Proteomics: A Laboratory Manual*, CSHL Press. Cold Spring Harbor, N.Y.
11. Galvani, M., Hamdan, M., Herbert, B., and Righetti, P.G., Alkylation kinetics of proteins in preparation for two-dimensional maps: a matrix assisted laser desorption/ionization-mass spectrometry investigation, *Electrophoresis*, 22, 2058–2065, 2001.
12. Mineki, R., Taka, H., Fujimura, T., Kikkawa, M., Shindo, N., and Murayama, K., In situ alkylation with acrylamide for identification of cysteinyl residues in proteins during one- and two-dimensional sodium dodecyl sulphate–polyacrylamide gel electrophoresis, *Proteomics*, 2, 1672–1681, 2002.
13. Luche, S., Diemer, H., Tastet, C., Chevallet, M., Van Dorsselaer, A., Leize-Wagner, E., and Rabilloud, T., About thiol derivatization and resolution of basic proteins in two-dimensional electrophoresis, *Proteomics*, 4, 551–561, 2004.
14. Hamdan, M., Bordini, E., Galvani, M., and Righetti, P.G., Protein alkylation by acrylamide, its N-substituted derivatives and cross-linkers and its relevance to proteomics: A matrix assisted laser desorption/ionization-time of flight-mass spectrometry study, *Electrophoresis*, 22, 1633–1644, 2001.
15. Hale, J.E., Butler, J.P., Gelfanova, V., You, J.S., and Knierman, M.D., A simplified procedure for the reduction and alkylation of cysteine residues in proteins prior to proteolytic digestion and mass spectral analysis, *Anal. Biochem.*, 333, 174–181, 2004.
16. Rüegg, U.T. and Rudinger, J., Reductive cleavage of cystine disulfides with tributylphosphine, *Methods Enzymol.*, 47, 111–116, 1977.

17. Herbert, B., Advances in protein solubilisation for two-dimensional electrophoresis, *Electrophoresis*, 20, 660–663, 1999.
18. Sebastiano, R., Citterio, A., Lapadula, M., and Righetti, P.G., A new deuterated alkylating agent for quantitative proteomics, *Rapid Commn. Mass Spectrom.*, 17, 2380–2386, 2003.
19. Witzmann, F.A., Clack, J.W., Geiss, K., Hussain, S., Juhl, M.J., Rice, C.M., Wang, C., *Electrophoresis*, 23, 2223–2232, 2002.
20. Reilly Industries, Inc., http://www.epa.gov/chemrtk/2vinylpy/c15018.pdf (last accessed June 2005).

13 Proteomic Approaches to the Study of Rheumatoid Arthritis

Mikaela Antonovici, Kumar Dasuri, Hani El-Gabalawy, and John A. Wilkins

CONTENTS

13.1 INTRODUCTION

One of the key objectives of the study of human diseases is to develop diagnostic and ultimately preventative or curative approaches. In order to achieve this goal, a clear understanding of the disease process and the associated systemic changes must be developed. Some of the realities that must be dealt with in clinical analysis include (1) the heterogeneity of patients (genetic and environmental) (2) the different stages of disease development, (3) the limited quantities and restricted sources of available material, and the (4) the frequent lack of availability of repeat samples of a given tissue. Thus a major goal is to develop approaches that maximize the information recovered from such potentially limiting and complex samples. Recently developed proteomic-based approaches provide capabilities, which can meet a number of these needs.[1–3]

Mass spectrometric–based proteomic techniques are highly sensitive and they can potentially deal with complex samples.[4] Using these approaches for the high content analysis of biological samples, hundreds to thousands of proteins can be identified in an undirected analysis of disease. The term undirected does not indicate a lack of hypothesis; rather, it reflects a broad-based analysis that is not necessarily directed at specified predetermined protein(s) or processes.

The types of information that can be acquired using these systems-wide approaches have relevance to a number of aspects of biomedical analysis. A detailed knowledge of the total proteomes of selected tissues or cells offers a basis for defining processes that may be involved in a particular pathology. At a more basic level the information acquired from compositional analysis offers insights into normal cellular processes and physiology. It should be recalled that we actually have relatively limited knowledge of the normal biology and biochemistry of most tissues. The recent development of a number of approaches for differential protein profiling allows for the comparison of normal and affected clinical samples.[5–7] These methods can be used to selectively identify differentially expressed proteins. These may function as biomarkers offering surrogate measures of disease progression or of response to therapy.[8–11] Since many tissues are not readily accessible on a repeat basis, these types of markers may provide noninvasive methods of monitoring process or response to therapy.[12] Alternatively, tissue or cell type restricted proteins may be useful as targets for novel therapies. While it has been suggested that patterns of protein analysis, independent of their identity, may be useful markers, it is clear that their identities will be essential for a based understanding of any disease.

13.2 RHEUMATOID ARTHRITIS

Rheumatoid arthritis (RA) is a disease characterized by inflammation and erosions of the cartilage of joints. The synovial membrane, which is responsible for maintaining normal joint function can become infiltrated with lymphocytes and undergo hypertrophic changes that result in membrane thickening and invasion of the adjacent articular cartilage with the subsequent destruction of the cartilage.[13] The disease is of unknown etiology and it is often diagnosed at a relatively late stage. Thus the focus of most clinically based studies has of necessity been on the later aspects of disease rather than the inductive events. The functional and compositional changes associated with RA are varied including inflammatory infiltrates, which may organize into secondary lymphoid follicles[14] and hyperplasia of the synovial membrane,[15] as well as a number of extraarticular manifestations.

The sources of clinical samples for analysis can derive from blood, synovial fluid, or synovial biopsy of the affected joint. It is also possible to obtain larger amounts of synovial tissues at the time of joint replacement. In the latter case it must be kept in mind that such tissues are likely to be more representative of late stage disease and as such may not provide insight into the early events of the disease process. However, these tissues can provide useful information about disease progression and the processes associated with joint destruction. There have been recent suggestions that the disease may have several simultaneous pathologies occurring in many patients, further highlighting the need to understand erosive and inflammatory events.[16]

One of the most studied components of the synovial membrane is the fibroblast-like synoviocytes (FLS). These cells display marked functional and phenotypic changes in arthritis.[17–19] Transplantation of cartilage explants from RA patients into immunodeficient mice has demonstrated that tissues containing only stromal cells can sustain activity suggesting a role in the rheumatoid process. However, at this time, relatively little is known about the overall proteome and biology of these cells.

The synovial fluid is a viscous material that fills the joint cavity, providing a medium for nutrient transport as well as providing structural properties that impact on joint lubrication and compression.[20] The fluid, which is normally highly viscous and restricted in composition, decreases in viscosity and approaches plasma level complexity in RA. Because this fluid also contains the products of local cellular and tissue synthesis in rheumatoid arthritis, the fluid represents an important clinical source of material to define pathologic processes in this disease.

This chapter describes some of our more recent proteomic approaches to characterizing the components of the rheumatoid joint.

13.3 METHODS AND MATERIALS

13.3.1 Clinical Samples and Cell Culture

All samples were obtained with informed consent from RA patients according to protocols approved by the local Ethics Review Board.

Synovial fluids samples were cleared of cellular and particulate material and frozen at −70°C until required. Synovial tissues were obtained from RA patients undergoing total knee arthroplasty. The cells were released from the dissected synovial tissue by digestion with collagenase (1 mg/ml) and hyaluronidase (0.05 mg/ml) (Sigma Chemicals, St. Louis) in Hank's buffer (ICN Biomedicals Costa Mesa, Calif.) for 1 to 2 hours at 37°C, washed with modified Dulbecco's Modified Eagle Medium (DMEM) medium, and cultured overnight in complete media at 37°C in a humidified 10% CO_2 environment. The adherent cells were cultured in fresh medium.[21] Once the cell layers were confluent, they were trypsinized and subcultured.

13.3.2 Two-Dimensional Polyacrylamide Gel Electrophoresis (2D PAGE) Analysis of FLS

Confluent cell layers were washed once with Hank's buffered saline and released with trypsin. The cells were collected by centrifugation, washed twice with phosphate-buffered saline (PBS) and once with isotonic sucrose solution (0.35 M). The cell pellet was dissolved in a sample buffer containing 7 M urea, 2 M thiourea, 4% 3-[(3-Cholamidopropyl)dimethylammonio]-1-propanesulfonate (CHAPS), 0.3%

(w/v) Bio-lyte ampholytes (pH 3–10), 75 mM dithiothreitol, and complete protease inhibitor cocktail (Roche Molecular Biochemicals Pleasanton, Calif). Protein levels were determined using a modified RC DC protein assay kit (BioRad Laboratories Hercules, Calif.)

Preparative 2D PAGE was performed on 1 mg total cellular protein dissolved in sample buffer. Immobilized pH gradient strips (17 cm, pH 3–10, nonlinear) were rehydrated overnight with sample in an IEF protean cell (BioRad) at 50 V. Electrofocusing was carried at 9000 V as the upper limit for a total 60 kV hours. Prior to analysis in the second dimension, separated proteins in the strips were reduced for 20 minutes at room temperature with 50 mM Tris (pH 8.8), 6 M urea, 2% sodium dodecyl sulfate (SDS), 20% glycerol, and 2% (w/v) dithiothreitol, and then alkylated with same buffer containing 2.5% (w/v) iodoacetamide for 20 minutes.

Second-dimension electrophoresis was carried on 12% SDS-PAGE gels (18.5 cm × 20 cm, 1 mm) (25 mA/gel at 20°C) using the PROTEAN II XL system (BioRad). Gels were fixed and stained using colloidal Coomassie blue G250 (Pierce Biotechnology, Rockford, Ill.). Gels were scanned and documented with Phoretix image analysis software (Perkin Elmer Life Sciences Inc., Boston).

13.3.3 Mass Spectrometry (MS) and Protein Identification

Spots were manually excised, destained, and in-gel digested with trypsin.[22] Peptides were extracted from the gel pieces with 25 mM ammonium bicarbonate containing 0.1% trifluoroacetic acid and 40% acetonitrile. The extracts were lyophilized and dissolved in 10 μl of 0.1% trifluoroacetic acid and 10% acetonitrile. Samples were mixed with an equal volume (0.5 μl) of 16% dihydroxybenzoic acid in 50% acetonitrile, deposited on a matrix-assisted laser desorption ionization (MALDI) target, and air dried.

13.3.4 MCE Fractionation of Synovial Fluid

Synovial fluid (1 ml) was reduced and alkylated according to the manufacturer's protocol (Proteome Systems Woburn, Mass.) and precipitated with acetone at room temperature. The pellet was resuspended in 5 ml of MCE chamber solution. The fractionation was performed in a multicompartment electrolyzer (MCE) using six immobilized pH membranes (IPM) as follows: pH 5, 5.5, 6, 6.5, 7, and 8. The sample was loaded in a pH 5.5–6 chamber. A three-phase method was applied: (1) 100 V for 4 hours, limit on power (1 W); (2) linear ramp from 100 to 1500 V for 4.5 kWh, limit on current (0.8 mA); (3) 1500 V for 7 hours, limit on power (1 W). Total running time was 16 hours and 44 minutes.

13.3.5 Chromatographic Separation

All seven fractions were collected, and urea was precipitated by decreasing the temperature to 4°C. Acetone precipitation was used for four of the fractions: pH <5, pH 5–5.5, pH 7–8, and pH >8. No pellet was obtained for the first fraction. Fractions of 50 μl, 300 μl, and 150 μl of pH 5.5–6, pH 6–6.5, and pH 6.5–7 were diluted 5× with

50 mM ammonium bicarbonate. All samples were digested with trypsin at 37°C overnight. The digest was dried down in a vacuum centrifuge.

Fractions pH 5.5–6, 6–6.5, and 6.5–7 were each desalted on an Agilent 1100 Series high-performance liquid chromatography (HPLC) (Agilent Technologies, Wilmington, Del.). Samples were injected onto an in-house packed SCX column (PolySulfoethyl A matrix), 0.125 × 100 mm and eluted after 15 minutes with a linear gradient of 0 to 100% 400 mM ammonium bicarbonate in 25% acetonitrile, 0.05% formic acid in 10 minutes, at a flow rate of 120 μl/min.

All five digested fraction were further separated on an Agilent 1100 Series HPLC. Samples were injected onto a C18 column, 5 μm of 0.150 × 150 mm in-house packed, with a linear gradient of 1 to 46.7% acetonitrile, 0.1% trifluoroacetic acid, in 35 minutes.

13.3.6 SDS-PAGE of MCE Fractions

Fractions of 100 μl, 600 μl, and 300 μl of pH 5.5–6, pH 6–6.5, and pH 6.5–7 were precipitated with UPPA reagent from Pierce. The pellet was solubilized in 5 μl sample buffer and 5 μl 1.5 M Tri-HCl, pH 8.8. Preparation for fractions with pH 5–5.5, 7–8, and over 8 was as described below. The samples were run on a 4–12% Bis-Tris gradient gel from Invitrogen. The running buffer was 3-(N-Morpholino)-propanesulfonic acid (MOPS), and SeeBlue Plus 2 was used as a marker.

13.3.7 MS

Mass spectra of the HPLC-separated tryptic fragments were acquired using a MALDI QqTOF mass spectrometer, developed at the Time of Flight laboratory, operated in single-MS or tandem MS (MS/MS) mode as described by Krohkhin et al.[23]

13.3.8 Protein Identification

Peak assignments were made using Knexus automation software (Proteometrics Canada Winnipeg, Canada), and peak lists were analyzed using the Global Proteome Machine (an open source protein identification system that uses the Tandem,[24] available at http://www.thegpm.org/) to search the ENSEMBL human genome protein translation (NCBI 34b). Default Global Proteome Machine (GPM) settings were used for the analysis (see supplementary material for details). All of the proteins reported as being identified were determined to have an expectation value $E < 3.3 \times 10^{-3}$. For notational simplicity, expectation values will be reported throughout as base10 logarithms, log (E).

13.4 RESULTS AND DISCUSSION

13.4.1 Analysis of FLS

Low-passage cultured cell lines are often used as surrogates of their *in vivo* counterparts because of the capacity to increase the cell numbers for analysis and reduce cellular heterogeneity relative to primary isolates. A consideration in the use of such lines as surrogates of their *in vivo* counterparts relates to their comparative properties with fresh isolates. Recent results suggest that cultures of synovial fibroblasts below four passages display similar phenotypes to primary isolates.[25] The cells in the present studies were only cultured up to passage four for this reason.

The 2D separation of the FLS resulted in a well-resolved mixture of proteins with >800 spots detectable by staining with colloidal Coomassie blue. We have also used gels stained with a silver-based method to detect proteins. However, in our experiences the Coomasie staining is preferred because there is much better correlation between the staining intensity and protein levels. Also, there is relatively little difference in the sensitivity of the two methods.

We have routinely used MALDI-based MS for this type of study because of the ability to archive and reanalyze samples of interest. Also the ease of multiple sample application makes the approach particularly appealing for this type of analysis. We set a fairly high confidence threshold for protein identifications (i.e., expectation value $<10^{-3.3}$) in order reduce the risk of false identifications. In those cases where novel gene products were identified, MS/MS was performed to confirm the identity of the protein.

A total of 368 spots of varying intensities were selected for analysis. Approximately 70% ($n = 254$) of these spots were identified by peptide finger printing. However, only 192 distinct proteins were identified. This reduced number of unique identifications of proteins relative to the number of spots analyzed derives from the fact that individual proteins undergo posttranslational modifications. This results in a single protein giving rise to multiple distinct spots. This is highlighted in Figure 13.1B where the beta actin and lamin A/C series are expanded to demonstrate these changes. Clearly this information may be relevant in those cases where charges in posttranslational modifications may be associated with disease-related alterations in protein function or activity. However, this property leads to redundancy in sampling of a given gene product thus reducing the efficiency of identifications.

Examples of identified proteins are listed in Table 13.1. The enzyme uridine diphosphoglucose dehydrogenase is involved in the synthesis of hylauronic acid. This is a component of the joint fluid that is essential for the maintenance of joint fluid viscosity. Histochemical methods have demonstrated the presence of this enzyme in the synovial lining and in cultured FLS.[26,27] Similarly a number of the identified proteins have been identified as auto antigens in RA (e.g., colligin, BiP, and Hcgp39).[28,29] Recently there have also been descriptions of the animal type lectins, galectin 1 and 3 in the synovial tissues of RA patients, and these were readily demonstrable in the present studies.[30,31] These observations suggest that the biosynthetic properties of the cultured cells display some similarity with cells in the rheumatoid tissue.

Several proteins of unknown function or description were also identified. The identification of these might be viewed as the discovery aspect of this type of analysis. Interestingly a relatively abundant protein identified as the product of chromosome 19 ORF 10 was also observed in the FLS-derived materials. The function of this protein is unknown at present. This identification has been confirmed on several samples by both MS and MS/MS. The fact that a previously unknown protein can

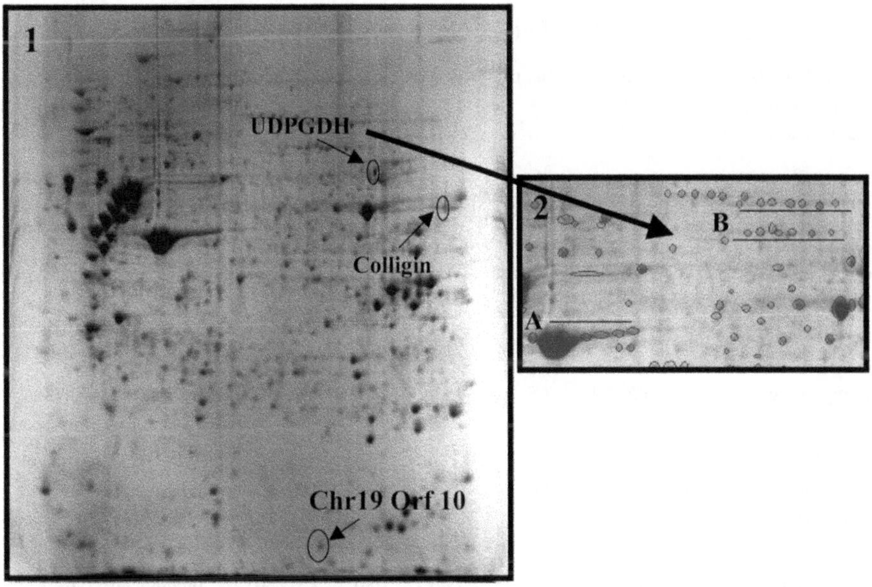

FIGURE 13.1 Total cell lysates of cultured FLS cells were resolved on a pH 3–10 IPG strips and separated by SDS-PAGE in the second dimension. The gels were stained with colloidal Comassie blue, and individual spots were excised and processed for in-gel digestion and MALDI MS analysis. (1) The positions of colligen, UDPGDH, Chr19 Orf10 are highlighted by the circles and arrow; (2) an enhanced view of the gel indicates examples of series of the same proteins undergoing posttranslational modification; (A) beta actin; (B) lamin A/C series.

be expressed at such significant levels highlights the value of detailed compositional analysis. This gene has been cloned and expressed allowing for the generation of reagents for functional and distributional analysis. Clearly such identifications provide the basis for a better understanding of these proteins and for a more complete characterization of synovial cell functions.

13.4.2 Analysis of Synovial Fluid

A major restriction of many analytical methods on clinical samples is the limited load capacities of separation steps. This means that low-abundance proteins will not be well represented in analysis of unfractionated materials. A number of approaches have been developed to deal with this problem based on the depletion of high-abundance proteins such as albumin and immmunoglobulin. Methods include antibody or affinity-based separations or prefractionations.[32] One of the limitations of affinity depletion methods for very high abundance proteins such as albumin or immunoglobulin relates to the high concentrations (mM) that must be removed. Highly specific antibody methods lack the capacity because of the near stoichiometry between the amounts of antibody and the immunosorbent. This limits the

TABLE 13.1 Proteins Identified in FLS

Protein ID	Gi No.	Obs. pI	Obs. MW	Cov.	Expect.	Function
Beta galactoside soluble lectin	gi|227920|	4.8	14	81	1.3×10^{-6}	Calcium binding/cell signaling/cell cycle regulation
Galectin 3	gi|28071074|	8.3	28.3	74	7.2×10^{-10}	Inflammatory/Stress Response/Antioxidant/Protease
Colligin-1	gi|4502597|	7.9	47.4	52	1.8×10^{-6}	Inflammatory/Stress Response/Antioxidant/Protease
Lymphocyte antigen 6 complex	gi|15306356|	7.5	16.5	53	4.6×10^{-5}	Inflammatory/Stress Response/Antioxidant/Protease
Manganese superoxide dismutase	gi|443133|	7.6	25.1	96	2.9×10^{-7}	Inflammatory/Stress Response/Antioxidant/Protease
Peroxiredoxin 1	gi|4505591|	8.2	24.6	80	2.3×10^{-13}	Inflammatory/Stress Response/Antioxidant/Protease
UDP-glucose dehydrogenase	gi|4507813|	7.1	52.4	66	3.5×10^{-13}	Inflammatory/Stress Response/Antioxidant/Protease
Cathepsin D preproprotein	gi|4503143|	5.4	28.9	43	7.6×10^{-7}	Inflammatory/Stress Response/Antioxidant/Protease
Collagen alpha 2(VI) chain precursor	gi|27808647|	5.6	72.6	11	1.3×10^{-7}	Cytoskeletal/Cell Adhesion
Inorganic pyrophosphatase	gi|11056044|	5.7	33.5	26	9.6×10^{-7}	Metabolic
BiP protein	gi|6470150|	6.6	31.7	20	3.6×10^{-7}	Inflammatory/Stress Response/Antioxidant/Protease
39 kDa synovial protein	gi|29726259|	6.9	25	31	3.3×10^{-8}	Inflammatory/Stress Response/Antioxidant/Protease
Cathepsin D preproprotein	gi|4503143|	5.9	29.2	50	7.0×10^{-11}	Inflammatory/Stress Response/Antioxidant/Protease

Note: Representative examples of identified proteins related to synovial function or inflammatory responses. The protein identities, GI numbers, observed pI and MW, percent coverage of protein sequence observed and expectation values are given for the examples.

utility for depleting anything but microliter quantities of serum. Such quantities are unlikely to be useful for the detection of very low abundance species. Other methods lack the selectivity of the immunological approaches. Thus increasing the possibility of simultaneously removing other proteins.

We have been examining the use of IsoelectrIQ^2 multichambered electrolyser (MCE Proteone Syster) as a means of fractionating the samples and eliminating high-abundance proteins. The underlying principle is that proteins are separated in solution between membranes with immobilized ampholytes of fixed pH values.[33] These provide a barrier for proteins with pIs between the pH values of the flanking membranes. The potential advantages relate to the large volumes and protein loads that the instrument can process. This affords the means to acquire significant levels of low-abundance proteins. Additionally the availability of membranes with narrow pH increments provides, at least in principal, the possibility of subfractionating the high-abundance components so as to isolate minor components that share comparable, but not identical, physicochemical properties. The system was originally developed as a preparatory step for two-dimensional gel analyses. However we have been examining the feasibility of using the MCE as a prefractionation step for single or multidimensional liquid chromatography. The rationale for such an approach relates to the ease of sample processing and the subsequent capacity to carry out a variety of high-resolution chromatographic steps on-line or off-line for subsequent MS analysis.

A problem encountered using this approach relates to the large volumes (5 ml) of the fractions that are generated during the MCE. As a preparatory step for subsequent 2D analysis the samples are precipitated using acetone and resuspended in buffers for isoelectric focusing. However these buffers contain chaotropes, detergents, and salts that are not compatible with downstream liquid chromatography (LC) and MS. Thus we sought other approaches for processing.

Samples of MCE fractions were diluted and directly digested with trypsin and desalted on a strong cation exchanger, and the peptides were eluted in ammonium bicarbonate buffer. The samples were then separated on a C18 micro LC column either off-line for MALDI or on-line for electrospray ionization (ESI) based MS/MSMS. Figure 13.2 outlines the procedural flow of the separation process.

Peak identification and mass assignments were generated using an automated program Knexus (Proteometrics Canada). The peak lists were submitted as .dta files for analysis using X! TANDEM TANDEM Spectrum Modeler. This bioinformatics component matches tandem mass spectra with peptide sequences for protein identification using a statistical significance algorithm very similar in concept to that used for the sequence similarity algorithm BLAST. The public user interface (and source code) for TANDEM is accessible at the Global Proteome Machine (http://www.thegpm.org/).

The SDS-PAGE analysis of reduced and alkylated synovial fluid demonstrated that there were several species that predominate in the fluid (Figure 13.3). Albumin and immunoglobulin are in such high excess that they result in gel distortion, making it difficult to detect minor species. Excision of the bands and in gel digestion provided sufficient material to identify the majority of visible bands by MALDI MS as derivatives of albumin and immunoglobulin heavy and light chains.

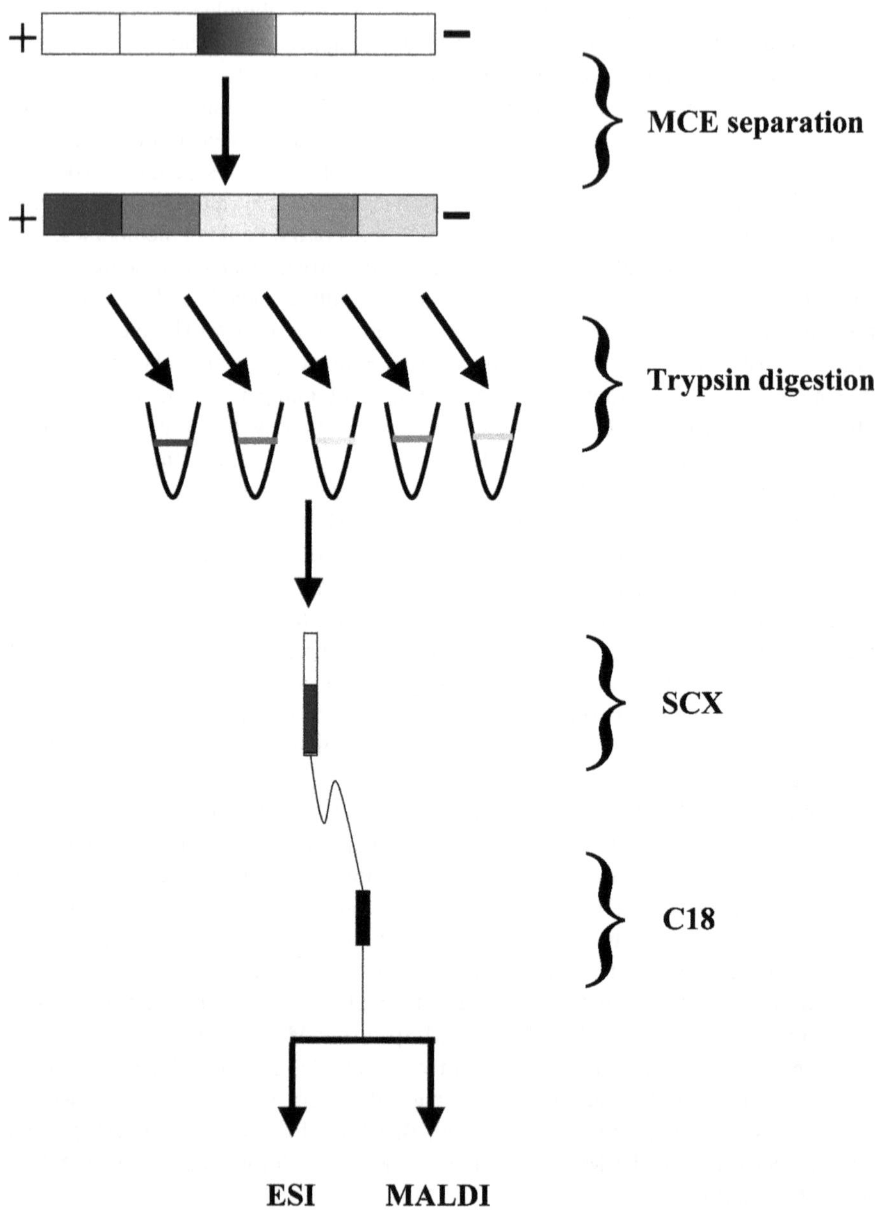

FIGURE 13.2 Flow chart of MCE-based synovial fluid analysis. Samples are separated by in solution isoelectric focusing using MCE. The individual fractions are digested with trypsin and desalted on a strong cation exchange column. The peptides are eluted and fractionated on a μLC C18 column for subsequent MS analysis.

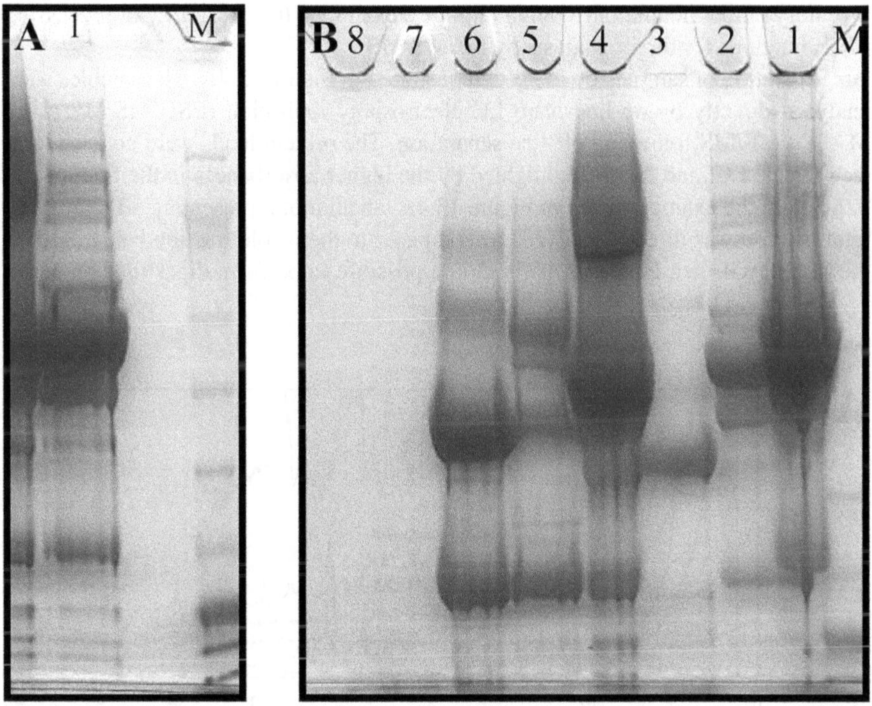

FIGURE 13.3 (A) SDS-PAGE analysis of reduced and alkylated synovial fluid. Albumin and immunoglobulins represent the predominant species. (B) SDS-PAGE analysis of MCE-fractionated synovial fluid. M, molecular weight markers; (1) untreated synovial fluid; (2) reduced and alkylated starting material, MCE fractions; (3) pH 5.0–5.5; (4) pH 5.5–6.0; (5) pH 6.0–6.5; (6) pH 6.5–7.0; (7) pH 7.0–8.0; (8) cathode pH >8.0. Note the absence of albumin in the pH 5.0–5.5 and pH 6.0–6.5 fractions.

A series of narrow pH membranes of 0.5 pH increments from 5.0 to 7.0 were used to fractionate the synovial fluid, and the samples were analyzed by SDS-PAGE. The 5.0–5.5 chamber contained virtually all of the albumin as detected by SDS-PAGE (Figure 13.3, compare lanes 2 and 3). In contrast the immunoglobulin chains were more broadly distributed over several fractions with the majority of protein located in fraction corresponding to pH 6.5–7.0. The immunoglobulins are expected to show a much broader pI distribution because of their well-recognized microheterogeneity. These results suggest that agents such as protein A or G might be a more effective way of removing immunoglobulins while albumin was very effectively removed using the MCE.

The MCE fractionation clearly resulted in the enrichment of a number of proteins that were either not visible or minor in the starting material (compare all lanes with lane 1 in Figure 13.3). These samples could readily be processed for in-gel digestion and MS-based peptide fingerprinting. As our interest is in obtaining a comprehensive inventory of the rheumatoid synovial fluid we examined the possibility of processing MCE fractions for direct analysis by LC MS/MS.

Separation of samples by LC is demonstrated in Figure 13.4A. The samples were analyzed directly by on-line nano LC electrospray ionisation (ESI) MS/MS or by MALDI MS/MS following off-line separation. The protein loads were not standardized for the LC, and this is highlighted by the higher absorbances in the fraction pH 5.0–5.5. In the example given in Figure 13.4A, an aliquot representing ~0.2% of the total fraction was directly analyzed as compared to the whole fraction being used on the SDS-PAGE gel. However, it should be possible to perform digestions on entire

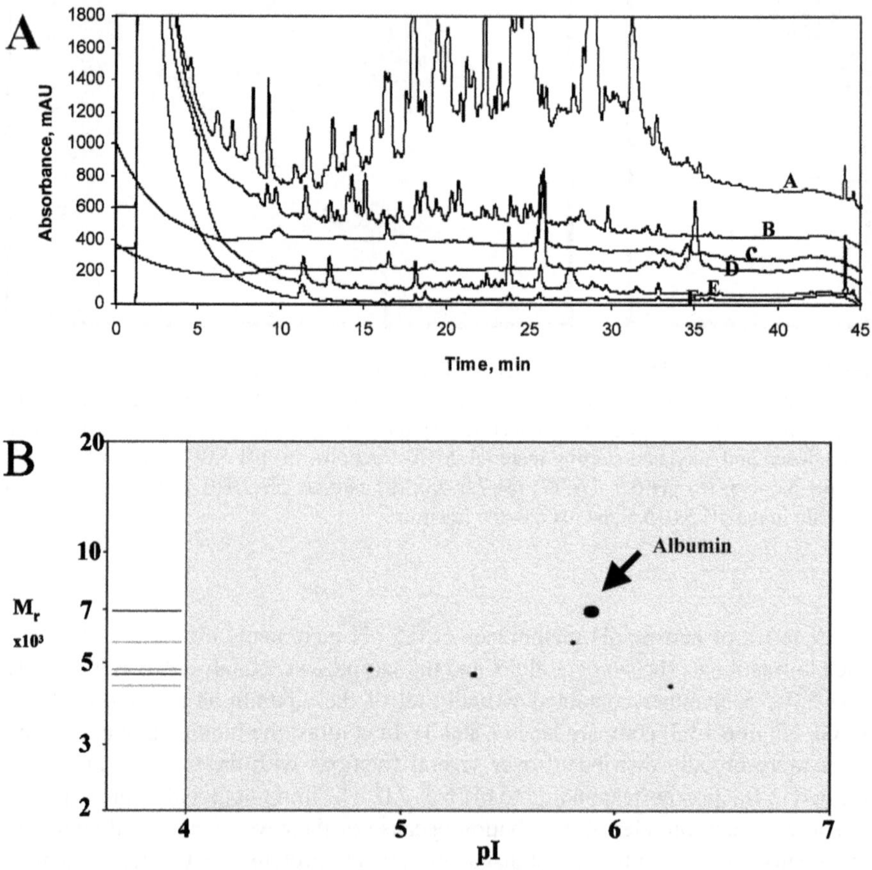

FIGURE 13.4 (A) Chromatograms of μLC separations of synovial fluid and MCE fractions. A, pH 5.5–6.0; B, Reduced and alkylated synovial fluid; C, pH >8.0; D, pH 7.0–8.0; E, pH 6.5–7.0; F, pH 6.0–6.5; (B) modified GPM gel view of identified proteins in the MCE pH 5.5–6.0 fraction. The spot intensities reflect the abundance of fragments identified for the indicated proteins. Program is publicly accessible at the Global Proteome Machine (http://www.thegpm.org/).

fractions with subsequent simultaneous desalting and concentration of the entire fraction on a cation exchange column.

Analysis of the pH 5.0–5.5 fraction demonstrated that albumin was the major protein in this fraction (Figure 13.4B). Furthermore the analysis suggested that this was the predominant protein. The other proteins detected in this fraction have predicted pI values that were consistent with the pH properties of the membranes flanking this fraction. It is worth noting that there is often a lack of correlation between predicted and observed pI values; thus, the former is never used as a constraint for protein identifications.

13.5 CONCLUSION

The application of proteomic approaches to the analysis of clinical samples for the analysis of human tissues and disease offers the opportunity to identify novel proteins that may serve as biomarkers or therapeutic targets. However, the identification of minor proteins will require high-capacity methods for the depletion of abundant proteins. Combinations of approaches based on selected affinity or high-resolution separations such as protein A/G and MCE offer the potential for a proteomic analysis of the rheumatoid process.

REFERENCES

1. Aebersold, R. and Mann, M., Mass spectrometry-based proteomics, *Nature*, 422, 198–207, 2003.
2. Hanash, S., Disease proteomics, *Nature*, 422, 226–232, 2003.
3. Pusch, W., Flocco, M.T., Leung, S.M., Thiele, H., and Kostrzewa, M., Mass spectrometry-based clinical proteomics, *Pharmacogenomics*, 4, 463–476, 2003.
4. Patterson, S.D. and Aebersold, R.H., Proteomics: the first decade and beyond, *Nat. Genet.*, 33 (Suppl.), 311–323, 2003.
5. Ong, S.E., Foster, L.J., and Mann, M., Mass spectrometric-based approaches in quantitative proteomics, *Methods*, 29, 124–130, 2003.
6. Regnier, F.E., Riggs, L., Zhang, R., Xiong, L., Liu, P., Chakraborty, A., Seeley, E., Sioma, C., and Thompson, R.A., Comparative proteomics based on stable isotope labeling and affinity selection, *J. Mass Spectrom.*, 37, 133–145, 2003.
7. Stein, R.C. and Zvelebil, M.J., The application of 2D gel-based proteomics methods to the study of breast cancer, *J. Mammary Gland. Biol. Neoplasia*, 7, 385–393, 2002.
8. Gunn, L. and Smith, M.T., Emerging biomarker technologies, *IARC Sci. Publ.* 437–450, 2004.
9. Lau, A.T., He, Q.Y., and Chiu, J.F., A proteome analysis of the arsenite response in cultured lung cells: evidence for *in vitro* oxidative stress-induced apoptosis, Biochem. J., 382, 641–650.
10. Colburn, W.A. and Lee, J.W., Biomarkers, validation and pharmacokinetic-pharmacodynamic modelling, *Clin. Pharmacokinet.*, 42, 997–1022, 2003.
11. Wulfkuhle, J.D., Paweletz, C.P., Steeg, P.S., Petricoin, E.F., III, and Liotta, L., Proteomic approaches to the diagnosis, treatment, and monitoring of cancer, *Adv. Exp. Med. Biol.* 532, 59–68, 2003.

12. Schaub, S., Rush, D., Wilkins, J., Gibson, I. W., Weiler, T., Sangster, K., Nicolle, L., Karpinski, M., Jeffery, J., and Nickerson, P., Proteomic-based detection of urine proteins associated with acute renal allograft rejection, *J. Am. Soc. Nephrol.* 15, 219–227, 2004.
13. Firestein, G. S., Evolving concepts of rheumatoid arthritis, *Nature*, 423, 356–361, 2003.
14. Magalhaes, R., Stiehl, P., Morawietz, L., Berek, C., and Krenn, V., Morphological and molecular pathology of the B cell response in synovitis of rheumatoid arthritis, *Virchows Arch.* 441, 415–427, 2002.
15. Lacey, D., Sampey, A., Mitchell, R., Bucala, R., Santos, L., Leech, M., and Morand, E., Control of fibroblast-like synoviocyte proliferation by macrophage migration inhibitory factor, *Arthritis Rheum.* 48, 103–109, 2003.
16. Kirwan, J. R., The synovium in rheumatoid arthritis: evidence for (at least) two pathologies, *Arthritis Rheum.* 50, 1–4, 2004.
17. Gay, S., Kuchen, S., Gay, R. E., and Neidhart, M., Cartilage destruction in rheumatoid arthritis, *Ann. Rheum. Dis.*, 61 (Suppl 2), ii87, 2002.
18. Muller-Ladner, U., Gay, R. E., and Gay, S., Activation of synoviocytes, *Curr. Opin. Rheumatol.*, 12, 186–194, 2000.
19. Yamanishi, Y. and Firestein, G. S., Pathogenisis of rheumatoid arthritis: the role of synoviocytes, *Rheum. Dis. Clin. North Am.*, 27, 355–371, 2001.
20. Berumen-Nafarrate, E., Leal-Berumen, I., Luevano, E., Solis, F. J., and Munoz-Esteves, E., Synovial tissue and synovial fluid, *J. Knee Surg.* 15, 46–48, 2002.
21. Hitchon, C., Wong, K., Ma, G., Reed, J., Lyttle, D., and El Gabalawy, H., Hypoxia-induced production of stromal cell-derived factor 1 (CXCL12) and vascular endothelial growth factor by synovial fibroblasts, *Arthritis Rheum.* 46, 2587–2597, 2002.
22. Dasuri, K., Antonovici, M., Chen, K., Wong, K., Standing, K., Ens, W., El Gabalawy, H., and Wilkins, J. A., The synovial proteome: analysis of fibroblast-like synoviocytes, *Arthritis Res. Ther.* 6, R161–R168, 2004.
23. Krokhin, O. V., Cheng, K., Sousa, S. L., Ens, W., Standing, K. G., and Wilkins, J. A., Mass spectrometric based mapping of the disulfide bonding patterns of integrin alpha chains, *Biochemistry*, 42, 12,950–12,959, 2003.
24. Craig, R. and Beavis, R. C., TANDEM: matching proteins with tandem mass spectra, *Bioinformatics*, 20, 1466–1467, 2004.
25. Zimmermann, T., Kunisch, E., Pfeiffer, R., Hirth, A., Stahl, H. D., Sack, U., Laube, A., Liesaus, E., Roth, A., Palombo-Kinne, E., Emmrich, F., and Kinne, R. W., Isolation and characterization of rheumatoid arthritis synovial fibroblasts from primary culture—primary culture cells markedly differ from fourth-passage cells, *Arthritis Res.* 3, 72–76, 2001.
26. Edwards, J. C., Wilkinson, L. S., and Pitsillides, A.A., Palisading cells of rheumatoid nodules: comparison with synovial intimal cells, *Ann. Rheum. Dis.*, 52, 801–805, 1993.
27. El Gabalawy, H., King, R., Bernstein, C., Ma, G., Mou, Y., Alguacil-Garcia, A., Fritzler, M., and Wilkins, J., Expression of N-acetyl-D-galactosamine associated epitope in synovium: a potential marker of glycoprotein production, *J. Rheumatol.*, 24, 1355–1363, 1997.
28. Corrigall, V. M., Bodman-Smith, M. D., Fife, M. S., Canas, B., Myers, L. K., Wooley, P., Soh, C., Staines, N. A., Pappin, D. J., Berlo, S. E., van Eden, W., van Der, Z. R., Lanchbury, J. S., and Panayi, G. S., The human endoplasmic reticulum molecular chaperone BiP is an autoantigen for rheumatoid arthritis and prevents the induction of experimental arthritis, *J. Immunol.*, 166, 1492–1498, 2001.

29. Hakala, B.E., White, C., and Recklies, A.D., Human cartilage gp-39, a major secretory product of articular chondrocytes and synovial cells, is a mammalian member of a chitinase protein family, *J. Biol. Chem.*, 268, 25,803–25,810, 1993.
30. Rabinovich, G.A., Rubinstein, N., and Toscano, M.A., Role of galectins in inflammatory and immunomodulatory processes, *Biochim. Biophys. Acta*, 1572, 274–284, 2002.
31. Ohshima, S., Kuchen, S., Seemayer, C.A., Kyburz, D., Hirt, A., Klinzing, S., Michel, B.A., Gay, R.E., Liu, F.T., Gay, S., and Neidhart, M., Galectin 3 and its binding protein in rheumatoid arthritis, *Arthritis Rheum.*, 48, 2788–2795, 2003.
32. Righetti, P.G., Castagna, A., Herbert, B., Reymond, F., and Rossier, J.S., Prefractionation techniques in proteome analysis, *Proteomics*, 3, 1397–1407, 2003.
33. Righetti, P.G., Wenisch, E., Jungbauer, A., Katinger, H., and Faupel, M., Preparative purification of human monoclonal antibody isoforms in a multi-compartment electrolyser with immobiline membranes, *J. Chromatogr.*, 500, 681–696, 1990.

14 Immunoglobulin Patterns in Health and Disease

Ingrid Miller and Marcia Goldfarb

CONTENTS

14.1 INTRODUCTION

Immunoglobulins (Igs) (antibodies) are highly complex and variable proteins, generated by the humoral immune system to a vast array of antigens. The humoral immune system is a flexible system of synthesis utilizing basic peptide segments with multiple variations. Perturbations can occur at many and various points causing formation of aberrant peptides and proteins. Two-dimensional electrophoresis (2DE), due to its ability to separate proteins by isoelectric point and molecular size, is a high-resolution method predestined for investigation of complex questions. It allows clear separation of Ig chains and classes, and thus is well suited for the study of Ig patterns in various body fluids and in different conditions (physiological or pathological) for research and diagnostic purposes.

14.2 IMMUNOGLOBULINS AND DISORDERS

14.2.1 Humoral Immune System

Immunoglobulin molecules are made up of four polypeptides, two identical heavy and two identical light chains. The heavy chains identify the five Ig classes; G, A, M, D, and E. Light (L-) chains can be one of two classes: kappa (κ) or lambda (λ). Further heavy-chain differences in primary amino acid composition divide classes into subclasses: $IgG_{1,2,3,4}$ and $IgA_{1,2}$, giving the molecules some different functions. There are also allotypic differences with subclasses 1,2,3 having Gm specificities. The term Gm indicates a sequence on G, IgG heavy chain, which is under genetic control and acts as a phenotypic marker, m. IgA subclasses 1,2 also have phenotypic markers, termed Am. IgG is the most abundant in human serum. It can pass the placenta in humans, and the infant is born with a concentration of the mother's IgG. IgG (except for subclass 4) binds complement strongly and enhances phagocytosis. IgA, the second highest concentrated, protects the mucosal surfaces. In plasma, it is usually a monomer. In secretions, it is a dimer joined by a peptide termed the J chain and has a peptide, secretory piece attached, which is part of the complex that transported the molecule to the mucosal surface. IgM appears early in an immune response and circulates in serum as a pentamer with a J chain. IgD appears to be mainly a receptor, along with IgM, on naïve B-lymphocytes. Only very low concentrations of IgE are present in serum. IgE appears on mast cells and is an effector of acute inflammatory reactions, causing allergic symptoms. In addition, the protein sequence of the variable portion of the Ig protein, which gives the molecule an epitope specificity (i.e., the antigen it recognizes), gives it an idiotype classification. This idiotype epitope can be recognized by other Igs. Antibodies of different classes that recognize the same antigenic epitope may have the same idiotype.[1]

Immunoglobulins are secreted by plasma cells, which have developed through an intricate series of cellular events. B-cells, so called because of development in the bone marrow, go through multiple stages before becoming mature plasma cells secreting antibody. Naïve B-cells coexpress IgD and IgM on the surface, and the cells have the ability to switch production of Ig class. B-cell response to antigenic stimulation is polyclonal involving the proliferation of multiple cells whose surface receptors recognize epitopes on the offending antigen. Clonal expansion of a single plasma cell or its precursor generates a monoclonal protein that is the common mark of plasma cell disorders.

14.2.2 Disorders

Plasma cell disorders, or dyscrasias, identified in a zone electrophoresis of serum as a heavy electrophoretically restricted band called an M-component, or paraprotein, are a heterogeneous group of B-cell immune system disorders. In many instances the causes remain unclear and the disease may or may not be clinically malignant. An early classification[2] lists dyscrasias as shown in Table 14.1.

14.2.2.1 Multiple Myeloma

Multiple myeloma is characterized by lytic bone lesions, an increase of abnormal plasma cells in the bone marrow, and a monoclonal protein in serum or urine, or both.[3] Patients may present with bone pain, anemia, spontaneous fractures, and recurrent infections. Approximately, 50% of patients have an IgG paraprotein (M-component) and 25% an IgA. Serum levels of the polyclonal Igs may be significantly reduced. Bence Jones proteins, which are free light chains, are detected in urine in approximately half of patients, and about 20% of patients have only Bence Jones proteinuria. There is also a nonsecretory myeloma (1% of patients) where no paraprotein is detectable in serum or urine and is only seen by immunostaining of plasma cells. M-proteins appear to be abnormal but are simply excessive quantities of normal Ig idiotypes. Monoclonal antibody specificity in humans has been associated with a wide variety of bacterial antigens, including streptolysin O, staphylococcal protein, *Klebsiella* polysaccharides, and *Brucella*.[4] However, most antigens are unknown. IgD myeloma (about 2% of cases) is a very serious finding with survival time after diagnosis approximately 14 months. Light-chain type in 90% of IgD cases is λ, whereas this is true in only 30% of IgG and IgA myelomas. Patients can have a normal concentration of Igs even when an M-protein is present. In IgD myeloma, the serum levels of total protein and monoclonal protein are usually not very high.[2] The finding of a monoclonal λ band by electrophoresis in the absence of a detectable monoclonal heavy chain band may suggest an IgD myeloma.

14.2.2.2 Waldenström's Macroglobulinemia

The major clinical feature of this dyscrasia is serum hyperviscosity caused by high concentration of monoclonal IgM, aggregate formation, cryoprecipitation, and

TABLE 14.1 Plasma Cell Disorders

Primary Monoclonal Gammopathy
Multiple myeloma
Waldenström's macroglobulinemia
Solitary plasmacytoma
Amyloidosis
Heavy-chain disease
Various types of non-Hodgkins lymphoma
Chronic lymphocytic leukemia
Light-chain disease
Secondary Monoclonal Gammopathy
Cancer (nonlymphoreticular)
Monocytic leukemia
Heptobiliary disease
Rheumatoid disorders
Chronic inflammatory states
Cold agglutinin syndrome
Benign hyperglobulinemic purpura of Waldenström
Papular mucinosis
Immunodeficiency
Monoclonal Gammopathy of Undetermined Significance (MGUS)
Transient
Persistent
Disorders of Immunodeficiency
Hypogammaglobulinemia, genetic and acquired

Note: Adapted from Wells et al.[2]

autoimmune antibody activity.[3] IgM is generally in the pentameric form, but many patients have monomeric. It is suggested Waldenström's is produced by proliferation of cells of B-1 lineage that secrete low affinity, multispecific autoantibodies, and so-called natural antibodies to bacterial carbohydrates.

14.2.2.3 Solitary Plasmacytoma

Solitary plasmacytoma is an isolated plasma cell tumor that can occur in bone or extramedullary tissue (arising outside of bone marrow). These may be two different diseases, often discovered by a routine x-ray or a patient complaint of pain. M-protein is found in serum or urine, or both, in 63% of patients. Solitary plasmacytoma of bone frequently progresses to multiple myeloma and has a poorer prognosis. Extramedullary has a predominance of IgA monoclonal protein, and only 15% of patients evolve to multiple myeloma.[5]

14.2.2.4 Amyloidosis

Amyloidosis, a heterogeneous group of diseases, is the effect of deposition of fibrils in tissue. The nature of the fibril protein is used to characterize the disease; most of these are not diseases of Igs. Solubilization and sequencing of fibrils identified that certain complexes are made up of polymerized fragments of light chain, mainly the variable region and an 8000 kD protein A. This disease is termed amyloidosis AL. Suggested mechanisms for deposition include catabolism by macrophages of deposited antigen-antibody complexes, *de novo* synthesis of light chains with reduced solubility, genetic deletions in light-chain gene producing a fragment of reduced solubility, separate synthesis of discrete regions of the light chain. Deposition is found in the gastrointestinal tract, liver, kidneys, adrenals, and spleen. Renal involvement is often a poor prognostic sign. The diagnosis is hard to establish, and detection of a monoclonal antibody in serum or urine is a helpful finding for making an early diagnosis, but tissue documentation by Congo Red staining in biopsy material is critical.

14.2.2.5 Heavy-Chain Diseases

Heavy-chain diseases are rare diseases characterized by an M-protein composed of incomplete heavy chains, either γ, μ, or α, with α as the most prevalent. Large amounts are excreted in urine, but serum concentration is usually low. N-terminal sequence is normal, but there are deletions from the variable domain through most of the C_H1 region. The chains lack the structure required to form cross-links to light chains, possibly due to a gene rearrangement defect during B-cell development.[3,6] A truncated α heavy chain with no light chains is expressed in mucosa-associated lymphoid tissue in immunoproliferative small intestinal disease (IPSID), thought to be caused by *Campylobacter jejuni.*[7]

14.2.2.6 Monoclonal Gammopathy of Undetermined Significance

The term monoclonal gammopathy of undetermined significance (MGUS) indicates the presence of a monoclonal protein in persons not presenting with clinical evidence of plasma cell disorders. The earlier use of the word benign in relation to these disorders was misleading because the patient may remain asymptomatic or may develop into symptomatic multiple myeloma, macroglobulinemia, amyloidosis, or other plasma cell disorders. Monoclonal gammopathy of undetermined significance occurs in 3% of persons over 70 years of age. The interval from time of recognition of the M-protein to diagnosis of clinical symptoms can range from 2 to 29 years. An important risk factor for progression to disease is the concentration of serum M-protein. The presence of a monoclonal light chain in urine also suggests multiple myeloma, although very small concentrations may be normal.[4,5,8]

14.2.2.7 Immunodeficiency

Antibody deficiency disorders comprise a spectrum of diseases, some congenital and some acquired. Disorders range from complete absence of all classes of Ig (hypo) to selective deficiency of a single class. The degree of symptoms found in patients is dependent on the degree of deficiency.

14.3 METHODS TO INVESTIGATE IMMUNOGLOBULIN PATTERNS (AS EXEMPLIFIED FOR SERUM)

14.3.1 One-Dimensional Separations

Classical methods for the detection of gammopathies have been immunoelectrophoresis and zone electrophoresis. In zone electrophoresis serum proteins are separated in several bands, almost exclusively on the basis of their surface charge, on a (theoretically) inert support without additional sieving effects of the matrix (nowadays mainly cellulose acetate strips or agarose gels).[9–11] Immunoglobulins show γ-globulin mobility (IgG), but depending on the class can appear also in the α_2- or β-zone: either as a general increase of the peak (for polyclonal Ig) or as a sharp and narrow band or spike (for monoclonal Ig, the so-called M-component). Whereas zone electrophoresis allows densitometric quantification of albumin and the single globulin fractions, immunoelectrophoresis is able to identify the Ig classes on the basis of the immunologic reaction with the respective antibodies.[2,9] In routine electrophoresis, nowadays immunoelectrophoresis is usually replaced by immunofixation after high-resolution electrophoresis. Gammopathies are detected by their specific reaction with appropriate antibodies and by band sharpness.[10,11]

In the last ten years, capillary zone electrophoresis has proven useful in serum and plasma protein electrophoresis, giving very similar separation patterns to those of gel-based methods.[12–14] Therefore, the method can also be routinely used for the detection of paraproteins[15,16] and is one of the recommended methods.[10,11]

For further investigation on the patient's pathological Ig pattern, isoelectric focusing,[17,18] sodium dodecyl sulfate–polyacrylamide gel electrophoresis (SDS-PAGE), and 2DE may give additional important details which can be helpful for diagnosis.

Figure 14.1 shows one-dimensional SDS-PAGE of serum specimens of several patients with monoclonal gammopathies compared to normal serum. With this method, not all cases of paraproteins can be seen as additional distinct bands in the L and H

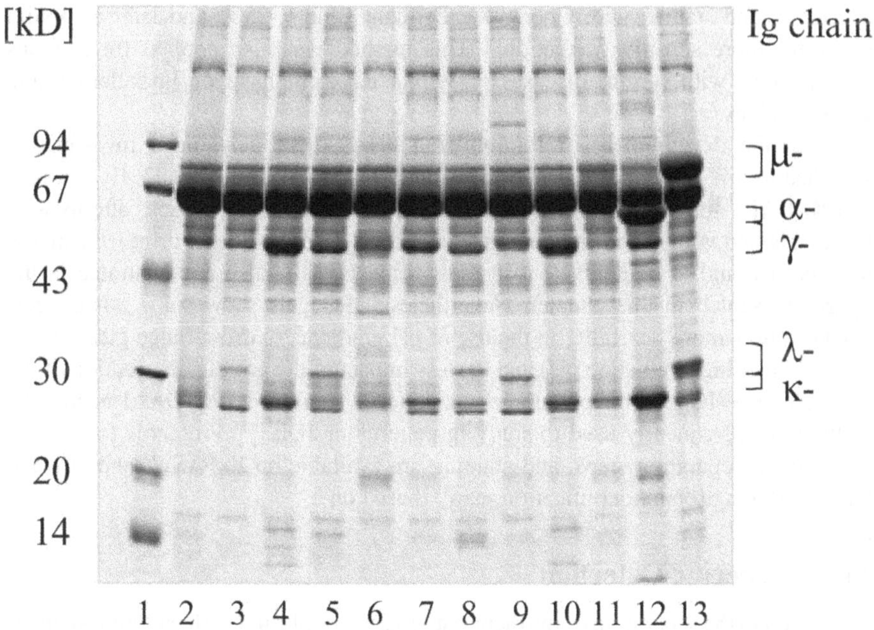

FIGURE 14.1 SDS-PAGE of myeloma serum samples. SDS-PAGE according to Laemmli[121] on gels of 140 × 140 × 1.5 mm (separation gel: T, 10–15%; C, 2.7%; stacking gel: T, 5%; C, 2.7%); stained with Coomassie Brilliant Blue R-250. The positions of the Ig chains are indicated on the right side. Samples: lane 1, molecular weight marker (LMW, Amersham Biosciences, Uppsala, Sweden); lanes 2–13, serum samples, corresponding to 0.5 μl serum each; lane 2, healthy volunteer; lanes 3–13, myeloma patients; lane 3, λ chain only; lane 4, multiple κ chains, polyclonal heavy chains; lane 5, λ chain only (see also Figure 14.9); lane 6, increased Igs; lane 7, κ-IgG monoclonal; lane 8, κ and λ chains only; lane 9, λ-IgG monoclonal (see also Figure 14.3C); lane 10, multiple κ chains, polyclonal heavy chains, same patient as in lane 4 (few weeks later); lane 11, κ-IgM monoclonal (see also Figure 14.5C); lane 12, triclonal κ-IgM/IgA; lane 13, biclonal κ-IgM with μ_S-chain (see also Figure 14.6).

range. Especially the μ-chain stays undetected when not present in high amounts, as it comigrates with transferrin.

14.3.2 2DE

From the early beginnings of 2DE on, serum protein patterns in health and disease were among the topics of interest because of the possible clinical and diagnostic relevance and the easy availability of serum samples. The first 2DE maps published for human serum proteins were from the Anderson group[20] and were established with the ISO-DALT system. The first-dimensional IEF was performed under reducing and denaturing conditions in tube gels in the presence of carrier ampholytes. After equilibration, the tube gels were placed on top of vertical SDS slab gels, run overnight, and stained with Coomassie Brilliant Blue R-250[20] and later also with silver for more sensitive detection.[21] This method was also applied for samples from patients with various diseases[22–25] with special emphasis on Ig patterns[26] and gammopathies.[27,28]

With the development of immobilized pH-gradient gels, most investigators switched to this new type of first-dimensional separation (IPG-DALT). Evaluations of IPG-DALT for the study of gammopathies and comparison of gel patterns with those obtained with carrier ampholytes have been described[29] (for urine).[30] Patterns have been found to be equivalent in most respects, as is also demonstrated in the figures presented in this chapter. Nevertheless, IPG gels show some advantages: (*a*) they allow more flexibility in the use of pH gradients (narrow-range gradients or ranges extending to alkaline pH) and (*b*) less impact of the protein distribution of the sample on the shape of the pH gradient is noticed. Therefore, IPG-DALT is now the method most frequently used to study Ig patterns in 2DE.[16,31] Protocols for sample preparation, separation steps, and staining are available on ExPASy[32] and also the human plasma protein identification map[33] (based on[25, 34–36]).

14.3.3 Specific Detection

In addition to the separation of unfractionated serum or plasma with pattern visualization by Coomassie Brilliant Blue and silver staining, immunoblotting and detection with specific antibodies have proven very helpful for identification of Ig classes.[26,30] For special applications, enrichment and purification of Igs have been performed on immobilized staphylococcal protein A[21,37] or protein G[38] or by affinity chromatography with specific antibodies.[38,39] In critical cases, especially when overlap occurs with high-abundance serum proteins like albumin and transferrin, only this purification step allows resolution of the paraprotein. In addition, by combination of these different affinity methods, overall Ig and antigen-specific antibody patterns (e.g., in course of a vaccination) may be compared.[40,41]

14.4 SERUM

14.4.1 Physiological Patterns

The prerequisite for the study of pathological patterns is the knowledge of physiological patterns and thus the setting up of a reference protein map of a healthy population. Since proteins are separated under denaturing and reducing conditions, heavy chains are well separated from light chains. Heavy chains of different Ig classes are detected in nonoverlapping areas of the gels. Positions of IgG single chains were already identified,[20] and those of the other Igs followed.[22,26,27] This resulted in a map with the locations of heavy and light chains of most of the Ig classes and subclasses,[26] as shown in Figure 14.2. In contrast to those early publications, Figure 14.2 shows a close-up of an IPG-DALT gel, the separation of serum from a healthy donor. Positions of the main Ig chains are indicated by boxes. Each Ig chain is characterized by a series of spots, displaying charge and size microheterogeneities

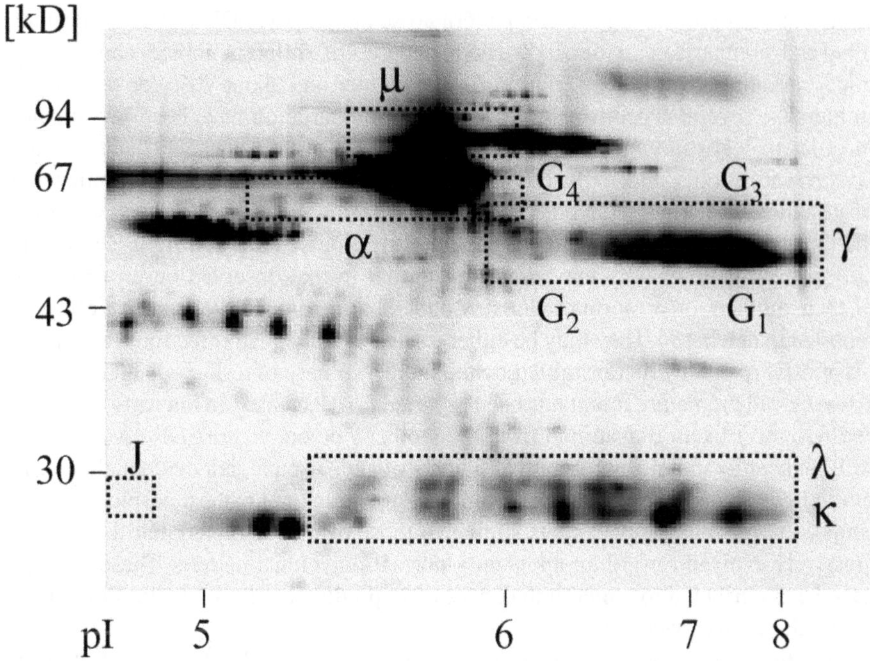

FIGURE 14.2 2DE of normal serum with the positions of single Ig light and heavy chains indicated. IPG-DALT with IPG pH 4–10 nonlinear (NL) over a distance of 10 cm as a first dimension and SDS-PAGE (as described in Figure 14.1) as a second; silver staining with a modified Heukeshoven method.[42] Sample amount corresponding to 0.5 μl serum (from a healthy individual). The positions of the heavy and light Ig chains of different classes and subclasses are indicated by boxes, as well as the J chain, which combines the monomers of dimeric IgA and pentameric IgM.

(mainly due to posttranslationational modifications, proteolytic cleavage, or genetic polymorphism). Polyclonal heavy chains are observed as fuzzy zones without distinct individualizable spots. According to Tracy et al.,[26] IgG heavy chains of different isotypes are located on specific areas of the gels. In contrast to that, Vu et al.[31] found no correlation between IgG isotype and position of the respective heavy chain on the gel. Nevertheless, both agree about the general gel location of γ chains. μ and also α chains are not very easily detectable in healthy persons. Polyclonal light chains appear as clustered nondiscrete spots, forming cloudy zones with unevenly distributed density over a wide pH and kD range. The apparent molecular mass of κ light chains is typically lower than that of λ light chains.

Patterns of healthy adult individuals have been given on numerous occasions, mainly as a reference when studying pathological processes.[20,21,24,25,27,33,34,43–44]

Igs are proteins with high diversity due to their properties and main functions in immune response and defense of the individual organisms. Therefore, the question about variability of the light-chain pattern arose quite early. 2DE was used for mapping and comparison of the light-chain patterns of different inbred and outbred mice, revealing 20 to 30% variation for the former and about 70% for the latter.[46] In humans, several aspects of Ig development and variation have been investigated in order to learn more about the humoral immune system in health and disease. One research topic was early human development.[47,48] From the eighteenth week of gestation, maternal IgG could be detected in growing concentrations, and from the thirty-eighth week, IgA and IgM could also be detected.[47] The light-chain clonality pattern from healthy newborns was similar to that observed on protein maps of their mothers or of normal adults. Monoclonal gammopathies have rarely been reported in newborns. They may be either related to congenital infections (e.g., toxoplasmosis) or passively transmitted by the mother. A case of a single λ-IgG spot in a 30-week-old premature infant has been reported.[49] It decreased in intensity over time and was most likely transmitted from the mother's blood, where 2DE revealed it as well. In healthy infants of 1 month to 5 years of age, discrete, but evident alterations of the light-chain patterns occurred: first a restricted clonal pattern developed (with single, sharp, well-resolved spots similar to those of monoclonals), but then it progressively evolved toward an apparently normal polyclonal pattern. These findings have been attributed to clonal imbalances of Ig production during maturation of the humoral immune system.[48]

Another interesting aspect of physiological importance, the diversity of antigen-specific antibodies, was studied in human adults.[40,41] Antibody production was induced by vaccination with tetanus toxoid and the overall as well as the specific Ig response followed over time. Differentiation of Ig populations was possible by a combination of protein G- and antigen-specific affinity chromatography, and comparison of Ig light-chain patterns in 2DE. The antigen-specific antibodies developed were polyclonal with a superimposed oligoclonality that was donor dependent and did not change upon time. Their light-chain diversity was unaffected by boosting and unrelated to the intensity of the response.[41]

14.4.2 Pathological Conditions

14.4.2.1 Monoclonal Gammopathies

Monoclonal Igs (monoclonal antibodies [Mabs]) are secretory products of single neoplastic plasma cell clones but are not always as homogeneous as one would expect. This was already recognized in studies applying one-dimensional isoelectric focusing (e.g., for testing of murine or human Mabs secreted by hybridoma technology).[50,51] The Mabs gave sets of multiple dense and several less prominent bands, although the pH range was narrower than in focus of polyclonal (< 0.6 units vs. > 2 units).[51] Microheterogeneity was attributed to mutations, amidation and deamidation of amino acids, variable glycosylation, and other posttranslational modifications,[50] but focusing patterns of intracellular and secreted antibodies may differ.[51]

In 2DE, monoclonal heavy chains are detected as sets of multiple well-resolved spots characterized by charge microheterogenenity. They are usually located at the same positions as the corresponding polyclonal heavy chains of the respective Ig class (Figure 14.2). A monoclonal light chain from a single monoclonal Ig is usually characterized by one dominant and well-defined single spot but can also disseminate in more than one spot. The isoelectric point (pI) of the light chain may vary over a wide range (from 5–11).[29]

Early 2DE studies (ISO-DALT on small size-gels) with sera of myeloma patients helped to establish the positions of the Ig single chains in the overall serum protein pattern and brought first insight into glycosylation sites.[27,52] Neuraminidase treatment of serum or purified myeloma proteins made spots of original Ig α, μ and δ chains shift markedly toward more alkaline pIs and confirmed the hypothesis that sialic acid is largely responsible for their charge microheterogeneity. Light and γ-chains were not influenced by this treatment, suggesting sterical hindrance, amidation and deamidation, or other modifications.[28,52] Light-chain patterns of patients may not change over months and years.[28] Not all changes in the light-chain region are related to Igs: the appearance of high levels of C-reactive protein (in severe inflammation), serum amyloid P or pro-apolipoprotein A-I may be observed (see Figure 12 of Tissot and Spertini[29]).

The usefulness of 2DE as an aid in the analysis of the clonality of Igs in different clinical situations was established by a large number of studies.[26,28,29,31,38,52,53] Results about distribution and appearance of certain disorders vary but may depend largely on the groups of patients investigated and the methods used. For instance, Harrison et al.[28] reported a 76% charge heterogeneity of light chains, but Tissot et al. reported only 29%[38,54] (i.e., a very different percentage of samples with multiple L-spots). Often, 2DE can reveal much more complex patterns than expected from routine electrophoresis: suspected monoclonal antibodies may turn out to have single aberrant chains (heavy or light), to be polyclonally increased or to contain truncated spot chains[53] or unusually large molecules. One example is given by Goldfarb,[55] a myeloma with a 100 kD monoclonal IgG. The protein, with a very acidic κ chain (pI < 5) as revealed in 2DE, was purified in a nondenaturing isoelectric focusing run on the Rotofor; only then the aberrant heavy chain was found and further characterized.

Figure 14.3 shows different examples for Mabs of the IgG type (A and B with alkaline κ chain; C with a moderately acidic λ chain). The patients represented in A and C belong to the IgG_1 type, patient B is IgG_3 with a clearly visible IgG_1 spot train at the position below. When the Mab is present in large amounts as in Figure 14.3A, additional fragments, possibly degradation products, become visible. This spot chain γ′ has also been found by several other investigators, especially in context with gamma heavy-chain disease,[6,56] and in those cases it has been identified to contain the Fc terminus of the respective γ chain. It is present as IgG γ_S intermediate chain also in the plasma map in[36] and in SWISS-2DPAGE.[33] Similarly, this map also shows products with higher molar mass quite frequently found in hypergammaglobulinemic sera, probably incompletely reduced and reaggregated IgG single chains or half molecules.

In a critical review about the analysis of Igs by 2DE,[29] Tissot and Spertini pointed out the main problems in detecting monoclonal γ chains: superposition of polyclonal

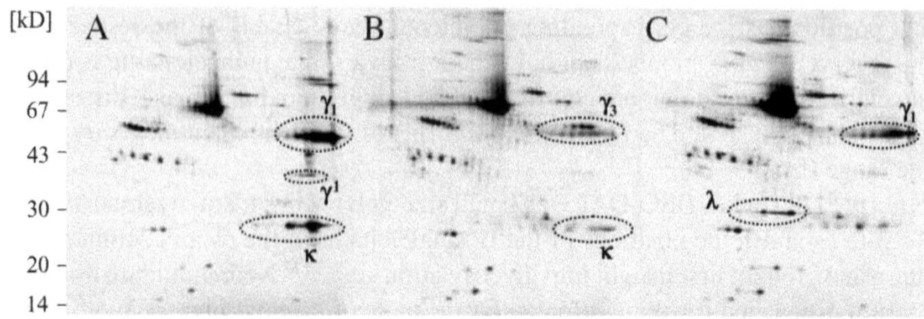

FIGURE 14.3 Sera of patients with monoclonal gammopathies of IgG type. 2DE as in Figure 14.2, corresponding to 0.2 μl serum each. Patients showed paraproteins (A) κ-IgG_1; (B) κ-IgG_3; and (C) λ-IgG_1.

γ chains and the basic pI of some monoclonal γ chains. At that time, they investigated Ig serum patterns by either ISO-DALT or IPG-DALT, but both in a broad pH range. In a later paper by the same group,[31] sample testing was extended also in the alkaline pH range: in 52% of the investigated monoclonal IgGs (pretested by immunofixation), the monoclonal γ chain was found by using IPG pH 3–10 in the first dimension, and additional 8% in the IPG range of pH 6–11; 4% were hidden by polyclonal γ chains, and for 36% no monoclonal heavy chain was found. In contrast to this, in 84% of the samples, monoclonal light chains were located. The study comprised 73 samples, displaying monoclonal light chains of pI 5 to 7.2 and γ chains of pI 6 to 9 (the majority of subtype γ_1). Both types of chains were highly heterogeneous in their electrophoretic behavior, and the authors were not able to find a classification based on their electrophoretic properties.

Examples for monoclonal IgA of the κ type are shown in Figure 14.4 and Figure 14.10C. In Figure 14.4, the α chain is rather close below the albumin chain, with almost the same pI; similarly, the light chain is adjacent to the apolipoprotein A-I spots. The IgA isoform in Figure 14.10C is slightly more acidic, of higher concentration, and displays a very prominent κ chain. Small amounts of dimerized light chain are visible (this phenomenon has already been described by Harrison[28]).

Figure 14.5 shows patterns from patients with different types of IgM gammopathies. Close-ups were taken from the μ chain, (located close to the large albumin spot) and from the light-chain region. It is a general finding that secretion of a Mab may downregulate the normal Ig production.[53] Patients represented in Figure 14.5A and 14.5C differ not only in the pI of the light and μ chain, but also in the presence of the J chain: it is quite marked in the patient represented in Figure 14.5C. The conclusion is that the patient in Figure 14.5C produces intact IgM pentamers, but the patient in Figure 14.5A probably produces monomeric IgM. It is not always easy in 2DE gels to distinguish basic monoclonal μ chain spots (of pI 6.1–6.8) and transferrin, as they often overlap (this has also been described by Tissot and Spertini[29]).

Figure 14.6 actually would belong to the next section (biclonal gammopathies), but is discussed here, as the sample shows some interesting features that occur also

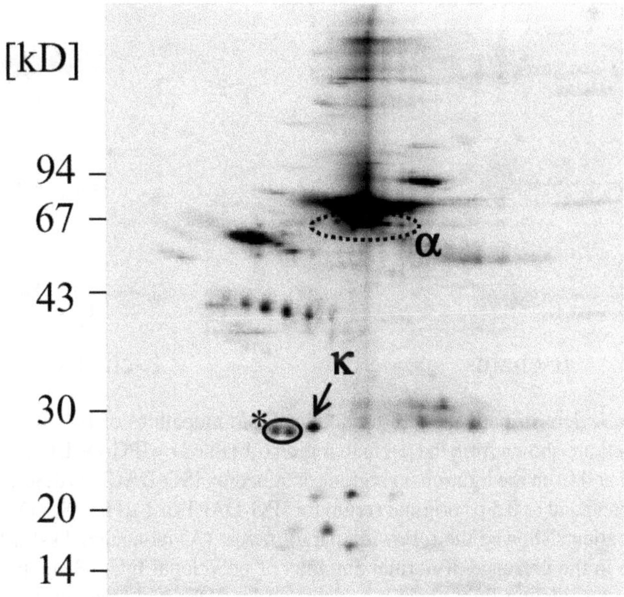

FIGURE 14.4 Serum of patient with monoclonal gammopathy of IgA type. 2DE as in Figure 14.2, corresponding to 0.2 μl serum. Patient sample with small amounts of κ-IgA (single chains indicated). The two spots of apolipoprotein A-I, adjacent to the κ spot, are marked with an asterisk.

in single Mabs. First, the serum displays two light chains that have both been typed by immunoblotting to be κ. Second, there is one intact μ chain at the same height as transferrin with spot chains almost fusing, and a second one, called μ_S (or intermediate μ_S), of much smaller molar mass. The first finding, a κ chain with a mobility in the λ-chain region, is not so unusual. Size variations in both light-chain types may occur due to heavy glycosylations or extended peptides.[57,58] Tracy et al.[26] reported that the predictive value for correct identification of κ and λ by relative mobility was 83 and 69%, respectively. The same authors also described a μ chain of about 48 kD, about 25 kD shorter than usual, and identified its class by immunoblotting. This truncated form was called the μ_S chain.[37] The same double spot row at 45–48 kD was found in purified preparations of polyclonal IgM and also an additional polypeptide associated with IgM, the latter especially in cryoglobulin preparations.[38,59] It is not clearly identifiable in the respective region of Figure 14.6 but should be present as two small spots at the lower left of the μ_S chain (pI 5.45, 44 kD). The protein was later identified as Spα (CD5 antigen-like),[60] a member of the scavenger receptor cysteine-rich superfamily of proteins that have important functions in the regulation of the immune system.[61]

In normal serum, polyclonal δ chains are present only in small amounts and comigrate with serum albumin; they are therefore not detectable. Monoclonal IgD is a rare paraprotein and is often not found in routine testing (if the panel of antibodies

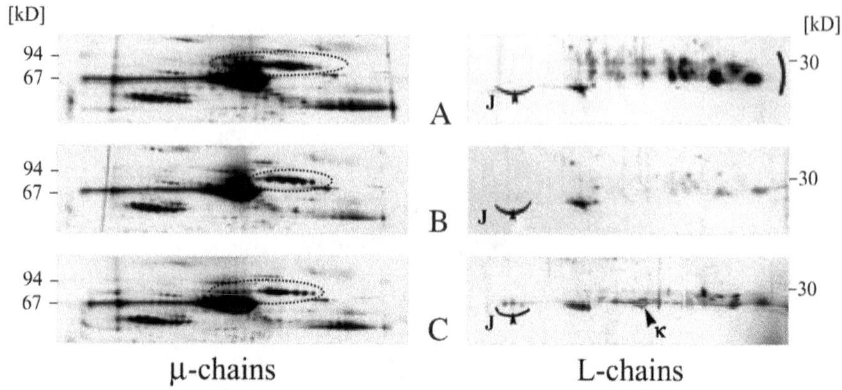

FIGURE 14.5 Sera of patients with monoclonal gammopathies of IgM-type. Close-ups from 2DE gels are shown from the μ-chain region (obtained by IPG-DALT, as described in Figure 14.2) and from the light-chain region (obtained by ISO-DALT, as described below). Samples correspond to 0.5 μl original serum for IPG-DALT or 1 μl for ISO-DALT, respectively. The patients showed the following paraproteins: (A) monoclonal κ-IgM (probably monomeric) in the presence of normal amounts of polyclonal IgG; (B) monoclonal IgM where no corresponding light chain was detected by 2DE, with marked downregulation of the IgG production; and (C) monoclonal κ-IgM (pentameric) in the presence of polyclonal IgG. ISO-DALT system: 2DE performed as described by Goldfarb[53] (tube gels of 1.5 mm diameter, CA pH 3–10, SDS-PAGE in 140 × 160 × 1.5 mm gels with T = 10–20%; silver staining).

does not include anti-IgD) but diagnosed as light-chain disease, which causes similar symptoms. In most reports λ chains prevail over κ chains. Figure 14.7 gives an example of a κ-IgD. Spertini et al.[39] described four cases of λ-IgD. The δ chain may not be discernable in complete serum, especially in the case of overlap with albumin. Affinity purification overcomes the problem and allows correct identification and treatment.[39]

IgD has been usually described as a multiple spot chain[39,52] (faintly seen also in Figure 14.7). This size heterogeneity (difference in molar mass approximately 3 kD) also persists in neuraminidase treatment, despite the pI shift of the spots due to sialic acid loss.[52]

Figure 14.7A and 14.7B show detailed views of the δ-chain region from ISO-DALT and IPG-DALT gels of the same sample. Patterns are very similar. As a horizontal method, IPGs offer the advantage that the sample application point can be chosen depending on the sample and the separation task (here near the anode). In this way, the risk of artifacts (e.g., smearing of the large albumin spots) can be reduced.

14.4.2.2 Biclonal and Oligoclonal Gammopathies

Biclonal gammopathy is characterized by two monoclonal Igs present in the same sample. They may either be produced by a single B-cell clone (differences due to

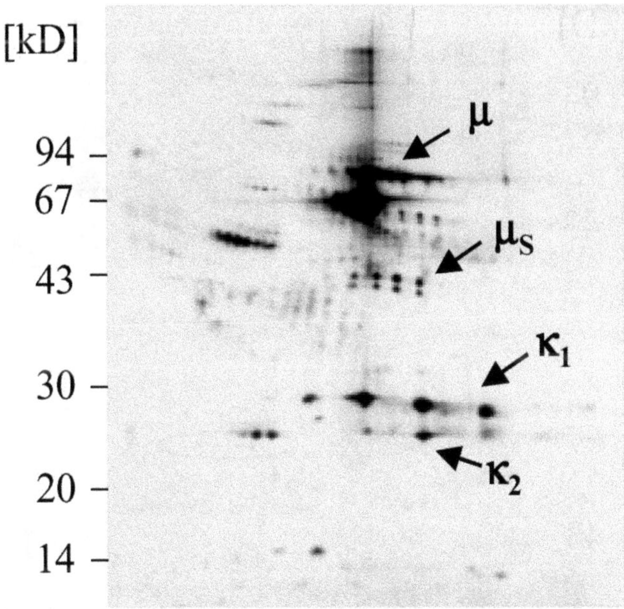

FIGURE 14.6 Serum of patient with biclonal gammopathy of κ-IgM-type. 2DE as in Figure 14.2, corresponding to 0.2 μl serum. Patient serum sample shows two different κ chains, an intact μ chain, and a μ_S chain (truncated form); single chains indicated. The light chains were typed by immunoblotting.

class switch) or by distinct cell populations. This depends on which kind of B-cell is involved, pre-switch B-cells or more differentiated plasma cells.[29] The origin of the biclonality largely influences the patterns, as reflected in Figure 14.6 and by images presented in literature: differences in heavy chain,[26] light chains,[53] and charge microheterogeneity of both chains.[38] As illustrated, biclonals are more easily detected with 2DE than with routine procedures; but—because 2DE works under denaturing and reducing conditions—one cannot in a single step assign the light chains correctly to the heavy chains.[26]

In routine techniques, oligoclonal gammopathies may be incorrectly diagnosed as monoclonal gammopathies if there is one predominating Mab. In 2DE, light-chain patterns are detected as a limited number of well-distinguishable spots. Oligoclonality may reflect T-B cell imbalance or selective antigenic pressures in different diseases, a limited repertoire of B-cell clones (e.g., in the development of the immune system; see section 14.4.1), or excessive B-cell proliferation.[29,54] An example of a time-course study in a patient with allogenic bone marrow transplantation is given:[54] once the allogenic transferred bone marrow began to produce detectable levels of Igs, the formerly polyclonal light-chain pattern changed to oligoclonal due to preferentially expanded Ig-producing B-cell clones, and only in course of time

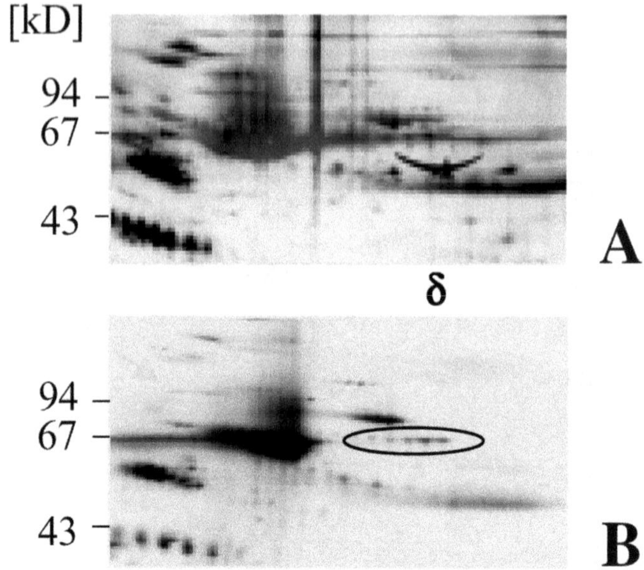

FIGURE 14.7 Serum of patient with monoclonal gammopathy of IgD-type; comparison of ISO-DALT and IPG-DALT. Close-ups from 2DE gels: (A) 2DE as in Figure 14.5 (ISO-DALT), corresponding to 1 μl serum; (B) 2DE as in Figure 14.2 (IPG-DALT), corresponding to 0.2 μl serum (anodic sample application). Patient sample with small amounts of κ-IgD (heavy chain indicated).

the distinct spot pattern became less defined. The development seemed similar to the development in children[47] and might reflect a recapitulation of B-cell development. Another time-course study involved a child with Wiskott-Aldrich syndrome,[25,62] where a change between monoclonal and oligoclonal patterns was noticed in the course of the disease.

14.4.2.3 Other Gammopathies

Agammaglobulinemia, hypogammaglobulinemia, and hypergammaglobulinemia are terms that refer to the amount of Igs present in the patient's serum. Except for the first term, they do not indicate the origin and clonality of the Igs. A pattern of agammaglobulinemia, complete absence of Igs, was shown by Hochstrasser and Tissot.[25] As already discussed in the previous section, Ig clonality can be studied best by inspection of the light-chain pattern. Massive polyclonal stimulation can lead to the appearance of numerous distinct clones in severe infection or autoimmune diseases such as B hepatitis or Sjögren's syndrome.[25] Besides this oligoclonal pattern, where just a limited number of Ig-secreting B-cell clones are preferentially expanded, there are also light-chain patterns with a wide spread of numerous indiscrete spots (polyclonal) or—as a minority—multiple small and well-delineated spots. Patterns look different depending on the patient and the disease, suggesting that different mechanisms are implicated in the production of hypergammaglobulinemia in humans.[54]

14.4.2.3.1 Cryoglobulins

Cryoproteins are defined as proteins precipitating reversibly at low temperatures. Most frequently, the precipitates contain Igs and are therefore called cryoglobulins. Cryoglobulins are associated with a wide variety of diseases such as various B-cell disorders and autoimmune diseases, or in association with viral, bacterial, or parasitic infections. Care has to be taken upon sample collection (prewarmed equipment); for further characterization, cryoglobulins can be isolated easily by precipitation in the cold and subsequent washing. Cryoglobulins are classified according to Brouet et al.[63] into type I monoclonal Ig, type II both monoclonal and polyclonal Igs, type III polyclonal Igs with or without other serum proteins. Identification of the different Ig classes is possible by immunoelectrophoresis, immunofixation, or immunoblotting after one-dimensional separations, but results are not always clearly interpretable.[64] 2DE offers more possibilities to discern monoclonal and polyclonal components and differentiation among Ig classes and subclasses. With the help of this method, an additional class type II–III (complexes of polyclonal IgG with a mixture of polyclonal and monoclonal IgM) was found.[59] A study on 335 samples[65] showed that the majority of cryoglobulins was of mixed origin, belonging to class II, II–III, or III. One sample of cryofibrinogen was found. In hepatitis C virus infection, a correlation between serum rheumatoid factor activity and the virus genotype has been described with cryoglobulin clonality.[66]

14.4.2.3.2 Immune Complexes

Immune complexes are formed every time an antibody meets a corresponding antigen, and generally they are phagocytozed either directly or upon reaction with the complement system and thus removed from circulation. Formation of circulating immune complexes in body fluids is therefore a normal part of the humoral immune defense against soluble antigens, irrespective of their nature. Occasionally, immune complexes persist in the body and are detected as elevated levels in body fluids of patients with infectious, neoplastic, or autoimmune diseases. 2DE has been used to detect the antigenic components of immune complexes in human plasma or serum.[67] Intact immune complexes were purified by taking advantage of their high molar mass and affinity to immobilized protein A or G. Then they were dissociated into their components because of the reducing and denaturing conditions of 2DE. Thus, the enzyme CK-BB could be detected in patients that showed macro creatine kinase.[67] The method serves also for detection and identification of yet unknown antigenic components in immune complexes, and with a modified protocol it is applicable for cerebrospinal fluid and urine.[68,69]

14.4.2.3.3 Cold Agglutinins

Cold agglutinins are autoantibodies, most frequently of the IgM isotype, that are capable of agglutinating erythrocytes at temperatures below 37°C. A study on 25 patients[29] showed that they can be either monoclonal or polyclonal and often occur in viral infections.

14.5 URINE

14.5.1 Patterns of Healthy Individuals

Urine is the filtration product of the kidneys, and its protein content in healthy individuals is minimal (protein concentrations exceeding 0.1–0.15 mg/ml have been defined as pathological).[70] Its composition in regard to salts and proteins may vary in time considerably, being largely influenced by factors such as liquid intake, exercise, and time.

Several classes of proteins can be found in urine: (*a*) normal serum and plasma proteins and their fragments; (*b*) proteins released by the kidney itself during urine production (e.g., Tamm-Horsfall protein); (*c*) proteins from other parts of the urogenital tract (they may be sex dependent), such as from tissues and cells. In case of disease and also tumor, antigens, products of viral and bacterial infection as well as proteins that leak in from some other parts of the body, may be found. First experiments with 2DE of urinary proteins were started almost at the same time as serum 2DE,[71] but in the beginning these experiments were hampered by the low protein concentration (at that time, gels were usually stained with Coomassie). Different concentration steps after desalting by dialysis or gel filtration were established.[72] Even after the introduction of silver-staining methods, concentration of urinary proteins was still a matter of concern, especially regarding whether different sample treatment gave different protein patterns. Proteins were precipitated with trichloroacetic acid and acetone or with dyes (Coomassie G250[30,73,74]) or concentrated with ultrafiltration devices.[75] First, urine patterns with identification of the major proteins, also including IgG heavy and light chains, were published;[76] later, they were extended and more refined.[77–79] The 2DE methods used were the same as for serum, either the ISO-DALT or the IPG-DALT system. By studying urinary protein patterns, markers for diseases or exposures were sought (e.g., different types of cancer,[79–81] cadmium exposure[82]). Today, a urine IEF database exists at the homepage of the Danish Centre of Human Genome Research,[83] established for the search for markers in bladder cancer. The number of identified urinary protein spots was further increased in a recent publication by combining chromatography-based prefractionation with 2DE.[84]

Figure 14.8 shows the urine protein pattern of a healthy volunteer (pattern differences and sex-specific proteins have been described[85]). In comparison to serum, some new proteins, especially acidic ones, appear, and high molar mass proteins are reduced. Some IgG is present, but no other Igs.

14.5.2 Pathological Patterns

In healthy individuals, the renal glomeruli restrict filtration of plasma proteins and the renal tubules reabsorb and degrade most filtered protein such that only trace amounts of proteins (in the range of 50–150 mg/day[86]) appear in the urine. In renal disease, the level of protein in the urine tends to be persistently elevated, and the protein composition is modified in a characteristic manner, depending on the type of disorder. Renal tubular damages (e.g., pyelonephritis, drug-induced tubulopathies, and heavy metal toxicity) result in excretion of mainly low molar mass proteins (β_2-microglobulin, lysozyme,

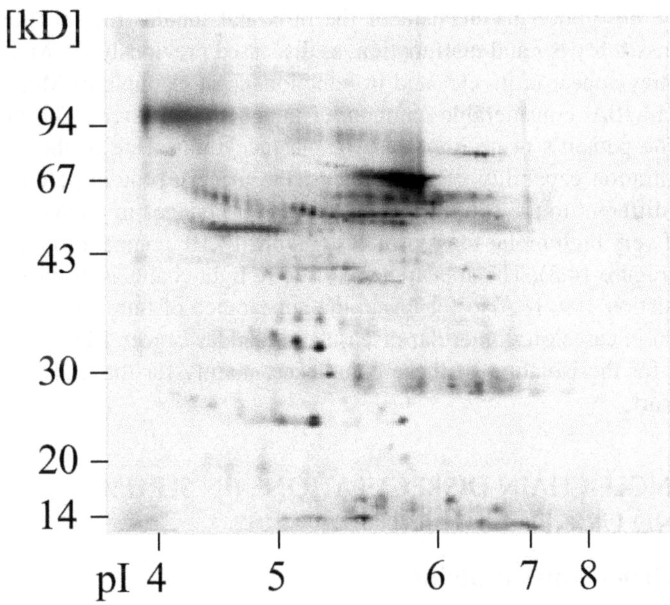

FIGURE 14.8 Urine of a healthy female volunteer. 2DE as in Figure 14.2. The appropriate volume of native urine was dialyzed against ammonium acetate and lyophilized to obtain 12 μg of urinary protein (as determined with a Coomassie G-250 binding assay[87]).

retinol-binding protein, α_1-microglobulin, and α_1-acid-glycoprotein); and the overall protein concentration is usually low and not necessarily higher than in physiologic proteinuria. Renal glomerular disease (glomerulonephritis, nephrosis, glomerulosclerosis, and IgA nephropathy) results in urine containing increased amounts of albumin and high molar mass proteins, mainly IgG. In severe cases, larger proteins may also be found in the urine (e.g., α_2-macroglobulin, IgA, IgM). This finding is usually accompanied by massive changes in protein distribution of serum (e.g., a marked drop of albumin levels) because the protein loss via the kidneys cannot be completely compensated by increased protein production in the liver.[87]

The study of molar mass distribution of excreted proteins by nonreducing SDS-PAGE has been described as a useful method to diagnose renal disorders.[88] Marker proteins and protein patterns have been defined for the different types of disorders or for mixed forms.[70,87] Examples for urinary 2DE patterns of patients with tubular and glomerular damage are shown in.[74,89] With increasing impairment of the glomeruli, the urinary protein pattern becomes more and more similar to serum, and the urine-specific proteins decrease in their relative percentage until they become hardly detectable.[89] Kidney damage in the form of glomerulonephritis occurs in most patients with monoclonal gammopathies.

Urinary Ig levels increase in cases of prerenal overflow proteinuria, when abnormally high plasma levels of low molar mass proteins (e.g., free Ig light chains) exhaust the absorption mechanism of the proximal tubules (see section 14.6.1). Another possibility is renal malfunction, as discussed previously.[70,87] Mabs or polyclonal Ig may appear (e.g., elevated in infections). An example of Mabs is shown in Figure 14.10A: considerable amounts of intact IgA and free light chains were found in the patient's urine, in a case of plasmacytoma. Due to the disease and reduced filtration capability of the kidneys, the urinary protein pattern does not look very different to the serum pattern, except for changes in the Mab levels and the lack of very high molar mass proteins (Figure 14.10; compare to physiological pattern in Figure 14.8). The appearance of free Ig light chains is discussed in more detail in section 14.6.1. Also for urine, the appearance of immune complexes has been shown in cases of glomerular disease and bladder cancer. Methods have been developed for the isolation of these complexes, mainly for the 2DE study of the antigenic part.[68,86]

14.6 SINGLE-CHAIN DISREGULATIONS (IN SERUM AND URINE)

14.6.1 Bence Jones Proteins

Beuce Jones proteins (BJP) are monoclonal Ig light chains originally detected in urine of patients with paraproteinemic disorders.These free monoclonal, kappa or lambda, Ig chains are named after Henry Bence Jones, a British pathologist who first described these proteins in a scientific paper in 1847.[90] Later investigations have found BJP in sera, also, where it is hader to detech. The presence and concentration of BJP in sera (proteinemia) or urine (proteinuria) have different diagostic significance.[91]

Small amounts of free light chains are present in the serum and urine of healthy individuals: serum values up to 19 mg/l for κ and 34 mg/l for λ; in urine up to 5 mg/l for κ and 3 mg/l for λ.[91–93] As these figures indicate, in serum, free λ chains prevail, but urine has more free κ chains.[94] This is attributed to differences in renal clearance.[93,94] The presence of elevated amounts of BJP in serum is a sign of asynchronous synthesis of Ig heavy chains so that excess light chains appear. Monoclonal BJP often accompany gammopathies, especially multiple myeloma, where they are found in about 50% of the cases in humans.[4] In these cases, their appearance in patients' urine in larger amounts is usually an example of overflow proteinuria, that is, a spillover of excess free monoclonal light chains into the urine. Besides free light chains, urine may also contain intact Mab in variable amounts.

Bence Jones proteins appear mainly as monomers or dimers (of about 22 or 44 kD, respectively), but fragments, tetramers, or different glycated forms have also been reported for urine. The two light-chain types κ and λ show a different tendency for polymerization: in serum as well as in urine, κ prevail as monomers and λ in most cases as dimers.[95]

The classical way to detect BJP in urine was the heating of the sample to 56–60°C: BJP precipitate but dissolve when the solution is boiled; the precipitate reappears upon cooling.[90,96] There have also been attempts to find urinary free light chains by immunofixation or by using κ–λ ratios.[94] Recently, polyclonal and monoclonal antibodies have been developed that can selectively detect free light chains only (of either isotype), as they are directed against light-chain parts hidden in intact Ig molecules.[93,97] The assays developed are now specific and sensitive enough also to detect free light chains in serum, and in this way, BJPs are also found in cases that have been reported as "nonsecretory multiple myeloma".[91]

As a routine electrophoretic method, for serum or urine, zone electrophoreses followed by immunofixation methods are in use. Research papers have mainly dealt with detection of BJP in urine. BJP and urinary Ig versus normal protein patterns have been studied in SDS-PAGE and 2DE, and sometimes also in isoelectric focusing. Patterns were either stained for overall protein or after blotting specifically for Ig.[30,74,76,95,98] A 2DE gel of a serum sample with BJP is shown in Figure 14.9. The free light chains are of λ type with a very acidic pI.

Especially for urine, SDS-PAGE has been used in nonreducing format in order to determine the protein size without destroying the subunit structure (except for molecule interactions already disturbed by the presence of SDS). 2DE has always been used in its classical form (reductive and denaturing). Only a recent study has shown the benefits of nonreducing 2DE, which allows study of free monomeric and dimeric single chains and intact Mabs in parallel.[42] Figure 14.10 exemplifies this with paired

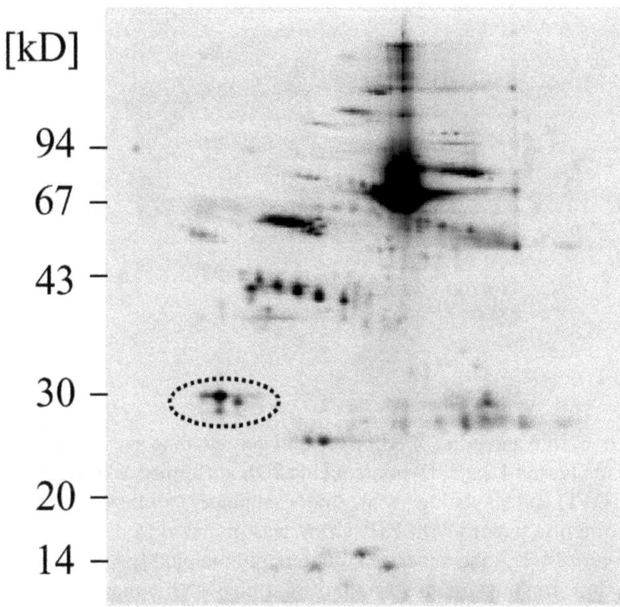

FIGURE 14.9 Patient with BJP in serum. 2DE as in Figure 14.2, corresponding to 0.2 μl serum. Patient sample with very acidic monoclonal λ chain (indicated).

samples from a patient with plasmacytoma. Serum and urine were analyzed in parallel: classical 2DE in serum reveals monoclonal IgA of the κ type (Figure 14.10C); in urine the same light-chain type is found in large quantities, as is some of the monoclonal α chain (Figure 14.10A). There are some spots in the 44 kD region that suggest light-chain dimers. From the pattern one may conclude that there must be free urinary light chains due to the gross differences in distribution of single chains. Nonreducing 2DE (Figure 14.10B, D) reveals the presence of intact monoclonal IgA in serum and, in minor amounts, in urine and major concentrations of monomeric and dimeric κ chains in urine (and probably one spot of dimers in serum). For comparison with healthy control patterns see Figure 14.8.

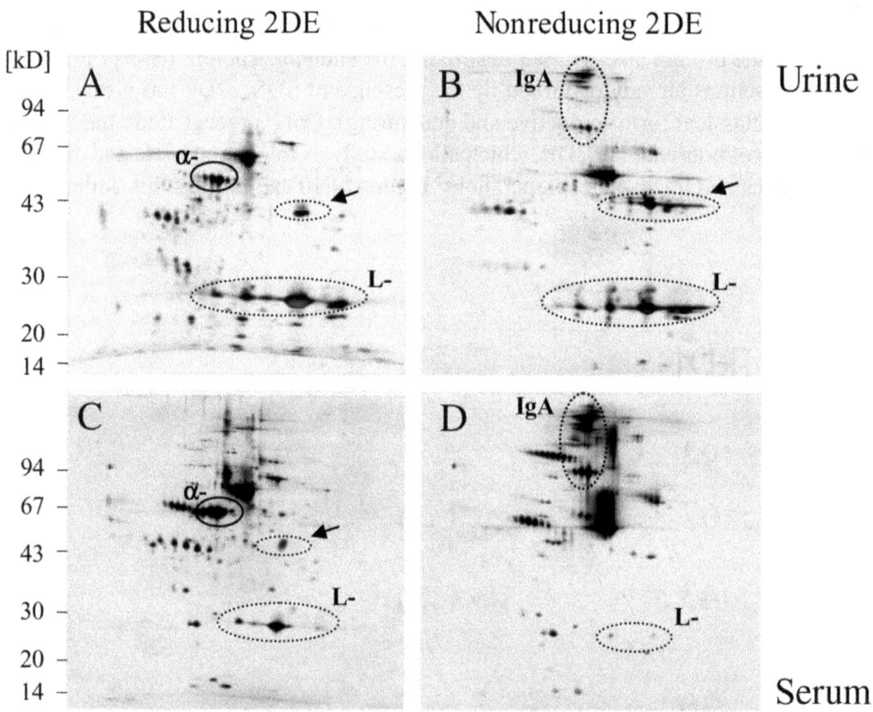

FIGURE 14.10 Serum and urine of BJP patient under reducing and nonreducing conditions. (A, C) 2DE as in Figure 14.2; (B, D) nonreducing 2DE performed as in Figure 14.2, but omitting reducing (DTT) and alkylating agents (iodoacetamide) (for method see[42]). (A, B) Urine and (C, D) serum of a patient with BJP. Urine was prepared as described in Figure 14.8, serum as in Figure 14.2. Appearance of complete monoclonal IgA molecules (IgA, in different polymeric forms), α chains (α-), and light chains (L-, monomeric); dimeric light chains are indicated with an arrow. Note that in comparing patterns with and without reduction, even monomeric proteins may shift in position. One example is serum albumin, which has been shown to change apparent Molecular weight (Mr) and pI depending on pH, denaturants, and other additives.[99,100]

As for serum Mabs, the pI of BJP chains is dependent on the B-cell clone it derives from and may vary over a very large pH range. It seems likely that not only the size or type of binding between the single light chains but also the pI may influence the processing of the protein in the kidney. Some authors think that alkaline BJP are more nephrotoxic than acidic ones (for further details see[95]).

Bence Jones proteins are also causative agents in the progression of renal failure and the development of systemic disease, especially when they are deposited extracellularly, leading to amyloidosis AL. Approximately 15–20% of monoclonal light chains appear to be amylogenic, that is, they may precipitate as fibrillar material resembling amyloid. No specific amino acid sequence common to all amyloidogenic light chains has yet been identified.[29] Detection of BJP is therefore important for diagnosis of myeloma and unclear cases of proteinuria and the follow-up of monoclonal gammopathies. Bradwell et al.[91] have reported that the half-life of free light chains is much shorter than that of intact Igs (2–4 hours vs. 20–25 days for IgG), and changes in their levels may give earlier hints for progression of the disease or effects of chemotherapy. The authors favor measuring BJP in serum, as BJP excretion in urine may be strongly affected by the reabsorptive capacity of the renal tubules.[91,96]

14.6.2 Heavy-Chain Disease

Heavy-chain diseases of all three major Ig isotypes have been described, in serum and in urine (for the latter, depending on the size of the molecules and the state of the kidney). Often fragments of the heavy chains are found, mainly at 30 to 36 kD, but no light chains. Single α chains are more frequent than γ chains. Blangarin et al.[56] have described a γ case and characterized the found 34 kD fragment in 2DE and immunoblotting as reactive to anti-Fc but not to anti-Fab. Tissot et al.[6] isolated the protein G-binding fraction of a serum sample from a patient with monoclonal gamma-chain disease and found several truncated γ chains besides the normal set of polyclonal IgG chains. Also, in the main fragment the γ Fc region could be detected.

14.7 OTHER BODY FLUIDS

The major part of proteins in all fluid compartments originates from plasma. The fluids can be considered as an ultrafiltrate of blood plasma in a first approximation. This is true for cerebrospinal fluid, separated from blood by the blood-brain barrier, for urine (see section 14.5), and for synovial fluid, lymph, exudates, and transudates. Plasma proteins normally are synthesized and excreted by the liver and lymphocytes. Protein synthesis, however, also occurs in the brain and different glands, which makes the composition of the respective body fluids specific.[101]

14.7.1 Cerebrospinal Fluid (CSF)

Cerebrospinal fluid (CSF), or spinal fluid, is a fluid surrounding the brain and spinal cord. Most of its proteins derive from serum, but CSF-specific proteins have also been found (e.g., prostaglandin D2 synthase). The actual proportion of serum proteins depends on the functional state of the blood-brain barrier, which is defective in several neurological diseases. Therefore, the study of its proteins and their alteration may yield useful information on that type of diseases.[101] The first 2DE maps of CSF of healthy individuals were established very early[102,103] and refined over the years.[33,101,104–107]

The main Ig present in CSF is IgG. In normal conditions, it originates from the blood; there is no antibody production within the central nervous system. Only up to 20% of CSF proteins are intrathecally synthesized, but the major fraction is blood derived.[108] In case of inflammatory neurological diseases such as multiple sclerosis, additional oligoclonal Ig bands or spots appear that are present in CSF but not in serum. One-dimensional isoelectric focusing for detection of CSF-specific oligoclonal Ig bands is used for diagnosis of multiple sclerosis.[108–110] Basic research in this field has been performed with 2DE: main differences in CSF patterns of healthy individuals and patients with various neurological diseases have been detected in the Ig light-chain pattern. Changes were of qualitative and quantitative nature; additional light chains were oligoclonal. Especially for multiple sclerosis, the overall CSF pattern of patient samples suggested that Igs were produced from cells within the central nervous system, as no additional spots were found proving leakage of a damaged blood-brain barrier.[111,112]

In patients with multiple sclerosis, also circulating immune complexes have been isolated from CSF, and glial fibrillary acidic protein has been found in the 2DE patterns.[69]

14.7.2 Saliva, Tears, and Milk

Saliva, tears, and milk all contain secretory IgA as a main Ig, and in considerably high amounts. Secretory IgA is a dimer joined by a J chain with a secretory piece attached. The secretory piece is a fragment of the poly-Ig receptor molecule, which helps transport the IgA across the mucosa.

Saliva composition has been studied in 2DE, using immunoblotting[113] and mass spectrometry as methods for protein identifications.[114,115] Interest in this fluid has increased because of its easy accessibility and potential use for monitoring general health and diagnosis of disease. For 2DE, whole saliva is not an easy sample, as it contains mucins and other high-molar-mass glycoproteins. Secretory IgA has been found a major defense factor in saliva, playing an important role in the protection against enteropathogenic bacteria and viruses.[114]

The protein composition of tears seems to vary depending on the way used to induce them or to collect the samples.[116] 2DE with nondenaturing first-dimensional separation as well as classical 2DE have been described, reporting albumin, secretory IgA, lactoferrin, lysozyme,[25,116] and zinc-α_2-glycoprotein[117] as main proteins in healthy persons. In patients with conjunctivitis, haptoglobin and IgG appear and increase with progression of the disease.[116]

Protein patterns of human milk, from colostrum to mature milk, have been studied and found to vary diurnally and individually. Milk contains caseins and whey proteins (mainly serum proteins). Secretory IgA is a major protein component, especially in colostrum (3–8 days postpartum), whereas IgG is barely detectable.[118,119]

14.7.3 Effusions and Lavage Fluids

In cases of disease, ascitic fluid accumulates and shows a protein distribution very similar to physiological or pathological plasma, with disease-specific changes. Two additional fibrinogen fragments have been found. Ig pattern is correlated with disease diagnosis and severity: massive IgA increase has been described in liver cirrhosis, and IgM and IgG were elevated in acute and chronic infection with polyclonal stimulation; in other malignancies, oligoclonal patterns occurred.[120] Hochstrasser and Tissot include an image of a patient's pleural fluid.[25]

Several studies deal with the composition of nasal and bronchoalveolar lavage fluids with or without occupational exposure or as a comparison of smokers to nonsmokers. IgA levels were higher in nasal fluids than in bronchoalveolar lavage fluids.[121–124]

Hochstrasser and Tissot[25] examine other body fluids in health and disease and also present 2DE examples.

14.8 CONCLUSIONS AND PERSPECTIVES

In a recent review on immunoglobulinopathies, Tissot et al.[16] gave a critical evaluation of the methods used for the study of gammopathies. The authors compared and evaluated routine and specialized methods. Whereas they found methods such as immunofixation (after serum electrophoresis on agarose gels) preferable for routine clinical applications, they still hold 2DE important for particular clinical situations and for research. This is not only true for serum and plasma, but also for other body fluids, since most of them contain Igs either originating from serum or due to local antibody production. Modifications in the distribution and clonality of total or antigen-specific Ig may reflect normal, as well as abnormal, immune events and need to be evaluated carefully. Due to its high resolution, 2DE is a powerful tool enabling us to study the diversity of the body's immune repertoire at the protein level, especially when combining data from reducing and nonreducing formats with sample prefractionation and postelectrophoretic immunologic and mass spectrometric identifications. Despite all efforts, the functioning of the immune system is not yet resolved in all its details. Apart from Igs, detection of polypeptides associated with immunological events may enhance further understanding, as may the parallel investigations on the secretory and intracellular level. Studies of particular diseases or patient cases are similarly helpful, as are those of mouse strains with selected immunodeficiency (e.g., SCID-mice).[41] New results and insights may then find their applications in clinical diagnosis—not necessarily as 2DE, but probably as specialized routine tests such as ELISA or other specific assays.

REFERENCES

1. Widmann, F.K. and Itatani, C.A., *An Introduction to Clinical Immunology and Serology*, 2nd ed., F. A. Davis Company, Philadelphia, 1998, 473 pp.
2. Wells, J.V., Isbister, J.P., Ries, C.A. Hematologic diseases, in *Basic & Clinical Immunology*, 6th ed., Stites, D.P., Stobo, J.D., Wells, J.V., Eds., Appleton & Lange, Norwalk, Conn., 1987, pp. 386–419.

3. Roitt, I.M. and Delves, P.J., *Essential Immunology*, 10th ed., Blackwell Science Ltd, Oxford, U.K., 2001, 481 pp.
4. Kyle, R.A., The monoclonal gammopathies, *Clin. Chem.*, 40, 2154–2161, 1994.
5. The International Myeloma Working Group, Criteria for the classification of monoclonal gammopathies, multiple myeloma and related disorders: A report of the International Myeloma Working Group, *Br. J. Haematol.*, 121, 749–757, 2003.
6. Tissot, J.-D., Tridon, A., Ruivard, M., Layer, A., Henry, H., Philippe, P., and Schneider, P., Electrophoretic analyses in a case of monoclonal γ chain disease, *Electrophoresis*, 19, 1771–1773, 1998.
7. Lecuit, M., Abachin, E., Martin, A., Poyart, C., Pochart, P., Suarez, F., Bengoufa, D., Feuillard, J., Lavergne, A., Gordon, J.I., Berche, P., Guillevin, L., and Lortholary, O., Immunoproliferative small intestinal disease associated with *Campylobacter jejuni, New Engl. J. Med.*, 350, 239–248, 2004.
8. Kyle, R.A. and Lust, J.A., Monoclonal gammopathies of undetermined significance, *Semin. Hematol.*, 26, 176–200, 1989.
9. Stites, D.P., Clinical laboratory methods for detection of antigens & antibodies, in *Basic & Clinical Immunology*, 6th Ed., Stites, D.P., Stobo, J.D., Wells, J.V., Eds., Appleton & Lange, Norwalk, Conn., 1987, 241–284.
10. Keren, D.F., Alexanian, R., Goeken, J.A., Gorevic, P.D., Kyle, R.A., and Tomar, R.H., Guidelines for clinical and laboratory evaluation of patients with monoclonal gammopathies, *Arch. Pathol. Lab. Med.*, 123, 106–107, 1999.
11. Keren, D.F., Procedures for the evaluation of monoclonal immunoglobulins, *Arch. Pathol. Lab. Med.*, 123, 126–132, 1999.
12. Chen, F.-T.A. and Sternberg, J.C., Characterization of proteins by capillary electrophoresis in fused-silica columns: Review on serum protein analysis and application to immunoassays. *Electrophoresis*, 15, 13–21, 1994.
13. Katzmann, J.A., Clark, R., Wiegert, E., Sanders, E., Oda, R.P., Kyle, R.A., Namyst-Goldberg, C., and Landers, J.P., Identification of monoclonal proteins in serum: A quantitative comparison of acetate, agarose gel, and capillary electrophoresis, *Electrophoresis*, 18, 1775–1780, 1997.
14. Wang, H.-P. and Liu, C.-M., Separation and identification of human serum proteins with capillary electrophoresis, *Clin. Chem.*, 38, 963–964, 1992.
15. Jenkins, M.A., Kulinskaya, E., Martin, H.D., and Guerin, M.D., Evaluation of serum protein separation by capillary electrophoresis: Prospective analysis of 1000 specimens, *J. Chromatogr. B*, 672, 241–251, 1995.
16. Tissot, J.-D., Vu, D.-H., Aubert, V., Schneider, P., Vuadens, F., Crettaz, D., and Duchosal, M.A., The immunoglobulinopathies: From physiopathology to diagnosis, *Proteomics*, 2, 813–824, 2002.
17. MacNamara, E.M. and Whicher, J.T., Electrophoresis and densitometry of serum and urine in the investigation and significance of monoclonal immunoglobulins, *Electrophoresis*, 11, 376–381, 1990.
18. Gallo, P., Sidén, A., and Tavolto, B., Isoelectric Focusing Combined with Anion Exchange Chromatography for Investigation of Normal and Monoclonal Immunoglobulins, in *Electrophoresis '86: Proceedings of the 5th Meeting of the International Electrophoresis Society*, Dunn, M.J., Ed., London, 1986, pp. 498–501.
19. Laemmli, U.K., Cleavage of structural proteins during the assembly of the head of bacteriophage T4, *Nature*, 227, 680–685, 1970.
20. Anderson, L. and Anderson, N.G., High resolution two-dimensional electrophoresis of human plasma proteins, *Proc. Natl. Acad. Sci. USA*, 74, 5421–5425, 1977.

21. Tracy, R.P., Currie, R.M., and Young, D.S., Two-dimensional gel electrophoresis of serum specimens from a normal population, *Clin. Chem.*, 28, 890–899, 1982.
22. Jellum, E. and Thorsrud, A.K., Clinical applications of two-dimensional electrophoresis, *Clin. Chem.* 28, 876–883, 1982.
23. Marshall, T. and Williams, K.M., The simplified technique of high resolution two-dimensional polyacrylamide gel electrophoresis: Biomedical applications in health and disease, *Electrophoresis*, 12, 461–471, 1991.
24. Tissot, J.-D., Schneider, P., James, R.W., Daigneault, R., and Hochstrasser, D.F., High-resolution two dimensional protein electrophoresis of pathological plasma/serum, *Appl. Theor. Electrophoresis*, 2, 7–12, 1991.
25. Hochstrasser, D.F. and Tissot, J.-D., Clinical application of high-resolution two-dimensional polyacrylamide gel electrophoresis, in *Advances in Electrophoresis*, Chrambach, A., Dunn, M.J., Radola, B.J., Eds., VCH, Weinheim, Germany, 1993, vol. 6, pp. 267–375.
26. Tracy, R.P., Currie, R.M., Kyle, R.A., and Young, D.S., Two-dimensional gel electrophoresis of serum specimens from patients with monoclonal gammopathies, *Clin. Chem.*, 28, 900–907, 1982.
27. Latner, A.L., Marshall, T., and Gambie, M., A simplified technique of high resolution two-dimensional electrophoresis: serum immunoglobulins, *Clin. Chim. Acta*, 103, 51–59, 1980.
28. Harrison, H.H., Patient-specific microheterogeneity patterns of monoclonal immunoglobulin light chains as revealed by high resolution, two-dimensional electrophoresis, *Clin. Biochem.* 25, 235–243, 1992.
29. Tissot, J.-D. and Spertini, F., Analysis of immunoglobulins by two-dimensional electrophoresis, *J. Chromatogr. A*, 698, 225–250, 1995.
30. Williams, K.M., Williams, J., and Marshall, T., Analysis of Bence Jones proteinuria by high resolution two-dimensional electrophoresis, *Electrophoresis*, 19, 1828–1835, 1998.
31. Vu, D.-H., Schneider, P., and Tissot, J.-D., Electrophoretic characteristics of monoclonal immunoglobulin G of different subclasses, *J. Chromatogr. B*, 771, 355–368, 2002.
32. Swiss-Prot Protein Knowledgebase, SWISS-2DPAGE, http://us.expasy.org/ch2d/protocols (accessed August 2005).
33. Swiss-Prot Protein Knowledgebase, SWISS-2DPAGE, http://us.expasy.org/ch2d/publi/inside1995.html (accessed August 2005).
34. Golaz, E., Hughes, G.J., Frutiger, S., Paquet, N., Bairoch, A., Pasquali, C., Sanchez, H.-C., Tissot, J.-D., Appel, R.D., Walzer, C., Balant, L., and Hochstrasser, D.F., Plasma and red blood cell protein maps: Update 1993, *Electrophoresis*, 14, 1223–1231, 1993.
35. Hughes, G.J., Frutiger, S., Paquet, N., Ravier, F., Pasquali, C., Sanchez, J.-C., James, R., Tissot, J.-D., Bjellqvist, B., and Hochstrasser, D.F., Plasma protein map: An update by microsequencing, *Electrophoresis*, 13, 707–714, 1992.
36. Sanchez, J.-C., Appel, R.D., Golaz, O., Pasquali, C., Ravier, F., Bairoch, A., and Hochstrasser, D.F., Inside SWISS-2DPAGE database, *Electrophoresis*, 16, 1131–1151, 1995.
37. Anderson, N.L., Tracy, R.P., and Anderson, N.G., High-resolution two-dimensional electrophoretic mapping of plasma proteins, in *The Plasma Proteins: Structure, Function, and Genetic Control*, 2nd ed., Putnam, F.W., Ed., Academic Press, Orlando, 1984, vol. 4, 221–270.
38. Tissot, J.-D., Hochstrasser, D.F., Spertini, F., Schifferli, J.A., and Schneider, P., Pattern variations of polyclonal and monoclonal immunoglobulins of different isotypes

and analyzed by high-resolution two-dimensional electrophoresis, *Electrophoresis*, 14, 227–234, 1993.
39. Spertini, F., Tissot, J.D., Dufour, N., Francillon, C., and Frei, P.C., Role of two-dimensional electrophoretic analysis in the diagnosis and characterization of IgD monoclonal gammopathy, *Allergy*, 50, 664–670, 1995.
40. Layer, A., Tissot, J.-D., Schneider, P., and Duchosal, M.A., Micropurification and two-dimensional polyacrylamide gel electrophoresis of immunoglobulins for studying the clonal diversity of antigen-specific antibodies, *J. Immunol. Methods*, 227, 137–148, 1999.
41. Layer, A., Tissot, J.-D., Schneider, P., and Duchosal, M.A., The diversity of antigen-specific antibodies in humans and in two xenochimeric SCID mouse models, *Electrophoresis*, 21, 2463–2475, 2000.
42. Miller, I., Teinfalt, M., Leschnik, M., Wait, R., and Gemeiner, M., Nonreducing two-dimensional gel electrophoresis for the detection of Bence Jones proteins in serum and urine, *Proteomics*, 4, 257–260, 2004.
43. Daufeldt, J.A. and Harrison, H.H., Quality control and technical outcome of ISO-DALT two-dimensional electrophoresis in a clinical laboratory setting, *Clin. Chem.*, 30, 1972–1980, 1984.
44. Harrison, H.H., Ober, C., Miller, K.I., and Elias, S., High-resolution two-dimensional electrophoretic survey of serum protein genetic types in Schmiedeleut hutterites, *Am. J. Hum. Biol.*, 3, 639–646, 1991.
45. Pieper, R., Gatlin, C.L., Makusky, A.J., Russo, P.S., Schatz, C.R., Miller, S.S., Su, Q., McGrath, A.M., Estock, M.A., Parmar, P.P., Zhao, M., Huang, S.-T., Zhou, J., Wang, F., Esquer-Blasco, F., Anderson, N.G., Taylor, J., and Steiner, S., The human serum proteome: Display of nearly 3700 chromatographically separated protein spots on two-dimensional electrophoresis gels and identification of 325 distinct proteins, *Proteomics*, 3, 1345–1364, 2003.
46. Anderson, N.L., High resolution two-dimensional electrophoretic mapping of immunoglobulin light chains, *Immunol. Lett.*, 2, 195–199, 1981.
47. Tissot, J.-D., Hohlfeld, P., Forestier, F., Tolsa, J.-F., Hochstrasser, D.F., Calame, A., Plouvier, E., Bossart, H., and Schneider, P., Plasma/serum protein patterns in human fetuses and infants: A study by high-resolution two-dimensional polyacrylamide gel electrophoresis, *Appl. Theor. Electrophoresis*, 3, 183–190, 1993.
48. Tissot, J.-D., Hohlfeld, P., Hochstrasser, D.F., Tolsa, J.-F., Calme, A., and Schneider, P., Clonal imbalances of plasma/serum immunoglobulin production in infants, *Electrophoresis*, 14, 245–247, 1993.
49. Tissot, J.-D., Schneider, P., Hohlfeld, P., Tolsa, J.-F., Calame, A., and Hochstrasser, D.F., Monoclonal gammopathy in a 30 weeks old premature infant, *Appl. Theor. Electrophoresis*, 3, 67–68, 1992.
50. Hamilton, R.G., Roebber, M., Reimer, C.B., and Rodkey, L.C., Isoelectric focusing-affinity immunoblot analysis of mouse monoclonal antibodies to the four human IgG subclasses, *Electrophoresis*, 8, 127–134, 1987.
51. Wenisch, E., Reiter, S., Hinger, S., Steindl, F., Tauer, C., Jungbauer, A., Katinger, H., and Righetti, P.G., Shifts of isoelectric points between cellular and secreted antibodies as revealed by isoelectric focusing and immobilized pH gradients, *Electrophoresis*, 11, 966–969, 1990.
52. Latner, A.L., Marshall, T., and Gambie, M., Microheterogeneity of serum myeloma immunoglobulins revealed by a technique of high resolution two-dimensional electrophoresis, *Electrophoresis*, 1, 82–89, 1980.

53. Goldfarb, M., Two-dimensional electrophoretic analysis of immunoglobulin patterns in monoclonal gammopathies, *Electrophoresis*, 13, 440–444, 1992.
54. Tissot, J.-D., Schneider, P., Hohlfeld, P., Spertini, F., Hochstrasser, D.F., and Duchosal, M.A., Two-dimensional electrophoresis as an aid in the analysis of the clonality of immunoglobulins, *Electrophoresis*, 14, 1366–1371, 1993.
55. Goldfarb, M., Use of Rotofor in two-dimensional electrophoretic analysis: Identification of a 100 kDa monoclonal IgG heavy chain in myeloma serum, *Electrophoresis*, 14, 1379–1381, 1993.
56. Blangarin, P., Deviller, P., Kindbeiter, K., and Madjar, J.-J., Gamma heavy chain disease studied by two-dimensional electrophoresis and immuno-blotting techniques, *Clin. Chem.*, 30, 2021–2025, 1984.
57. Bouvet, J.P., Liacopoulos, P., Pillot, J., Banda, R., Tung, E., and Wang, A.C., Three M-components IgA lambda + IgG kappa n + IgG kappa h in one patient (DA): Lack of shared idiotypic determinants between IgA and IgG, and the presence of an unusual kappa H chain of 30,000 M.W., *J. Immunol.*, 125, 213–220, 1980.
58. Bouvet, J.P., Pillot, J., and Liacopoulos, P., Human myeloma light chains with increased molecular weight: High frequency among lambda chains, *Mol. Immunol.*, 20, 397–407, 1983.
59. Tissot, J.-D., Schifferli, J.A., Hochstrasser, D.F., Pasquali, C., Spertini, F., Clement, F., Frutiger, S., Paquet, N., Hughes, G.J., and Schneider, P., Two-dimensional polyacrylamide gel electrophoresis analysis of cryoglobulins and identification of an IgM-associated peptide, *J. Immunol. Methods*, 173, 63–75, 1994.
60. Tissot, J.-D., Sanchez, S.-C., Vuadens, F., Scherl, A., Schifferli, J.A., Hochstrasser, D.F., Schneider, P., and Duchosal, M.A., IgM are associated to Spα (CD5 antigen-like), *Electrophoresis*, 23, 1203–1206, 2002.
61. Gebe, J.A., Kiener, P.A., Ring, H.Z., Li, X., Francke, U., and Aruffo, A., Molecular cloning, mapping to human chromosome 1 q21-q23, and cell binding characteristics of Spα, a new member of the scavenger receptor cysteine-rich (SRCR) family of proteins, *J. Biol. Chem.*, 272, 6151–6158, 1997.
62. Tissot,J.D.,Schneider,P.,Pelet,B.,Frei,P.C.,andHochstrasser,D.,Mono-oligoclonal production of immunoglobulin in a child with the Wiskott-Aldrich syndome, *Br. J. Haematol.*, 75, 436–438, 1990.
63. Brouet, J.C., Clauvel, J.P., Danon, F., Klein, M., and Seligmann, M., Biological and clinical significance of cryoglobulins: A report of 86 cases, *Am. J. Med.* 57, 775–788, 1974.
64. Musset, L., Diemert, M.C., Taibi, F., Thi Huong Du, L., Cacoub, P., Leger, J.M., Boissy, G., Gaillard, O., and Galli, J., Characterization of cryoglobulins by immunoblotting, *Clin. Chem.*, 38, 798–802, 1992.
65. Tissot, J.-D., Invernizzi, F., Schifferli, J.A., Spertini, F., and Schneider, P., Two-dimensional electrophoretic analysis of cryoproteins: A report of 335 samples, *Electrophoresis*, 20, 606–613, 1999.
66. Antonescu, C., Mayerat, C., Mantegani, A., Frei, P.C., Spertini, F., and Tissot, J.-D., Hepatitis C virus (HCV) infection: Serum rheumatoid factor activity and HCV genotype correlate with cryoglobulin clonality, *Blood*, 92, 3486–3487, 1998.
67. Wiederkehr, F., Büeler, M.R., and Vonderschmitt, D.J., Chromatographic and electrophoretic studies of circulating immune complexes in plasma, *J. Chromatogr.* 566, 77–87, 1991.
68. Wiederkehr, F., Büeler, M.R., and Vonderschmitt, D.J., Analysis of circulating immune complexes isolated from plasma, cerebrospinal fluid and urine, *Electrophoresis*, 12, 478–486, 1991.

69. Wiederkehr, F., Wacker, M., and Vonderschmitt, D.J., Analysis of immune complexes of cerebrospinal fluid by two-dimensional gel electrophoresis, *Electrophoresis*, 10, 473–479, 1989.
70. Bianchi-Bosisio, A., D'Agrosia, F., Gaboardi, F., Gianazza, E., and Righetti, P.G., Sodium dodecyl sulphate electrophoresis of urinary proteins, *J. Chromatogr.*, 569, 243–260, 1991.
71. Anderson, N.G., Anderson, N.L., and Tollaksen, S.L., Proteins of human urine. I. Concentration and Analysis by Two-Dimensional Electrophoresis, *Clin. Chem.*, 25, 1199–1210, 1979.
72. Anderson, N.G., Anderson, N.L., Tollaksen, S.L., Hahn, H., Giere, F., and Edwards, J., Analytical techniques for cell fractions. XXV. Concentration and two-dimensional electrophoretic analysis of human urinary proteins, *Anal. Biochem.*, 95, 48–61, 1979.
73. Marshall, T. and Williams, K.M., Recovery of protein by Coomassie Brilliant Blue precipitation prior to electrophoresis, *Electrophoresis*, 13, 887–888, 1992.
74. Marshall, T. and Williams, K., Two-dimensional electrophoresis of human urinary proteins following concentration by dye precipitation, *Electrophoresis*, 7, 1265–1272, 1996.
75. Marshall, T., Centriprep ultrafiltration for fractionation of serum and urinary proteins before electrophoresis, *Clin. Chem.*, 39, 1558, 1993.
76. Edwards, J.J., Tollaksen, S.L., and Anderson, N.G., Proteins of human urine. III. Identification and two-dimensional electrophoretic map positions of some major urinary proteins, *Clin. Chem.*, 28, 941–948, 1982.
77. Büeler, M.R., Wiederkehr, F., and Vonderschmitt, D.J., Electrophoretic, chromatographic and immunological studies of human urinary proteins. *Electrophoresis*, 16, 124–134, 1995.
78. Harrison, H.H., The "Ladder Light Chain" or "Pseudo-Oligoclonal" pattern in urinary immunofixation electrophoresis (IFE) studies: A distinctive IFE pattern and an explanatory hypothesis relating it to free polyclonal light chains, *Clin. Chem.*, 37, 1559–1564, 1991.
79. Rasmussen, H.H., Orntoft, T.F., Wolf, H., and Celis, J.E., Toward a comprehensive database of proteins from the urine of patients with bladder cancer, *J. Urol.*, 155, 2113–2119, 1996.
80. Edwards, J.J., Anderson, N.G., Tollaksen, S.L., von Eschenbach, A.C., and Guevara, J., Proteins of human urine. II. Identification by two-dimensional electrophoresis of a new candidate marker for prostatic cancer, *Clin. Chem.*, 28, 160–163, 1982.
81. Tollaksen, S.L. and Anderson, N.G., Two-dimensional electrophoresis of human urinary proteins in health and disease, in *Electrophoresis '79: Proceedings of the Second International Conference on Electrophoresis*, Radola, B.J., Ed., Walter de Gruyter, Berlin, 1980, pp. 405–414.
82. Marshall, T., Williams, K.M., and Vesterberg, O., Unconcentrated human urinary proteins analysed by high resolution two-dimensional electrophoresis with narrow pH gradients: Preliminary findings after occupational exposure to cadmium, *Electrophoresis*, 6, 47–52, 1985.
83. Danish Centre of Human Genome Research, http://proteomics.cancer.dk/ (last accessed August 2005)
84. Pieper, R., Gatlin, C., McGrath, A.M., Makusky, A.J., Mondal, M., Seonarain, M., Field, E., Schatz, C.R., Estock, M.A., Ahmed, N., Anderson, N.G., and Steiner, S., Characterization of the human urinary proteome: A method for high-resolution display

of urinary proteins on two-dimensional electrophoresis gels with a yield of nearly 1400 distinct protein spots, *Proteomics*, 4, 1159–1174, 2004.
85. Büeler, M.R.M., Charakterisierung und Identifizierung von Proteinen in Human-Urin, Ph.D. thesis, University Zuerich, 1991, 168 pp (in German).
86. Weber, M.H., Urinary protein analysis, *J. Chromatogr.*, 429, 315–344, 1988.
87. Bradford, M.M., A rapid and sensitive method for the quantitation of microgram quantities of protein utilizing the principle of protein-dye binding, *Anal. Biochem.*, 72, 248–254, 1976.
88. Schiwara, H.-W., Hebell, T., Kirchherr, H., Postel, W., Weser, J., and Görg, A., Ultrathin-layer sodium dodecyl sulfate-polyacrylamide gradient gel electrophoresis and silver staining of urinary proteins, *Electrophoresis*, 7, 496–505, 1986.
89. Tracy, R.P., Young, D.S., Hill, H.D., Cutsforth, G.W., and Wilson, D.M., Two-dimensional electrophoresis of urine specimens from patients with renal disease, *Appl. Theor. Electrophoresis*, 3, 55–65, 1992.
90. Kyle, R.A., Multiple myeloma: An odyssey of discovery, *Br. J. Haematol.*, 111, 1035–1044, 2000.
91. Bradwell, A.R., Carr-Smith, H.D., Mead, G.P., Harvey, T.C., and Drayson, M.T., Serum test for assessment of patients with Bence Jones myeloma, *Lancet*, 361, 489–491, 2003.
92. Abe, M., Goto, T., Kosaka, M., Wolfenbarger, D., Weiss, D.T., and Solomon, A., Differences in kappa to lambda (κ:λ) ratios of serum and urinary free light chains, *Clin. Exp. Immunol.*, 111, 457–462, 1998.
93. Bradwell, A.R., Carr-Smith, H.D., Mead, G.P., Tang, L.X., Showell, P.J., Drayson, M.T., and Drew, R., Highly sensitive, automated immunoassay for immunoglobulin free light chains in serum and urine, *Clin. Chem.*, 47, 673–680, 2001.
94. Levinson, S.S. κ/λ Index for confirming urinary free light chain in amyloidosis AL and other plasma cell dyscrasias, *Clin. Chem.*, 37, 1122–1126, 1991.
95. Marshall, T. and Williams, K.M., Electrophoretic analysis of Bence Jones proteinuria, *Electrophoresis*, 20, 1307–1324, 1999.
96. Bradwell, A.R., *Serum Free Light Chain Assays*, The Binding Site Ltd, Birmingham, U.K., 2003, 136 pp.
97. Nakano, T. and Nagata, A., ELISAs for free light chains of human immunoglobulins using monoclonal antibodies: Comparison of their specificity with available polyclonal antibodies, *J. Immunol. Methods*, 275, 9–17, 2003.
98. Norden, A.G., Fulcher, L.M., and Flynn, F.V., Immunoglobulin light-chain immunoblots of urine proteins from patients with tubular and Bence-Jones proteinuria, *Clin. Chim. Acta*, 166, 307–315, 1987.
99. Gianazza, E., Galliano, M., and Miller, I., Structural transitions of human serum albumin: An investigation using electrophoretic techniques, *Electrophoresis*, 18, 695–700, 1997.
100. Miller, I. and Gemeiner, M., Peculiarities in electrophoretic behaviour of different albumins, *Electrophoresis*, 14, 1312–1317, 1993.
101. Büeler, M.R., Wiederkehr, F., and Vonderschmitt, D.J., Two-dimensional gel electrophoresis of proteins in body fluids, in *Two-Dimensional Electrophoresis: Proceedings of the International Two-Dimensional Electrophoresis Conference*, Endler, A.T., Hanash, S., Eds., VCH, Weinheim, Germany, 1989, pp. 192–205.
102. Goldman, D., Merril, C.R., and Ebert, M.H., Two-dimensional gel electrophoresis of cerebrospinal fluid proteins, *Clin. Chem.*, 26, 1317–1322, 1980.
103. Merril, C.R. and Harrington, M.G., "Ultrasensitive" silver stains: Their use exemplified in the study of normal human cerebrospinal fluid proteins separated by two-dimensional electrophoresis, *Clin. Chem.*, 30, 1938–1942, 1984.

104. Yun, M., Wu, W., and Harrington, M., Human cerebrospinal fluid protein database: Edition 1992, *Electrophoresis*, 13, 1002–1013, 1992.
105. Raymackers, J., Daniels, A., De Brabandere, V., Missiaen, C., Dauwe, M., Verhaert, P., Vanmechelen, E., and Meheus, L., Identification of two-dimensionally separated human cerebrospinal fluid proteins by N-terminal sequencing, matrix-assisted laser desorption/ionization—mass spectrometry, nanoliquid chromatography-electrospray ionization-time of flight-mass spectrometry, and tandem mass spectrometry, *Electrophoresis*, 21, 2266–2283, 2000.
106. Yuan, X., Russell, T., Wood, G., and Desiderio, D.M., Analysis of the human lumbar cerebrospinal fluid proteome, *Electrophoresis*, 23, 1185–1196, 2002.
107. Sickmann, A., Dormeyer, W., Wortelkamp, S., Woitalla, D., Kuhn, W., and Meyer, H.E., Toward a high resolution separation of human cerebrospinal fluid, *J. Chromatogr. B*, 771, 167–196, 2002.
108. Sindic, C.J., Van Antwerpen, MP., and Goffrette, S., The intrathecal humoral immune response: Laboratory analysis and clinical relevance, *Clin. Chem. Lab. Med.*, 39, 333–340, 2001.
109. Andersson, M., Alvarez-Cermeno, J., Bernardi, G., Cogato, I., Fredman, P., Fredriksen, J., Fredrikson, S., Gallo, P., Grimaldi, L.M., Gronning, M., Keir, G., Lamers, K., Link, H., Magalhaes, A., Massaro, A.R., Öhman, S., Reiber, H., Rönnbäck, L., Schluep, M., Schuller, E., Sindic, C.J.M., Thompson, E.J., Trojano, M., and Wurster, U., Cerebrospinal fluid in the diagnosis of multiple sclerosis: A consensus report, *J. Neurol. Neurosurg. Psychiatry*, 57, 897–902, 1994.
110. Verbeek, M.M., de Reus, H.P.M., and Weykamp, C.W., Comparison of methods for the detection of oligoclonal IgG bands in cerebrospinal fluid and serum: Results of the Dutch quality control survey, *Clin. Chem.*, 48, 1578–1580, 2002.
111. Harrington, M.G., Merril, C.R., Goldman, D., Xu, X., McFarlin, D.E., Two-dimensional electrophoresis of cerebrospinal fluid proteins in multiple sclerosis and various neurological diseases, *Electrophoresis*, 5, 236–245, 1984.
112. Wiederkehr, F., Ogilvie, A., and Vonderschmitt, D.J., Two-dimensional gel electrophoresis of cerebrospinal fluid immunoglobulins, *Electrophoresis*, 7, 89–95, 1986.
113. Beeley, J.A. and Khoo, K.S., Salivary proteins in rheumatoid arthritis and Sjögren's syndrome: One-dimensional and two-dimensional electrophoretic studies, *Electrophoresis*, 20, 1652–1660, 1999.
114. Ghafouri, B., Tagesson, C., and Lindahl, M., Mapping of proteins in human saliva using two-dimensional gel electrophoresis and peptide mass fingerprinting, *Proteomics*, 3, 1003–1015, 2003.
115. Vitorino, R., Lobo, M.J.C., Ferrer-Correira, A.J., Dubin, J.R., Tomer, K.B., Domingues, P.M., and Amado, F.M.L., Identification of human whole saliva protein components using proteomics, *Proteomics*, 4, 1109–1115, 2004.
116. Mii, S., Nakamura, K., Takeo, K., and Kurimoto, S., Analysis of human tear proteins by two-dimensional electrophoresis, *Electrophoresis*, 13, 379–382, 1992.
117. Glasson, M.J., Molloy, M.P., Walsh, B.J., Willcox, M.D.P., Morris, C.A., and Williams, K.L., Development of mini-gel technology in two-dimensional electrophoresis for mass-screening of samples: Application to tears, *Electrophoresis*, 19, 852–855, 1998.
118. Anderson, N.G., Powers, M.T., and Tollaksen, S.L., Proteins of human milk. I. Identification of major components, *Clin. Chem.*, 28, 1045–1055, 1982.
119. Goldfarb, M., Two-dimensional electrophoretic analysis of human milk proteins, *Electrophoresis*, 10, 67–70, 1989.

120. Toussi, A., Paquet, N., Huber, O., Frutiger, S., Tissot, J.-D., Hughes, G.J., and Hochstrasser, D.F., Polypeptide marker and disease patterns found while mapping proteins in ascitis, *J. Chromatogr.*, 582, 87–92, 1992.
121. Lindahl, M., Ståhlbom, B., and Tagesson, C., Two-dimensional gel electrophoresis of nasal and bronchoalveolar lavage fluids after occupational exposure, *Electrophoresis*, 16, 1199–1204, 1995.
122. Lindahl, M., Ståhlbom, B., Svartz, J., and Tagesson, C., Protein patterns of human nasal and bronchoalveolar lavage fluids analyzed with two-dimensional gel electrophoresis, *Electrophoresis*, 19, 3222–3229, 1998.
123. Noël-Georis, I., Bernard, A., Falmagne, P., and Wattiez, R., Database of bronchoalveolar lavage fluid proteins, *J. Chromatogr. B*, 771, 221–236, 2002.
124. Ghafouri, B., Ståhlbom, B., Tagesson, C., and Lindahl, M., Newly identified proteins in human nasal lavage fluid from non-smokers and smokers using two-dimensional gel electrophoresis and peptide mass fingerprinting, *Proteomics*, 2, 112–120, 2002.

15 Difference Gel Electrophoresis (DIGE)

Mustafa Ünlü and Jonathan Minden

CONTENTS

15.1 INTRODUCTION

15.1.1 Of Proteomes and Gels

Proteomics is a recently coined term that refers to the field of study of the proteome, defined for the first time in 1994 as "the PROTEin products of a genOME".[1] Since its introduction, the term has found widespread acceptance across the breadth of biological research. A simple search of the Medline database indicated an almost exponential increase in the number of published articles about proteomics since 1995 (Figure 15.1). The increase continues unabated to the present.

The attractiveness of proteomics derives from its promise to uncover changes in global protein expression accompanying many biologically relevant processes, such as development and tumor genesis. Since proteins are the effector molecules that carry out most cellular functions, studying proteins directly has clear advantages over and (at least in theory) achieves results that go beyond those accessible by genomic analyses. Spatiotemporal patterns of protein expression are very complex and dynamic due to fluctuations in abundance. This complexity is increased multiple-fold by the ability of proteins to be functionally affected through posttranslational modifications.

In order to detect changes in global protein expression, the methodology employed must be able to generate and compare snapshots of the entire protein component of an organism, cell, or tissue type. Two-dimensional gel electrophoresis (2DE) is a key separation technique for complex polypeptide mixtures because it offers the highest practical resolution in protein fractionation. This resolution power derives from orthogonally combining two separations based on two independent parameters: isoelectric focusing (IEF) separates proteins based on charge, and sodium dodecyl sulfate–polyacrylamide gel electrophoresis (SDS-PAGE) separates by size. Since separation is achieved under two independent parameters, charge and size, and no dilution effects are incurred, 2DE outperforms all other current separation techniques and yields a 2D gel on which hundreds to a few thousand proteins are fractionated

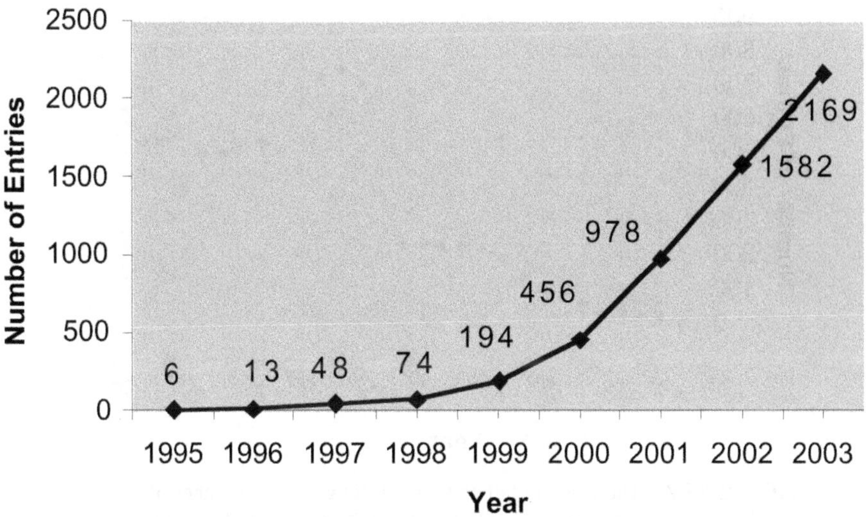

FIGURE 15.1 The meteoric rise of a trend. The number of articles published each year, with the phrase PROTEOM,* according to a Medline search.

and displayed. Traditionally, comparison of two or more gels then allows for the detection of the protein species differing between the compared states of the organism, cell, or tissue type under investigation.

2DE was first described simultaneously by several groups in 1975.[2–4] Unfortunately, using 2DE to compare protein extracts to each other has not been very fruitful.[5] It is instructive to examine the number of published articles on 2DE in the database and then to compare these with the results for proteomics (Figure 15.2). The numbers indicate that as a technique that has been around for 25 years and promised much, 2D gels have not delivered, primarily because of unavoidable systemic irreproducibilities between different gels. The other salient shortcoming is incomplete proteome display due to the poor performance of 2DE with basic and hydrophobic proteins. Some progress has been made with the latter,[6] but most efforts have been concentrated on surmounting the more serious irreproducibility problem. Given that all conventional protein detection methods are nonspecific, running different gels in order to compare different samples has remained a requirement. Thus, developments until the late 1990s focused primarily on physically improving the 2DE process[7] and also on computational approaches to better compare several gels.[8] The results of these efforts were at best partially successful, also having the side effect of increased material and time investment in a technique that was already considered to be difficult.

15.1.2 DIGE

We developed a detection system that differentiates between two different samples without introducing differences on its own, in order to be able to run two different samples on the same 2D gel.[9] DIGE uses two fluorescent dyes to label two protein

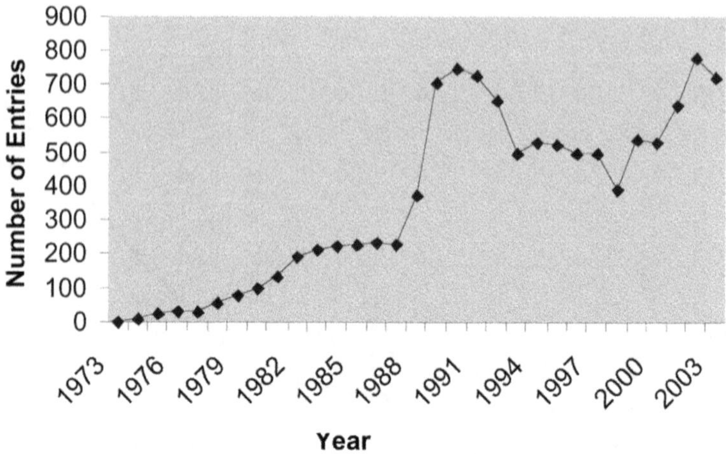

FIGURE 15.2 The rise and fall and rise of 2D gels. The number of articles published each year with the words TWO and DIMENSIONAL and GEL and ELECTROPHORESIS, according to a Medline search.

mixtures. The dyes enable samples to be differentiated from each other on the same gel. Dye design was nontrivial: no electrophoretic mobility differences should be imparted to otherwise identical proteins that were differentially tagged; and issues of sensitivity and photostability needed to be properly addressed. We succeeded in this approach and unquestionably demonstrated the reproducibility and feasibility of our modified technique. Thus, in DIGE, every identical protein in one sample superimposes with its differentially labeled counterpart in the second sample, eliminating the need to compare different gels and the problems associated with that approach.

15.1.3 Of Proteomes and Dynamic Range of Expression

Application of mass spectrometry to in-gel enzymic digests protein spots from 2D gels facilitates the rapid and sensitive identification of proteins of interest.[10] Previously the only way to identify candidate spots was by slow and less-sensitive methods such as coexpressing or coelectrophoresing known proteins, and Western blotting or N-terminal sequencing by Edman degradation. An in-depth look at how mass spectrometry has developed and its applications to proteome research is beyond the scope of this article; however, suffice it to state that mass spectrometry, in conjunction with genome sequencing projects, has brought higher throughput and new levels of sensitivity to the field, literally allowing the word *proteomics* to be invented.

Widespread use of mass spectrometry to analyze large numbers of spots from 2D gels has also brought out another shortcoming of 2DE. Articles on proteome-wide analyses of *Saccharomyces cerevisiae* pointed out that no spots from silver-stained 2D gels could be identified with low-abundance,[11,12] estimates of protein abundance have been linked to codon bias,[13] and the completion of the yeast genome sequence has enabled protein spot abundance to be coupled with quantitative transcript-based analyses.

The conclusion of these recent studies[14] is that there is an inherent limitation in 2D gels, namely that of maximum load compared to the range of protein abundances that precludes low-abundance protein identification therefrom. It would appear from these conclusions that in order to be useful in detecting any protein with less than moderate abundance, 2DE must be applied to subfractionated or otherwise enriched samples. In fact, it has been suggested that the most productive proteomics approaches should bypass the 2DE step and depend on liquid chromatography (LC) based methods for separation and isotope-coded affinity tags measured by mass spectrometry for quantitation.[15]

15.1.3.1 Good Night, Sweet Gel?

Despite the preliminary results reported from several new promising techniques currently in development that do not require prefractionation of complex protein mixtures by gels, such as ICAT technology,[15] the advantages of separating a mixture of thousands of proteins (such as whole cell lysates) are obvious. LC-based fractionation suffers even more from the enormous dynamic range of protein expression present in very complex biological samples. Furthermore, LC-based separation is often done at the peptide level (a must when inline LC to mass spectrometry is being attempted), which multiplies the complexity of the sample and spreads the higher abundant signal throughout the separation time.

On the other hand, 2DE achieves its remarkable resolution by separating whole proteins via two independent parameters, charge and size. This means that differentially modified versions of proteins are often separated and may be detected and quantitated with respect to each other. Furthermore, dilution effects during the electrophoresis procedure are minimal compared to column chromatography, thus enabling 2DE to outperform multidimensional chromatography based approaches in separation.

We feel that the abundance limitation is not insurmountable, but rather more a problem of the level of detection associated with the staining techniques employed and losses caused by postelectrophoretic handling prior to mass spectrometry. By affording facile and reproducible detection of differences, DIGE has eliminated the need for running multiple gels. The fact that all proteins are already fluorescently labeled takes sensitivity of detection to lower levels than was affordable. Also, gels do not need to be stained, cutting down the number of postelectrophoretic handling steps. Direct excision from differentially labeled gels is possible with a device that incorporates imaging and excision abilities in the same instrument. Thus, coupling the fluorescent label to automated spot cutting and digestion will also minimize losses associated with handling.

15.2 USAGE GUIDE

This is a detailed how-to guide for performing DIGE experiments using the GE Healthcare IPGPhor II with the IPGPhor Manifold for the first dimension and SE660 for the second.

15.2.1 General Notes

All references to H_2O should be read as double distilled H_2O unless stated otherwise.

In recipes, the information given in each line corresponds to final concentration, the name of ingredient, and the amount used, in that order.

15.2.2 Recipes, Apparatus, and Chemicals for Sample Solubilization and Labeling

40% Methylamine in water, urea, thiourea, DTT, and 3-[(3-Cholamidopropyl)dimethylammonio]-1-propanesulfonate (CHAPS) were obtained from Sigma-Aldrich (Milwaukee, Wisc.). N 2-Hydroxyethlypiperazine-N'-ethanesulfonic acid (HEPES) was from Fisher Scientific (Pittsburgh, Penn.).

15.2.2.1 Lysis Buffer

8 M urea	24.0 g

(Alternatively, 6 M urea and 2 M thiourea may be used.*)

6 M urea	18.0 g
2 M thiourea	7.6 g

Make up to 40 ml with high-performance liquid chromatography (HPLC) quality H_2O and dissolve the urea.

2% CHAPS (American Bioanalytical Natick, Mass.)	1.0 g
10 mM DTT	500 µl of 1 M stock or 0.077 g solid
10 mM NaHEPES, pH 8.5	5.0 ml of 100 mM stock†‡

Make up to 50.0 ml with HPLC-quality H_2O and store at −80°C in 1 to 1.5 ml aliquots.

15.2.2.2 Labeling Solution

We typically label samples in lysis buffer with no further modification. Note that we do not add IPG Buffers to this solution.§

* A mixture of urea and thiourea has been reported to aid in the solubilization of membrane proteins.[6]

† Never warm a protein lysate in urea; always endeavor to keep it at least on ice, if not frozen. Minimize the time samples spend away from −80°C. This is because at high temperature and pH urea spontaneously breaks down to yield cyanate, which modifies lysine residues and leads to carbamylated charge trains in the IEF dimension. Low temperature slows down but does not stop this process.[16]

‡ Usually the pH of HEPES is adjusted with KOH; however, if SDS is used in any subsequent step (such as running lysate directly on SDS-PAGE), KDS will be formed, which is insoluble in water. Use NaOH. When making up this buffer, make sure to add the HEPES last (see note †).

§ The presence of IPG Buffers in the Lysis buffer interferes with labeling since some contain amines that react with and inactivate the dye. Similarly, the presence of any other primary amine containing compound (such as Tris) should be avoided.

15.2.2.3 Quenching Solution

5 M methyl amine in 100 mM NaHEPES, pH 8.0. Dissolve 2.38 g HEPES in 38.8 ml of 40% methyl amine aqueous solution. Slowly add approximately 60 ml concentrated HCl with stirring. Cool the solution on ice, and measure the pH as the HCl is added until the pH reaches 8.0.

15.2.3 IEF

The IEF equipment is a IPGPhor II from GE Healthcare (Peapack, N.J.), with the DryStrip kit installed. The teflon membranes are from the YSI (Yellow Springs, Ohio) model 5793 standard membrane kit for oxygen electrodes. Light paraffin oil was from obtained from GE Healthcare.

15.2.3.1 Rehydration Buffer

2% CHAPS	0.4 g
8 M urea	9.6 g (Same composition as lysis buffer above*)
(Alternatively, 6 M urea or 2 M thiourea may be used)	
2 mM acetic acid	2.7 µl of glacial acetic acid (~17 M)
10 mM DTT	200 µl of 1 M stock or 0.031 g solid

Make up to 20 ml with H_2O; store at 4°C.†

15.2.3.2 Equilibration Buffer I

1 × Stacking Gel Buffer	25 ml of 4 × solution
1% SDS	20 ml of 10% SDS solution
8.7% Glycerol	10 ml of 87% solution
5 mM DTT	500 µl of 1 M solution or 0.076 g

Bring up to 100 ml with H_2O
Store at 4°C‡

15.2.3.3 Equilibration Buffer II

Same as above, except 2% iodoacetamide (add 2 g) replaces DTT, and a trace amount of Bromophenol blue is added.

15.2.4 Sodium Dodecyl Sulfate–Polyacrylamide Gel Electrophoresis (SDS-PAGE)

Sodium dodecyl sulfate (SDS) was obtained from Fisher Scientific. The gradient maker, acrylamide, and bis-acrylamide of the highest purity were obtained from Bio-Rad

* See first footnote in section 15.2.2.1.

† As in the second footnote in section 15.2.2.1, there is a danger of cyanate formation in this buffer. It is thus preferable to add the acetic acid at the same time as the urea. The lower pH will slow down the breakdown process. IPG Buffer will need to be added just before rehydration according to the manufacturer's instructions.

(Hercules, Calif.). The gel equipment was a Hoefer SE-660 18 × 24 cm apparatus from GE Healthcare. The 0.2 μm filters were from Nalgene (Rochester, N.Y.).

15.2.4.1 4 × Resolving Gel Buffer

36.3 g Tris
Add 150 ml H_2O
Adjust to pH 8.6 with 6 N HCl
Make up to 200 ml with H_2O
Filter sterilize through a 0.2 μm filter and store at 4°C

15.2.4.2 30% Monomer Solution

60.0 g acrylamide
1.6 g bis-acrylamide
Make up to 200 ml with H_2O
Filter sterilize and store as above

15.2.4.3 4 × Stacking Gel Buffer

3.0 g Tris
Add 40 ml H_2O
Adjust pH to 6.8 with 6N HCl
Make up to 50 ml with H_2O
Filter sterilize as above and store at −20°C

15.2.4.4 Light Gradient Gel Solution

10% acrylamide	8.25 ml of 30% monomer solution
0.375 M Tris	6.25 ml of 4 × resolving gel buffer
H_2O	10 ml
0.1% SDS	250 μl of 10% SDS solution

Add these right before pouring the gel:

Ammonium Persulfate (APS)	82.5 μl of 10% stock solution
N,N,N',N'-Di-(dimethylamin.) ethane TEMED	8.25 μl

15.2.4.5 Heavy Gradient Gel Solution

15% acrylamide	12.25 ml of 30% monomer solution
0.375 M Tris	6.25 ml of 4 × resolving gel buffer
H_2O	3.8 ml
0.1% SDS	250 μl of 10% SDS solution
Sucrose	3.75 g

Add these right before pouring the gel:

APS	82.5 μl of 10% stock solution
TEMED	8.25 μl

15.2.4.6 Stacking Gel Solution

3.5% acrylamide	400 μl of 30% monomer solution
0.175 M Tris	800 μl of 4 × stacking gel solution
H_2O	2.0 ml
0.1% SDS	33 μl of 10% SDS solution

Add these two before pouring the gel:

APS	16.7 μl of 10% stock solution
TEMED	1.7 μl

15.2.5 Imager Specifications

We use a home-built device to acquire fluorescence images of DIGE gels.[17] The cooled CCD camera is a 16-bit, series 300 model purchased from Photometrics/ Roper Scientific (Tucson, Ariz.). It is fitted with a standard 105 mm macro lens from Nikon, available from most photographical suppliers. Two 250 W quartz-tungsten-halogen lamps from Oriel (Stratford, Conn.) are used for illumination. Single bandpass excitation (2.5 cm diameter) filters from Chroma Technology (Brattleboro, Vt.), are used to excite 545 ± 10 nm and 635 ± 15 nm for Cy3 and Cy5, respectively. A multi-wavelength bandpass emission filter from Chroma Technology is used to image the gels at 587.5 ± 17.5 nm and 695 ± 30 nm for Cy3 and Cy5, respectively. The imager housing was constructed in-house from black plexiglass. An automated gel cutter is built as an integral component of the fluorescent gel imager. Image acquisition and spot excision is semiautomated and is controlled by a Silicon Graphics Inc. (Mountain View, Calif.) O2 computer workstation.

15.2.5.1 Destain Solution

1% Acetic acid	10 ml of glacial acetic acid
40% Methanol	400 ml
H_2O	590 ml

15.3 METHOD

15.3.1 Sample Solubilization

15.3.1.1 Introduction and General Notes

Since the number of different cells or tissue types that will be typically analyzed by 2DE is quite large, it is hard to describe a single protocol that will be applicable for all cases. Most samples have merely needed lysis buffer to be added to extract protein. Some examples of this kind are almost all prefractionated or prepurified proteins, human brain slices, *Drosophila* testes, mouse embryonic genital ridges, and pellets of *Escherichia coli*. Some preparations have required only slightly more vigorous disruption (i.e., with a ground glass homogenizer). Other preparations have required even more severe extractions, such as using sonication or glass beads for efficient disruption. Yeast and some cultured cells are examples of the latter. Each sample, if it does not simply lyse upon buffer addition, will require an empirical approach to determine the most efficient preparation method.

In general, all samples should be kept as cold as possible. All steps leading to and including lysis should be performed on ice. As soon as lysis is complete, samples may be stored indefinitely at −80°C. Repeated thawing and refreezing does not seem to have a deleterious effect. Do note that the lysis buffer contains urea.*† Following are three sample extraction protocols that may be used as a basis for developing further protocols:

Tissue Cultured 3T3 Mouse Fibroblasts

- Grow cells in a 150 mm Petri dish
- Wash three times with 5 ml culture medium without serum
- Add 2 ml medium to plate
- Remove cells by scraping and transfer to 15 ml conical tube
- Repeat the above with an additional 2 ml, transfer to the same conical tube
- Centrifuge for 5 minutes at 5000*g*
- Discard supernatant and resuspend cells in 1 ml medium
- Transfer cells to a 1.5 ml tube on ice
- Centrifuge for 5 seconds at maximum speed in a tabletop microfuge
- Remove supernatant
- Vortex lightly to loosen pellet
- Add 50 μl of lysis buffer for every 10^6 cells and vortex
- Centrifuge in the cold room for 5 to10 minutes at 13,000*g*
- Store at −80°C

This extract is expected to yield between 4 and 8 mg/ml protein.

15.3.1.2 *Drosophila* Embryos

Embryos are dechorinoted and observed in eggwash solution under a dissection microscope in order to determine their age. Embryos that are at the right stage are removed and rinsed once with ethanol to remove the eggwash solution. They are immediately transferred into ethanol over dry ice. Several days of collections may be accumulated, with the embryos being kept in ethanol at −80°C. After accumulating enough embryos (assume a yield of 1–1.5 μg protein yield per embryo), transfer the embryos into a cold ground glass homogenizer on ice, add 1 μl of lysis buffer per embryo and briefly homogenize. Spin in a tabletop microfuge for 5 to 10 seconds,

* See second footnote in section 15.2.2.1.

† We typically do not use protease inhibitors. The combination of the lysis buffer with its reducing ability, the chaotropic effects of the urea and the surfactant, and the cold temperature seems to inactivate proteolytic activity. We also do not perform any steps requiring room temperature or protein activity (such as the DNAse and RNAse treatment found in some protocols). Furthermore, the presence of the inhibitors may sometimes interfere with the fluorescent labeling.

take the supernatant, and discard any solid material. Store at −80°C. This extract should have between 1 and 2 mg/ml protein.

15.3.1.3 *S. Cerevisiae*

- Obtain a 250 ml yeast culture of 0.5 optical density $(OD)^{600}$
- Centrifuge the culture at 5000*g* for 15 minutes at 4°C
- Decant the supernatant
- Wash the cells as follows:
  - Suspend the cells in 5 ml of 100 mM NaHEPES, pH 8.0
  - Centrifuge the cells for 10 minutes at 5000*g* at 4°C*
  - Remove the supernatant
  - Resuspend in 5 ml of 100 mM NaHEPES, pH 8.0 and repeat the above steps once
- Suspend the pellet of washed cells in 750 μl of lysis buffer
- Transfer to a 15 ml conical tube containing 1 g acid washed glass beads
- Vortex for 3 minutes in the cold room
- Spin down and transfer the supernatant to fresh tubes
- Store the extract at −80°C. This extract has an expected protein concentration between 1 and 4 mg/ml.

15.3.1.4 Sample Labeling

Measure the protein concentration in the extracts with the method of choice (Bradford, etc.). In order to obtain matched fluorescence images, use equal amounts of protein for each sample. Anywhere between 50 and 250 μg of protein per sample may be loaded on a single 13–24 cm IEF strip, making the total maximum load 0.5 mg when two samples are being compared. There is very little to no loss of resolution even at the highest level of loading, and there is no reason not to try to achieve it. In fact, the greater load may allow for the detection of lower-abundance proteins and permit a better chance of success in the eventual identification of the spots of interest. Dye-dependent changes are rarely seen when the labeling reaction is done correctly. However, errors in labeling or quenching are often indicated by many dye-dependent changes.

- Bring the desired amount of each sample to up to 48 μl with lysis buffer†
- Add 1 μl of the appropriate dye to each sample‡
- Incubate on ice for 15 minutes
- Add 1 μl of quenching solution, followed by 0.5 μl IPG Buffer solution

* It is preferable to use a swing-out rotor to get a more compact pellet and minimize loss of yeast cells. We use a bench-top centrifuge with a swing-out TechnoSpin rotor from Sorvall Instruments.

† The sample cups (see section 15.3.2) on the IEF gel have about 100 μl maximum capacity. However, if necessary, more volume can be handled by ordering more sample cup holder bars separately from the Dry strip kit and used to spread one sample between several cups. Because IEF is a focusing technique, sample does not necessarily need to be all applied in exactly the same spot.

‡ The dye synthesis is detailed elsewhere.[9] The dyes are commercially available from GE HealthSciences.

- Incubate on ice for 30 minutes
- Immediately load and run on the first dimension

15.3.2 IEF

15.3.2.1 Introduction and General Notes

This is arguably the most complex and problematic step in the whole procedure. Notable difficulties include, but are not limited to, sample leakage, gel sparking, and insufficient or incomplete focusing. The first-dimension gels are purchased as dry strips and come in a variety of sizes and pH ranges. The strips of choice are from GE Healthcare; however, good results may also be obtained by using strips by other vendors. In all cases, the instructions that come with the gels should be followed in general; however, do note that some of the solutions and procedures recommended by GE Healthcare have been modified here. When there's a conflict, use this version for the best results.

15.3.2.2 Rehydration

The strips are stored at –20°C. Bring them to room temperature before opening the package to prevent condensation from forming on them.

Follow the manufacturer's instructions for rehydrating the strips. Note that IPG Buffer will need to be added to the rehydration buffer just before use. The amount of rehydration buffer one needs to prepare at this step depends on the size and number of the gels in each experiment. For 11cm strips, use 250 μl buffer, for 18 cm strips, use 340 μl, and for 24 cm strips, use 450 μl buffer per strip. To the appropiate amount of rehydration buffer add 4 μl of IPG Buffer per 750 μl on ice.

Level the swelling tray, then add the appropriate volume of rehydration buffer to each lane, loading it all at one end. Remove the strip from its package and remove the thin protective plastic covering the gel side of the strip. While manipulating the strips, it is best to hold them at either end where there is no IEF gel on the plastic backing membrane. Lay the strip on the buffer solution in swelling tray gel-side down, taking care to not trap any air bubbles under the gel. Once the gel strip is placed on the buffer, slide the strip back and forth to be sure that the rehydration buffer is evenly distributed across the gel.

Close the lid on the swelling tray and place it into a large plastic bag (or any other container) containing some wet paper towels. This will retard evaporation of the buffer during rehydration. Seal the bag and leave it at room temperature overnight. Gels will swell to 0.5 mm thickness.*

15.3.2.3 Setting Up and Running the First Dimension

Set-up the IPGPhor II with the cup-loading IPGPhor Manifold and level it. Clean the white tray and align it correctly on the electrode plate. Add 110 ml light mineral oil, covering all 12 lanes even if only running a few.

* rehydration should continue for a minimum of 8 hours. The gels should not be used after 24 hours in the rehydration solution.

Place the gel strips into the lanes, one gel per lane, gel side up. Be sure strip is correctly placed within the lane (see manufacturer's instructions) and completely covered with oil.

Place the sample cups so that they are not directly over protrusions in the lane walls and are only as close to the end of the gel so that there is still room for wick overlap. It is preferable to place the cups close to the acidic end of the gels. GE Healthcare gels are conveniently marked with a + sign at the acidic end. Snap the cups firmly into place.

Prewet the wicks on a plastic surface using 150 μl of water per wick. Place two wicks per gel, one at each end. At the acidic end of the strip, place the wick so that it touches the feet of the sample cup.

Apply the samples into each cup, placing the pipette tip under the surface of the oil as you deliver sample.

Place the acidic electrode with diagonal lines toward you so that it contacts both the wicks and the gold cathode plate at the outside edges. The edges of the electrode should be ~1 mm from the cup feet. Snap it into place and tighten the cams at the edges. Place the basic electrode similarly, having it contact the wicks and the anode plate.

Close the lid and select an IET protocol. There are charts of suggested protocols in the GE Healthcare literature. A sample protocol (for 11 cm 3-11 NL strips) is shown below:

a. No rehydration time (since that was already done)
b. Temp. set to 20°C and current to 50 uA per strip
c. Step, 500 V, 1 hour
d. Grad, 1000 V, 1 hour
e. Grad, 6000 V, 2 hours
f. Step, 6000 V, 30 minutes

Begin the focusing run and observe the current. As the first step executes, there should be at least some current. If there is no current, pause the run and check that the electrodes are placed correctly.

15.3.2.4 Equilibration of IEF Gels

At the end of the IEF run, equilibrate the gels in Equilibration Buffer I for 15 minutes, rinse briefly with H_2O to remove DTT, and equilibrate in Buffer II for another 15 minutes. After equilibration, gels should be either run immediately on the second dimension or stored at −80°C. They will last almost indefinitely at −80°C.*

15.3.3 SDS-PAGE

15.3.3.1 Assembling the Gel Cassettes

Rinse the glass plates with H_2O and then 95% EtOH and air dry. Rinse clamps and spacers in H_2O and dry. Make sure the insides of clamps are dry and the spacers are free of particulate debris.

* A small disposable petri dish makes a great receptacle for equilibration. Wrap the gel around the inside of the dish, with the gel side pointing in. The dishes may also be marked to easily keep track of the identities of multiple gels.

Align glass plates and spacers so that all edges are flush, especially at the bottom. Place a large thumb screw clamp toward the top and a small one at the bottom on each of the long sides of the glass plates. Make sure the bottom of the cassette is sticking out slightly (1–2 mm) from the clamps. This drives the gel cassette into the stand and seals the bottom when the cams are inserted and twisted in opposite directions. If desired, check seal by squirting water on the outside of sandwich assembly.

Assemble the equipment for pouring, with the gradient maker on a stir plate, over the cassette(s).*

15.3.3.2 Pouring the 10–15% Gradient Gels

Pour the heavy solution in front chamber of gradient maker, add stirbar
Pour light solution in back chamber of gradient maker, add stirbar balancer
Open the mixing channel and the front stopcock at the same time
Pour gel and overlay with *n*-butanol
Allow gel to polymerize (about 30 minutes)
Pour off *n*-butanol, rinse top of gel with H_2O twice, and pour about 0.5 cm of stacking gel
Overlay with *n*-butanol again
Allow to polymerize for at least eight hours†

15.3.3.3 Setting Up and Running the Second Dimension

Microwave 10 ml of autoclaved 1% low melting agarose in 1 × stacking gel buffer until it starts to boil. (We also add a trace amount of Bromophenol blue.) Place an IEF strip on top of the second-dimension gel plastic side down, acidic side facing left. Then rotate the IEF strip so that the plastic backing is parallel to the face of the back gel plate and slide it down vertically while making sure the plastic side, not the gel side, is contacting the glass plate.‡

Push the strip down until it is in firm contact with the top of the SDS gel. Add the agarose until it barely covers the top of the strip. Make sure to add evenly from both sides and to avoid air bubbles. Burst bubbles with gel loading pipettor tips after the agarose hardens.

Make sure to mark the second-dimension gel with sample ID.

Electrophorese at 4°C. For a run of about 8 to 10 hours, use 20 mA per gel at constant voltage. For a run of about 16 hours, use 8 to 10 mA per gel.

* Before starting to pour the gradient gels, make sure the gradient maker chamber is free of polyacrylamide pieces. The tube and mixing channel can easily be blocked by small pieces of gel. Working fast, four gels can be poured one after the other. No washing of the chamber in between gels is necessary.

† The eight hours is to allow the polymerization reaction to go to completion. This time should not be shortened to less than about four hours, due to potential side chain and N-terminal modification by acrylamide. Gels may be allowed to polymerize as long as overnight provided that the butanol layer is increased so that it does not dry out. We have also stored gels at 4°C for up to 24 hours.

‡ If the IEF gel has dried, this may cause a problem as the gel has a higher tendency to stick to the glass and tear. Should this start happening, try wetting the IEF strip a little with Equilibration Buffer II.

15.3.4 Image Acquisition and Analysis

15.3.4.1 Introduction and General Notes

The layout of the imaging system is vital for the success of DIGE. The technique has unique requirements, so all the hardware and most of the software for image acquisition was built from scratch. Our system is experimental and transitory in nature. At the time of the writing, the apparatus is under constant revision and development and is not commercially available in its current form. Thus, rather than give the exact minutiae of the image acquisition and analysis process, we list the nature and aim of the operations that are performed to arrive at the final result. Hopefully, this will assist those who are interested in either building or acquiring their own imaging system.

At the time of writing of this manuscript, commercially available DIGE imagers are also available as well. For those who might consider either building or acquiring an imager, we list the absolute minimum requirements that the hardware must meet:

1. In addition to the obvious requirement that the gel not physically move during imaging, no changes in protein spot position due to optically induced deformations should occur while switching between channels.
2. The imaging hardware needs to be sensitive enough to detect minimally labeled proteins. Since a cooled, coupled charge device (CCD) had been used previously to image in gel fluorescently labeled proteins,[1,18] we decided to use a scientific grade CCD camera. Without cooling, CCD cameras do not have the requisite sensitivity of low-noise capabilities.
3. The imaging cabinet must be light-tight, and illumination must be filtered through the appropriate filter sets.
4. It is highly preferable that the imaging device have an integrated cutting piece. This allows for the rapid and accurate excision of spots of interest with minimal handling.

In the current incarnation of the DIGE imager, the gels are placed flat on a black plexiglass surface at the bottom of the cabinet. The camera is mounted vertically over the gel at about 30 cm away. Illumination is provided by two halogen lamps with fiber-optic leads mounted on top of the cabinet at $\pm 60°$ incident angles to the bottom to provide an even field of illumination. The standard image acquired by the imager is a 4×4 cm square made up from 256×256 pixels, each storing 65,000 gray levels as unsigned short integers.

15.3.4.2 Image Acquisition

After the second-dimension run, the gels are removed from their cassettes and incubated in Coomassie destain solution for a minimum of one hour. This removes surfactants and salts and allows the proteins to be fixed in the gel. Gels are then equilibrated in and placed under 1% acetic acid in H_2O during imaging. Two images are acquired from the central region of the gel, one with each excitation filter. A few spot intensities are compared and used to normalize the acquisition times for the two channels in tile mode. Tile mode is where the entire gel is imaged into a single file by stitching together thirty 4 × 4 cm squares to generate one 20 × 24 cm image. Two such images are generated, one for each channel. Each image corresponds to one of the dyes and thus represents one of the samples that was run on the gel. Acquisition times typically vary between 10 and 180 seconds per square, which translates to 10 to 180 minutes total acquisition time per gel.

15.3.4.3 Image Analysis

The two images from a single gel are inverted and then normalized to each other so that the most abundant spots appear at the same level of intensity. The images are then converted to byte format and placed into a two-frame Quicktime movie. Playing this movie in a continuous loop allows for the visual detection of differences. The images are normalized at several different grayscale levels. Normalizing for high values allows for the detection of the abundant protein changes and normalizing at lower values covers the lower-abundant protein changes.

In addition, composite images are also submitted to the SExtractor program (a freeware image analysis program used in astronomy applications), which detects spots and creates a measurement mask to be applied to the two original image fragments. The mask outlines an area of the gel that contains a protein spot. This is used to define the area in which the pixel values were integrated. The SExtractor program also performs a localized background estimation to determine the base values across the mask area. The output from the image analysis is the sum of pixel values in the mask area less the background sum.

REFERENCES

1. Wasinger, V.C., Cordwell S.J., Cerpa-Poljak, A., Yan, J.X., Gooley, A.A., Wilkins, M.R., Duncan, M.W., Harris, R., Williams, K.L., and Humphery-Smith, I., Progress with gene-product mapping of the Mollicutes: *Mycoplasma genitalium*, *Electrophoresis*, 16, 1090–1094, 1995.
2. Klose, J., Protein mapping by combined isoelectric focusing and electrophoresis in mouse tissues: A novel approach to testing for induced point mutations in mammals, *Humangenetik*, 26, 231–243, 1975.
3. O'Farrell, P.H., High resolution two-dimensional electrophoresis of proteins, *J. Biol. Chem.*, 250, 4007–4021, 1975.
4. Scheele, G.A., Two dimensional gel analysis of soluble proteins: Characterization of guinea pig exocrine pancreatic proteins, *J. Biol. Chem.*, 250, 5375–5385, 1975.
5. Alban, A., David, S.O., Bjorkesten, L., and Andersson, C., A novel experimental design for comparative two-dimensional gel analysis: Two-dimensional difference gel electrophoresis incorporating a pooled internal standard, *Proteomics*, 3, 36–44, 2003.
6. Rabilloud, T., Adessi, C., Giraudel, A., and Lunardi, J., Improvement of the solubilization of proteins in two-dimensional electrophoresis with immobilized pH gradients, *Electrophoresis*, 18, 307–316, 1997.
7. Görg, A., Postel, W., and Günther, S., The current state of two-dimensional electrophoresis with immobilized pH gradients, *Electrophoresis*, 9, 531–546, 1998.

8. Wilkins, M.R., Hochstrasser, D.F., Sanchez, J.C., Bairoch, A., and Appel, R.D., Integrating two-dimensional gel databases using the Melanie II software, *Trends Biochem Sci.*, 12, 496–497, 1996.
9. Ünlü, M., Morgan, M.E., and Minden, J.S., Difference gel electrophoresis: A single gel method for detecting changes in protein extracts, *Electrophoresis*, 18, 2071–2077, 1997.
10. Shevchenko, A., Jensen, O.N., Podtelejnikov, A.V., Sagliocco, F., Wilm, M., Vorm, O., Mortensen, P., Shevchenko, A., Boucherie, H., and Mann, M., Linking genome and proteome by mass spectrometry: Large-scale identification of yeast proteins from two dimensional gels, *Proc. Natl. Acad. Sci.*, 25, 14, 440–414, 445, 1997.
11. Futcher, B., Latter, G.I., Monardo, P., McLaughlin, C.S., and Garrels, J.I., A sampling of the yeast proteome, *Mol. Cell. Biol.*, 11, 7357–7368, 1999.
12. Gygi, S.P., Rochon, Y., Franza, B.R., and Aebersold, R., Correlation between protein and mRNA abundance in yeast, *Mol. Cell. Biol.*, 3, 1720–1730, 1999.
13. Coghlan, A. and Wolfe, K.H., Relationship of codon bias to mRNA concentration and protein length in *Saccharomyces cerevisiae, Yeast*, 12, 1131–1145, 2000.
14. Gygi, S.P., Corthals, G.L., Zhang, Y., Rochon, Y., and Aebersold, R., Evaluation of two-dimensional gel electrophoresis based proteome analysis technology, *Proc. Natl. Acad. Sci.*, 97, 9390–9395, 2000.
15. Gygi, S.P., Rist, B., Gerber, S.A., Turecek, F., Gelb, M.H., and Aebersold, R., Quantitative analysis of complex protein mixtures using isotope-coded affinity tags, *Nat. Biotechnol.*, 10, 994–999, 1999.
16. Hagel, P., Gerding, J.J.T., Fieggen, W., and Bloemendal, H., Cyanate formation in solutions of urea, *Biochim. Biophys. Acta*, 243, 366–379, 1971.
17. Gong, L., Puri, M., Ünlü, M., Young, M., Robertson, K., Viswanathan, S., Krishnaswamy, A., Dowd, S.R., and Minden, J.S., *Drosophila* ventral furrow morphogenesis: A proteomic analysis, *Development*, 131, 643–656, 2003.
18. Urwin, V.E. and Jackson, P., Two-dimensional polyacrylamide gel electrophoresis of proteins labelled with the fluorophore monobromobimane prior to the first-dimensional isoelectric focusing: Imaging of the fluorescent protein spot patterns using a cooled charge-coupled device, *Anal. Biochem.*, 209, 57–62, 1993.

16 Principles and Challenges of Basic Protein Separation by Two-Dimensional (2D) Electrophoresis

Anton Posch, Aran Paulus, and Mary Grace Brubacher

CONTENTS

16.1 INTRODUCTION

Since the pioneering work of Klose[1] and O'Farrell,[2] two-dimensional (2D) electrophoresis has become a key technology in proteome analysis. It provides the high resolving power necessary to separate complex protein samples prior to protein identification via mass spectrometry methods. In the early publications from Klose's group, around 300 protein spots in a mouse tissue sample could be separated. With many refinements over the last three decades, 2D gels are now capable of separating up to 5000 proteins in a single run. 2DE is the only analytical technique capable of such resolving power. It is somewhat surprising that so far all technological attempts to achieve protein separation in a single and preferably automated step have not been successful.

The first 2D gels used a carrier ampholyte (CA) generated pH gradient for the isoelectric focusing (IEF). However, the setup suffered from pH gradient instability, a phenomenon called cathodic drift, which resulted in poor reproducibility and made interlaboratory comparisons virtually impossible.[3,4]

A major difference in both the performance and the popularization of the 2D gel approach was made with the introduction of immobilized pH gradients (IPGs) for IEF in 1982 as an alternative to CA-generated pH gradients.[5] IPGs provide a significantly higher protein load capacity and can separate up to milligram quantities of protein for micropreparative purposes.[6] High protein loads are essential for the separation and identification of low-abundance proteins using mass spectrometry.

Although IPGs have modernized IEF in terms of resolution, reproducibility, and user-friendliness, some problems remain. In particular, the separation of basic proteins remains challenging, and the needs of the life science community in respect to this class of compounds are not adequately addressed. The analysis of polypeptides with isoelectric point (pI) values above 9 has always been a problem for the first step of 2D gel analysis, independent of whether carrier ampholytes or IPGs are used.

A theoretical analysis based on genomic data and their translation into protein sequence data reveals that there should be a considerable number of proteins with basic pIs in most if not all organisms.[7] However, when a proteome is separated and analyzed, very few basic proteins are identified. Furthermore, only a small number of proteomics laboratories are capable of routinely using basic IPGs in their work.

Here we provide an overview on the principles of IEF using IPGs, the difficulties of running basic IPG gels, and the most recent developments to develop robust IPGs in the basic range.

16.2 PRINCIPLES OF IPGS

As briefly discussed in the introduction, pH gradients formed with carrier ampholytes are subject to instability (cathodic drift) during extended focusing runs.[8] This phenomenon is not fully understood but is characterized by a drift of the gradient toward the cathode and is accompanied by acidification at the anode. This results in flattening of the gradient in the neutral pH region and ultimately a loss of alkaline proteins. IPGs were developed to overcome these limitations of carrier ampholytes.

16.2.1 Chemistry of IPGs

IPGs are generated by copolymerization of acrylamide monomers and acrylamide derivatives. These derivatives contain modifications on the amino group that improve buffering capability and are called acrylamido buffers. They are characterized by the general structure CH_2=CH–CO–NH–R, where R represents either a carboxylic or an amino group. An acrylamido buffer is defined by its pK value. The first publications on IPG technologies worked with a total of six acrylamido compounds with pK values at pH 3.6, 4.6, 6.2, 7.0, 8.5, and 9.3. Theoretical and experimental considerations led to formulations and protocols for casting IPGs spanning ranges of pH 4 to pH 10. Later on it became possible to expand the pH range by using additional acrylamido buffers with p*K* values of 0.8, 10.4, and above 12. This allowed the preparation of IPGs with very acidic and basic pH gradients as well.[9–11]

16.2.2 Casting of IPG Gels

IPGs are generated through a polymerization process. Because the pH gradient is formed by localized charges, the ratio of different acrylamide buffers is varied in a gradient mixer prior to the polymerization reaction. In practice, two different polymerization solutions, each containing acrylamide, bisacrylamide, acrylamido buffers, and catalysts, are mixed with a gradient mixer and poured into a cassette lined on one side with a plastic backing for support. After polymerization, the polymerized sheets are washed to remove residual catalysts and monomers, dried, and for 2D electrophoresis applications, cut into 3 mm wide strips of appropriate length. Hence the term IPG strip.

Unlike conventional IEF with carrier ampholytes, the polymerized gel strips now contain covalently bound carboxylic and amino groups with buffering capacity. Also contrary to carrier ampholytes, IPGs exhibit no moving charges upon application of an electrical field. In theory, IPGs, with their constant pH gradient, should allow steady-state focusing and eliminate cathodic drift. This should result in establishment of highly reproducible protein patterns, which can extend into the basic pH range.

16.2.3 Types of IPG Gels for 2D Electrophoresis

With the help of computational approaches, a vast number of pH gradients, both in range and shape gradient types (linear, convex, etc), have been created.[12] Generally they can be divided into three categories, namely micro range (0.1–1.0 pH units), narrow range (2–4 pH units), and wide range (5–8 pH units). The required resolution for the particular protein sample will determine which range should be used. Wide gradients (e.g., pH 3–10) are useful to get an overview of an unknown protein sample with respect to its pI distribution. If the goal is the resolution of the maximal number of protein spots, two strategies can be pursued: either using wide gels (e.g., 24 cm) or using multiple overlapping micro or narrow IPGs (zoom-in gels).

Although protocols for casting IPG gels are extensively described in the literature, most laboratories commonly use commercially available IPG strips. The commercial products are made with strict control using automated procedures. Therefore, they offer high reproducibility both when compared between laboratories as well as over project time. This and the convenience and hassle-free use are why most laboratories favor commercial strips over homemade IPGs.

16.2.4 Sample Application

IPGs either from a commercial source or homemade are supplied as dried gel strips that must be rehydrated to their original thickness of 0.5 mm prior to use. The rehydration solution typically contains additives such as urea, thiourea, detergents, reducing agents, and carrier ampholytes. The rehydration process takes about 8 to 12 hours, typically at room temperature. There are two popular sample application methods: cup loading and in-gel rehydration.

16.2.4.1 Cup Loading

Specially designed cups with a capacity up to 150 μL (up to ~800 μg) are used to load samples into rehydrated IPG strips (Figure 16.1). In theory, the actual application point should not be critical because IEF with IPGs is a true equilibrium technique. In practice, however, the sample is usually applied at one of the pH extremes (anodic or cathodic), where most of the proteins are charged. Under these conditions, the migration of the sample into the IPG matrix is simplified, minimizing the risk of sample loss due to precipitation. Cup-loading devices are well suited for analytical protein loads, but the limited volume of commercial cups may hamper preparative applications.

16.2.4.2 In-Gel Rehydration

The need to rehydrate IPG strips prior to their use led to a simplification of the sample application technique first published by Rabilloud et al.[13] During the rehydration step, Rabilloud and coworkers loaded the protein samples over the whole IPG gel, thus avoiding the risk of sample loss by precipitation in sample cups. This technique is popular because it is easy and convenient. The sample volume to be applied is only restricted by the volume required for proper IPG strip rehydration. For example, 350 μl is the rehydration sample volume used for 17 cm IPG strips, which allows higher protein loads. Also, highly diluted samples may be analyzed without further concentration steps.

Interestingly, the way a protein sample is applied to the IPG strip affects the quality of basic protein separation. Initially the protein sample was applied by sample cups at either end of the strip, preferably at the anode. Since this method suffered

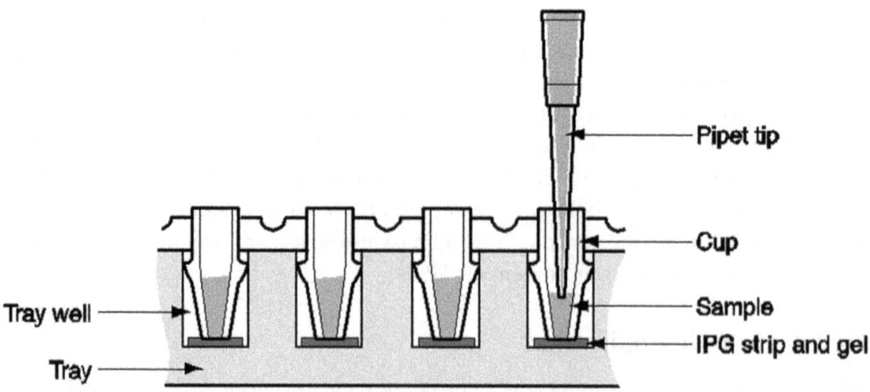

FIGURE 16.1 Side view of focusing tray with IPG strips, cups, and sample. Courtesy of Bio-Rad laboratories.

from leakage of the sample cups and difficult handling, Rabilloud et al.[13] developed the in-gel rehydration method, which has been successfully demonstrated in a number of publications.[14–17]

However, when sample application by in-gel rehydration is used at pH extremes, basic protein separation is not achieved. Therefore, for basic proteins with high pH IPG strips, cup loading at the anode seems to be the method of choice (Figure 16.2). Barry et al.[18] compared cup loading with the alternative passive and active rehydration methods to identify the advantages one loading method has over another. Replicate 2D gels from each loading method were quantitatively evaluated for gel-to-gel reproducibility using pH 6–11 IPG strips and semipreparative protein loads (300 μg). They stated that cup loading was far superior for pH 6–11 separations than active or passive rehydration methods. Cup loading consistently produced the greatest number of detectable spots, the best spot matching efficiency (56%), lowest spot quantity variation (28% coefficient of variation), and the best-looking gels qualitatively.

16.3 OVERCOMING DIFFICULTIES WITH BASIC PROTEIN SEPARATIONS

The theory and methodology of working with IPGs are well understood.[19] Applying the theoretical knowledge led Righetti and coworkers to recognize that extreme pH ranges pose challenges both in preparing and running IPG strips. As already discussed in section 16.2.1, the synthesis of very basic acrylamido buffers with p*K* values of 10.4 and greater than 12 is a requirement for the production of very alkaline IPG gels (e.g., pH 10–12). However, early attempts to run basic IPG gels polymerized with novel basic acrylamide buffers under standard conditions resulted in poor

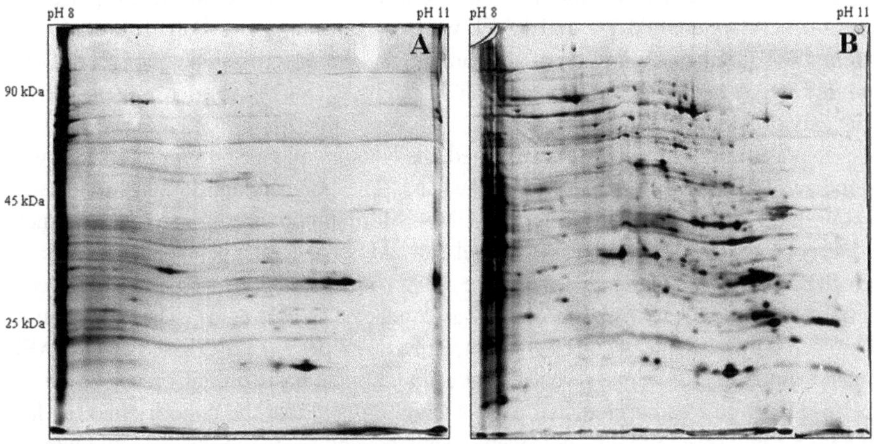

FIGURE 16.2 Effect of sample loading method on resolution. pH 8–11 acrylamide gels loaded by (A) standard rehydration methods; and (B) anodic cup loading. Courtesy of Wiley-VCH Publishing.[7]

focusing. This was explained by the occurrence of endosmotic flow (EOF), which can contribute to poor focusing due to water transport from the cathode to the anode. Water flow can slow protein migration in the opposite direction, leading to incomplete focusing. In addition, dehydration of the IPG strip at the cathode can reduce the pore size of the gel, leading to slower protein migration and incomplete focusing.[20] Solutions to this problem include replacing the paper wick, changing the wick during the run, and adding organic modifiers such as glycerol, isopropyl alcohol, or methyl cellulose to the rehydration/sample solution[21] (see section 16.3.2). However, EOF is not the only hurdle to the separation of basic proteins. Others are discussed further in a later section.

16.3.1 Instability of the Acrylamide Matrix

Under extremely acidic and basic conditions, acrylamide is subject to hydrolysis and forms acrylic acid, which may enhance EOF. Trying to increase hydrolytic stability, Chiari et al.[22] developed alternative, N-substituted acrylamide derivatives where hydrolysis of the amido group is sterically hindered. These efforts resulted in new matrices for electrophoretic separations, such as dimethylacrylamide (DMAA) and *N*-acryloylaminoethoxyethanol (AAEE). Compared to polyacrylamide (PAA), they offer the following advantages: (*a*) strong resistance to alkaline hydrolysis (AAEE appears to be 500 times more stable compared to polyacrylamide), (*b*) greater porosity, due to the higher molecular weight of the monomers, and for AAEE, (*c*) high hydrophilicity (not for DMAA). Görg et al.[21] successfully used DMAA to create alkaline IPGs (pH 8–12, 9–12 and 10–12) for 2D separations of ribosomal and nuclear proteins.

Molloy et al.[7] extended the work of Görg and coworkers by investigating IPG matrix effects on 2D protein separation of *Caulobacter crescentus* outer membrane proteins under standard conditions.[7] For the preparation of alkaline IPGs (pH 8–11, 17 cm), acrylamide, DMAA, and AAEE were used as monomers. Polyacrylamide gels were 4% T (total concentration of acrylamide) and 3% C (cross-linker concentration), while DMAA and AAEE IPGs were 5% T and 3% C, each with an average buffering capacity of 2.9 mequiv/pH/l.

Interestingly, for each gel matrix tested, the outer membrane protein sample showed good resolution with well-defined spot shape. The polyacrylamide and AAEE gels showed less protein spot streaking than DMA. More importantly, toward the extremely basic portion of the DMAA and AAEE gels, an additional 20 polypeptides were detected compared to the standard polyacrylamide gel. A similar study on basic IPGs for the separation of *Helicobacter pylori* proteins was conducted by Bae et al.[23] They compared a commercially available pH 6–11 gradient with a novel pH 9–12 IPG based on an AAEE matrix. Spot identifications revealed that even the most basic proteins resolved on the pH 6–11 2D gels had a predicted pI less than 10.0 and that the majority had pIs less than 9.5. Only seven protein spots were visible beyond pH 9.5 in this gel, although the presence of significantly more proteins in this region was theoretically expected. With the pH 9–12 IPG, a noteworthy improvement was obtained, and 15 more spots with a pI above 9.5 could be detected and characterized by mass spectrometry.

The aforementioned results of Görg, Molloy, and Bae indicate that the use of alternative monomers such as DMAA and AAEE results in better alkaline protein separations compared to the reference polyacrylamide gel. However, there is still an apparent lack of proteins with predicted pI >10. An alternative approach to the examination of extremely alkaline proteins is a combination of prefractionation liquid-phase IEF and both sodium dodecyl sulfate–polyacrylamide gel electrophoresis (SDS-PAGE) followed by liquid chromatography–tandem mass spectrometry posttryptic digest.[23]

16.3.2 Running Conditions for Basic IPG Strips

The enhanced stability of DMMA and AAEE gel matrices under extreme alkaline conditions is only one among many approaches to improve separation power and resolution of basic IPG gels. To obtain highly resolved 2D patterns in the alkaline range, a series of optimization steps related to sample application, additives, and running conditions are necessary.

Gelfi et al.[24] addressed the EOF problem with the use of conductivity quenchers. The anode-directed water flow discussed above is easy to observe during the IEF step when a moistened cathodic paper bridge dries out and sticks to the gel, whereas an anodic paper bridge remains well wetted. A sorbitol gradient (0–10% from anode to cathode) during gel rehydration reduces EOF and quenches the high conductivity of the IPG gel at the cathodic end during IEF (pH 10–11). Using this method, excellent focusing results have been reported for lysozyme (pI 10.55), So-6 (a leaf protein, pI 10.49), cytochrome c (pI 10.45) and ribonuclease (pI 10.12).

The separation of histones by Righetti et al.[25] represents an important milestone in the development of basic IPG focusing. Using the same strategy of EOF quenching, they used a nonlinear IPG, pH 10–12, to separate histone fractions (II-AS, VI-S, VII-S, and VIII-S) with pI values between 11 and 12. A further improvement in the separation of histones and ribosomal proteins was achieved by Görg et al.[21] by introducing isopropanol and glycerol as EOF quenchers. The addition of 10 to 16% isopropanol and 10% glycerol to the reswelling solution enhanced the quality of the 2D pattern of ribosomal proteins significantly. It is noteworthy that both Righetti and Görg covered the IPG strips with silicone oil during IEF to act as a barrier to the surrounding atmosphere.

Encouraged by these promising results, other groups tried similar approaches to separate basic proteins from different sources. Although all of them generated 2D maps of apparently good quality, they independently reported a less obvious problem: the appearance of extra spots and spot trains, with the number and intensity of the spots varying randomly from gel to gel. Further investigations revealed that dithiothreitol (DTT) may be the main culprit for this phenomenon; DTT is typically used in the sample solubilization solution and IPG strip rehydration buffer to cleave disulfide bonds and maintain them in a reduced state. As already outlined in Chapter 8 (this volume), DTT is deprotonated and negatively charged at alkaline pH (Figure 16.3). As DTT migrates toward the anode, the cathode is depleted of DTT, which causes a reoxidation of thiol groups and the formation of interchain and

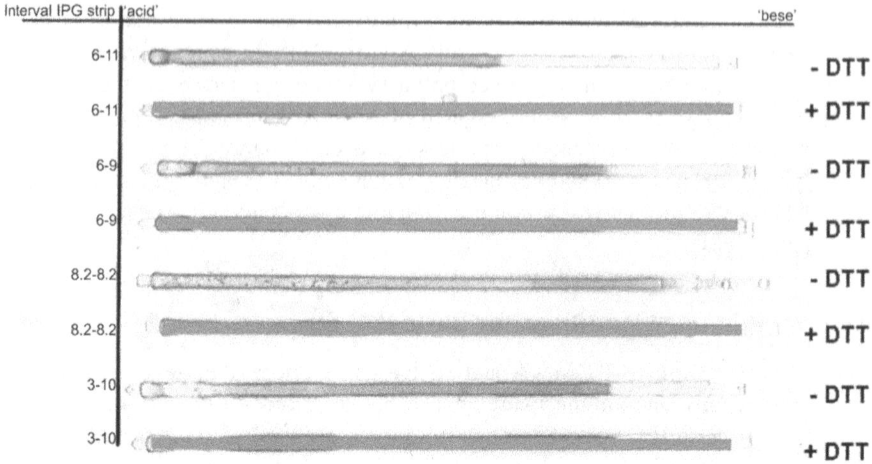

FIGURE 16.3 IPG strips treated with Ellman's reagent. Different IPG strips were rehydrated in 7 M urea, 2 M thiourea, 4% CHAPS, 2.5% DTT, 10% isopropanol, 5% glycerol, 2% IPG buffer (pH 6–11), focused for 34 kVh (IPG 6–9 and 6–11) or for 80 kVh (IPG 3–10 and 6.2–8.2) and treated with Ellman's reagent as described. The dark shaded area on the anodic side of the strip indicates the presence of free thiol groups, and the light shaded areas show the positions where DTT has been depleted. Because no sample was used, the only source of free thiol groups is DTT (marked –DTT). The experiment was repeated under the same conditions, but with the introduction of the same buffer (3.5% DTT) at the cathodic end of the strip with the paper-bridge method. Ellman's reagent was applied on the strip and demonstrated that DTT was present on the entire length of the strip (marked +DTT). Courtesy of Wiley-VCH Publishing.[31]

intrachain disulfide bonds. The following four remedies have been recommended to overcome the limitations of DTT as reducing agent in basic range IPG strips:

- Soaking extra electrode paper strips in rehydration buffer with 3.5% DTT and placing them at the cathodic end of the IPG strip[26]
- Performing IEF in the presence of hydroxyethyl disulfide (HED)[27,28]
- Replacing DTT by uncharged reducing agents such as tributylphosphine (TBP)[29]
- Reduction and alkylation of proteins prior to isoelectric focusing during sample preparation[30]

Use of a DTT-soaked electrode paper at the cathode has not gained wide acceptance, although it has worked successfully in some studies.[31] The experimental protocol is complicated, and protein loss can occur because of liquid exudation resulting from electroosmotic pumping introduced by DTT transport during focusing.[27] The same group suggested first-dimension IEF in an excess of HED. An excess of HED would drive the conversion of all thiol groups between the protein cysteine and one half of HED according to the following equilibrium:

$$\text{Prot–S}^- + \text{HO–CH}_2\text{–CH}_2\text{–S–S–CH}_2\text{–CH}_2\text{–OH} \Leftrightarrow$$

$$\text{Prot–S–S–CH}_2\text{–CH}_2\text{–OH} + \text{HO–CH}_2\text{–CH}_2\text{–S}^-.$$

This approach was successfully used for the separation of basic human brain proteins by Pennington et al.[28]

The replacement of DTT by the uncharged reducing agent tribuylphosphine has also attracted many researchers and has definitely improved the overall focusing quality in many 2D electrophoresis experiments. In contrast to DTT, short chain phosphines are chemically more difficult to handle because they are volatile, toxic, and highly flammable in concentrated stock solutions. However, commercial suppliers have developed TBP formulations that are safe for shipping and laboratory use.

Yet another approach to separate basic proteins in 2D gels is reduction and alkylation prior to IEF as part of sample preparation.[30] This is potentially a breakthrough for 2D electrophoresis and will be discussed in more detail in the next section.

16.3.3 Reduction and Alkylation Prior to 2D Electrophoresis

A number of streaking and artifactual spot problems in 2D electrophoresis of basic proteins are associated with disulfide bond formation in the protein sample. Reduction and alkylation of thiol groups has long been practiced in protein chemistry to prevent the reformation of intramolecular and intermolecular disulfide bonds. Alkylation agents such as iodoacetamide or acrylamide have been successfully used to chemically modify the thiol group, preventing disulfide formation.

However, the standard procedure[32] currently adopted in proteomics protocols includes a reduction step prior to the IEF step, followed by a second reduction and alkylation step during equilibration after IEF but prior to the SDS-PAGE step. Unfortunately, not every cysteine is blocked using this procedure. Herbert et al.[30] demonstrated using matrix-assisted laser desorption and ionization–time of flight (MALDI-TOF) mass spectrometry that failure to reduce and alkylate proteins prior to both electrophoretic steps results in a large number of spurious spots in the alkaline pH region due to scrambled disulfide bridges between random polypeptide chains. As a consequence, they suggest a sample preparation protocol using TBP as reducing and iodoacetamide as alkylating reagent. Figure 16.4 demonstrates how dramatically the reduction and alkylation of proteins prior to IEF can improve the separation of basic proteins, in this case using a commercial kit.

It is important to recognize that the alkylation reaction is pH and time dependent, regardless of the alkylation agent used. Therefore, the reaction conditions must be optimized and carefully controlled to obtain a high alkylating efficiency and specificity as well as to avoid unwanted side reactions such as the alkylation of lysines.[33,34] Alternative alkylating reagents like maleimide derivates[35] are currently

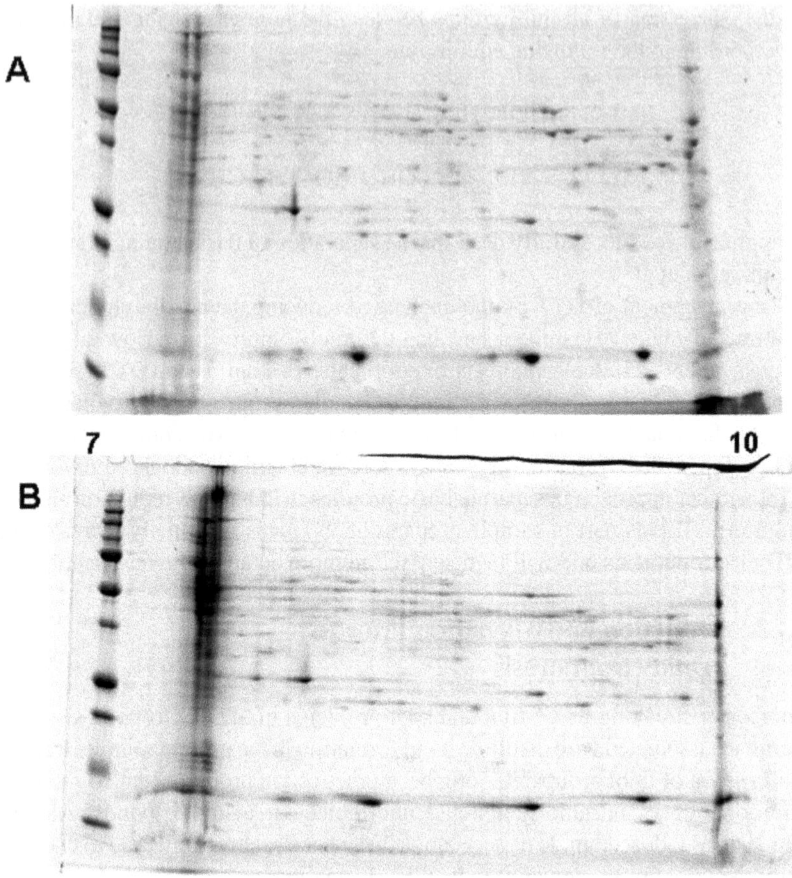

FIGURE 16.4 Improved 2D results with basic protein separation from mouse liver extract after treatment with the ReadyPrep Reduction-Alkylation Kit (Bio-Rad). Both samples were applied by cup loading onto 11 cm pH 7–10 IPG strips and both samples contained DTT reducing agent in the rehydration and sample buffer. (A) Treated with the kit; (B) untreated.

under investigation, and substantial progress in the field of the alkylation of protein thiols for the benefit of IEF in the alkaline range can be expected in the near future.

16.4 CONCLUDING REMARKS

Broad-range 2D electrophoresis is an excellent tool to separate and display more proteins than any other analytical method and therefore allows a first look at the protein composition of the biological material of interest. On closer inspection, the proteins are not uniformly distributed over the molecular weight and pI range used for separation.

In fact, in most gels the basic range above pH 9 resembles a *terra incognita*, a white map with only a few spots. However, the analysis of genomic data, now available for a number of model organisms, including humans, suggests a sizable number of protein structures with a pI of 10 or higher. Why are they not found in 2D gels, and what has stopped researchers from their discovery?

Limitations of 2D gels for separation of hydrophobic membrane proteins have been well documented and understood, whereas the mysteries of basic proteins are only now being unraveled. A first obstacle is manufacturing of a suitable immobilized matrix at high pH, which requires the chemical synthesis of suitable acrylamido buffer compounds with pIs above 10 and their incorporation into a base-stable matrix. Still, at pH 11 or 12, a strong EOF is difficult to suppress and is detrimental to a separation step relying on steady-state conditions. Clearly, the addition of highly viscous additives and the use of pH-stable base matrix compounds are helping.

However, not only the physics and chemistry of the separation system are responsible for the poor results with basic proteins so far. The high pH environment has a profound influence on the sample composition, and we are only now learning that a careful sample preparation strategy is necessary to preserve the protein sample in its original state during analysis. Oxidation of disulfide groups from cysteines heavily influences tertiary protein structure and therefore migration and precipitation characteristics. A number of strategies to prevent the formation of disulfide bonds, including the use of DTT and alkylation of the resulting thiol groups, show promise.

Despite the advances in the last decade in the separation of basic proteins, an acceptable protocol has yet to be developed. A better understanding of the chemical environment of basic proteins and how it influences the separation conditions has helped. Improved resolution of basic proteins with very high pI values remains the biggest challenge. Further advances in sample preparation, in particular with respect to solubilization, adsorption of basic proteins to surfaces they come in contact with, and preservation of a completely denatured state of the proteins will help in the development of more robust IPG separations in the basic range.

REFERENCES

1. Klose, J., Protein mapping by combined isoelectric focusing and electrophoresis of mouse tissues: A novel approach to testing for induced point mutations in mammals, *Humangenetik*, 26, 231–243, 1975.
2. O'Farrell, P., High resolution two-dimensional electrophoresis of proteins, *J. Biol. Chem.*, 250, 4007–4021, 1975.
3. Corbett, J.M. et al., Positional reproducibility of protein spots in two-dimensional polyacrylamide gel electrophoresis using immobilised pH gradient isoelectric focusing in the first dimension: An interlaboratory comparison, *Electrophoresis*, 15, 1205–1211, 1994.
4. Blomberg, A. et al., Interlaboratory reproducibility of yeast protein patterns analyzed by immobilized pH gradient two-dimensional gel electrophoresis, *Electrophoresis*, 16, 1935–1945,1995.

5. Bjellqvist, B. et al., Isoelectric focusing in immobilized pH gradients: Principle, methodology and some applications, *J. Biochem. Biophys. Methods*, 6, 317–339, 1982.
6. Bjellqvist, B. et al., Micropreparative two-dimensional electrophoresis allowing the separation of samples containing milligram amounts of proteins, *Electrophoresis*, 14, 1375–1378, 1993.
7. Molloy, M.P. et al., Profiling the alkaline membrane proteome of *Caulobacter crescentus* with two-dimensional electrophoresis and mass spectrometry, *Proteomics*, 2, 899–910, 2002.
8. Garfin, D.E., Isoelectric focusing, *Methods Enzymol.*, 182, 459–77, 1990.
9. Chiari, M. et al., Synthesis of buffers for generating immobilized pH gradients. II. Basic acrylamide buffers, *Appl. Theor. Electrophor.*, 1, 103–107, 1989.
10. Chiari, M. et al., Synthesis of buffers for generating immobilized pH gradients. I. Acidic acrylamide buffers, *Appl. Theor. Electrophor.*, 1, 99–102, 1989.
11. Chiari, M. et al., Synthesis of an hydrophilic, pK 8.05 buffer for isoelectric focusing in immobilized pH gradients, *J. Biochem. Biophys. Methods*, 21, 165–172, 1990.
12. Celentano, F.C., Gianazza, E., and Righetti, P.G., On the computational approach to immobilized pH gradients, *Electrophoresis*, 12, 693–703, 1991.
13. Rabilloud, T., Valette, C., and Lawrence, J.J., Sample application by in-gel rehydration improves the resolution of two-dimensional electrophoresis with immobilized pH gradients in the first dimension, *Electrophoresis*, 15, 1552–1558, 1994.
14. Rouquie, D. et al., Construction of a directory of tobacco plasma membrane proteins by combined two-dimensional gel electrophoresis and protein sequencing, *Electrophoresis*, 18, 654–660, 1997.
15. Gorg, A. et al., Recent developments in two-dimensional gel electrophoresis with immobilized pH gradients: Wide pH gradients up to pH 12, longer separation distances and simplified procedures, *Electrophoresis*, 20, 712–717, 1999.
16. Cordwell, S.J., Basseal, D.J., and Humphery-Smith, I., Proteome analysis of *Spiroplasma melliferum* (A56) and protein characterisation across species boundaries, *Electrophoresis*, 18, 1335–1346, 1997.
17. Nouwens, A.S. et al., Complementing genomics with proteomics: The membrane subproteome of *Pseudomonas aeruginosa* PAO1, *Electrophoresis*, 21, 3797–3809, 2000.
18. Barry, R.C. et al., Quantitative evaluation of sample application methods for semipreparative separations of basic proteins by two-dimensional gel electrophoresis, *Electrophoresis*, 24, 3390–3404, 2003.
19. Righetti, P.G., *Immobilised pH Gradients: Theory and Methodology. Laboratory Techniques in Biochemistry and Molecular Biology*, Burdon, R.H., van Knippenberg, P.H., Eds., Amsterdam, Elsevier, 1990, vol. 20.
20. Chang, H., Brubacher, M.G., and Strong, W.B., Preventing horizontal streaks on 2-D gels, *BioRadiations*, 111, 56–57, 2003.
21. Gorg, A. et al., Very alkaline immobilized pH gradients for two-dimensional electrophoresis of ribosomal and nuclear proteins, *Electrophoresis*, 18, 328–337, 1997.
22. Chiari, M. et al., Towards new formulations for polyacrylamide matrices: *N*-acryloylaminoethoxyethanol, a novel monomer combining high hydrophilicity with extreme hydrolytic stability, *Electrophoresis*, 15, 177–186, 1994.
23. Bae, S.H. et al., Strategies for the enrichment and identification of basic proteins in proteome projects, *Proteomics*, 3, 569–579, 2003.
24. Gelfi, C. et al., Isoelectric focusing in immobilized pH gradients in the pH 10–11 range, *J. Biochem. Biophys. Methods*, 15, 41–48, 1987.

25. Righetti, P.G. et al., Steady-state two-dimensional maps of very alkaline proteins in an immobilized pH 10–12 gradient, as exemplified by histone types, *J. Biochem. Biophys. Methods*, 31, 81–91, 1996.
26. Gorg, A. et al., Two-dimensional polyacrylamide gel electrophoresis with immobilized pH gradients in the first dimension (IPG-Dalt): The state of the art and the controversy of vertical versus horizontal systems, *Electrophoresis*, 16, 1079–1086, 1995.
27. Olsson, I. et al., Organic disulfides as a means to generate streak-free two-dimensional maps with narrow range basic immobilized pH gradient strips as first dimension, *Proteomics*, 2, 1630–1632, 2002.
28. Pennington, K. et al., Optimization of the first dimension for separation by two-dimensional gel electrophoresis of basic proteins from human brain tissue, *Proteomics*, 4, 27–30, 2004.
29. Herbert, B.R. et al., Improved protein solubility in two-dimensional electrophoresis using tributyl phosphine as reducing agent, *Electrophoresis*, 19, 845–851, 1998.
30. Herbert, B. et al., Reduction and alkylation of proteins in preparation of two-dimensional map analysis: Why, when, and how? *Electrophoresis*, 22, 2046–2057, 2001.
31. Hoving, S. et al., Preparative two-dimensional gel electrophoresis at alkaline pH using narrow range immobilized pH gradients, *Proteomics*, 2, 127–134, 2002.
32. Gorg, A. et al., The current state of two-dimensional electrophoresis with immobilized pH gradients, *Electrophoresis*, 21, 1037–1053, 2000.
33. Galvani, M. et al., Protein alkylation in the presence/absence of thiourea in proteome analysis: A matrix assisted laser desorption/ionization-time of flight-mass spectrometry investigation, *Electrophoresis*, 22, 2066–2074, 2001.
34. Galvani, M. et al., Alkylation kinetics of proteins in preparation for two-dimensional maps: A matrix assisted laser desorption/ionization-mass spectrometry investigation, *Electrophoresis*, 22, 2058–2065, 2001.
35. Luche, S. et al., About thiol derivatization and resolution of basic proteins in two-dimensional electrophoresis, *Proteomics*, 4, 551–561, 2004.

17 Multidimensional Separation of Membrane Proteins

Susan Francis-McIntyre and Simon J. Gaskell

CONTENTS

17.1 INTRODUCTION

The molecular biology of membranes has been studied for many years.[1] Membrane proteins perform a range of functions from cell signaling to solute transport;[1,2] although well characterized biochemically, these functions have not been studied greatly using the methods of proteomics. In principle, proteomics approaches offer the opportunity to identify the proteins involved in particular cellular processes, to study the context within which the proteins are expressed, and to monitor protein expression and turnover in relation to the dynamics of the cellular environment. The structure of the cell membrane consists of a lipid bilayer with which proteins are associated. Membrane proteins can be broadly divided into two categories: integral (or intrinsic) membrane proteins, which are inserted in the lipid bilayer, and membrane associated (or peripheral) proteins, which are bound to the membrane through covalent or noncovalent interactions:

17.1.1 Integral Membrane Proteins

Integral or intrinsic membrane proteins contain hydrophobic transmembrane domains (TMDs), which pass through the lipid bilayer several times, anchoring the protein within the structure. Since water is absent in the membrane, peptide bonds in the bilayer form hydrogen bonds that are primarily intramolecular. This can be maximized if the polypeptide chain forms a tightly ordered secondary or tertiary structure as it crosses the bilayer.[2] Many integral membrane proteins have TMDs of 15 to 25 amino acids that span the lipid bilayer in the form of an α helix. Alternatively, porins, which are found in the outer membranes of bacteria, form β sheets organized into a β-barrel tertiary structures that can allow the passage of molecules across the membrane. Integral membrane proteins can be considered as true membrane proteins since they are embedded in the membrane itself. They will be present exclusively in membrane preparations and are the most challenging to characterize biochemically.

17.1.2 Membrane-Associated Proteins

This is a loosely defined category of all proteins associated with, but not actually embedded within, the membrane. Proteins may be attached to the bilayer by means of one or more covalently bound fatty acid chains or by a covalent linkage via an oligosaccharide to phosphatidylinositol in the outer lipid monolayer of the plasma membrane, as observed with glycosylphosphatidylinositol (GPI) anchor proteins. Other peripheral membrane proteins are attached to the membrane by noncovalent interactions with other membrane proteins. Membrane associated proteins are often arranged in large complexes at the surface of the membrane and may be enriched in cytosolic and membrane fractions depending on the strength of the interaction involved and the solubilization treatment favored.

17.1.3 The Need for Membrane Proteomics

It is estimated that 20 to 30% of all open reading frames (ORFs) encode integral membrane proteins containing at least two predicted transmembrane domains.[3] However, the proteomics literature available today does not reflect this, with membrane proteins being underrepresented in two-dimensional gel electrophoresis (2DE) experiments, mainly due to their poor solubilization.[4] Furthermore, Schwartz et al.[5] analyzed estimated isoelectric point (pI) values for the expression products of all predicted ORFs in a selection of fully sequenced genomes and proposed a link between pI and subcellular location. On this basis, the proportion of the proteome corresponding to membrane proteins is underestimated.

Membrane proteins can be recognized through sequence analysis software based on known structural motifs. Proteins comprising α-helical domains of 20 to 30 amino acids are easier to recognize than β sheets that are encoded in much smaller numbers of residues. However, proteins can be mislabeled as membrane proteins if the sequence prediction software wrongly assigns the cleavable signal sequence of a protein as a potential transmembrane domain.[6] Moreover, the limited biochemical characterization of many of these transmembrane proteins prevents further development in prediction software algorithms and further improvement in their reliability.[6] Furthermore, the grand average hydropathy index (GRAVY) score, which is often used as an indicator of the hydrophobicity of a protein, does not consider the number of TMDs that may be present and therefore cannot be used as a guide to protein behavior.[7,8]

In common with other proteins, membrane proteins are often modified by posttranslational modifications (PTMs) in order to perform specific cellular functions. Proteomics, especially a 2DE-based approach where proteins are resolved on the basis of pI and size, offers an insight into the nature of these modifications and may also indicate how these change in response to the local environment.

Here we present a review of current literature in the field of membrane proteomics using both the classical 2DE-based approach and alternative protein identification strategies using liquid chromatography coupled with mass spectrometry (LC-MS). We illustrate how these techniques have been applied in a specific area, fungal proteomics.

17.2 CLASSICAL TWO-DIMENSIONAL ELECTROPHORESIS (2DE) BASED METHODS OF MEMBRANE PROTEIN ISOLATION AND CHARACTERIZATION

There are numerous advantages to a gel-based approach to membrane proteomics. Significant technical advances mean that the techniques are relatively robust and easily accessible to many groups. The expression of one gene can result in many different protein isoforms by posttranslational modification of the expression product. 2DE offers at least a preliminary means of monitoring the subtle changes in molecular weight and pI that a protein may undergo during processing. Santoni et al. demonstrated the power of a 2DE-based approach studying the expression patterns of root and leaf plasma membrane extracts from *Arabidopsis thaliana*.[9] Several proteins, which were plasma membrane (PM) specific in one extract, displayed similar abundances in both the PM and cytosol fractions of the other material, illustrating how the local environment in which the protein exists affects its cellular role and determines whether the protein appears in the membrane or cytosolic fractions for a given cell extract. Modern technical advances such as differential gel electrophoresis (DIGE) provide a method of relative quantification of protein preparations,[10–12] allowing changes in protein abundance to be monitored and quantified.

During the last decade a number of new reagents have been developed that have improved protein solubility for 2DE and have therefore improved the technique's suitability for the analysis of membrane proteins; these developments have been reviewed by Santoni et al. and by Molloy.[7,13] However, no all-purpose reagents are available, and often the procedure used needs to be optimized for each application.[7,13,14] Nevertheless, simple changes in the constituents of the solubilization buffer used in extraction can vastly improve the number of protein spots seen by 2DE. Mechin et al.[15] compared buffers for their efficiency of solubilizing maize endosperm. The R2D2 buffer, based on the classic Rabilloud solubilization buffer[16] with modifications to include two reducing agents dithiothreitol (DTT) and Tris(2-carboxyethyl)phosphine hydrochloride (TCEP) in addition to 3-[(3-cholamidopropyl)dimethylammonio]-1-propanesulfonate (CHAPS) and caprylyl sulfobetaine (SB3–10) detergents improved protein solubilization and resolution for 2DE. This general solubilization buffer was found to give efficient solubilization for a wide range of extracts including yeast and maize mitochondria. Alternatively, the addition of a harsh detergent to a conventional extraction procedure can greatly improve protein solubilization and recovery as illustrated by Wang et al.[17] Here the addition of a buffer containing 2% sodium dodecyl sulfate (SDS) and 5% β-mercaptoethanol during phenol extraction of olive leaf proteins yielded more material than when either treatment was used separately.

17.2.1 Detergent Extraction Based Approaches

Detergent solubilization of integral membrane proteins requires a detergent to mimic the natural lipid environment in which the proteins of interest exist. When mixed, the hydrophobic ends of the detergent bind to the hydrophobic regions of the protein, thereby displacing the lipid molecules, bringing the membrane proteins into solution in the form of detergent-protein complexes. Santoni et al. proposed three important criteria which detergents must fulfill for the solubilization of proteins for 2DE.[7] Firstly, the lipid environment must be dissolved and remaining lipids should not interfere with the isoelectric focusing (IEF) process. Secondly, the membrane proteins must be extracted from the membrane in solution typically as a detergent-protein complex. Finally, these proteins must remain soluble during IEF, notably at their pI where their solubility is at a minimum. Unfortunately, the degree of mimicry of the detergent and the hydrophobic nature of the membrane vary and hence a range of detergents must be tested and the best mimic selected, based on the quality of the extracts obtained and the compatibility with downstream analysis procedures.

Phase-separation approaches exploit the fact that a number of detergents, such as the Triton series, have two transition temperatures in water at which they will separate from the aqueous phase to give a detergent-rich and a detergent-poor phase. Membrane proteins will favor the detergent-rich phase because this is most like their lipid-rich natural environment, and therefore such an approach offers a simple method of membrane protein enrichment.[7,12,18,19]

Rabilloud et al. have optimized the extraction conditions for normal or plasmodium-infected erythrocyte ghosts using the zwitterionic amidosulfobetaine detergents ASB-14, ASB-16, ØC6, C6BzO and C7BzO.[20] The aim was to solubilize and detect band III, a transmembrane protein with 12 putative transmembrane domains, not previously identified by 2DE using the standard CHAPS/SB3–10 solubilization buffer and thought to be heavily glycosylated. Initial analysis by 1DE suggested that solubilization buffers supplemented with 2% ASB-14 or 1% ASB-16 or 2% C7BzO resulted in the most efficient extraction of band III. When tested by 2DE, it was found that ASB-14 or C7BzO solubilization, in combination with urea/thiourea, were superior. Further analyses with other classical detergents such as Tween (polyoxyethylenesorbitan monolaurate derivatives), Brij-56, Triton X-100, and α- and β-dodecyl maltosides did not offer any further improvement in solubilization.[21]

In contrast, when the same detergent mixtures were applied to *A. thaliana* leaf membrane proteins, C7BzO proved more suited to the solubilization of plant material with Brij 56 being marginally better than ASB-14 for the detection of aquaporins and H+-ATPase.[21] During plant membrane preparation, the material was partially delipidated by treatment with Triton X-100, while in the preparation of red blood cell membranes no delipidation occurred. These results illustrate that the lipid content of the membrane extracts has a significant effect on the solubilization efficiency: detergents with an aromatic core (e.g., C8Ø) being inefficient at delipidation whist being effective at breaking aggregates. Linear detergents, such as ASB-14, have the opposite effects.[22]

Further assessment of detergent compatibility with the solubilization of two recombinant proteins, rat purinoceptor subtype $P2X_3$ ($rP2X_3$) and human histamine H2 receptor (hH2R), was carried out by Henningsen et al.[23] Five zwitterionic detergents, CHAPS, C8Ø, SB3–10, ASB-14, and ASB-16, were tested. For both proteins, C8Ø gave substantial improvements in solubility. It is interesting to note the contrast in efficiency of ASB-14 for the two proteins: for $rP2X_3$, ASB-14 gave poor solubility, while for hH2R, ASB-14 was second only to C8Ø in efficiency. This illustrates the need to assess a range of detergents and, ideally, use a range of solubilization buffers to maximize the number of proteins detected. Furthermore, a characteristic train of spots corresponding to $rP2X_3$, which could be visualized by western blotting, could not be detected by Colloidal Coomassie, silver, or Sypro ruby even when the sample concentration was increased 20-fold. These data suggest that sample solubilization and protein abundance are not the only factors governing protein detection and that even if proteins are sufficiently solubilized they may still not be detected using a gel-based approach with common staining techniques.

SDS is known to solubilize many proteins; however, its incompatibility with the first (IEF) stage of 2DE means that its use should be avoided unless the extracts obtained can be diluted sufficiently to minimize its detrimental effects. An alternative to SDS, lithium dodecyl sulfate (LDS), has been evaluated for compatibility with 2DE using *Haemophilus influenzae* extracts.[24] Solubilization by SDS, LDS, or guanidinium hydrochloride (Gndn HCl) resulted in the detection of several outer membrane proteins (OMPs) not previously seen using the standard urea/CHAPS based solubilization method. LDS was found to be an efficient solubilization agent leading to the observation during 2DE of highly abundant membrane proteins, with no significant spot stretching observed, in contrast to the results obtained on SDS treatment. It is thought that this detergent can be displaced by IEF-compatible reagents more readily than SDS and many therefore provide a substitute for classical SDS-based protein extraction methods.

Despite the increase in protein solubilization, a number of proteins such as lipoprotein D and OMP P5 were detected in developed gels at the IEF loading positions, suggesting that these proteins could not enter the gel and precipitated at this point. It is interesting to note that Fountoulakis et al.[24] report a decrease in the number of spots observed when IEF strips were rehydrated in protein solution compared to that of conventional cup loading, also observed by Barry et al.[25] In-gel sample rehydration should offer improved solubilization of hydrophobic proteins and circumvent problems of aggregation and protein loss commonly observed when cup loading is used.[13]

Based on the results obtained, Fountoulakis and Takacs conclude that two TMDs represent the limit of applicability of a 2DE-based approach. This is in contrast to findings of Rabilloud et al. who were able to detect the protein band III from red blood cells with 12 putative TMDs.[20,21] These findings[20,21,24] illustrate that the choice of detergent and its ability to maintain sample solubility, especially during IEF, plays a key role in the maximum number of TMDs detected.

17.2.2 Sequential Extraction Based Techniques

As a development of the single detergent extraction procedures described above, Molloy et al. devised an elegant sequential extraction approach to yield cellular proteins in a range of fractions.[26] This method makes use of the known limited solubility of membrane proteins in standard solubilizing reagents, using a range of solutions of increasing detergent, chaotrophe, and reducing agent strength to release proteins sequentially from the membrane. Moreover, by splitting the protein pool into several fractions and reducing sample complexity, more proteins can be visualized than by a single extraction alone, and therefore low-abundance proteins, enriched in a specific fraction, are more likely to be observed.

This technique was originally applied to the extraction and enrichment of membrane proteins from *Escherichia coli*[26] but has since been modified and marketed as a universal extraction reagent (e.g., SIGMA and Bio-Rad sequential extraction kits). Initially, cell extracts are solubilized in a conventional buffer containing 8 M urea and 4% CHAPS, which solubilizes the majority of cytosolic proteins. The insoluble proteins that remain are then treated with an enhanced reagent that contains a higher concentration of chaotrophes (urea and thiourea) in combination with two detergents, CHAPS and SB3–10. The reducing agent DTT is also replaced with tributylphosphine (TBP), which does not migrate during IEF and therefore is better at maintaining disulfide bridges in their reduced form.[27] Finally, the insoluble pellet is solubilized in a 1% SDS solution and boiled for 5 minutes. Analysis of this final fraction by 2DE demonstrated the presence of a number of proteins including OmpF, an outer membrane porin, although these gels were poorly focused.

Although the authors were able to identify most of the major protein components of the *E. coli* bacterial outer membrane including a number of proteins not previously seen by 2DE, none of these proteins were in the top 15% of GRAVY values[4] showing that there are still a number of issues in protein solubility which need to be addressed. However, such a sequential approach allows detergents, or other components of the extraction buffer, to be substituted as new reagents become available, and the tailoring of the approach taken to the properties of the proteins of interest.

Sequential and single detergent based extraction, using digitonin and Triton X-100, were compared for the enrichment of membrane proteins from a human alveolar type II cell line, with the success of each method assessed by the recovery of E-cadherin, an integral membrane protein with 1 TMD.[28] The single detergent based extraction method yielded the most membrane proteins, although a number of these proteins were also observed in the cytosolic fraction. In contrast, the sequential extraction approach gave a much cleaner separation of membrane proteins with no E-cadherin detected in the cytosolic fraction. Using these methods, a number of membrane proteins from plasma membranes, endoplasmic reticulum (ER), and mitochondrial membranes were identified and detected exclusively in membrane fractions; again, the proteins detected mainly had negative GRAVY values and low numbers of TMDs, suggesting that only weakly hydrophobic proteins were able to be solubilized by this method. Lehner et al. concluded that a sequential extraction approach, using CHAPS and SB3–10, offered a poorer yield compared to the digitonin and Triton X-100 detergent extraction. However, the sequential extraction approach gave extracts of higher purity and was better suited for small quantities of sample. Nevertheless, the authors suggest that if large quantities of sample are available, both methods should be applied as they complement each other in maximizing the number of proteins detected.

17.2.3 Organic Solvent Based Extraction

In an alternative approach to sequential extraction, Molloy et al. used a methanol and chloroform based extraction procedure to extract outer membrane proteins from *E. coli.*[29] The lower organic layer was dried and solubilized in buffer containing CHAPS/SB3–10 or ASB-14 and analyzed by IEF. A number of hydrophobic cytoplasmic, periplasmic, and hypothetical lipoproteins were identified, of which eight had not previously been detected by 2DE. This included malate dehydrogenase membrane-associated protein with a GRAVY value of +0.294. This technique was also effective applied to the isolation of chloroplast envelope proteins from *Spinacia oleracea.*[30,31] Santoni et al. combined Triton X-114 phase partitioning with organic solvent extraction to isolate proteins from *Arabidopsis* plasma membranes.[22] Here a methanol–chloroform ratio of 8:1 combined with ØC5 detergent solubilization prior to IEF was found to be the most effective.

In an alternative approach, a phase-separation system based on trifluoroethanol (TFE) and chloroform has been recently reported and applied to the extraction of integral membrane proteins from *E. coli.*[32] The extraction procedure used was similar to that for methanol–chloroform extraction[29] with a 2:1 ratio of TFE to chloroform being used. After centrifugation, three phases were obtained of which the upper TFE-rich phase and interface of insoluble material enriched for different proteins. Protein extracts were solubilized in buffer containing CHAPS and 50% TFE. On 2DE analysis, a number of OMPs and other membrane-associated proteins were identified that were not observed when a standard CHAPS/SB3–10 solubilization buffer was used without the inclusion of TFE. Both of these organic extraction procedures offer an alternative to the standard detergent based extraction and solubilization procedures. Trifluoroethanal has the added advantage of being compatible with IEF and can be added to the rehydration solution, thereby improving sample entry into the strip and maintaining protein solubility during focusing.

17.2.4 Membrane Stripping Procedures

Peripheral membrane proteins loosely associated with the membrane through noncovalent interactions can provide a major source of contamination in integral membrane preparations. In addition, there are nonintegral proteins anchored to the membrane via covalent bonds. These peripheral proteins can hinder the detection of low-abundance, more difficult-to-solubilize, integral proteins.

A simple method for removal of noncovalently bound proteins, and therefore to enrich for integral membrane proteins in a membrane protein preparation, is to apply a membrane-stripping procedure. One of the most common procedures is treatment with aqueous sodium carbonate. This simple, effective treatment offers a method of converting sealed membrane vesicles into flat sheets, thereby releasing the contents within.[33] This technique has been applied to the study of the outer membrane proteins of *E. coli*.[34] Using this approach, the authors were able to identify 69% of the predicted integral OMPs in the pH 4–7, 10–80 KDa mass range studied and confirmed the presence of two proteins (YbiF and YeaF) that were previously defined as hypothetical. A number of low-abundance nonmembrane proteins were also identified, however, illustrating that not all peripheral membrane proteins can be removed by sodium carbonate treatment.

Two membrane-stripping procedures, sodium carbonate or 0.25% Triton X-100 and potassium bromide (KBr) treatment, were assessed using root or leaf extract plasma membrane proteins from *A. thaliana* by Santoni et al.[9] It was found that of the 52 functionally identified membrane proteins, only 6 were observed when the Triton X-100/KBr method was used compared with 20 proteins from sodium carbonate treatment. Further studies suggested that the use of sodium carbonate membrane stripping, preferably followed by a C8Ø-based solubilization buffer, offered the best results for membrane preparations,[35] especially after delipidation of the sample with SDS.[22] While KBr stripping treatment offers a higher stringency than carbonate treatment, there is an increased risk of losing genuine membrane proteins by this harsh treatment; therefore a milder carbonate-based approach offers an opportunity to remove the majority of weakly associated proteins while minimizing the loss of proteins of interest. Again, the choice of detergent used for solubilization plays an important role in protein detection; a larger number of spots were detected when the sample was solubilized with C8Ø-based solubilization buffer including the detection of H+ ATPase (10 TMDs) and water channel proteins.[22,35] It is interesting to note that when this buffer was used for total protein extraction without membrane-stripping treatments, no significant advantage was offered over the classical CHAPS-based IEF buffer.

17.3 LIQUID CHROMATOGRAPHY COUPLED WITH MASS SPECTROMETRY APPROACHES

Although 2DE offers many advantages for the characterization and quantification of proteins, it has a number of inherent disadvantages that limit its use for the separation of membrane proteins: the resolution of hydrophobic, alkaline proteins is often poor, and low-abundance proteins can be difficult to detect on gels.[13,36] Studies by

Pederson et al. suggest that low-abundance proteins, in the region of 10,000 copies per cell, make up 80% of the predicted yeast proteome and that therefore a 2 mg load protein is needed to detect a 50 kDa protein present at this expression level by Coomassie blue staining.[37] The apparent absence of expression of a low-abundance protein can result in false negative results in which the cellular role of a protein may be overlooked due to its expression levels being below the limit of detection. This is illustrated by the analysis of the *E. coli* outer membrane proteins by Molloy et al. where iron receptor proteins could only be identified when grown in iron-restrictive conditions that result in the upregulation of the expression of these proteins.[34] Unless many different environmental conditions are explored, it is apparent that many proteins could be missed when using 2DE as the basis for proteomics experiments. Subcellular fractionation and enrichment strategies can help overcome this limitation, but only if the proteins of interest are amenable to 2DE analysis.

Mass spectrometry (MS) for protein identification and characterization has become an integral part of proteomics.[38,39,40] In many cases MS, in the form of peptide mass fingerprinting (PMF) or tandem MS (MS/MS) peptide sequencing, has been used for the routine identification of proteins from 1DE and 2DE gel experiments (e.g., [30,31,41]). However, the digestion of proteins within the gel, and subsequent peptide extraction, often leads to poor peptide recovery yielding relatively small amounts of material for analysis by these methods. Often the standard in-gel digestion technique using trypsin is unsuitable for use for membrane proteins since recovery of the hydrophobic peptides generated can be difficult without the inclusion of octyl-β-glucopyranoside during peptide extraction or additional cleavage methods such as cyanogen bromide (CNBr) being employed.[41] Moreover, this is a time-consuming approach with protein separation and identification taking several days. Liquid chromatography (LC) based approaches can complement or provide an alterative to gel-based separation techniques.[42,43] Thus, an automated analysis of proteolytic peptides using a LC separation coupled with MS can offer a viable alternative to classical gel techniques, especially for large-scale protein identification and characterization.

17.3.1 1D LC-MS/MS Analysis Techniques

Minor components of a protein mixture from a gel band or spot are difficult to identify by MALDI PMF or MS/MS peptide sequencing experiments, with peptides missed simply because the sample is too complex. Complexity can be reduced by using a chromatographic separation prior to MS/MS analysis; typically, a reverse-phase separation is employed with each peptide being analyzed as it elutes from the column. The conditions required for peptide separation must be optimized to allow sufficient time for analysis of eluted peptides. Alternatively, total extracted membrane proteins can be digested in solution, with separation performed solely at the peptide, rather than protein, level by LC.

Combining detergent phase partioning, 1DE, and LC-MS/MS techniques, Elortza et al.[19] and Borner et al.[12] identified a number of glycosylphosphatidylinositol-anchored proteins (GPI-APs). The shave and conquer approach[19] couples the phase separation detergent extraction of membrane proteins with enzymatic treatment to release GPI-APs into the aqueous phase through the hydrolysis of phosphatidylinositol. *A. thaliana* and Human HeLa cell raft-enriched membrane extracts were prepared, and two-phase detergent extraction with Triton X-114 was performed.[18] The detergent phase was recovered and treated with phosphatidylinositol phospholipase C (PI-PLC); the proteins released into the aqueous phase were separated by1DE. Protein bands were then excised, digested, and analyzed by nano LC-MS/MS. For human HeLa cell extracts, 17 proteins were identified in the aqueous phase, of which 6 were identified as GPI-APs.[19] Further studies of *A. thaliana* membrane proteins identified 64 proteins, of which 44 were predicted to be GPI-anchor proteins (18% of the predicted GPI-anchor proteome). Furthermore, ω sites, where the GPI group is attached to anchor the protein to the membrane, could be identified in most of these proteins. Twenty contaminant proteins present were secreted proteins or membrane proteins with at least one TMD and represented minor components in the aqueous phase.[19]

Borner et al.[12] adopted a slightly different approach combining the 1DE and LC-MS/MS techniques with 2D DIGE to identify proteins enriched in the aqueous phase after PI-PLC treatment.[12] Such a modification-specific approach to proteomics, choosing an enzyme treatment to release a specific protein subtype, allows the systematic identification and characterization of posttranslationally modified proteins. The release of proteins into the aqueous (detergent-poor) phase leaves the detergent-rich phase (containing the majority of membrane proteins) free to be used for alternative enrichment and separation procedures, maximizing the amount of data generated from a single sample.

A further example of modification specific proteomics is the study of phosphorylated membrane proteins from *A. thaliana* by Nuhse et al.[44] Here a membrane shaving approach was applied to expose and release phosphorylated proteins. Plasma membranes, isolated by phase partitioning, are usually orientated with the cytoplasmic domains of membrane proteins on the inside of the vesicles. These vesicles can be inverted with almost complete efficiency by treatment with Brij58. Therefore, the normally hidden domains of these proteins become accessible for protease digestion. On applying this technique, the membrane was digested with trypsin, and the resulting complex mixture separated by strong anion exchange (SAX) chromatography yielding several fractions of peptides. The phosphorylated peptides were then separated from these fractions using immobilized metal ion affinity chromatography (IMAC) before being analyzed by nano LC-MS/MS. In the absence of an SAX prefractionation step, very few phosphopeptides could be observed in the protein mixture. In contrast, over 90% of the peptides identified by this combined SAX and IMAC approach were phosphopeptides, illustrating the success of the technique for reducing sample complexity and enriching for phosphorylated peptides. Peptides from a number of proteins known to be involved in signal transduction were identified, many with previously uncharacterized sites of phosphorylation.

This technique allows the large-scale, automated analysis and identification of phosphopeptides and potentially the site of phosphorylation. Such an approach is unlikely to be successful using conventional 1DE or 2DE approaches because insufficient enrichment of the phosphopeptides is achieved.

Zhao et al. described a novel method of cell surface profiling, combining biotin labeling and 1DE and nano LC-MS/MS technology.[45] Cell surface proteins from human lung cancer cells were biotinylated and recovered by incubation with streptavidin magnetic beads. After washing with potassium chloride (high salt) and sodium carbonate (high pH) to remove cytosolic proteins, the sample was treated with SDS before being analyzed by1DE. The separated proteins were then excised, digested, and analyzed by nano LC-MS/MS. The combination of 1DE and nano LC-MS/MS led to the identification of many proteins from each single gel band, and using this approach, 526 integral membrane proteins were identified from the gel. However, the method was biased against extremely hydrophobic proteins: only 10% of the identified integral membrane proteins contained seven or more TMDs. This could be due to protein precipitation and reduced solubility leading to poor entry into the SDS-PAGE gel or difficulty in extraction of the peptides post digestion. A number of cytosolic proteins were also identified, suggesting that further enrichment for membrane proteins may be required. Nevertheless, biotin-directed affinity purification (BDAP) offers a versatile method for isolation and enrichment of membrane proteins and can be combined with a number of downstream techniques such as 2DE, isotope coded affinity tag (ICAT), or shotgun proteomics approaches.

As demonstrated by Molloy et al., protein extraction with methanol offers improved solubilization of membrane proteins for 2DE.[29] Methanol, unlike many other denaturants such as detergents or urea, is compatible with tryptic digestion. It can therefore be included in the digestion buffer to ensure that proteins remain solubilized during digestion and do not precipitate out of solution. Similar properties are offered by the acid labile surfactant RapiGest, as demonstrated by the complete solubilization and digestion of bacteriorhodopsin.[46,47]

The benefits of using methanol in the extraction of membrane proteins is illustrated by a number of studies by Blonder et al.[48,49] and Gosh et al.[50] Initial experiments used sodium carbonate treated membrane proteins from *Deinococcus radiodurans* extracted using 60% methanol prior to digestion and analysis by micro LC-MS/MS.[48] Significant membrane protein enrichment was observed using this method compared to membrane proteins solubilized in ammonium bicarbonate alone. To illustrate the versatility of this technique, Smith and coworkers sought to develop a method of protein extraction, solubilization, and digestion that was applicable to a range of cell types.[48] Intact purple AQ4 membrane proteins from *Halobacterium haloblum* were extracted and digested with trypsin in 50 mM ammonium bicarbonate with 60% methanol. Although the presence of methanol reduces the activity of trypsin significantly, the digestion was sufficient for the generation of peptides for micro LC-MS/MS analysis using this method. A major component of the purple membrane, bacteriorhodopsin (GRAVY, 0.723), was characterized. The protein has three predicted TMD trypsin cleavage sites; evidence for two of these sites was obtained using this method. Forty additional purple membrane proteins were also identified, over half of which were considered hydrophobic, three quarters of the proteins being previously recognized to function within the membrane. This approach was also applied to the characterization of human epidermis plasma membrane enriched fractions, identifying 117 proteins, illustrating the applicability of such a technique to different sample types.

The combination of affinity isolation and micro LC-MS/MS was also exploited by Goshe et al. to achieve enrichment of membrane proteins and therefore reduced sample complexity and improved detection of low-abundance proteins.[50] Biotinylation of cysteine resides was followed by affinity chromatography and micro LC-MS/MS of the eluted peptides. Cysteine labeled peptides accounted for 72% of all the peptides identified. A total of 40 integral inner membrane proteins were identified, of which 15 proteins had three or more TMDs. The identification of an iron–sulfur binding protein with six cysteine-labeled peptides illustrated that this approach is able to identify multiple cysteine residues originating from the same protein. The limitation of this technique, as with all cysteine-labeling strategies, is that cysteine is a low-abundance amino acid and therefore proteins with few or no cysteine residues will be missed by such a procedure. However, the use of a protein labeling strategy introduces the opportunity to quantify differences in protein expression levels and abundance.

17.3.2 Shotgun Proteomics: 2D LC-MS/MS Approaches

The term shotgun proteomics is often used to refer to the high-throughput automated identification of proteins via analysis of the total proteolytic peptides derived from a complex protein mixture or proteome. This generally involves a 2D separation, separating peptides using two different LC columns to offer increased peptide separation and resolution. The strategy is not limited by the constraints in pI, molecular weight, or protein abundance that apply to 2DE, making such an approach ideal for membrane protein analysis. It should be noted, however, that this method is usually directed toward protein recognition (via database searching) rather than full characterization of individual proteins and their modifications. Moreover, such an approach generates vast amounts of data that must be analyzed and verified to ensure correct protein identification and to minimize false positive results.

A 2D LC-MS/MS approach to peptide separation, termed multidimensional protein identification technology (MudPIT), was applied by Washburn et al. to identify proteins from *Saccharomyces cerevisiae* on a large scale.[51] The soluble fraction obtained after yeast cell lysis was removed and digested with Lys-C and trypsin. Separate samples of the insoluble proteins were washed with phosphate-buffered saline (PBS) either once or extensively; each was then treated with 90% formic acid and cyanogen bromide (CNBr) and subsequently digested with Lys-C and trypsin. The three digested fractions were then analyzed by 2D LC-MS/MS, generating data from 5540 peptides representing 1484 proteins. The insoluble fractions, particularly the heavily washed fraction, were enriched with integral and membrane-associated proteins. In total, 72 of the 231 predicted membrane-associated proteins were identified. The use of formic acid aids membrane solubilization, with the exposed regions of the membrane, typically the loop regions of transmembrane proteins, being cleaved by CNBr treatment. Furthermore, this approach can provide an insight into the topology of a protein. For example, PMA1, a H^+ transporting P-type ATPase, has ten predicted TMDs. Using this approach, 13 peptides were observed, of which 10 were from the loop region between TMDs 4 and 5 with one peptide spanning TMD5 itself, suggesting that this TMD is not embedded in the bilayer.

In an alternative approach, Peng et al. used an off-line combination of two LC separations prior to MS/MS.[52] Here the fractions eluted from the first LC separation were collected prior to transfer to the second-dimension separation and analysis by MS/MS. There are a number of technical advantages to this type of method of analysis, which allows the user to assess fractions of interest and reanalzye them at a later date if necessary. In a single 2D LC-MS/MS analysis of yeast lysate, 1504 proteins were identified.

Further to the work of Washburn et al.,[51] Wu et al.[53] combined sodium carbonate high pH treatment with the use of a nonspecific protease, proteinase K, with 2D LC-MS/MS to analyze rat membrane proteins. The use of proteinase K results in many unique and potentially overlapping peptides for each protein, thereby increasing the sequence coverage obtained. Using brain homogenates, containing membrane and cytosolic proteins, 1610 proteins were identified with >95% confidence. Of these, 454 proteins were predicted to have TMDs. The presence of many overlapping peptides from the same protein allowed the identification and location of some PTMs. While less than 20% sequence coverage was observed for most proteins, 24 modifications present in 18 membrane proteins were identified, providing structural information on the proteins' interaction with the membrane. For example, sequence coverage of approximately 26% was obtained for aquaporin, mainly at the C-terminal end of the protein. Five peptides were identified that overlapped into the predicted C-terminal TMD, suggesting that the domain is exposed and not embedded within the bilayer as predicted. Two previously reported sites of phosphorylation could not be confirmed due to poor sequence coverage; however, a previously unidentified site of phosphorylation was identified with 98% confidence. In a further experiment, Golgi membranes isolated from rat liver were subjected to repeated high pH and proteinase K treatment. By comparing the single and multiple treatments, information on the relative topology and localization of a protein could be deduced. However, such an approach requires high sequence coverage, and therefore a further enrichment step may be required to optimize the detection of minor components.

Although SDS interferes with MS analysis, it can be used to solubilize proteins prior to tryptic digestion and then removed during preparative LC separation of peptides prior to analysis by MALDI MS/MS.[54] Furthermore, the use of a 2D LC-MS/MS approach sufficiently reduces the concentration of SDS to allow ESI-MS/MS analysis, a technique more sensitive to the presence of impurities. Both of these shotgun proteomics approaches were able to identify over 100 membrane proteins from Trition X-100 extracted human HT29 cells.[54]

17.4 FUNGAL PROTEOMICS EXAMPLES

As discussed in section 17.2, changes in the detergent composition of the solubilization buffer can greatly improve the number of spots detected by 2DE. Harder et al. have demonstrated that preboiling cell extracts with SDS prior to sample dilution in urea/thiourea/CHAPS lysis buffer increased the solubilization of high molecular weight proteins.[55] Since SDS disrupts noncovalent interactions between proteins, these extracts are more likely to include peripheral membrane proteins.

In a novel approach to study the components of the cell wall of *S. cerevisiae*, Pardo et al. monitored the proteins secreted by regenerating protoplasts.[56] Protoplasts were formed by treating cells with glusulase, completely removing the cell wall, in an isotonic medium. The cells were then allowed to regenerate the cell wall, with the cells and the medium being analyzed by 2DE to assess the components and mechanisms involved. Western blotting and N-terminal sequencing were used to confirm the presence of known cell wall proteins. Using this method, a number of GPI-anchored proteins were detected, together with several heat-shock proteins, members of the PIR family of cell wall proteins, and a number of glycolytic enzymes thought to be present on the cell surface. Eg1, an exo-β-1,3-glucanase that has been implicated in cell wall β-glucan assembly, was also identified. This type of approach removes the need to breakdown and solubilize the cell wall structure, and provides an insight into the timing of expression of the different types of proteins involved and allows a mechanism for the formation of the cell wall to be proposed.

Later studies by the same group[57] took this analysis a stage further using MS to perform a large-scale investigation of cell wall regeneration with analysis of 2DE separated proteins by MALDI, nano ESI-MS/MS, or on-line LC-MS/MS. This approach allowed a much more detailed study of the proteins involved with many more proteins being identified, including a number of unknown function and so-called moonlighting proteins, whose function depends on their location and environment, and which were not previously thought to be involved in cell wall construction.

The benefits of a 2DE-based sequential extraction approach to membrane proteomics are illustrated in the study of the change in composition of the cell wall in *Candida albicans* as the cells change from yeast to filamentous forms.[58,59] *C. albicans* cell wall extracts were washed with decreasing salt concentrations to remove any extracellular or cytosolic protein contaminants before being treated to sequential release of cell wall components. Cell surface associated and other loosely attached membrane proteins were released by SDS extraction. The remaining insoluble pellet was washed and treated with sodium hydroxide (releasing Pir-CWPs and CEPs directly linked to β-1,3-glucan via alkali-labile bonds) or digested with β-1,3-glucanase (releasing mannoproteins attached to β-1,3-glucan, via β-1,6-glucan, by a phosphodiester bridge). Remaining extracts from the latter treatment were then treated with exochitinase to release GPI-CWP attached to chitin through a β-1,6-glucan moiety. Extracts from all treatments were then solubilized with standard IEF buffers and separated by 2DE, with spots of interest digested and analyzed by MALDI-TOF/TOF MS.

This sequential extraction procedure demonstrated that different expression profiles could be observed for each cell type and gave an insight into the proteins involved in cell wall remodeling and organization. It also provided an understanding of the mechanisms by which mannoproteins were incorporated, assembled, and retained in the cell wall, and their interaction with wall polysaccharides. Although this approach generates a large number of cell wall subfractions, the removal of cytosolic material, protein enrichment, and the breakdown of cell walls into components allow the release and detection of low-abundance integral membrane proteins previously overlooked in whole cell preparations. In an alternative approach, Vediyappan et al. used extraction with aqueous ammonium carbonate containing 1% β-mercaptoethanol to monitor expression changes in cell surface associated membrane proteins from *C. albicans* when grown in the presence of glucose or galactose.[60]

Mrsa et al. described the characterization of phosphorylated cell wall components of *S. cerevisiae* using a range of biochemical techniques.[61] The authors used negative ion mode ESI-MS to provide structural analysis of phosphorylated trisaccharides and disaccharides from cell wall extracts treated with SDS and TFA. Using this approach, a number of oligosaccharide structures were characterized that were previously only identified in soluble glycoproteins.

Pedersen et al. have attempted to address the problem of detection of low-abundance *S. cerevisiae* membrane proteins by combining advances in prefractionation with 2DE analysis.[37] The benefits of such an approach can be seen by comparing the standard total membrane protein approach versus the prefractionation approach. For total protein extracts separated on a broad pH range gel, the components of 88 spots were identified, corresponding to the products of 42 genes. None of these proteins contained TMDs, and only three proteins were classed as low-abundance proteins (using the codon adaption index score of <0.2 to indicate low abundance). When isolated membrane preparations were studied and when solution IEF was employed (using a multicompartment electrolyzer) to recover an alkaline protein fraction, 2DE analysis and MS of in-gel digests yielded 780 identifications corresponding to 323 genes. From these results, 9 proteins were classified as low abundance, and of the 264 genes identified as encoding membrane proteins, 105 were integral membrane proteins containing TMDs. Since the solubilization solution used was the same as that for total protein analysis, these results suggest that poor solubility is not the sole explanation for poor detection of membrane constituents of a total protein fraction and removal of the large excess of abundant and soluble proteins facilitates the recovery and detection of membrane proteins. Protein separation could be further enhanced by the use of narrow range IEF strips (1 pH unit), which allow higher protein loads and therefore the opportunity to detect lower-abundance proteins.

The powerful combination of 2DE and biochemical and mass spectrometric techniques for subcellular proteomics is illustrated by the characterization of the mitochondrial proteome of *S. cerevisiae* by Sickmann et al.[62] Both purified mitochondria and mitochondrial-associated proteins isolated by trypsinisation or salt extraction were analyzed by 2DE or nana LC-MS/MS, or both. Using a 2DE-based approach, 109 proteins were identified, of which 102 had been previously localized to the mitochondria. In an alternative approach, the extracts were treated with four different proteases (trypsin, chymotrypsin, Glu-C, or subtilisin) before being separated and analyzed by nano LC-MS/MS. By pooling the data obtained, a total of 750 proteins were identified, representing over 90% of the predicted mitochondrial proteome of *S. cerevisiae*. Of the proteins identified, 436 were known mitochondrial proteins with the other proteins being associated with other cellular compartments (dual localization) or of unknown function.

This combined analysis approach illustrates how several techniques can be employed together to maximize the amount of data obtained from a single sample. However, such an approach is not only time-consuming but requires an array of data-handling solutions to manage the vast (in excess of 15 million) numbers of mass spectra obtained. It is interesting to note that even with such a large quantity of data it is not possible to generate a truly complete proteome since some proteins are expressed only in particular conditions or may escape detection by the methods used. Nevertheless, the product of such an analysis serves as a database for future reference for the characterization of novel proteins and their functions.

17.5 CONCLUSIONS

The study of membrane proteins is one of the most challenging areas of proteomics, with the proteins being difficult to isolate and solubilize for analysis. By employing prefractionation techniques and improving the solubilization conditions to mimic the natural environment in which these proteins exist, these proteins become easier to identify and characterize using proteomics techniques.

Both gel-based and LC-MS/MS based approaches have a number of inherent advantages and disadvantages, and therefore the choice of method must be based on a number of factors. Where solubilization of membrane proteins for 1DE or 2DE can be achieved (using a variety of new agents), the gel approaches, with or without Western blotting, can provide indications of protein identity that can be substantiated by MS analyses of in-gel digests. Detailed characterization (e.g., to define the nature and location of posttranslational modifications) almost invariably requires the use of combined LC-MS/MS. Extension of the latter method by incorporation of two dimensions of LC separation provide an alternative to gel-based approaches by analyzing proteolytic digests of protein mixtures without prior gel separation. This may be advantageous in avoiding the limitations of applicability of the gel methods.

The approach adopted by Borner et al.[12]—combining phase-partitioning, targeted enzymatic release of the proteins of interest, 2DE DIGE analysis of enriched fractions, and LC-MS/MS of 1DE separated extracts—illustrates how several techniques can be used in concert to provide a detailed characterization of the proteins present in a membrane extract.

Combining gel and MS analyses with biochemical techniques enables further characterization of the proteome by identifying changes in protein expression and posttranslational modification as the local environment changes.[58,61] While this can be a time-consuming approach, the data obtained provide a detailed characterization of the proteins expressed in a specific set of conditions and their subcellular roles and localization. These data can then act as a reference map for further analyses as other techniques are applied or the conditions modified to induce different expression patterns.

ACKNOWLEDGMENTS

The authors would like to thank Dr. Caroline Evans and Dr. Isabel Riba for their suggestions during preparation of this manuscript. Research within the Michael Barber Centre for Mass Spectrometry is funded by the Biotechnology and Biological Sciences Research Council (BBSRC).

REFERENCES

1. Singer, S.J., Some early history of membrane molecular biology, *Annu. Rev. Physiol.*, 66, 1–27, 2004.
2. Alberts, B., Gray, D., Lewis, J., Raff, M., Roberts, K., Watson, J.D., *Molecular Biology of the Cell*, 3rd ed., Garland Publishing, Oxford, U.K., 1994. AQ5
3. Wallin, E. and von Heijne, G., Genome-wide analysis of integral membrane proteins from eubacterial, archaean, and eukaryotic organisms, *Protein Sci.* 7, 1029–1038, 1998.
4. Wilkins, M.R., Gasteiger, E., Sanchez, J.C., Bairoch, A., Hochstrasser, D.F., Two-dimensional gel electrophoresis for proteome projects: The effects of protein hydrophobicity and copy number, *Electrophoresis*, 19, 1501–1505, 1998.
5. Schwartz, R., Ting, C.S., and King, J., Whole proteome pI values correlate with subcellular localizations of proteins for organisms within the three domains of life, *Genome Res.*, 11, 703–709, 2001.
6. Chen, C.P. and Rost, B., State-of-the-art in membrane protein prediction, *Appl. Bioinformatics*, 1, 21–35, 2002.
7. Santoni, V., Molloy, M., and Rabilloud, T., Membrane proteins and proteomics: Un amour impossible? *Electrophoresis*, 21, 1054–1070, 2000.
8. Kyte, J. and Doolittle, R.F., A simple method for displaying the hydropathic character of a protein, *J. Mol. Biol.*, 157, 105–132, 1982.
9. Santoni, V., Doumas, P., Rouquie, D., Mansion, M., Rabilloud, T., and Rossignol, M., Large scale characterization of plant plasma membrane proteins, *Biochimie*, 81, 655–661, 1999.
10. Shaw, J., Rowlinson, R., Nickson, J., Stone, T., Sweet, A., Williams, K., and Tonge, R., Evaluation of saturation labelling two-dimensional difference gel electrophoresis fluorescent dyes, *Proteomics*, 3, 1181–1195, 2003.
11. Tonge, R., Shaw, J., Middleton, B., Rowlinson, R., Rayner, S., Young, J., Pognan, F., Hawkins, E., Currie, I., and Davison, M., Validation and development of fluorescence two-dimensional differential gel electrophoresis proteomics technology, *Proteomics*, 1, 377–396, 2001.
12. Borner, G.H., Lilley, K.S., Stevens, T.J., and Dupree, P., Identification of glycosylphosphatidylinositol-anchored proteins in *Arabidopsis*: A proteomic and genomic analysis, *Plant Physiol.* 132, 568–577, 2003.
13. Molloy, M.P., Two-dimensional electrophoresis of membrane proteins using immobilized pH gradients, *Anal. Biochem.*, 280, 1–10, 2000.
14. Herbert, B., Advances in protein solubilisation for two-dimensional electrophoresis, *Electrophoresis*, 1999, 20, 660–663.
15. Mechin, V., Consoli, L., Le Guilloux, M., and Damerval, C., An efficient solubilization buffer for plant proteins focused in immobilized pH gradients, *Proteomics*, 3, 1299–1302, 2003.
16. Rabilloud, T., Adessi, C., Giraudel, A., and Lunardi, J., Improvement of the solubilization of proteins in two-dimensional electrophoresis with immobilized pH gradients, *Electrophoresis*, 18, 307–316, 1997.
17. Wang, W., Scali, M., Vignani, R., Spadafora, A., Sensi, E., Mazzuca, S., and Cresti, M., Protein extraction for two-dimensional electrophoresis from olive leaf, a plant tissue containing high levels of interfering compounds, *Electrophoresis*, 24, 2369–2375, 2003.
18. Bordier, C., Phase separation of integral membrane proteins in Triton X-114 solution, *J. Biol. Chem.*, 256, 1604–1607, 1981.

19. Elortza, F., Nuhse, T.S., Foster, L.J., Stensballe, A., Peck, S.C., and Jensen, O.N., Proteomic analysis of glycosylphosphatidylinositol-anchored membrane proteins, *Mol. Cell. Proteomics*, 2, 1261–1270, 2003.
20. Rabilloud, T., Blisnick, T., Heller, M., Luche, S., Aebersold, R., Lunardi, J., and Braun-Breton, C., Analysis of membrane proteins by two-dimensional electrophoresis: Comparison of the proteins extracted from normal or *Plasmodium falciparum*-infected erythrocyte ghosts, *Electrophoresis*, 20, 3603–3610, 1999.
21. Luche, S., Santoni, V., and Rabilloud, T., Evaluation of nonionic and zwitterionic detergents as membrane protein solubilizers in two-dimensional electrophoresis, *Proteomics*, 3, 249–253, 2003.
22. Santoni, V., Kieffer, S., Desclaux, D., Masson, F., and Rabilloud, T., Membrane proteomics: Use of additive main effects with multiplicative interaction model to classify plasma membrane proteins according to their solubility and electrophoretic properties, *Electrophoresis*, 21, 3329–3344, 2000.
23. Henningsen, R., Gale, B.L., Straub, K.M., and DeNagel, D.C., Application of zwitterionic detergents to the solubilization of integral membrane proteins for two-dimensional gel electrophoresis and mass spectrometry, *Proteomics*, 2, 1479–1488, 2002.
24. Fountoulakis, M. and Takacs, B., Effect of strong detergents and chaotropes on the detection of proteins in two-dimensional gels, *Electrophoresis*, 22, 1593–1602, 2001.
25. Barry, R.C., Alsaker, B.L., Robison-Cox, J.F., and Dratz, E.A., Quantitative evaluation of sample application methods for semipreparative separations of basic proteins by two-dimensional gel electrophoresis, *Electrophoresis*, 24, 3390–3404, 2003.
26. Molloy, M.P., Herbert, B.R., Walsh, B.J., Tyler, M.I., Traini, M., Sanchez, J.C., Hochstrasser, D.F., Williams, K.L., and Gooley, A.A., Extraction of membrane proteins by differential solubilization for separation using two-dimensional gel electrophoresis, *Electrophoresis*, 19, 837–844, 1998.
27. Herbert, B.R., Molloy, M.P., Gooley, A.A., Walsh, B.J., Bryson, W.G., and Williams, K.L., Improved protein solubility in two-dimensional electrophoresis using tributyl phosphine as reducing agent, *Electrophoresis*, 19, 845–851, 1998.
28. Lehner, I., Niehof, M., and Borlak, J., An optimized method for the isolation and identification of membrane proteins, *Electrophoresis*, 24, 1795–1808, 2003.
29. Molloy, M.P., Herbert, B.R., Williams, K.L., and Gooley, A.A., Extraction of *Escherichia coli* proteins with organic solvents prior to two-dimensional electrophoresis, *Electrophoresis*, 20, 701–704, 1999.
30. Ferro, M., Salvi, D., Brugiere, S., Miras, S., Kowalski, S., Louwagie, M., Garin, J., Joyard, J., and Rolland, N., Proteomics of the chloroplast envelope membranes from *Arabidopsis thaliana*, *Mol. Cell. Proteomics*, 2, 325–345, 2003.
31. Ferro, M., Salvi, D., Riviere-Rolland, H., Vermat, T., Seigneurin-Berny, D., Grunwald, D., Garin, J., Joyard, J., and Rolland, N., Integral membrane proteins of the chloroplast envelope: Identification and subcellular localization of new transporters, *Proc. Natl. Acad. Sci. USA*, 99, 11,487–11,492, 2002.
32. Deshusses, J.M., Burgess, J.A., Scherl, A., Wenger, Y., Walter, N., Converset, V., Paesano, S., Corthals, G.L., Hochstrasser D.F., and Sanchez, J.C., Exploitation of specific properties of trifluoroethanol for extraction and separation of membrane proteins, *Proteomics*, 3, 1418–1424, 2003.
33. Fujiki, Y., Hubbard, A.L., Fowler, S., and Lazarow, P.B., Isolation of intracellular membranes by means of sodium carbonate treatment: application to endoplasmic reticulum, *J. Cell Biol.*, 93, 97–102, 1982.

34. Molloy, M.P., Herbert, B.R, Slade M.B., Rabilloud, T., Nouwens, A.S., Williams, K.L., and Gooley, A.A., Proteomic analysis of the *Escherichia coli* outer membrane, *Eur. J. Biochem.*, 267, 2871–2881, 2000.
35. Santoni, V., Rabilloud, T., Doumas, P., Rouquie, D., Mansion, M., Kieffer, S., Garin, J., and Rossignol, M., Towards the recovery of hydrophobic proteins on two-dimensional electrophoresis gels, *Electrophoresis*, 20, 705–711, 1999.
36. Gygi, S.P., Corthals, G.L., Zhang, Y., Rochon, Y., and Aebersold, R., Evaluation of two-dimensional gel electrophoresis-based proteome analysis technology, *Proc. Natl. Acad. Sci. USA*, 97, 9390–9395, 2000.
37. Pedersen, S.K., Harry, J.L., Sebastian, L., Baker, J., Traini, M.D., McCarthy, J.T., Manoharan, A., Wilkins, M.R., Gooley, A.A., Righetti, P.G., Packer, N.H., Williams, K.L., and Herbert, B.R., Unseen proteome: Mining below the tip of the iceberg to find low abundance and membrane proteins, *J. Proteome Res.*, 2, 303–311, 2003.
38. Chalmers, M.J. and Gaskell, S.J., Advances in mass spectrometry for proteome analysis, *Curr. Opin. Biotechnol.*, 11, 384–390, 2000.
39. Godovac-Zimmermann, J. and Brown, L.R., Perspectives for mass spectrometry and functional proteomics, *Mass Spectrom. Rev.* 20, 1–57, 2001; Aebersold, R. and Mann, M., Mass spectrometry–based proteomics, *Nature*, 422, 198–207, 2003.
40. van Montfort, B.A., Canas, B., Duurkens, R., Godovac-Zimmermann, J., and Robillard, G.T., Improved in-gel approaches to generate peptide maps of integral membrane proteins with matrix-assisted laser desorption/ionization time-of-flight mass spectrometry, *J. Mass Spectrom.* 37, 322–330, 2002.
41. Shevchenko, A., Jensen, O.N., Podtelejnikov, A.V., Sagliocco, F., Wilm, M., Vorm, O., Mortensen, P., Shevchenko, A., Boucherie, H., and Mann, M., Linking genome and proteome by mass spectrometry: Large-scale identification of yeast proteins from two dimensional gels, *Proc. Natl. Acad. Sci. USA*, 93, 14,440–14,445, 1996.
42. Romijn, E.P., Krijgsveld, J., and Heck, A.J., Recent liquid chromatographic-(tandem) mass spectrometric applications in proteomics, *J. Chromatogr A.*, 1000, 589–608, 2003.
43. Wu, C.C. and Yates, J.R., III, The application of mass spectrometry to membrane proteomics, *Nat. Biotechnol.*, 21, 262–267, 2003.
44. Nuhse, T.S., Stensballe, A., Jensen, O.N., and Peck, S.C., Large-scale analysis of in vivo phosphorylated membrane proteins by immobilized metal ion affinity chromatography and mass spectrometry, *Mol. Cell. Proteomics*, 2, 1234–1243, 2003.
45. Zhao, Y., Zhang, W., Kho, Y., and Zhao, Y., Proteomic analysis of integral plasma membrane proteins, *Anal. Chem.*, 76, 1817–1823, 2004.
46. Yu, Y.Q., Gilar, M., and Gebler, J.C., A complete peptide mapping of membrane proteins: A novel surfactant aiding the enzymatic digestion of bacteriorhodopsin, *Rapid Commn. Mass Spectrom.*, 18, 711–715, 2004.
47. Yu, Y.Q., Gilar, M., Lee P.J., Bouvier, E.S., and Gebler, J.C., Enzyme-friendly, mass spectrometry-compatible surfactant for in-solution enzymatic digestion of proteins, *Anal. Chem.*, 75, 6023–6028, 2003.
48. Blonder, J., Goshe, M.B., Moore, R.J., Pasa-Tolic, L., Masselon, C.D., Lipton, M.S., and Smith, R.D., Enrichment of integral membrane proteins for proteomic analysis using liquid chromatography-tandem mass spectrometry, *J. Proteome Res.*, 1, 351–360, 2002.
49. Blonder, J., Conrads, T.P., Yu, L.R., Terunuma, A., Janini, G.M., Issaq, H.J., Vogel, J.C., and Veenstra, T.D., A detergent- and cyanogen bromide-free method for integral membrane

proteomics: Application to *Halobacterium* purple membranes and the human epidermal membrane proteome, *Proteomics*, 4, 31–45, 2004.
50. Goshe, M.B., Blonder, J., and Smith, R.D., Affinity labeling of highly hydrophobic integral membrane proteins for proteome-wide analysis, *J. Proteome Res.*, 2, 153–161, 2003.
51. Washburn, M.P., Wolters, D., and Yates, J.R., III, Large-scale analysis of the yeast proteome by multidimensional protein identification technology, *Nat. Biotechnol.*, 19, 242–247, 2001.
52. Peng, J., Elias, J.E., Thoreen, C.C., Licklider, L.J., and Gygi, S.P., Evaluation of multidimensional chromatography coupled with tandem mass spectrometry (LC/LC-MS/MS) for large-scale protein analysis: The yeast proteome, *J. Proteome Res.*, 2, 43–50, 2003.
53. Wu, C.C., MacCoss, M.J., Howell, K.E., Yates, J.R., III, A method for the comprehensive proteomic analysis of membrane proteins, *Nat. Biotechnol.*, 21, 532–538, 2003.
54. Zhang, N., Li, N., and Li, L., Liquid chromatography MALDI MS/MS for membrane proteome analysis, *J. Proteome Res.*, 3, 719–727, 2004.
55. Harder, A., Wildgruber, R., Nawrocki, A., Fey, S.J., Larsen, P.M., and Gorg, A., Comparison of yeast cell protein solubilization procedures for two-dimensional electrophoresis, *Electrophoresis*, 20, 826–829, 1999.
56. Pardo, M, Monteoliva, L., Pla, J., Sanchez, M., Gil, C., Nombela, C., Two-dimensional analysis of proteins secreted by *Saccharomyces cerevisiae* regenerating protoplasts: A novel approach to study the cell wall, *Yeast*, 15, 459–472, 1999.
57. Pardo, M., Ward, M., Bains, S., Molina, M., Blackstock, W., Gil, C., and Nombela, C., A proteomic approach for the study of *Saccharomyces cerevisiae* cell wall biogenesis, *Electrophoresis*, 21, 3396–3410, 2000.
58. Pitarch, A., Sanchez, M., Nombela, C., and Gil, C., Sequential fractionation and two-dimensional gel analysis unravels the complexity of the dimorphic fungus *Candida albicans* cell wall proteome, *Mol. Cell. Proteomics*, 1, 967–982, 2002.
59. Pitarch, A., Sanchez, M., Nombela, C., and Gil, C., Analysis of the *Candida albicans* proteome. I. Strategies and applications, *J. Chromatogr. B. Analyt. Technol. Biomed. Life Sci.*, 787, 101–128, 2003.
60. Vediyappan, G., Bikandi, J., Braley, R., and Chaffin, W.L., Cell surface proteins of *Candida albicans*: Preparation of extracts and improved detection of proteins, *Electrophoresis*, 21, 956–961, 2000.
61. Mrsa, V., Ecker, M., Strahl-Bolsinger, S., Nimtz, M., Lehle, L., and Tanner, W., Deletion of new covalently linked cell wall glycoproteins alters the electrophoretic mobility of phosphorylated wall components of *Saccharomyces cerevisiae, J. Bacteriol.*, 181, 3076–3086, 1999.
62. Sickmann, A., Reinders, J., Wagner, Y., Joppich, C., Zahedi, R., Meyer, H.E., Schonfisch, B., Perschil, I., Chacinska, A., Guiard, B., Rehling, P., Pfanner, N., and Meisinger, C., The proteome of *Saccharomyces cerevisiae* mitochondria, *Proc. Natl. Acad. Sci. USA*, 100, 13,207–13,212, 2003.

18 Structural Approaches in Glycoproteomics

Heidi Geiser, Cristina Silvescu, and Vernon Reinhold

CONTENTS

18.1 THE DEVELOPING FIELD OF GLYCOPROTEOMICS

When genomes are sequenced, the information is immediately usable for bioinformatics, microarrays, and many other applications. Genomes can be compared and genes manipulated knowing only the DNA sequence. The function of DNA and the one-way translation of data were rarely in question. From this simple understanding the significance of the proteome rose to astonishing prominence and became the research objective of many with goals and dreams, "...capable of identifying up to 1 million proteins a day...."[1] But much has changed over the last few years; histone tails are modified by acetylation, phosphorylation, and methylation, all affecting the local chromatin structure and transcriptional regulation of adjacent genes.[2] Micro RNAs (about 22 nucleotides in length) add further complexity, producing transcripts that form imperfect hairpin structures and guide RNA silencing.[3,4] Alternative pre-mRNA splicing is another mechanism for regulating gene expression in higher eukaryotes. Recent estimates indicate 30% of the primary transcripts of the human genome are subject to alternative splicing showing specific spatial/temporal patterns during normal development. In complex genes, alternative splicing generates dozens or even hundreds of different mRNA isoforms from a single transcript. It has been argued that nonprotein coding RNA,[5] including microRNAs, siRNAs, and intronic RNA, are part of a vast regulatory mechanism, and phenotypic variations between species (and individuals) results from differences in the control architecture, not the proteins themselves.

Proteomics is now projecting a similar divergent story. The amino acid sequence alone is insufficient to describe the function of a protein. Beyond folding and conformation, which can be difficult to predict, arise a growing multitude of new considerations: alternative activities at different sites, binding of metal cofactors, cleavage of signal sequences, phosphorylation, glycosylation, and a smaller number of other post-translational modifications (PTMs). These embellishments make protein sequencing and understanding functional relationships an ever-challenging endeavor. However, PTMs are vital to activity. Glycosylation specifically, with but a small set of monomers, generates a profusion of structures that are rarely brought to full understanding. Countering these demanding biochemical problems has been a steady assault of new techniques, instruments, and significant methodological changes that require continuous vigilance to the literature. This chapter summarizes some of those more recent techniques.

18.1.1 Significance of Molecular Glycosylation

The abundant and pleiotropic modifications introduced by carbohydrate residues affect cell signaling, trafficking, and cell-cell adhesion as well as the physical properties of the protein itself, such as solubility and stability.[6] On review of the Swiss-Prot database,[7] it has been predicted that over half of all human proteins are glycosylated.[8] When single pathways of glycan biosynthesis are blocked by mutation, humans suffer numerous developmental disorders (seizures, gastrointestinal problems, skin lesions, impaired vision, and psychomotor and mental retardation), reflecting a profusion of altered function when such changes are expressed through all glycosylated products.[9] Even within a single protein, individual glycans contribute to different metabolic activities, as recently illustrated for a human protease.[10] In this case all five N-linked glycosylation sites were found to be utilized, and at one site (Asn-286), a dual role of oligosaccharide folding and lysosomal targeting was observed.

18.1.2 Basic Glycan Structures

In a recent survey,[11] a total of 13 different monosaccharides and 8 amino acids were involved in linkages to proteins providing a total of 41 unique bonds. The majority of eukaryotic conjugations, however, involve a much smaller set of two different

2-acidamidohexoses, GlcNAc and GalNAc. The former structures are defined as N-linked glycans that attached to the amide group of asparagine via a beta linkage, (Figure 18.1A). Glycans that are alpha linked through GalNAc to the hydroxyls of serine and threonine moieties are defined as O-linked (Figure 18.1B). Site occupancy for both glycan types is an important metabolic measure, but characterized rarely. A peptide sequence for glycosyl attachment is known only for N-linked glycans, (-Asn-X-Ser/Thr-, where X can be any amino acid except proline). This is defined as the consensus site, and a dolichol-bound oligosaccharide ($Glc_3Man_9GlcNAc_2$) is transferred *en bloc* to this site to initiate protein N-glycosylation. This glycan subsequently loses two terminal glucose residues leaving a product ($Glc_1Man_9GlcNAc_2$-) that serves as a chaperone receptor insuring native protein conformations; a quality-control component indigenous to glycoprotein biogenesis. Misfolded proteins are exported to the cytoplasm and degraded by proteasomes through a mechanism known as ER-associated degradation (ERAD). Native structures are trimmed of the remaining glucose moiety and exported to the Golgi. These glycans are partially trimmed of mannose and reglycosylated with a different set of resident glycosyltransferases in a time- and location-dependent manner, providing the hybrid and complex glycans found in many tissues.

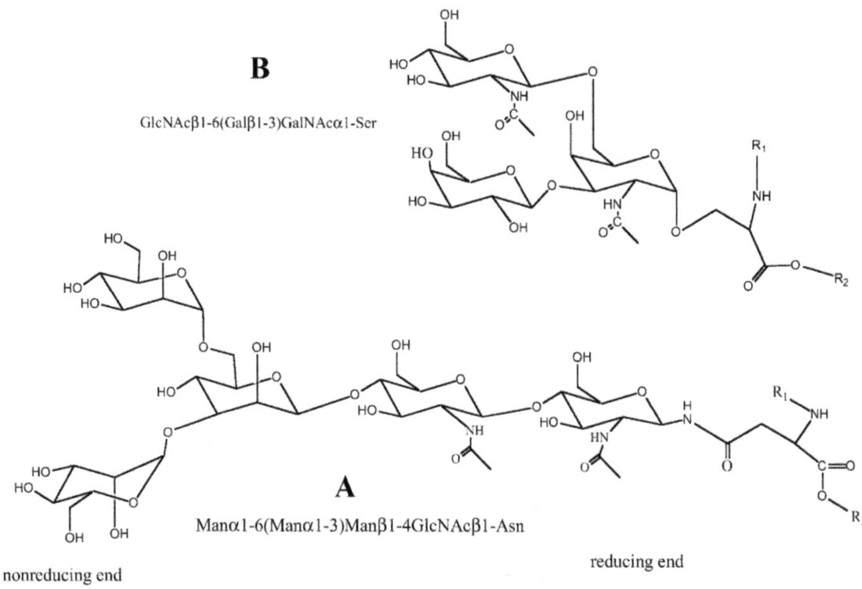

FIGURE 18.1 Two common protein-carbohydrate linkage structures and core regions: (A) GlcNAc N-linked to amide of asparagine; (B) GalNAc O-linked to threonine or serine moieties. Extensions to these core regions define glycan type. N-linkage of GlcNAc occurs at the defined peptide sequence Asn-Xaa-Ser/Thr, where Xaa can be any amino acid except proline. This has been defined as the consensus sequence.

O-linked glycans, with 2-acidamidogalactose at their reducing terminus, are synthesized by stepwise addition of monomer units.[12,13] Antennae extension with a β3-linked galactose initiates a family of structures defined as Core 1 (Figure 18.2A). A second group of structures, defined as Core 2 (Figure 18.2B), increments the Core 1 structure by the addition of 2-acetamidoglucose in a β6 linkage. Cores 3 and 4 alter these structures by replacing the β3-linked galactose with a β3-linked 2-acetamidoglucose residue (Figure 18.2C and 2D, respectively). The most common aspect of all glycosylation is diversity in composition, topology, and function. Glycosylation at any N- or O-linked site is protein dependent and varies with a host of temporal and environmental conditions.[14] Beyond the cores (Figure 18.1 and Figure 18.2), product glycans are frequently heterogeneous at any one site as well as different between sites. These modifications provide a distribution of structures termed glycomers.[15] N-linked glycans are classified into three groups based on their monomer composition: complex, hybrid, and high mannose. All three contain the same core structure (Figure 18.1A). High mannose glycans, as the name suggests, contain only mannose residues attached to the core. Complex glycans contain fucose and other sugar residues. Hybrid glycans contain both high mannose and complex chains. A comparable nomenclature has developed for O-linked glycans that are defined as cores based on branching from the single GalNAc linked to the protein, (Figure 18.2).

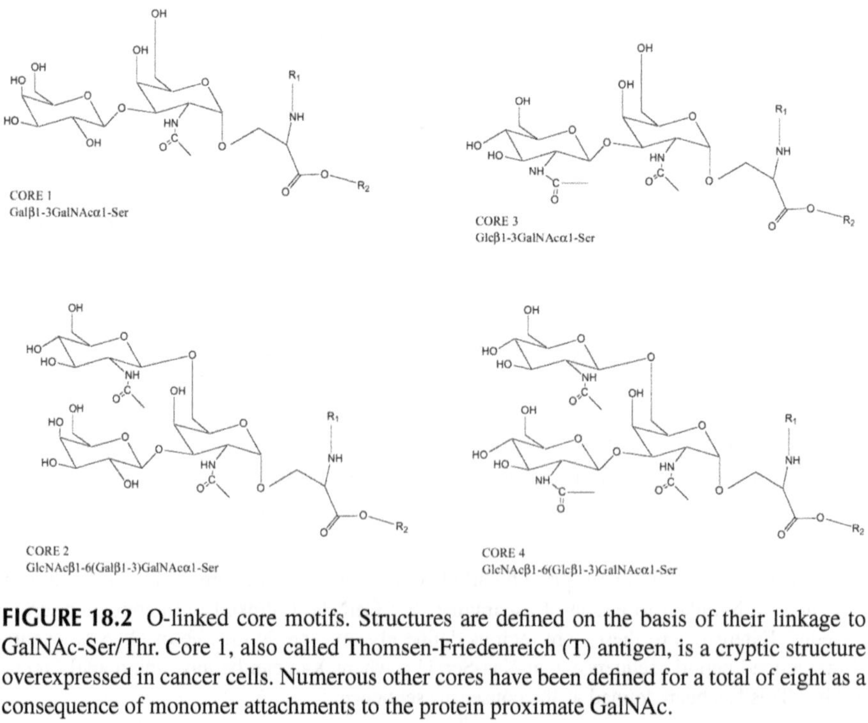

FIGURE 18.2 O-linked core motifs. Structures are defined on the basis of their linkage to GalNAc-Ser/Thr. Core 1, also called Thomsen-Friedenreich (T) antigen, is a cryptic structure overexpressed in cancer cells. Numerous other cores have been defined for a total of eight as a consequence of monomer attachments to the protein proximate GalNAc.

18.2 SEPARATION BY TWO-DIMENSIONAL GEL ELECTROPHORESIS

Global analysis of gene expression via a cross-section of bioanalytical techniques for resolving, identifying, and quantifying proteins remains the major health-related task of our generation.[16] Beyond purification, glycoprotein samples require a number of steps for a complete characterization, and these details are briefly discussed below. In essence the steps require an independent sequence of the protein and the heterogeneous glycan and the linkage site of conjugation.

For obtaining purified samples, the technique of choice has been two-dimensional polyacrylamide gel electrophoresis (2D PAGE).[17–20] The ability to separate thousands of proteins coupled with the advantage of differential display, simplicity, cost, and fairly precise quantification has made 2D gels very popular. The hope is that the buffers chosen and the cell rupture techniques selected provide all the proteins of interest. For analytical purposes, altering any of these parameters to enhance protein yield or number of gel spots observed provides a fundamental start for acquiring a tissue glycoproteome. In the similar way, comparative 2D gel studies form the molecular basis for monitoring the dynamic changes of a biological process. In this regard it must be assumed that all components are represented and visualized, reflecting gene expression.

18.2.1 Limitations

Despite the exceptional analytical power of gels, a number of constraints are apparent.[21] Some of these difficulties are reflected in protein dynamics, representation of all proteins (e.g., solubility), lower and upper limits of molecular weight (e.g., chemokines), and the imposed difficulties of posttranslational modifications such as glycosylation and phosphorylation that require a different set of analytical tools. The most challenging problem to consider remains resolution of nascent proteins from those terminally translocated.[22] Current strategies remain blind to cellular events when protein translocation is a component of function, and such techniques as differential analysis will fail to identify changes.

Prefractionation of proteins prior to gel electrophoresis, described in more detail in a previous chapter, can improve detection of low-abundance proteins. Several companies manufacture devices for separating proteins in solution according to their isoelectric point. Examples are the Bio-Rad Rotofor and the Proteome Systems IsoelectrIQ. Following separation, each fraction can be run on a 2D gel using a narrow range IPG strip. Alternatively, prefractionation may take advantage of differing protein solubilities,[21] and commercial kits are available from several companies for this type of fractionation as well. For example, the ProteoPrep Universal Extraction Kit from Sigma generates a soluble/cytoplasmic protein fraction and a membrane protein fraction, and the ReadyPrep Sequential Extraction Kit from Bio-Rad generates three fractions based on protein solubility. An advantage for including a prefractionation step for glycoconjugates is to increase the concentration of low-abundance proteins because glycan stains are not as sensitive as protein stains, and small sample loading achieves better resolution.

18.2.2 Methods

Protein 2D gel spots are typically excised from the gel, washed, and in-gel digested with a protease. For maximal recovery of products, each step requires repeated washing, drying, and swelling of gel pieces between chemical or enzymatic treatments. Peptides are then extracted using aqueous and organic mixtures at acidic or basic conditions and prepared for peptide mass fingerprinting or *de novo* sequencing, or both. The combined techniques of 2D gels and MS identifies only the most abundant proteins even though many more can be visualized by silver staining. Thus, doubts have been raised about the efforts currently devoted to implementing the 2D gel Matrix-assisted laser desorption ionization (MALDI) fingerprint strategy, and new technologies have been reported to compensate for the shortcomings mentioned. In this regard, efficient gel transfer of separated products has been a problem, and a capillary extraction apparatus has been demonstrated for its rapid and effective transfer from polyacrylamide gels to a fused-silica capillary.[23] Alternative strategies, such as on-probe in-gel digestion, has provided improved sensitivity and coverage, especially for large hydrophobic peptides.[24] Decreasing sample manipulation has shown improved sensitivities and coverage using combined 96-well microplates connected via tiny capillaries. The first microplate was used for gel destaining, reduction/alkylation, dehydration, and digestion, and the second for extraction and cleanup. A connecting capillary contained C18 reversed-phase modified monolithic silica rods. Peptides were eluted directly from the solid-phase extraction plate onto the MALDI sample support.[24] Developments of this type will extend the utility and sustain the applications of 2D gels.

18.3 GLYCOPROTEIN DETECTION ON 2D GELS

Identification of glycoproteins on gels or blots is based on one of two general methods. One method utilizes lectin binding as immunoblots to antibodies and has limited specificity. The other scheme is based on periodate oxidation of sugar cis-hydroxyl groups and converting them to aldehydes. The aldehydes are then conjugated to a variety of reagents including digoxigenin, fluorescence, biotin, or acidic fuchsin dye.

18.3.1 Methods

These procedures have varying sensitivities and ease of use. An oxidation-conjugation of fuchsin dye method (Gel-Code Glycoprotein Stain, Pierce, Rockford, Ill.) can stain 0.625 ng of avidin or 0.16 μg of horseradish peroxidase according to product literature, for example. Some stains are compatible only with gels or blots, others with both. Few have sensitivity sufficient to detect 2D gel spots. It is difficult to describe the sensitivities of dyes in terms of glycoprotein concentration, as the degree of glycosylation affects staining intensity as much as abundance of the glycoprotein. Roche (Roche Diagnostics Corp., Indianapolis, Ind.) offers a staining kit, the DIG Glycan Detection Kit, which labels glycoproteins either in solution prior to sodium dodecyl sulfate–polyacrylamide gel electrophoresis (SDS-PAGE) or after the gel has been blotted to a membrane. This kit employs periodate oxidation followed by digoxigenin labeling. Digoxigenin is detected using an antibody conjugated to alkaline phosphatase.

After staining, glycoprotein bands appear dark gray. This method is used for blots, not gels, and has a reported sensitivity between 10 (α_1-acid glycoprotein) and 250 ng (carboxypeptidase Y). Molecular Probes has developed a more sensitive, fluorescent stain, Pro-Q Emerald, which also utilizes periodate oxidation chemistry to stain gels and blots.[25,26]

One traditional glycoprotein identification method detects electroblotted proteins that bind to ConA.[27,28] Blots are probed similar to traditional immunoblots, using ConA instead of an antibody. Proteins separated by SDS-PAGE are electroblotted to nitrocellulose membranes. The blot is blocked with excess nonglycosylated protein in a saline buffer and then incubated with ConA conjugated to biotin in the blocking solution. After washing to remove excess ConA, the blot is incubated in streptavidin conjugated to horseradish peroxidase, then washed again. To detect the peroxidase bound to Con A, the blot is incubated in 4-chloro-1-naphthol and hydrogen peroxide. The protein bands stain blue in less than ten minutes.

18.4 GLYCOPROTEIN ANALYSIS

For sensitivity and specificity, mass spectrometry (MS) is unsurpassed in its ability to unravel a protein sequence. Overall the techniques for sequencing have continuously improved, but the more significant contribution to sequence analysis has been the adaptability of MS to meet newer challenges of sensitivity and specificity. Comparable progress in the characterization of molecular glycosylation has been lacking. Glycan structures are largely reported devoid of interresidue linkage and anomericity, and monomers are identified for the most part on the basis of established motifs. Enzymes, when utilized, may define monomer substrate and linkage but fail to differentiate site of origin when multiple antenna are involved. Comparative analysis, internal standards, or orthogonal methods are seldom reported and few procedures have acquired a routine status. A glycoprotein structure is usually approached by glycan release, component separation, and independent sequence analysis of each conjugate. Consensus sites are generally identified within glycopeptides following proteolytic cleavage and a study of peptide mass fingerprint (PMF) data.

18.4.1 Protein Sequencing

Two different methodological approaches are usually considered for sequencing the protein, a top-down and a bottom-up strategy. The most popular bottom-up approach includes 2D gel separations, proteolysis, and MS peptide pattern recognition. This approach, while comprehensive, requires multiple chemical and instrumental considerations. An alternative bottom-up approach, not discussed here, uses reversed phase high-performance liquid chromatography (HPLC) MS with off-line MS/MS, or N-terminal amino acid sequencing to obtain sequence tags for identification. The significantly different top-down approach has introduced an ionization method for intact proteins, tandem MS/MS, and exact mass measurement. The higher resolving power provides an accurate molecular weight for each target protein, a critical piece of information unattainable in the bottom-up characterization. Recent reports, however, have altered these general considerations in very significant ways. Leading the way has been electron capture dissociation (ECD)[29–31] adaptable to Fourier transform ion cyclotron resonance (FTICR) instruments with high magnetic fields, and more recently, the exciting developments of ion–ion chemistry for less-expensive instruments including electron transfer dissociation (ETD).[32] This latter advance provides for the first time low-energy electrons with high-ionization efficiencies for nonergodic radical bond cleavage. The radical initiated fragmentation is bond specific, homogeneous, abundant in large polypeptides, and dominated by N–C bond cleavage, preferentially giving N-terminal (C-type) fragments (Figure 18.3).[33] These products are in contrast to the more sequence-specific, vibrationally induced CO–N backbone fragments that produce b and y ions. Most interestingly, both ECD and ETD fragments have been found to retain labile PTM groups, such as O-, and N-glycosylation[34,35] and phosphorylation[36,37] and in all cases far superior to traditional MS/MS. Providing ion–ion reaction chemistry for radical ergodic backbone fragmentation, combined with quadrupole ion-trap MS, offers a fast and efficient way to analyze intact proteins, without heavy reliance on initial proteolysis, peptide mass fingerprints, and most important, costly magnetic analyzers. It is important to note that although some sequence information can be observed using ECD for glycans, it seriously lacks the details appropriate for a full structural understanding.

18.4.2 MS Profiles of Intact Glycoproteins: Top Down

There are different strategies within the top-down approach that expose new interests for proteome sequencing. While protein mass is fundamentally informative, it alone is generally not useful in identifying an *a priori* unknown protein due to post-translational processing. However, in combination, partial protein sequence information and intact protein mass can provide a level of insight not available from either piece of information alone. The top-down sequence of whole glycoprotein ions has

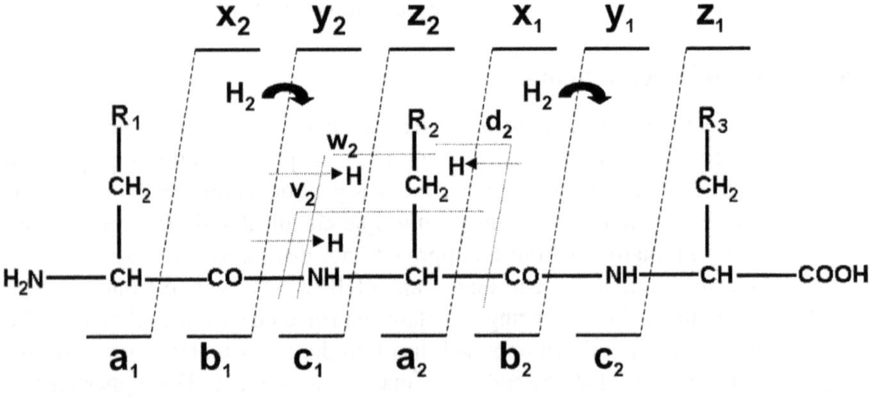

FIGURE 18.3 Peptide fragments frequently observed as a consequence of CID. Modified from Biemann.[33]

been reported, employing electrospray ionization, collision-induced dissociation, and ion–ion proton-transfer reactions in a quadrupole ion trap mass spectrometer.[38] Ribonuclease B and its N-linked high mannose glycomers were compared with the unglycosylated analog, ribonuclease A, to determine the differential influence of ionization. Facile gas-phase fragmentation was noted to occur along the protein backbone at the C-terminal of aspartic acid residues and at the N-terminus of proline, depending on the precursor ion charge state. No deglycosylation of the N-linked sugar occurred presumably due to effective competition from the facile amide bond cleavage channels that protect the N-linked glycosidic bond from cleavage. This would appear to provide an identification method for N-linked consensus sites and even further inquiry into the glycan structure; however, no details beyond composition were provided. Simplifying the precursor ion population by the application of a relatively high amplitude single-frequency pulse at 1000 Hz lower than the secular frequency was necessary to eject the high-mannose-containing ions from the trap.[38]

18.4.3 MS Profiles of Intact Glycoproteins: Bottom Up

Electrospray ionization (ESI) and MALDI MS profile analysis of glycoprotein PMF samples has always offered problems due to the poor ionization efficiency

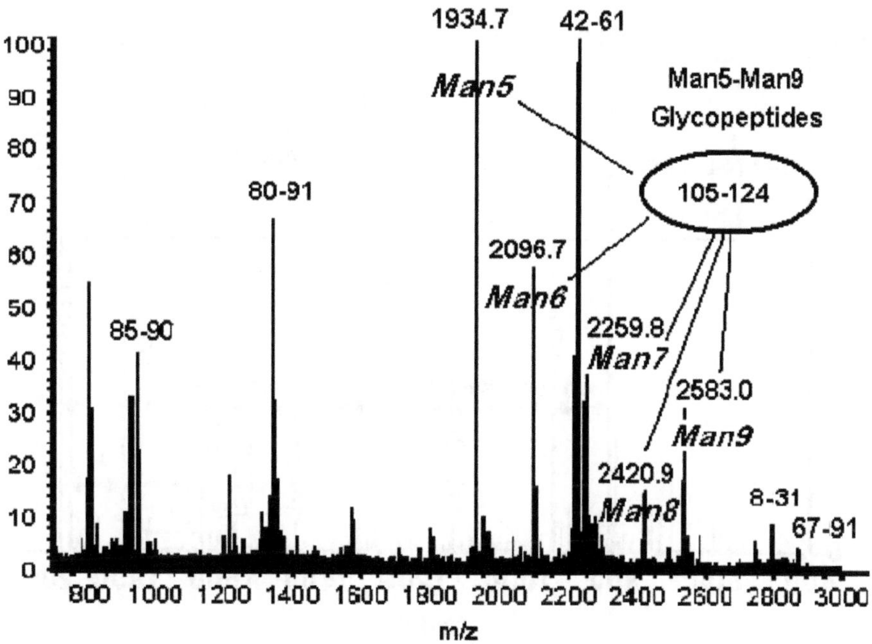

FIGURE 18.4 AP MALDI-IT TOF of PMF RNAase B glycoprotein. Major ions represent glycopeptide fragments. Ion distribution was found to be consistent with the known glycomers abundance, Man_5–Man_9. All ions are $(M + H)^+$. Procedures and instrument setup kindly provided by Dr. Koichi Tanaka, Shimadzu, Kyoto.

of glycopeptides. This constraint appears to be resolved with atmosphere pressure MALDI (Axima-QIT, Shimadzu, Kyoto) as first reported by Dr. Koichi Tanaka (ASMS abstract, Nashville, TN, 2004) and repeated in our laboratory with an identical modification (Center for Structural Biology, UNH) (Figure 18.4). This peptide-glycopeptide fingerprint profile coverage was approximately 90% of the total protein and provided base-line abundance for the glycohexapeptide known to carry the Man5 glycomer, *m/z* 1934.8. The 162 atomic mass unit (amu) increments to the hexapeptide (SRNLTK), for the Man6, Man7, Man8, and Man9 were readily identified at *m/z* 2096.9, 2259.9, 2421.0, and 2583.1, respectively. Subsequent selection of the glycopeptide base ion, *m/z* 1934.8, and MS^2 analysis with the API-MALDI-IT instrument provided a spectrum showing neutral loss of hexose monomers continuing the core GlcNAc residue (Figure 18.5).

18.4.4 Consensus Site Methods

Consensus site identification by releasing glycans in the presence of ^{18}O water,[39] is an effective way to identify former N-linked sites (site occupancy), although such strategies fail with blocked chitobiose (GlcNAc-GlcNAc) cores. The overall details

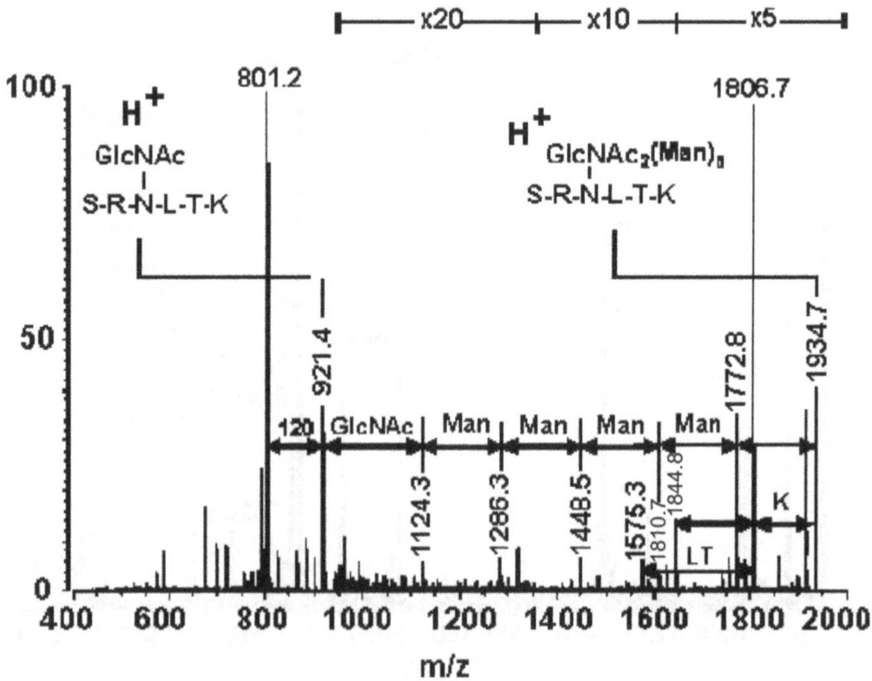

FIGURE 18.5 AP MALDI-IT TOF (Axima-QIT, Shimadzu, Kyoto) of *m/z* 1934.7 glycopeptide ion representing the base ion in Figure 18.4. MS^2 spectrum shows mannose neutral loss to core peptide linked GlcNAc. Procedures and instrument setup kindly provided by Dr. Koichi Tanaka, Shimadzu, Kyoto.

involve the following: (*a*) reduction, alkylation, and proteolysis; (*b*) enzymatic deglycosylation in the presence of ^{18}O-water; and (*c*) MS profile analysis (Figure 18.6). Glycopeptides rarely are observed in a PMF profile because they compete poorly for protons in a MALDI or ESI plume. A shift in the peptide ion isotope pattern indicates an ^{18}O-peptide. MS/MS analysis of these peptides can easily identify the consensus site. Alternatively, glycopeptides can be identified with HPLC and collision-induced dissociation (CID) monitoring of glycan fragment ions at *m/z* 163 (Hex), 204 (HexNAc), 292 (sialic acid), or 366 (HexHexNAc).

18.5 GLYCAN RELEASE TECHNIQUES

Characterization of glycoprotein samples, whether N-, or O-linked, should aspire to provide quantitative glycan release, stable products (peptide/glycan), and a marker of former glycan location. Stable products provide an opportunity to sequence each conjugate and identify the peptide site of glycosylation. Many aspects of these analytical concerns have been described, but they have not been described in a single congruent analysis. Chemical methods avoid the problems of enzymatic release that fail with blocked cores (N-glycans) and a number of chemical reagents have been described for both *N*-, and *O*-linked glycans including anhydrous hydrazine,[40,41] trifluoromethanesulphonic,[42] hydrogen fluoride,[43] and a variety of E2-elimination conditions using organic bases. Specific amino acid labeling of former O-glycan sites can be achieved via alkylamine addition to the generated alkene(s) following E2 base release. Labeling N-glycan sites can be approached enzymatically (endoglycosidase) in

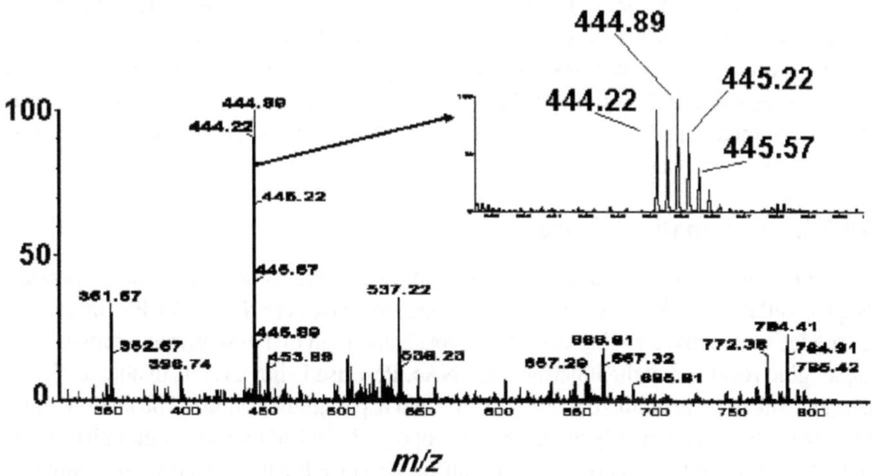

FIGURE 18.6 ESI-TOF MS profile of tryptic PMF obtained from mussel glycoprotein deglycosylated in the presence of O-18 water. Insert expanded mass range showing peptide a consensus site pattern with O-18 incorporated into aspartic acid. (Spectra provided by J. Huang, Thesis, Proteomic Study of a Novel Mussel Glycoprotein Involved in Biomineralization, UNH 2004.)

the presence of ^{18}O-water providing the ^{18}O-aspartyl peptide(s)[44] easily characterized by MS^2. Each reagent has specific limits and advantages and clearly should be carefully evaluated before assuming any quantitative relationships. Classically, O-linked glycans are effectively released under the E_2 conditions of sodium hydroxide coupled with *in situ* glycan reduction ($NaBH_4$).[45] The reduction to alditols stabilizes glycans against peeling,[46] but this approach is detrimental to the protein and peptides.[47] For simplicity, semiquantitative release, and salt-free products, anhydrous hydrazine has proven successful for releasing both the N-, and O-linked glycans.[48,49] Partial loss of N-acyl groups does occur, but this can be easily corrected by N-reacetylation (some originality may be lost, e.g., neuraminic acid N-glycolyl, O-acyl). When working with unknown glycans, chemical derivatives in conjunction with comparative MS profiling greatly assist structural understanding (e.g., sodium borohydride reduction of the reducing terminus, methyl ester formation of acids, desialylation, and deuteriomethylation). Clearly, there is great need for specific chemical release strategies that would provide N- and O-glycans separately in conjunction with protein stability. In that regard, there has been considerable interest in the milder conditions of O-glycan release provided by ammonia and alkylamines that would circumvent reduction and glycan peeling,[50,51] but the quantification and protein stability remains in question.

18.5.1 Methods

Chemical Release: Glycoprotein samples are thoroughly dried, the reagent added, and the samples are heated at 60°C to release O-linked glycans (a lower temperature than N-linked glycans). For release of both N- and O-linked oligosaccharides, samples are incubated in the dark for six hours at 90°C, cooled, and then dried to remove excess hydrazine.[48] Glycans are N-acetylated with saturated bicarbonate and acetic anhydride, desalted with Dowex, and purified from peptides and minor contaminants with microcrystalline cellulose tips. Product glycans are dried, desiccated, and MS profiled directly or methylated for improved sensitivity and greater structural detail when followed by MS^n analysis.

18.5.2 Enzymatic Release

Enzymes commonly used to cleave N-linked glycoproteins are the amidases peptide N-glycosidase F (PNGase F) and peptide N-glycosidase A (PNGase A). PNGase F has a broader specificity, cleaving complex, hybrid, and high mannose glycans between the asparagine residue and the innermost GlcNAc. PNGase F, however, is unable to cleave glycans linked to an amino or carboxyl terminal asparagine residue.[51] The enzyme fails to release glycans efficiently in the presence of α(1-3)-linked fucose,[52] although α(1-6)-linked fucosylated cores are cleaved readily. PNGase F can be used with a range of buffers but is sensitive to the presence of SDS. PNGase A is similar to PNGase F but is able to cleave core fucosyl residues linked α(1-3). It also fails to cleave when neuraminic acid is present. Other enzymes used for glycan cleavage are endoglycosidases. These have the disadvantage of leaving an *N*-acetylglucosamine residue attached to the asparagine(s) as well as a narrower range of specificity. This could be advantageous in defining glycan consensus site occupancy. Exoglycosidases that release a single sugar from the nonreducing end can have very high specificity. For example, (β(1-4)-galactosidase releases only that residue.[53] Digestion with a series of these enzymes can suggest sequence and linkage information, but complications regarding origin arise with multibranched structures, and kinetic release does not always follow the sequence order, making enzymological methods somewhat questionable for characterizing sequence.

18.5.3 Discussion

The above strategies complement sequence understanding, especially when supported by more exacting structural methods such as MS[54] and nuclear magnetic resonance.[55] In MS studies, sample derivatization brings many advantages, sample quantification[56] and detection sensitivity being the most pronounced. Derivatization, in conjunction with MS, defines chain termination, a component of molecular topology. Fragmentation, MS^n, unwraps connectivity exposing features defined as branching and linkage positions. Methylation is performed by dissolving dry samples in dimethylsulfoxide containing powdered sodium hydroxide.[57] Methyl iodide is added to the solubilized sample, and the products are extracted from the residue with chloroform after drying. The reaction is rapid and limitations (incomplete methylation) arise only because of analyte insolubility, and the methylation reaction can be repeated for complex samples.[58] To illuminate functional relationships of glycoproteins, glycan profiles may bring insight into their specific involvement. This function can be pursued very effectively by a study of gene knockouts (glycosyltransferase), and glycan MS profiles can be followed for confirmation.[59,60] To fully support these studies, a reproducible and complete recovery of all glycans is essential, a feature identified as coverage. This analytical measure is an appraisal of cell rupture, protein extraction, glycan release, and derivatization chemistry. For most glycan reports, the effort has been focused on qualitative relationships of analytes, and rarely is quantification properly evaluated. As an example, MS profiles have been considered for *Caenorhabditis elegans* as a glycome measure for the worm.[61] Although several groups have pursued such measurements, an absence of standard protocols has resulted in a wide disparity between laboratories.[62–67] When glycans are released en masse from a large set of proteins and then characterized, it is impossible to know the protein origin of specific glycans, and thus a subsequent functional relationship to protein structure. Strategies discussed in section 18.4 using both top-down and bottom-up MS techniques appear well positioned to clarify this relationship.

18.6 SEQUENCING RELEASED GLYCANS

Proteomics is an endeavor to understand gene function and characterize the structural relationships of the living cell. In proteomics, MS has become a powerful analytical technology to identify proteins by the analysis of peptides (PMF) with database relationships. In considerable contrast to proteomics, where global comprehensive studies are conducted in a systematic fashion, glycomics is still lost in multiple procedures looking mostly for qualitative answers and not quantitative results. A full understanding of cellular activity will, like proteomics, require comprehensive and systematic studies. Clearly, the application of glycomics to biology holds great potential for identifying the mechanisms that lead to development, malignancy, and subsequent therapeutic public health strategies. But these contributions are distant, and many fundamental analytical studies are lacking. While MS^2 is sufficient for determining the simple linear amino acid sequence, a full understanding of glycan structure (interresidue linkage, branching, anomer and monomer stereochemistry) requires diverse and orthogonal approaches. In that regard, we suggest complete characterization to mean knowing all aspects of structure in sufficient detail to identify a single structure appropriate to initiate. Anything short of such objectives would fail to match the biological challenge. Many analytical methods and instrumentation have been described to approach such goals. In terms of sensitivity and specificity, MS instrumentation is continuously improving and probably the most appropriate. Defining a congruent methodology to approach glycan sequencing is not clear. Thus, for this brief account of glycoproteomic sequencing, we provide a series of excellent reviews[68–70] and briefly describe a set of techniques we consider most appropriate for a comprehensive understanding of structural detail.

18.6.1 The Details of Glycan Structure by MS^n

Glycan fragments generated during MS^n disassembly of methylated precursors reveal features of their original structure as alterations in mass. As a simple example, disaccharide spectra are dominated by facile glycosidic cleavage allowing a topological differentiation of termini (Figure 18.7A), a feature exposed only with derivatized precursors. Although all disaccharides show spectral identity, an interesting contrast was observed with 1,2-anhydroglycans (products of gas-phase CID). In these analogs major differences in spectra were exhibited that included changes with interresidue linkage and monomer and anomer stereochemistry (Figure 18.7B). This CID specificity inherent in 1,2-anhydro analogs extends to monomers (Figure 18.8) and provided clear strategies to differentiate isobars in higher oligomers (Figure18.9, Figure 18.10, and Figure 18.11). A composition library of such products from known precursors is serving as a bottom-up strategy for the characterization of glycan disassembly end products.

To demonstrate these features, an MS profile (reduced and permethylated N-linked glycans) was prepared from ovalbumin (Figure 18.9, top left). The glycan ion at m/z 2167.8 was selected for detailed structural analysis by MS^n, which was shown to comprise four isobars following the disassembly path, MS^2 m/z 2167, MS^3 m/z 838, and MS^4 m/z 611. Detailed study of these spectra (Figure 18.10 and Figure 18.11) showed fragments that can only be rationalized on the basis of the isobars inserted in the figure.

Thus, an MS^n fragment ion is a mass composition with structural information. Tracking such changes through multiple disassembly steps provides a set of compiled clues that will define specific structures. These details were observed foremost with gas-phase collision products possessing a 1,2-anhydro reducing terminus,

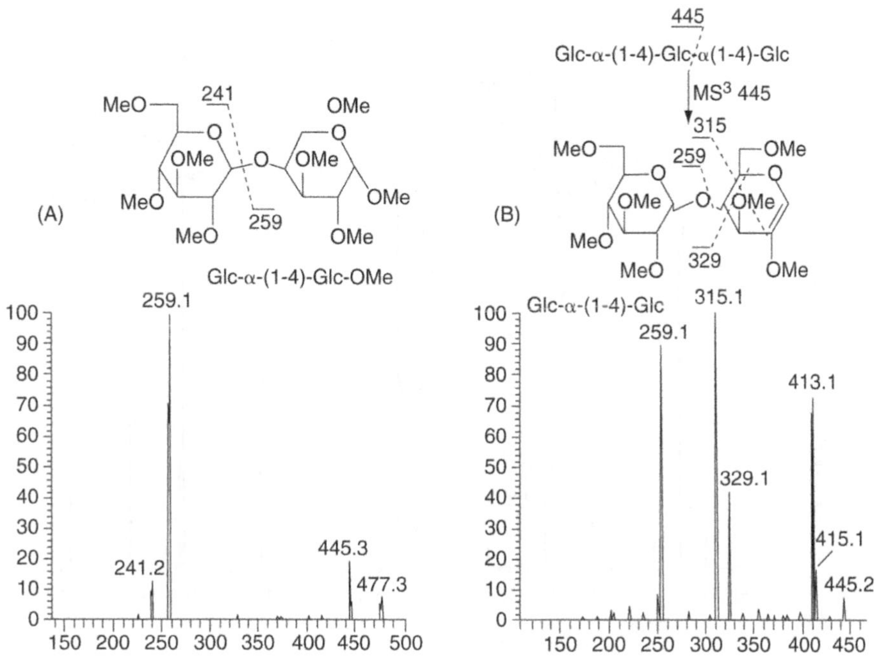

FIGURE 18.7 ESI-IT MS^2 of (A) methylated Glc-Glc methyldiglucoside; and (B) MS^2 of methylated trisaccharide to provide 1,2 Anhydro diglucoside. MS^3 of 1,2 anhydro product showed spectral details and differences unique in all 1,2-disaccharides studied.

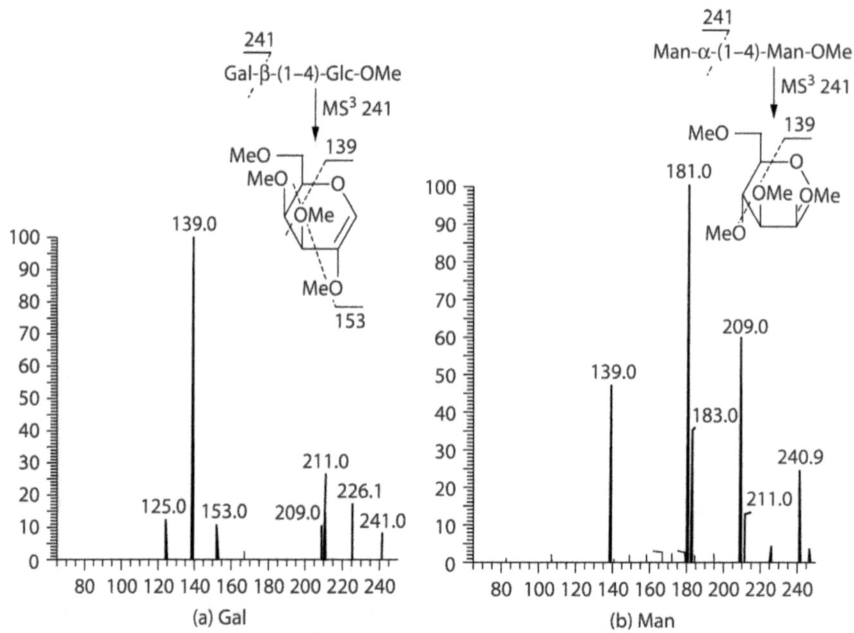

FIGURE 18.8 MS-CID of methylated disaccharides followed by isolation of (A) galactose; and (B) mannose product fragments, *m/z* 241. MS-CID and MS^3 of these products showed clear fragment ion and abundance specificity.

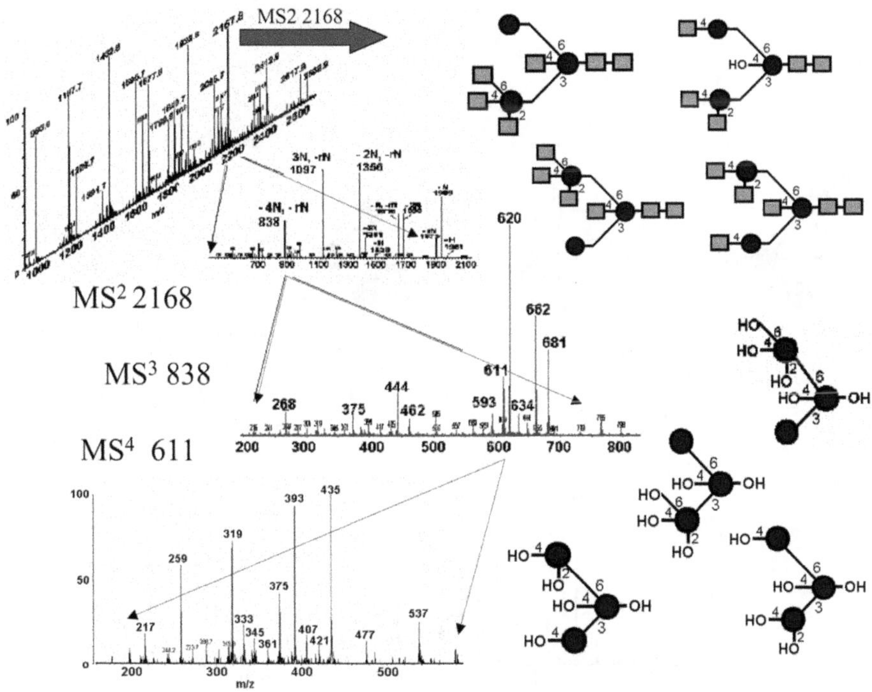

FIGURE 18.9 MALDI-IT-TOF MS profile of methylated and reduced N-linked glycans from the glycoprotein ovalbumin; 2,5-DHB was used as the matrix. The glycan at m/z 2167.8 was selected for detailed structural analysis by MS^n, which was shown to comprise four isobars following the disassembly path: MS^2, m/z 2167; MS^3, m/z 838; and MS^4, m/z 611. The squares in the modeled structure denote GlcNAc, while the circles denote Man. Facile neutral losses of GlcNAc residues result in open hydroxyl residues and position of former GlcNAc location in the isobar. Core fragments (Figure 18.10 and Figure 18.11) can only be rationalized on the basis of the structures inserted.

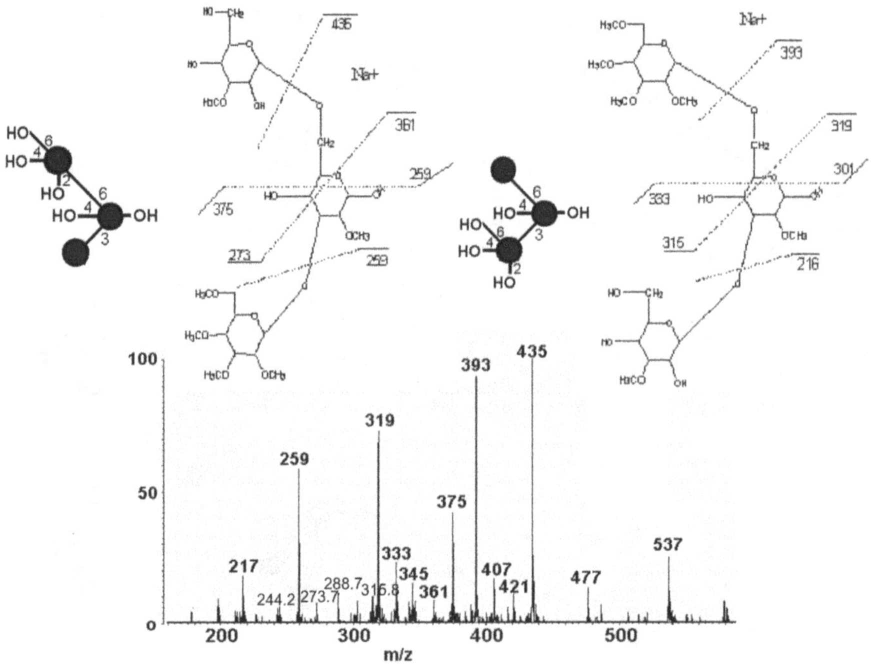

FIGURE 18.10 MALDI-IT-TOF MS profile of methylated and reduced N-linked glycans from the glycoprotein ovalbumin; 2,5-DHB was used as the matrix. MS^4 of selected fragment ion, *m/z* 611, was shown to comprise two isobars following the disassembly pathway: MS^2, *m/z* 2167; MS^3, *m/z* 838; and MS^4, *m/z* 611. Detailed study of this spectrum identified core fragments consistent with the structures inserted. Fundamental to this determination was the cross-ring cleavage of the central mannose, which resolved 6-linked and 3-linked isobars.

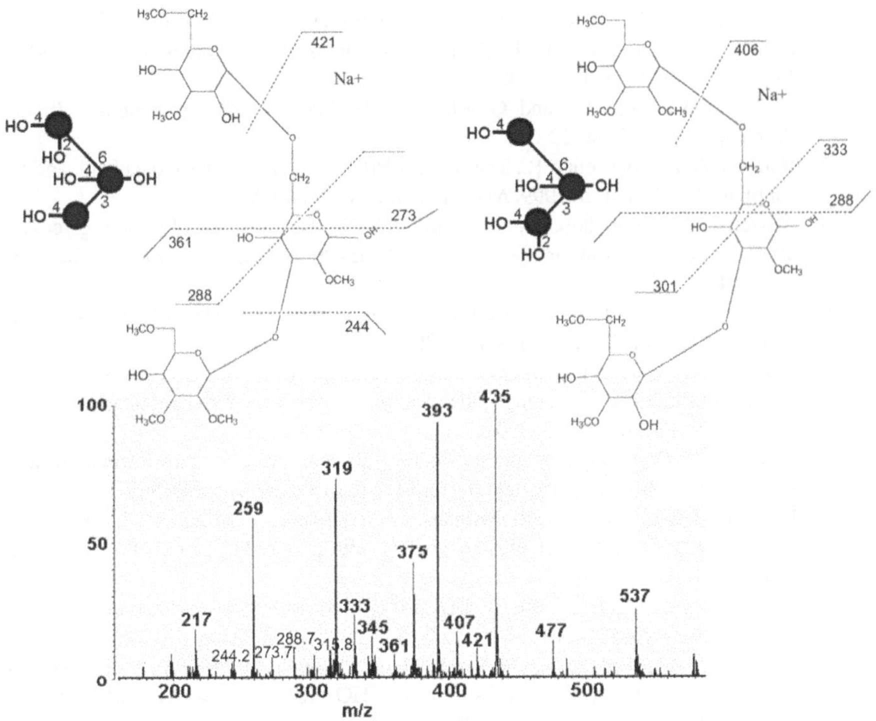

FIGURE 18.11 MALDI-IT-TOF MS profile of methylated and reduced N-linked glycans from the glycoprotein ovalbumin; 2,5-DHB was used as the matrix. The fragment ion, *m/z* 611, was selected for detailed structural analysis, which was shown to comprise two additional isobars following the disassembly path MS^2, *m/z* 2167; MS^3, *m/z* 838; and MS^4, *m/z* 611. Detailed study of this spectrum identified core fragments consistent with the structures inserted. Fundamental to this determination was the cross-ring cleavage of the central mannose, which resolved 6-linked and 3-linked isobars.

and this specificity we attribute to sodium nesting next to the pyran ring among methoxy-paired electrons that change with monomer stereochemistry thereby altering disassembly.

REFERENCES

1. Service, R.F., Proteomics: Can Celera do it again? *Science*, 287, 2136–2138, 2000.
2. Jenuwein, T. and Allis, C.D., Translating the histone code, *Science*, 293, 1074–1080, 2001.
3. Pfeffer, S., Zavolan, M., Grässer, F.A., Chien, M., Russo, J.J., Ju, J., John, B., Enright, A.J., Marks, D., Sander, C., and Tuschl, T., Identification of virus-encoded micro RNAs, *Science*, 304, 734–736, 2004.
4. Stortz, G., An expanding universe of noncoding RNAs, *Science*, 296, 1260–1263, 2002.

5. Mattick, J.S., The evolution of controlled multitasked gene networks: The role of introns and other noncoding RNAs in the development of complex organisms, *Mol. Biol. Evol.*, 18, 1611–1630, 2001.
6. Taylor, S.W., Fahy, E., and Ghosh, S.S., Global organellar proteomics, *Trends Biotechnol.*, 21, 82–88, 2003.
7. Bairoch, A. and Apweiler, R., The SWISS-PROT protein sequence data bank and its supplement TrEMBL in 1999, *Nucleic Acid Res.*, 27, 49–54, 1999.
8. Apweiler, R., Hermjakob, H., and Sharon, N., On the frequency of protein glycosylation, as deduced from analysis of the SWISS-PROT database, *Biochim. Biophys. Acta*, 1473, 4–8, 1999.
9. Grunewald, S., Matthijs, G., and Jaeken, J., Congenital disorders of Glycosylation: A review, *Pediatr. Res.*, 52, 618–624, 2003.
10. Wujek, P., Kida, E., Walus, M., Wisniewski, K.E., and Golabek, A.A., N-glycosylation is crucial for folding, trafficking, and stability of human tripeptidyl-peptidase I, *J. Biol. Chem.*, 279, 12,872–12,939, 2004.
11. Spiro, R.G., Protein glycosylation: Nature, distribution, enzymatic formation, and disease implications of glycopeptide bonds, *Glycobiology*, 12, 43R–56R, 2002.
12. Fukuda, M., Cell surface carbohydrates: cell type-specific expression, in *Molecular Glycobiology*, Fukuda, M., and Hindsgaul, O., Eds., Oxford University Press, Oxford, 1994, pp. 1–52.
13. Tsuboi, S. and Fukuda, M., Roles of O-linked oligosaccharides in immune responses, *BioEssays*, 23, 46–53, 2001.
14. Andersen, D.C., Bridges, T., Gawlitzek, M., and Hoy, C., Multiple cell culture factors can affect the glycosylation of Asn-184 in CHO-produced tissue-type plasminogen activator, *Biotechnol. Bioeng.*, 70, 25–31, 2000.
15. Rademacher, T.W., Parekh, R.B., and Dwek, R.A., Glycobiology, *Ann. Rev. Biochem.*, 57, 785–838, 1988.
16. Wilkins, M.R., Williams, K.L., Appel, R.D., and Hochstrasser, D.F., Eds., *Proteome Research: New Frontiers in Functional Genomic*, Springer, Berlin, 1997, pp. 1–243.
17. O'Farrell, P.H., High resolution two-dimensional electrophoresis of proteins, *J. Biol. Chem.*, 250, 4007–4021, 1975.
18. Righetti, P.G., Castagna, A., and Herbert, B., Prefractionation techniques in proteome analysis, *Anal. Chem.*, 73, 320A–325A, 2001.
19. Kuster, B., Wheeler, S.F., Hunter, A.P., Dwek, R.A., and Harvey, D.J., Sequencing of N-linked oligosaccharides directly from protein gels: In-gel deglycosylation followed by matrix-assisted laser desorption/ionization mass spectrometry and normal-phase high-performance liquid chromatography, *Anal. Biochem.*, 250, 82–101, 1997.
20. Hellman, U., Wernstedt, C., Gonez, J., and Heldin, C-H., Improvement of an "in-gel" digestion procedure for the micropreparation of internal protein fragments for amino acid sequencing, *Anal. Biochem.*, 224, 451–455, 1995.
21. Jacobs, D.I., van Rijssen, M.S., van der Heijden, R., and Verpoorte, R., Sequential solubilization of proteins precipitated with trichloroacetic acid in acetone from cultured *Catharanthus roseus* cells yields 52% more spots after two-dimensional electrophoresis, *Proteomics*, 1, 1345–1350, 2001.
22. Malinow, R. and Malenka, R.C., AMPA receptor trafficking and synaptic plasticity, *Ann. Rev. Neurosci.*, 25, 103–126, 2002.
23. Cooper, J.W., Gao, J., and Lee, C.S., Gel protein capillary extraction apparatus electronic protein transfer, *Anal. Chem.*, 74, 1182–1186, 2002.

24. Klenø, T.G., Andreasen, C.M., Kjeldal, H.O., Leonardsen, L.R., Krogh, T.N., Nielsen, P.F., Sørensen, M.V., and Jensen, O.N., MALDI MS peptide mapping performance by in-gel digestion on a probe with prestructured sample supports, *Anal. Chem.*, 76, 3576–3583, 2004.
25. Nissum, N., Schneider, U., Kuhfuss, S., Obermaier, C., Wildgruber, R., Posch, A., and Eckerskorn, C., In-gel digestion of proteins using a solid-phase extraction microplate, *Anal. Chem.*, 76, 2040–2045, 2004.
26. Steinberg, T., Top, K., Berggren, K., Kemper, C., Jones, L., Diwu, Z., Haugland, R., and Patton W., Rapid and simple single nanogram detection of glycoproteins in polyacrylamide gels and on electroblots, *Proteomics*, 1, 841–855, 2001.
27. Hart, C., Schulenberg, B., Steinberg, T., Leung, W., and Patton, W., Detection of glycoproteins in polyacrylamide gels and on electroblots using Pro-Q Emerald 488 dye, a fluorescent periodate Schiff-base stain, *Electrophoresis*, 24, 588–598, 2003.
28. Hawkes R., Identification of Concanavalin A-binding proteins after sodium dodecyl sulfate-gel electrophoresis and protein binding, *Anal. Biochem.*, 123, 143–146, 1982.
29. Dobos, K.M., Swiderek, K., Khoo, K.-H., Brennan, P.J., and Belisle, J.T., Evidence for glycosylation sites on the 45-kilodalton glycoprotein of *Mycobacterium tuberculosis, Infection Immunity*, 63, 2846–2853, 1995.
30. Zubarev, R.A., Kelleher, N.L., and McLafferty, F.W., Electron capture dissociation of multiply charged protein cations: A non-ergodic process., *J. Am. Chem. Soc.*, 120, 3265–3266, 1998.
31. Cargile, B.J., McLuckey, S.A., and Stephenson, J.L., Jr., Simplification of product ion spectra derived from multiply charged parent ions via ion/ion chemistry, *Anal. Chem.*, 73, 1277–1285, 2001.
32. Baba, T., Hashimoto, Y., Hasegawa, H., Hirabayashi, A., and Waki, I., Electron capture dissociation in a radio frequency ion trap., *Anal. Chem.*, 76, 4263–4266, 2004.
33. John, E.P., Syka, J.E.P., Coon, J.J., Schroeder, M.J., Shabanowitz, J., and Hunt, D.F., Peptide and protein sequence analysis by electron transfer dissociation mass spectrometry, *Proc. Natl. Acad. Sci.*, 101, 9529–9533, 2004.
34. Biemann, K., Nomenclature for peptide fragment ions (positive ions), *Methods Enzymol.*, 193, 886–887, 1990.
35. Mirgorodskaya, E., Roepstorff, P., and Zubarev, R.A., Localization of O-glycosylation sites in peptides by ECD in a Fourier transform mass spectrometer, *Anal. Chem.*, 71, 4431–4436, 1999.
36. Hå kansson, K., Cooper, H.J., Emmett, M.R., Costello, C.E., Marshall, A.G., and Nilsson, C.L., Electron capture dissociation and infrared multiphoton dissociation MS/MS of an N-glycosylated tryptic peptide to yield complementary sequence information, *Anal. Chem.*, 73, 4530–4536, 2001.
37. Stensballe, A., Jensen, O.N., Olsen, J.V., Haselmann, K.F., and Zubarev, R.A., Electron capture dissociation of singly and multiply phosphorylated peptides, *Rapid Commn. Mass Spectrom.*, 14, 1793–1800, 2000.
38. Shi, S.D.H., Hemling, M.E., Carr, S.A., Horn, D.M., Lindh, I., and McLafferty, F.W., Phosphopeptide/phosphoprotein mapping by electron capture dissociation mass spectrometry, *Anal. Chem.*, 73, 19–22, 2001.
39. Reid, G.E., Stephenson, J.L., and McLuckey, S.A., Tandem mass spectrometry of ribonuclease A and B: N-linked glycosylation site analysis of whole protein ions, *Anal. Chem.*, 74, 577–583, 2002.

40. Kuster, B. and Mann, M., ^{18}O-Labeling of N-glycosylation sites to improve the identification of gel-separated glycoproteins using peptide mass mapping and database searching, *Anal. Chem.*, 71, 1431–1440, 1999.
41. Takasaki, S., Misuochi, T., and Kobata, A., *Methods Enzymol.*, 83, 263–268, 1982.
42. Patel, T., Bruce, J., Merry, A., Bigge, C., Wormald, M., Jaques, A., and Parekh, R., Use of hydrazine to release in intact and unreduced form both N- and O-linked oligosaccharides from glycoproteins, *Biochemistry*, 32, 679–693, 1993.
43. Edge, A., Deglycosylation of glycoproteins with trifluoromethanesulphonic acid: Elucidation of molecular structure and function, *Biochem. J.*, 376, 339–350, 2003.
44. Axelsson, M.A.B., Hansson, E.M., Sikut, R., and Hansson, G.C., Deglycosylation by gaseous hydrogen fluoride of mucus glycoproteins immobilized on nylon membranes and in microtiter wells, *Glycoconjugate J.*, 15, 749–755, 1998.
45. Kuster, B. and Mann, M., ^{18}O-Labeling of N-glycosylation sites to improve the identification of gel-separated glycoproteins using peptide mass mapping and database searching, *Anal. Chem.*, 71, 1431–1440, 1999.
46. Carlson, D., Structures and immunochemical properties of oligosaccharides isolated from pig submaxillary mucins, *J. Biol. Chem.*, 243, 616–626, 1968.
47. White, C.C. and Kennedy, J.F., Base degradation, in *An Introduction to the Chemistry of Carbohydrates*, 3rd ed., Kennedy, J.F., Ed., Clarendon Press, Oxford, U.K., 1988, pp. 42–67.
48. Rademaker, G.J., Pergantis, S.A., Blok-Tip, L., Langridge, J.I., Kleen, A., and Thomas-Oates, J.E., Mass spectrometric determination of the sites of O-glycan attachment with low picomolar sensitivity, *Anal. Biochem.*, 257, 149–160, 1998.
49. Merry, A.H., Neville, D.C.A., Royle, L., Matthews, B., Harvey, D.J., Dwek, R.A., and Rudd P.M., Recovery of intact 2-aminobenzamide-labeled O-glycans released from glycoproteins by hydrazinolysis, *Anal. Biochem.*, 304, 91–99, 2002.
50. Royle, L., Mattu, T.S., Hart, E., Langridge, J.I., Merry, A.H., Murphy, N., Harvey, D.J., Dwek, R.A., and Rudd, P.M., An analytical and structural database provides a strategy for sequencing O-glycans from microgram quantities of glycoproteins, *Anal. Biochem.*, 304, 70–90, 2002.
51. Hanisch, F., Jovanovic, M., and Peter-Katalinic, J., Glycoprotein identification and localization of O-glycosylation sites by mass spectrometric analysis of deglycosylated/ alkylaminylated peptide fragments, *Anal. Biochem.*, 290, 47–59, 2001.
52. Maley, F., Trimble, R.B., Tarentino, A.L., and Plummer, T.H.J., Characterization of glycoproteins and their associated oligosaccharides through the use of endoglycosidases, *Anal. Biochem.*, 180, 195–204, 1989.
53. Tretter, V., Altmann, F., and Marz, L., Peptide-N4-(N-acetyl-beta-glucosaminyl) asparagine amidase F cannot release glycans with fucose attached $\alpha(1–3)$ to the asparagine-linked N-acetylglucosamine residue, *Eur. J. Biochem.*, 199, 647–652, 1991.
54. Kobata, A., Use of endo- and exoglycosidases for structural studies of glycoconjugates, *Anal. Biochem.*, 100, 1–14, 1979.
55. Harvey, D.J., Identification of protein-bound carbohydrates by mass spectrometry, *Proteomics*, 1, 311–328, 2001.
56. Bubb, W., NMR in the study of carbohydrates: Characterization of structural complexity, *Concepts Magn. Resonance, Part A*, 19A, 1–19, 2003.
57. Tao, W.A., and R. Aebersold,, Advances in quantitative proteomics via stable isotope tagging and mass spectrometry, *Curr. Opinion Biotechnol.*, 14, 110–118, 2003.
58. Ciucanu, I., and Kerek, F., A simple and rapid method for the permethylation of carbohydrates, *Carbohydr. Res.*, 131, 209–217, 1984.

59. Linsley, K., Chan, S-Y., Chan, S., Reinhold, B.B., and Reinhold, V.N., Detailed site specific characterization of *N*- and *O*-Linked glycans from erythropoietin (ES-MS/CID/MS), *Anal. Biochem.*, 219, 207–217, 1994.
60. Chen, S.H., Tan, J., Reinhold, V.N., Spence, A.M., and Schachter, H., UDP-N-acetylglucosamine: alpha-3-D mannoside beta 1,2-N-acetylglucosaminyltransferase I and UDP-N-acetylglucosamine: alpha-6-D-mannoside beta-1,2-N-acetylglucosaminyltransferase II in *Caenorhabditis elegans*, *Biochim. Biophys. Acta Gen. Subjects*, 1573, 271–279, 2002.
61. Schachter, H., Chen, S., Zhang, W., Spence A.M., Zhu, S., Callahan, J.W., Mahuran, D.J., Fan, X., Bagshaw, R.D., She, Yi-Min, Rosa, J.C., and Reinhold, V.N., Functional post-translational proteomics approach to study the role of N-glycans in the development of *Caenorhabditis elegans, Biochem. Soc. Symp.*, 69, 1–21, 2002.
62. Hirabayashi, J., Hayama, K., Kaji, H., Isobe, T., and Kasai, K-I., Affinity capturing and gene assignment of soluble glycoproteins produced by the nematode *Caenorhabditis elegans*, *J. Biochem.*, 132, 103–114, 2002.
63. Altman, F., Fabini, G., Ahorn, H., and Wilson, I.B.H., Genetic model organisms in the study of *N*-glycans, *Biochimie*, 83, 703–712, 2001.
64. Guerardel, Y., Balanzino, L., Maes, E., Leroy, E., Coddeville, B., Oriol, R., and Strecker, G., The nematode *C. elegans* synthesizes unusual *O*-linked glycans: identification of glucose-substituted mucin-type O-glycans and short chondroitin-like oligosaccharides, *Biochem. J.*, 357, 167–182, 2001.
65. Cipollo, J.F., Costello, C.E., and Hirschberg, C.B., The fine structure of *Caenorhabditis elegans N*-Glycans, *J. Biol. Chem.*, 277, 49,143–49,157, 2002.
66. Natsuka, S., Adachi, J., Kawaguchi, M., Nakakita, S-I., Hase, S., Ichikawa, A., and Ikura, K., Structural analysis of *N*-linked glycans in *C. elegans, J. Biochem.*, 131, 807–813, 2002.
67. Haslam S.M. and Dell, A., Hallmarks of *C. elegans N*-glycosylation: Complexity and controversy, *Biochimie*, 85, 25–32, 2003.
68. Fan, X.L., She, Y.M., Bagshaw, R.D., Callahan, J.W., Schachter, H., and Mahuran, D.J., A method for proteomic identification of membrane-bound proteins containing Asn-linked oligosaccharides, *Anal. Biochem.*, 332, 178–186, 2004.
69. Mechref, Y. and Novotny, M.V., Structural investigations of glycoconjugates at high sensitivity, *Chem. Rev.*, 102, 321–369, 2002.
70. Harvey, D.J., Matrix-assisted laser desorption/ionization mass spectrometry of carbohydrates, *Mass Spectrom. Rev.*, 18, 349–451, 1999.
71. Burlingame, A.L., Characterization of protein glycosylation by mass spectrometry, *Curr. Opinion Biotechnol.*, 7, 4–10, 1996.

[illegible]

19 Enrichment and Analysis of Glycoproteins in the Proteome

Nicole L. Wilson, Niclas G. Karlsson, and Nicolle H. Packer

CONTENTS

19.1 INTRODUCTION

Glycosylation is a posttranslational modification of high relevance for proteins in biological systems,[1] as is the alteration of these glycoproteins in disease and disease progression.[2] The analysis of glycoproteins can be often masked by more abundant nonglycosylated proteins (e.g., albumin, IgG, and fibrinogen in human plasma). This chapter describes methods of enriching biological samples for glycoproteins prior to one-dimensional (1D) and two-dimensional polyacrylamide gel electrophoresis (2D PAGE). We first describe methods using lectin and boronate affinity chromatography to enrich for subfractions of glycoproteins.[3] We then describe the separation and detection of these glycoproteins by 2D PAGE. In our laboratory we then use these gel-separated glycoproteins for the release of N- and O-linked oligosaccharides, separation of these by graphitized carbon liquid chromatography, and analysis by electrospray mass spectrometry. The interpretation of the resulting fragmentation mass spectra is then facilitated by use of our glycoinformatic tools.

19.2 LECTIN AFFINITY CHROMATOGRAPHY

Lectins are sugar binding proteins of nonimmune origin that are capable of specific recognition, and of reversible binding to, carbohydrate moieties of complex glycoconjugates without altering their covalent structures. Lectins, which were first described by Stillmark in 1888,[4] bind to cell surface glycoproteins, proteoglycans, and glycolipids. They have been used widely in medical and biological research due to the availability of a large variety of carbohydrate-specific lectins.[5] Lectins can be immobilized on a chromatographic support, such as agarose, and will selectively bind glycoproteins with different affinities for particular oligosaccharide epitopes. The specificity of the chromatography is dependent on both the specificity of the lectin and the experimental conditions used to bind and elute the carbohydrates. Lectin interactions are fragile biological complexes, and so the sample pH, detergents, and contaminants should be considered when using this method for glycoprotein enrichment. The immobilized lectin, under the appropriate conditions, will bind to the glycoconjugate, allowing the unbound residual material to be easily removed. The bound glycoconjugates can then be displaced by the addition of a competing oligosaccharide or simple carbohydrate.

Many lectins have been studied and shown to have various specificities for particular oligosaccharide epitopes. For example *Sambucus nigra* and leukoagglutin and the *Maackia amurensis* hemagglutinin have been used to detect α2–6 and α2–3 linked sialic acid, respectively.[6] These and others could be used in a similar approach to that described here, which uses the lectin from Jack fruit, jacalin, as an example. Other lectins that we have used, the conditions required, and the carbohydrate moieties that they recognize are summarized in Table 19.1. This table also includes a summary of the conditions used for the boronate and jacalin affinity chromatography, for which the methods are given below. The jacalin lectin binds galactose-containing oligosaccharides and has been identified as a general lectin for binding O-linked glycoproteins.[7]

It is important to note that lectin specificities are based on relative, rather than absolute, affinities for carbohydrate epitopes and as such should be used with the expectation of enrichment rather than identification of structural epitopes.

19.3 BORONATE AFFINITY CHROMATOGRAPHY

A unique class of chemical compounds that has many of the properties of lectins are boronates. Although boronate compounds have the disadvantage of not having the specificity of lectins, they have several advantages over lectins, such as increased stability and a general affinity for the majority of glycoproteins.[8]

TABLE 19.1 Summary of Lectin and Boronate Affinity Chromatography Conditions

Affinity Chromatography	Specificity	Equilibration Solution	Elution Solution	Wash Solution
Jacalin	Galβ(1-3) GalNAc (O-linked)	175 mM Tris HCl, pH 7.5	0.8 M galactose in 175 mM Tris, pH 7.5	—
Aminophenylboronic acid	Hexose, sialic acid	100 mM phosphate buffer, 0.15 M NaCl, pH 8.8	100 mM Tris, buffered to pH 9.3 with phosphoric acid	50 mM sodium acetate, pH 4.8
WGA	$(GlcNAc)_2$, NeuAc	175 mM Tris HCl, pH 7.5	0.8 M *N*-acetylglucosamine in 175 mM Tris, pH 7.5	—
Concanavalin A	α-Man and α-Glc	175 mM Tris HCl, pH 7.5	0.1 M methyl α-$_D$-mannopyranoside in 175 mM Tris HCl, pH 7.5	—

In order for boronate interaction to occur with oligosaccharides, the carbohydrate hydroxyl groups must be on adjacent carbon atoms in a coplanar conformation. Sugars such as fucose that possess cis-diol groups are prime targets for boronate affinity chromatography and enable fucose-containing glycoproteins to be separated from the nonglycosylated proteins present in a complex sample such as plasma. While 1,2-cis-diols (hexoses) form the strongest boronate ester bonds, 1,3-cis-diols (sialic acids) and trident interactions, with cis-inositol or triethanolamine, can occur.[8] In solution, in alkaline conditions (pH >8), the boronate ligand is ionized, converting it from its normal trigonal coplanar geometry to its tetrahedral boronate anion, which covalently interacts with the appropriate hydroxyl geometry, releasing two water molecules. This reaction can be reversed, releasing the bound molecule either under more acidic conditions or alternatively by displacing the relevant glycoprotein from the column with a soluble diol-containing agent such as sorbitol.

19.4 METHODS

19.4.1 Lectin Affinity Chromatography

Agarose-bound jacalin was purchased from Vector Laboratories (Burlingame, Calif.). Heat stable, cross-linked 4% agarose beads are used as the solid-phase matrix to which the lectins are covalently bound. Approximately 0.5 ml of a slurry of the agarose bound jacalin was packed into a HPLC column or into a gravity flow liquid chromatography column (GFLC). The column was equilibrated in 175 mM Tris

HCl at pH 7.5 (5 ml for the HPLC system at ~1 ml/min, or 2 ml for the GFLC). Approximately 300 μl of human plasma was buffered in 175 mM Tris HCl pH 7.5 and run through the agarose/jacalin column (the HPLC sample was washed onto the column using the Tris solution, and the GFLC- sample was loaded in a 1:1 ratio (v/v) of plasma to buffer, onto the column). The column was washed with 175 mM Tris HCl, pH 7.5 (HPLC, 5 ml; GFLC, 3 × 1 ml). The flow through from the washing of the sample onto the column and the following wash steps were collected and pooled. This fraction will contain the majority of the non-O-linked glycosylated proteins.

Glycoproteins bound to the jacalin were then eluted from the column using 0.8 M galactose in 175 mM Tris pH 7.5 (HPLC, 5 ml; GFLCm 3 × 1 ml). The proteins were ethanol precipitated using a 1:9 (v/v) ratio of sample to cold ethanol and then placed at 20°C for 30 minutes. The precipitated proteins were pelleted by centrifuging the sample at 4000 rpm for 10 minutes.

For 2D PAGE separation, the pellet was resuspended in an appropriate volume of 2D sample solubilization buffer (7 M urea, 2 M thiourea, 4% (w/v) CHAPS). The proteins were reduced and alkylated in a final concentration of 5 mM tributylphosphine (TBP) and 1 mM acrylamide. The reduced and alkylated proteins were passed through a 10 kDa molecular weight cut-off filter (Centricon Plus 20 Centrifugal Units, Millipore, Bedford, Mass.) using 2D sample solubilization buffer to remove any excess salt or other impurities that may affect the first-dimensional focusing of the proteins. A Bradford protein assay was used to determine the protein concentration of the samples.

19.4.2 Boronate Affinity Chromatography

Approximately 0.5 ml of the aminophenylboronic acid immoblized on 6% bedded agarose (Sigma) was packed into a HPLC or into a gravity flow column. The column was equilibrated with 100 mM phosphate buffer, 0.15 M NaCl, pH 8.8 (HPLC, 5 ml; GFLC, 3 × 1 ml). Approximately 300 μl of human plasma, which was buffered in 100 mM phosphate buffer, 0.15 M NaCl, pH 8.8, was run through the column (HPLC, washed onto the column in 5 ml; GFLC, 1:1 ratio of sample to buffer). The column was washed with 100 mM phosphate buffer, 0.15 M NaCl, pH 8.8 (HPLC-5 ml; GFLC, 3 × 1 ml). The flow through from the washing of the sample onto the column and the following wash steps were collected and pooled. This fraction contained the majority of the nonglycosylated proteins.

The glycoproteins bound to the boronate were eluted from the column using 100 mM Tris solution adjusted to pH 9.3 with phosphoric acid (HPLC, 5 ml; GFLC, 3 × 1 ml). The column was washed with 50 mM sodium acetate buffer, pH 4.8 (HPLC, 5 ml; GFLC, 3 × 1 ml). The column was then equilibrated with Tris-phosphate buffer. Eluted proteins were precipitated, pelleted, resuspended, and reduced and alkylated for 2D PAGE separation as described for the jacalin affinity chromatography.

19.4.3 2D PAGE

2D PAGE is used to separate proteins according to molecular size and charge. We prefer this technique for the glycosylation analysis of proteins because it not only concentrates and purifies the proteins, but it also has the potential to separate the protein glycoforms with differing charges in a 2D gel. This allows for the glycosylation of different proteins to be compared on the same gel from the one sample, and also for the glycosylation on the individual isoforms of the same protein to be compared. These glycoforms have been shown to have the potential to be used as diagnostic markers of disease.[9]

Using the protein assay results, the appropriate concentration (approximately 6 mg/ml for human plasma) of sample in 220 μl was used to rehydrate 11 cm, 4–7 pH IPG strips (Proteome Systems Ltd, Sydney, Australia), for approximately 5 to 6 hours until all the solution was absorbed into the strip. The sample volume varies with strip length and sample. The IPG strips were focused at 10,000 V, with a maximum current setting of 50 μA/strip, for approximately 75 kVH for an 11 cm strip. The strips were equilibrated in a solution containing 6 M urea, 50 mM Tris acetate pH 7.0, 2% (w/v) SDS, for approximately 20 minutes. Bromophenyl blue 1% (w/v) can be added to this solution so that a dye front can be seen while running the second dimension. The 11cm IPG strips were placed on top of the second-dimension gel (GelChip, Proteome Systems Ltd, Sydney, Australia), and the upper and lower chambers of the gel running tank were filled with running buffer (50 mM Tris, 50 mM Tricine, 0.1% w/v SDS). Gels were run at 50 mA per gel until the dye front ran off the bottom of the gels (approximately 1 to 1.5 hr). Gels were stained with either of the two staining protocols below.

19.4.4 Protein and Glycoprotein Staining

Both general protein and carbohydrate staining can be used on the gels produced from the lectin and boronate affinity chromatography of human plasma.

19.4.4.1 Proteins

The general protein stain GelChip Blue, a reformulation of colloidal Coomassie Blue (Proteome Systems Ltd), was used to stain the proteins immobilized in the gels.

Gels were removed from the cassettes and placed in individual containers. The gels are then covered with fixative solution for 30 minutes, placed in GelChip Blue according to manufacturer's instructions, and incubated overnight in a sealed container with gentle agitation. Maximum sensitivity will have been reached after this time, and no destaining is required. Gels can then be imaged and analysed or stored for later use.

Gels were scanned and the images were analyzed with ImagepIQ™ (Proteome Systems Ltd). Spots were automatically detected and were matched and aligned to each other allowing a overlaid image that enabled the protein content of the different fractions to be visualized.

Proteins were identified by tryptic digestion of the spots both before and after PNGase F enzymatic release of the N-linked oligosaccharides. Samples were desalted using μC_8 ZipTip columns (Millipore), and the peptides were eluted onto a hydrophobic

MALDI plate using 1 μl of matrix (2.5 mg/ml α-cyano-4-hydroxycinnamic acid) prior to MALDI-MS analysis. All mass spectra were acquired in positive reflectron mode with delayed pulse extraction. Internal mass calibration was performed, and an in-house peptide mass fingerprinting program (IonIQ, Proteome Systems Ltd) was used to identify the proteins using the SwissProt database.

19.4.4.2 Glycoproteins

Periodic acid/Schiff (PAS) staining is used as a general carbohydrate stain. Although this technique is not highly sensitive, it can visualize 1 to 10 μg of highly glycosylated proteins. There are now fluorescent glycoprotein stains on the market (Pro-Q® Emerald, Molecular Probes, or Glycoprotein detection kit, Sigma) based on the same chemistry. We have found PAS to be the most reliable stain for visualizing glycoproteins because nonspecific labeling seems to be a problem associated with the fluorescent stains. Periodic acid oxidizes vicinal diols of glycosyl residues to dialdehydes. The aldehydes are then allowed to react with fuchsin (Schiff's reagent) to form a Schiff's base that stains pink on a clear gel background. One disadvantage of all these methods of carbohydrate staining is that both the carbohydrates and the protein backbone to which the carbohydrates are attached are ixidized during the staining, making it difficult to conduct further analysis on the proteins in the stained gels.

The gels were removed from the cassettes after the second dimension separation and placed in individual containers. Glycoprotein staining using PAS was performed after fixation of the gels in a solution of 50% (v/v) methanol and 1% (v/v) acetic acid for at least 1 hour. The gels were then washed with water for 10 minutes. The gels were then incubated for 30 minutes in a solution of 1% (w/v) periodic acid and 3% (v/v) acetic acid before being washed with 3 × 5 minute water washes, and 2 × 10 minutes in 0.1% (w/v) sodium metabisulfite, 12 mM HCl. The gels were incubated for 30 minutes to overnight with Schiff reagent, depending on the degree of staining required to produce pink-stained proteins on the gel. The gels were washed thoroughly with 0.1% (w/v) sodium metabisulfite, 12 mM HCl. Approximately five washes should remove the majority of the background staining; however more could be required if the background staining is dark.

19.4.5 Glycosylation Analysis

Mass spectrometry has now reached a level at which the study of glycoproteins and the attached oligosaccharide structures can be undertaken on a protein separated by one-dimensional (1D) PAGE and 2D PAGE. We prefer the use of graphitized carbon as an LC chromatographic medium combined with MS for the separation and detection of released oligosaccharides in negative ion mode at high pH.[10] This approach reduces the amount of sample and preparation time needed since both acidic and neutral oligosaccharides can be detected sensitively without derivatization. The same method of analysis can be used for both PNGase F released N-linked and reductively β-eliminated O-linked oligosaccharides, with minimal handling of the sample. In addition, the development of nanoflow liquid chromatographic separation and nanospray ESI has allowed the sensitivity of detection to be increased to approximate that of MALDI-MS.[11] These combined technologies are now available to enable the large-scale analysis of the oligosaccharides on small amounts of gel-separated glycoproteins, which is difficult to achieve by analyzing the glycopeptides in a tryptic digestion of these spots. Glycopeptides are often not seen in tryptic digests because of their heterogeneity and low ionization potential in the presence of other peptides. By our approach, once the released oligosaccharide heterogeneity has been determined, it is then often possible to find the glycopeptides in a tryptic digestion of the protein by specifically selecting for ions of predetermined mass based on the known attached oligosaccharide masses.

19.4.5.1 N-Linked Oligosaccharides

N-linked oligosaccharides that attach to the asparagine residue on the glycoprotein backbone can be enzymatically released without destroying the amino acid backbone. As the proteins have been denatured in the SDS-PAGE separation, there is no need to add detergent to the incubation. N-Glycosidase F (PNGase F) (Roche Molecular Biochemicals, Mannheim, Germany), can be used to cleave the oligosaccharide from the protein producing the intact oligosaccharide while replacing the asparagine in the amino acid backbone with aspartic acid (+1 Da). This technique allows not only for the oligosaccharides to be analyzed, but also for the PMF analysis of the deglycosylated proteins after further digestion with trypsin. It is also possible by this approach for information to be gained on the sites of glycosylation due to the 1 Da peptide mass increase after PNGase F deglycosylation.

The released N-linked oligosaccharides can then be analyzed directly, or alternatively, they can be reduced prior to MS analysis. The reduction has been shown to direct the fragmentation in MS/MS for easier interpretation (unpublished data).

19.4.5.2 O-Linked Oligosaccharides

O-linked oligosaccharides are attached via the hydroxyl group of serine or threonine, or both, to the glycoprotein. There is no universal enzyme available that cleaves this linkage. Chemical β-elimination in a reducing environment is widely used to liberate and study glycoprotein O-linked oligosaccharide structures. The reduction of the reducing terminus of the eliminated oligosaccharide is necessary to prevent the released oligosaccharides from undergoing subsequent peeling and destruction. We use this release method following PNGase F digestion to allow analysis of both the N- and O-linked oligosaccharides to be performed from a single protein spot after 2D PAGE separation.[12]

19.4.6 LC-ESI-MS of Oligosaccharides

Negative ion LC-ESI-MS provides the best means, in our hands, for the analysis of both the N- and O-linked reduced oligosaccharides using graphitized carbon as a LC chromatographic medium at high pH.[10] This method not only allows for the analysis of both neutral and sialylated oligosaccharides simultaneously, but also the minimal handling reduces the sample losses and preparation time. Following separation MS/MS of the oligosaccharides, along with their chromatographic retention time, can produce both sequence and linkage structural information.

Peak lists obtained from the MS/MS data, including *m/z* and intensity data, as well as the oligosaccharide parent ion mass, can then be interpreted using GlycosidIQ™ (www.glycosuite.com, Proteome Systems Ltd), which compares the observed fragmentation masses with the predicted masses of the *in silico* fragmentation of oligosaccharides contained in the relational database, GlycosuiteDB (Proteome Systems Ltd).

19.5 RESULTS AND DISCUSSION

The experimental work flow that we use to probe samples for glycoproteins and then to release and analyze both the N- and O-linked oligosaccharides attached to the protein is shown in Figure 19.1. The following results are from the enrichment of the glycoproteins in human plasma using lectin affinity chromatography (jacalin) to isolate glycoproteins, which contain O-glycosidically linked oligosaccharides. The 2D PAGE separations of both the unbound and bound fractions and identification of the major proteins are shown in Figure 19.2. Both the flow through and retained fractions contain proteins that are unique to that fraction, illustrating the separation ability of this technique.

Several proteins appear on both gels, and glycoproteins are found in both fractions. The glycoproteins could be easily seen as trains of spots, presumably due to the differences in sialylation between glycoforms of the same protein. Together with the glycoproteins found in the bound fraction, a significant amount of Apolipoprotein A1 (Apo A1) was detected in the 2D PAGE protein map at approximately 30 kDa. Because Apo A1 has not been described as glycosylated and does not stain with PAS, its presence in the bound fraction must be explained by other reasons. A possible explanation is that since the sample is prepared under native conditions, Apo A1 may still have been associated with either proteins in the plasma, some of which are glycosylated. The majority of the glycoproteins in the retarded fraction have been reported to contain O-linked glycosylation (e.g., hemopexin, α-HS-glycoprotein, and inter-α trypsin inhibitor heavy chain H4). The jacalin column also isolated a subfraction of α-1-antitrypsin. This protein has not been reported to contain O-linked oligosaccharides, although it is substantially N-linked glycosylated. Whether jacalin is interacting with a subfraction of these N-linked oligosaccharides, or if there is a subfraction of α-1-antitrypsin that contains O-linked oligosaccharides, will need to be further investigated. In many cases we were able to detect proteins in the bound fraction that were not visible on 2D PAGE gels of unfractionated plasma. These proteins include hemopexin, which is a glycoprotein containing both N- and O-linked oligosaccharides, usually hidden under albumin in an unfractionated human plasma 2D PAGE separation.

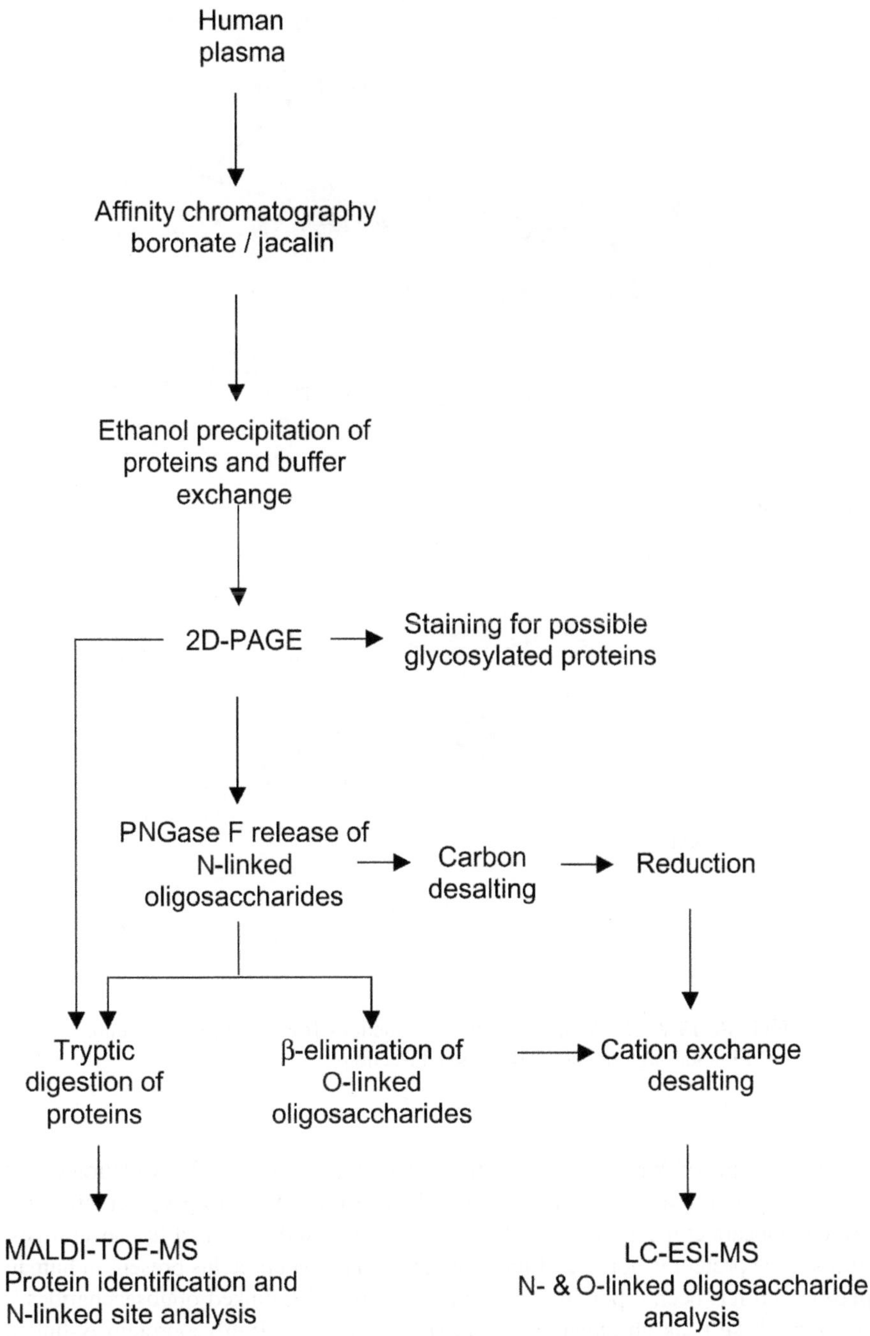

FIGURE 19.1 Experimental flow chart of glycoprotein enrichment, detection.

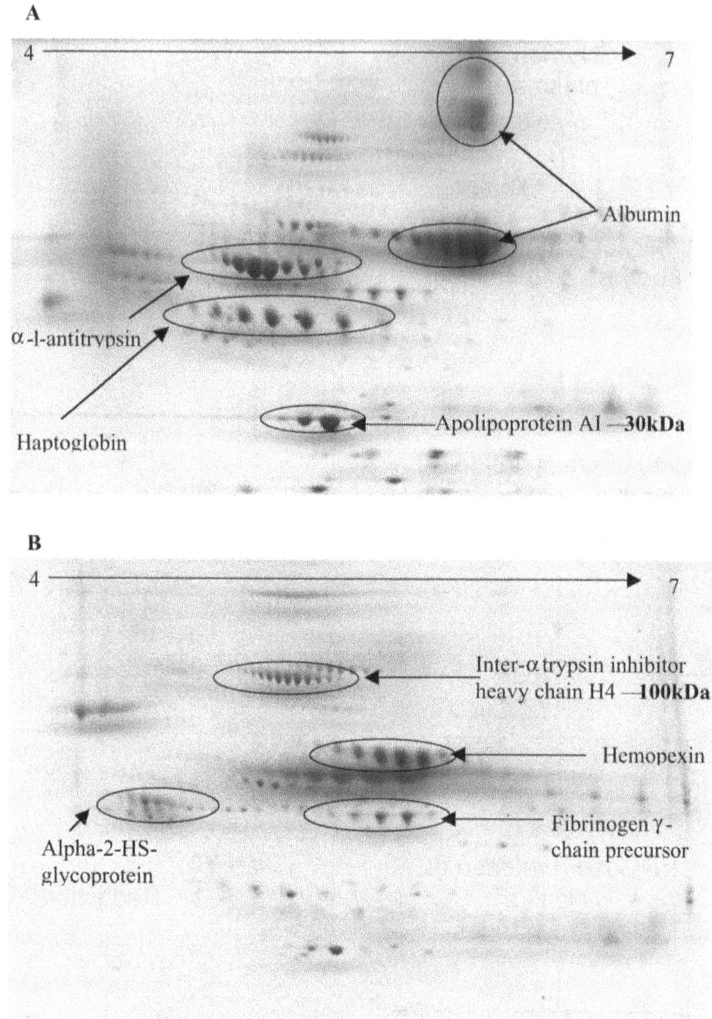

FIGURE 19.2 2D PAGE separation of fractions from the jacalin column. (A) Flow-through fraction; (B) retained fraction.

An example of our procedure for the further analysis of the glycosylation of gel isolated glycoproteins is shown for one of the isoforms of α_2-HS-glycoprotein. This protein appears at the acidic end of the plasma 2D PAGE map and has several isoforms. This protein is not one of the more abundant glycoproteins present in human plasma and has been reported to contain both N- and O-linked oligosaccharides.[13] A typical base peak chromatogram and spectrum of α_2-HS-glycoprotein N-linked oligosaccharides are shown in Figure 19.3. The N-linked oligosaccharides found on this isoform are typical of the oligosaccharides found throughout plasma proteins.

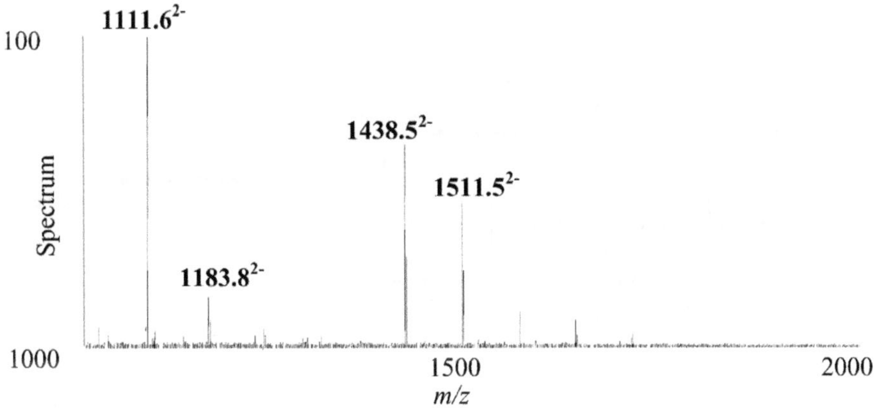

FIGURE 19.3 Base peak spectrum of N-linked oligosaccharides released from α_2-HS-glycoprotein gel spot.

The difference between the glycoforms has been shown to result from the relative amounts of different charged oligosaccharide structures on the protein.[12]

The MS/MS spectra of the peaks present in the base peak spectrum were analyzed using glycofragment mass fingerprinting GlycosidIQ software to predict the glycan structures.[14] Figure 19.4 shows the MS/MS spectra and corresponding structure of the $[M-2H]^{2-}$ ion with *m/z* of 1111.6 as assigned manually. The same structure was predicted by the bioinformatic tool GlycosidIQ as shown in Figure 19.5, which also gives a scoring based on quality of matching (segmentation score) and coverage (correspondence score) and links to the structure found in the GlycosuiteDB database. This structure corresponds to the N-linked structure previously found in α_2-HS-glycoprotein.[13] The other structures that were identified using their MS/MS spectra and the GlycosidIQ tool were not reported in the GlycosuiteDB database as attached to α_2-HS-glycoprotein but have been found to be attached to other human plasma proteins.[15]

The base peak spectrum of the O-linked oligosaccharides released using reductive β-elimination from the same isoform from which the N-linked oligosaccharides were enzymatically released is shown in Figure 19.6. The MS/MS spectra corresponding to the *m/z* 966.1^{1-} peak is shown in Figure 19.7. The daughter ions and their intensities were used along with the parent ion mass to obtain the predicted structure of the oligosaccharide using GlycosidIQ. This is illustrated in Figure 19.8, which shows the search results as well as the matched glycosidic fragments. Both the structure NeuAc(a2–3)Gal(b1–3)[NeuAc(a2–6)]GalNAc (*m/z* 966.1^{1-}) and NeuAc(a2–3)Gal(b1–3)GalNAc (*m/z* 675.2^{1-}) are commonly found on plasma O-linked glycoproteins.

The technique of using lectin affinity in combination with 2D PAGE allows the relatively large-scale analysis of oligosaccharides on glycoprotein isoforms and can

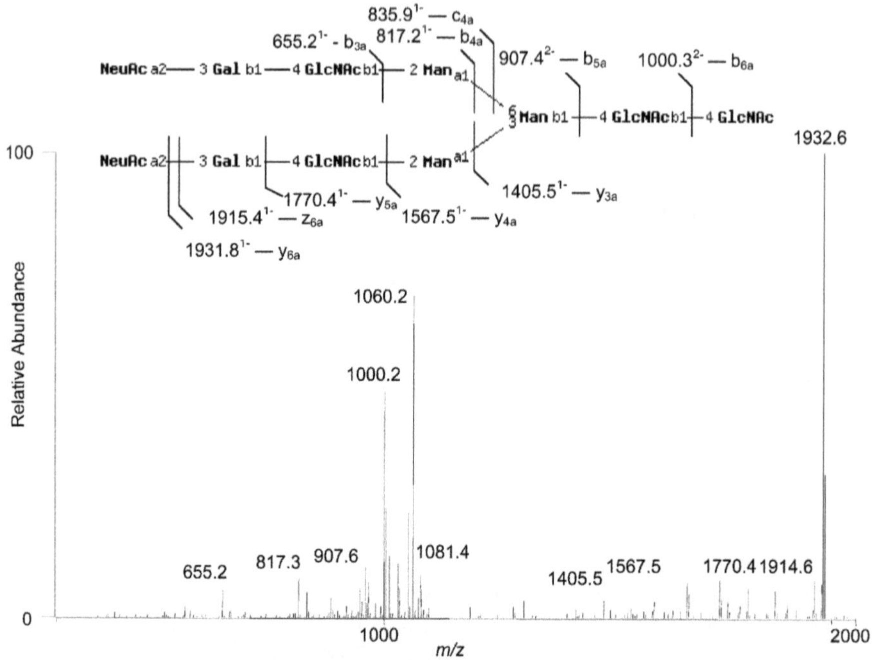

FIGURE 19.4 MS/MS spectrum and manually assigned structure of the N-linked oligosaccharide with $[M\text{-}2H]^{2-}$ ion at *m/z* 1111.6 in Figure 19.3 corresponding to a composition of $[Hex_5HexNAc_4NeuAc_2]$.

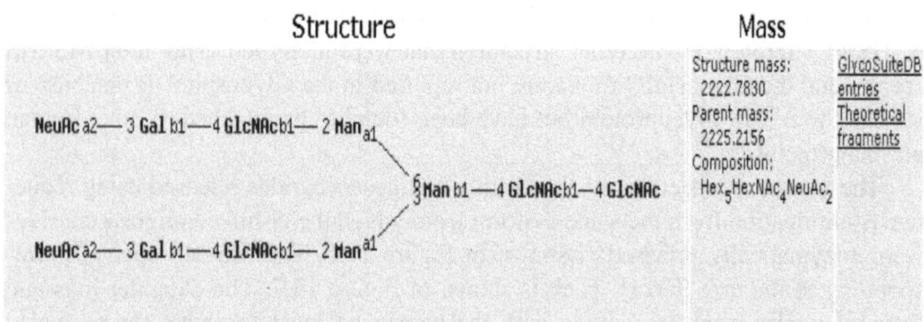

FIGURE 19.5 GlycosidIQ™ results from matching the MS/MS spectra of the $[M\text{-}H]^{2-}$ ion at *m/z* 1111.6 with the theoretical fragmentation of the GlycoSuiteDB database. The MS/MS fragmentation data for each doubly charged peak seen in the base peak spectra were searched using following parameters: parent ion error, 0.7 Da; ion mode, negative; adduct, H; spectrum error, 0.5 Da; reducing terminal derivatization, reduced.

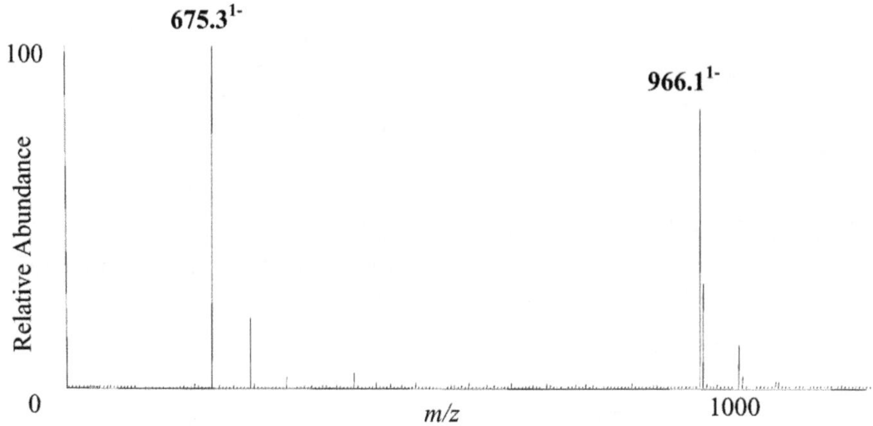

FIGURE 19.6 Base peak spectrum of O-linked oligosaccharides released from α_2-HS-glycoprotein gel spot using reductive β-elimination after PNGase F digestion. Spectrum obtained from the protein spot at the more basic end of the isoform chain.

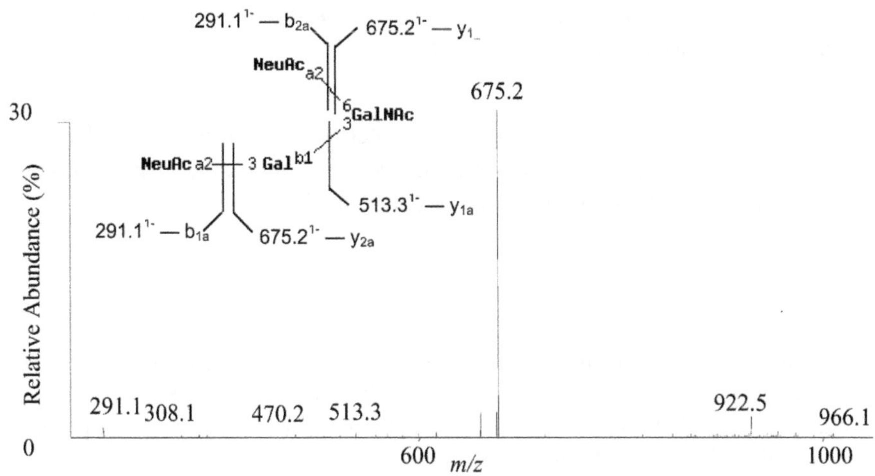

FIGURE 19.7 MS/MS spectrum and manually assigned structure of the O-linked oligosaccharide of the *m/z* 966.1 parent ion in Figure 19.6.

be used to expand the picture of protein glycosylation offered by 2D PAGE analysis. By combining either lectin or boronate affinity chromatography with the traditional 2D PAGE separation techniques we were able to prefractionate samples based on their oligosaccharide content rather than the more traditional molecular weight and isoelectric point fractionations.[16] Advances in MS and bioinformatics now allow the

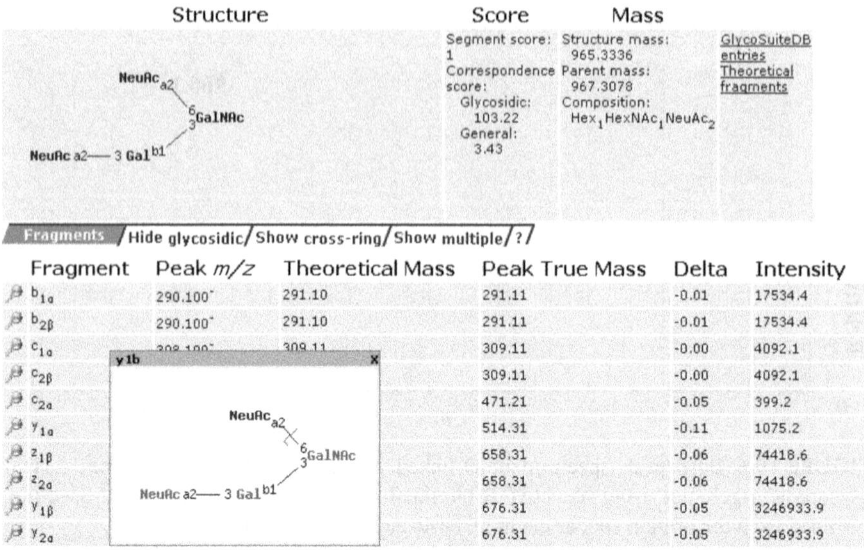

FIGURE 19.8 Screen shot of the results from GlycosidIQ of matching the MS/MS spectra of the $[M\text{-}H]^-$ ions at m/z 966.1 with the theoretical fragmentation of the GlycoSuiteDB database. These include the predicted attached oligosaccharide structures, with the list of matched fragments and their pictorial representation.

analysis of oligosaccharides present on glycoprotein isoforms from 2D PAGE gels,[12] and by combining this with the above enrichment techniques the glycosylation of previously undetected glycoproteins can be determined.

REFERENCES

1. Spiro, R.G., Protein glycosylation: Nature, distribution, enzymatic formulation, and disease implications of glycopeptide bonds, *Glycobiology*, 12, 43–56, 2002.
2. Schachter, H., Congenital disorders involving defective N-glycosylation of proteins, *Cell. Mol. Life Sci.*, 58, 1085–1104, 2001.
3. Hirabayashi, J. and Kasai, K., Separation technologies for glycomics, *J. Chromatogr B.*, 771, 67–87, 2002.
4. Stillmark, H., and Über, R., Ein giftiges Ferment aus den Samen von *Ricinus comm.* L. und einigen anderen Euphorbiaceen, M.D. Dissertation, University of Dorpat, Dorpat, 1888.
5. Gorelik, E., Galili, U., and Raz, A., On the role of cell surface carbohydrates and their binding proteins (lectins) in tumor metastasis, *Cancer Metastasis Rev.* 20, 245–277, 2001.
6. Brinkman-Van der Linden, E.C., Sonnenburg, J.L., and Varki, A., Effects of sialic acid substitutions on recognition by *Sambucus nigra* Agglutinin and *Maackia amurensis* Hemagglutinin, *Anal. Biochem.*, 303, 98–104, 2002.

7. Peumans, W.J., Avan Damme, E.J., Barre, A., and Rouge, P., Classification of plant lectins in families of structurally and evolutionary related proteins, *Adv. Exp. Med. Biol.*, 491, 27–54, 2001.
8. Yu Cai Li, Separation of Glycoproteins Using Shielding Boronate Affinity Chromatography, Doctoral Thesis, Department of Biotechnology, Lund University, 2001.
9. Hoffmann, A., Nimtz, M., and Conradt, H.S., Molecular characterisation of beta-trace protein in human serum and urine—A potential diagnostic marker for renal diseases, *Glycobiology*, 7, 499–506, 1997.
10. Kawasaki, N., Ohta, M., Hyuga, S., Hyuga, M., and Hayakawa, T., Application of liquid chromatography with tandem mass spectrometry to the analysis of the site-specific carbohydrate heterogeneity in erythropoietin, *Anal. Biochem.*, 285, 82–91, 2000.
11. Okafo, G., Burrow, L., Carr, S.A., Roberts, G.D., Johnson, W., and Camilleri, P., A coordinated high-performance liquid chromatographic, capillary electrophoretic, and mass spectrometric approach for the analysis of oligosaccharide mixtures derivatized with 2-aminoacridone, *Anal. Chem.* 68, 4424–4430, 1996.
12. Wilson, N.L., Schulz, B.L., Karlsson, N.G., Packer, N.H., Sequential analysis of *N*-and *O*-linked glycosylation of 2D-PAGE separated glycoproteins, *J. Proteome Res.*, 1, 521–529, 2002.
13. Watzlawick, H., Walsh, M.T., Yoshioka, Y., Schmid, K., and Brossmer, R. Structure of the N- and O-glycans of the A-chain of human plasma alpha-2HS-glycoprotein as deduced from the chemical composition of the derivatives prepared by stepwise degradation with exglycosidases, *Biochem.*, 31, 12,198–12,203, 1992.
14. Joshi, H.J., Harrison, M.J., Schulz, B.L., Cooper, C.A., Packer, N.H., and Karlsson, N.G., Development of a mass fingerprinting tool for automated interpretation of oligosaccharide fragmentation data, *Proteomics*, 4, 1650–1664, 2004.
15. Ogawa, H., Yoneda, A., Seno, N., Hayashi, M., Ishizuka, I., Hase, S., and Matsumoto, I., Structures of the N-linked oligosaccharides on human plasma vitronectin, *Eur. J. Biochem.*, 230, 994–1000, 1995.
16. Righetti, P.G., Castagna, A., Herbert, B., Reymond, F., Rossier, J.S., Prefractionation techniques in proteome analysis, *Proteomics*, 3, 1397–1407, 2003.

Part IV

Applications of High-Performance Liquid Chromatography

20 Proteomic Analyses Using High-Efficiency Separations and Accurate Mass Measurements

Jon M. Jacobs and Richard D. Smith

CONTENTS

20.1 INTRODUCTION

While the development of analytical techniques for detecting and identifying proteins has been an actively pursued area of research for many years, a recent and concerted reemergence of work in this area has resulted in an explosion of new developments that can be applied across multiple disciplines. The advent of numerous mass spectrometric and separation advances may be viewed as the technical driving force behind these developments, but other factors are perhaps equally crucial. These factors include the increasing availability of sequences for many organisms as a result of recent advancements in genome analysis, recognition that there are limitations to using sequence and transcriptome information alone, and the more apparent value of directly characterizing the biomolecules and pathways that actually drive cellular events. Direct identification of structural and functional proteins allows perturbations at the protein level to be linked to responses at the cellular level and provides a method for analyzing global proteomic changes in virtually any model system.

The field of proteomics, in particular, has benefited from numerous recent analytical advances that have provided increased sensitivity and dynamic range of proteins detected, as well as analysis throughput. Traditional approaches typically have involved separations using two-dimensional polyacrylamide gel electrophoresis (2D PAGE) in conjunction with a mass spectrometric (MS) component for protein identification.[1] These seminal techniques set the foundation for the advancing field of proteomics. The limitations associated with the gel-based techniques are well known and stem mostly from a lack of sufficient dynamic range needed for in-depth proteome coverage[2] and from the large amount of time needed for analysis, which dramatically limits throughput. Significant developments have been made in an attempt to address both these and other issues. With the advent of electrospray ionization, researchers can now couple multidimensional liquid chromatography (LC) separations with MS analysis. Combined with a bottom-up strategy, in which proteins are digested into smaller peptide fragments for identification, the LC-MS platform has proved highly successful for identifying peptides (and generally proteins) from complex samples.

This chapter discusses proteome analysis techniques that are based on high-resolution LC separations and high mass accuracy MS instrumentation, specifically the use of Fourier transform ion cyclotron resonance (FTICR) MS, and their application for high-throughput global identification of proteins. The simultaneously expanded sensitivity, dynamic range, and mass accuracy afforded by FTICR place this system at the forefront of analytical systems that can attain proteomic identifications from complex biological samples. We will begin the discussions by describing a high-throughput proteomics approach that capitalizes on the attributes of FTICR. The critically important supporting components that combine to provide a comprehensive proteomic analysis capability—electrospray ionization efficiency, pre-MS high efficiency LC separations, and downstream data processing methods—are discussed with the intent of providing insight into the process as a whole and to highlight analytical challenges that remain to be addressed. Because quantitative measurements are so important to understanding even the simplest biological systems, some of the quantitation techniques and approaches being developed and applied in our laboratory are also discussed. Several applications are exemplified prior to concluding this chapter.

20.2 HIGH-THROUGHPUT APPROACH FOR GLOBAL PROTEOMIC MEASUREMENTS

The level of proteome coverage achievable depends on both the separation quality and the MS platform being utilized. For proteomic analyses from complex peptide mixtures, the better the separations prior to tandem MS (MS/MS) analysis, the more complex a mixture that can be addressed.[3] The first implementation of a combined LC-MS approach for global proteomics was by Washburn et al.,[4,5] who utilized 2D LC separations that involved sequential step elutions of tryptically digested peptides from a strong cation exchange (SCX) resin followed by reversed-phase solvent gradient separations in combination with ion-trap MS/MS analysis. The resulting peptide dissociation information enabled identification of multiple peptides from a mixture and hence the parent proteins present.[6–12]

This shotgun approach to proteomics produced unprecedented initial coverage of the global yeast proteome; however, challenges and limitations are inherent with the technique. For example, throughput is limited since MS/MS measurements require that peptides be selected one at a time for analysis. Thus, while many peptides can be detected in a single first-stage mass spectrum, this approach requires that these peptides be individually selected for a second MS stage analysis. New linear ion trap MS instrumentation has improved the efficiency of MS/MS analyses, but one analytical cycle at a minimum is still needed to identify each peptide. In addition, complex proteomic samples often have large numbers of peptides that coelute from even the best multidimensional liquid separations. MS/MS analysis dictates that only a small subset of these peptides can be picked for the second-stage MS fragmentation and identification, which generally leads to a global undersampling problem when performing MS/MS experiments. This issue can be theoretically addressed by even more extensive pre-MS fractionation or separations chromatography to reduce the complexity of the mixture eluting to the MS at any time point, but only at the expense of significantly lower throughput, increased sample consumption, and likely increased specific losses of peptides. To overcome this MS/MS bottleneck, we developed a strategy that increases throughput by avoiding *routine* MS/MS measurements.

The proteomics strategy developed in our laboratory is a variation of the shotgun proteomics approaches. As with shotgun approaches, proteins are first cleaved into peptide fragments (e.g., by a specific proteolytic enzyme such as trypsin) after cell lysis and sample processing, and LC-MS/MS proteome analyses are employed to identify peptides. However, in our strategy, these multiple analyses need only be performed once for a particular biological system. The results from the initial LC-MS/MS analyses are used to create a look-up table that contains a characteristic accurate mass and LC separation elution time for each peptide; that is, a mass and time (MT) tag, which serves as a unique 2D marker for its subsequent identification. Once this look-up table (referred to as an MT tag database) is created, a future sample need only be analyzed by high-resolution capillary LC-FTICR to detect peptides with the same accurate mass and same elution time characteristics as in the MT tag database. This approach not only eliminates the need for routine MS/MS, but also provides both greater analytical sensitivity and increased analysis throughput.

The two parts of this overall proteome measurement strategy—(1) generating potential MT tags from extensive automated LC-MS/MS analyses and (2) performing high throughput LC-FTICR analyses that use the MT tag database in subsequent studies for peptide and protein identifications—are depicted in Figure 20.1. The peptide MT tags are assigned using conventional software tools (e.g., SEQUEST[13]) and

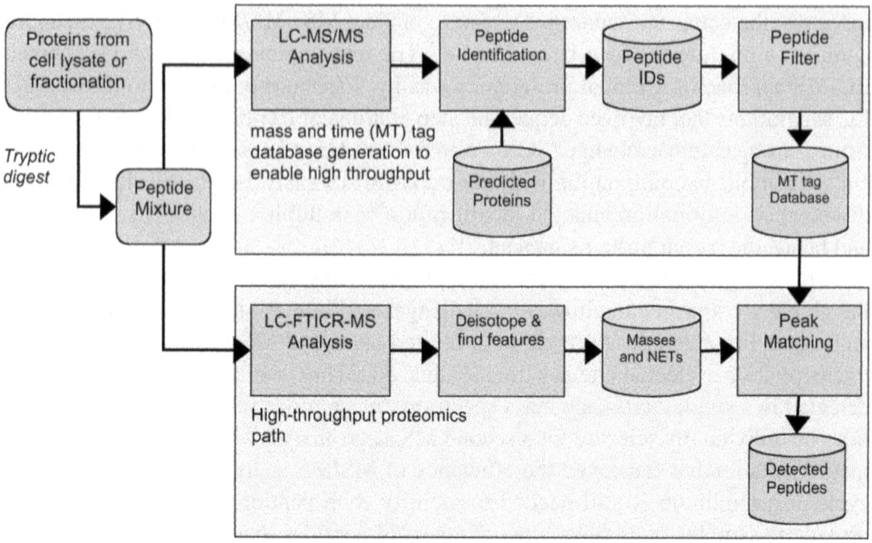

FIGURE 20.1 Basic schematic of the accurate mass and time (AMT) tag approach. The AMT approach consists of two main components: (1) Creation of an MT tag database using a LC-MS/MS peptide identification strategy; and (2) use of LC-FTICR-MS for high-throughput accurate mass measurements that will be compared against the MT tag database for the identification of peptides and creation of an AMT tag. Reproduced with permission from Ref.[14], copyright 2005 Elsevier.

reflect both the calculated accurate mass and the normalized LC elution time (to <2% uncertainty based on present approaches). Subsequent identification by LC-FTICR accurate mass measurements effectively validates an MT tag as an accurate mass and time (AMT) tag. This so-called AMT tag approach is similar to the identification of a protein spot in a 2D-PAGE study. Once a protein spot has been identified in a gel for a particular biological system, a spot at the exact same 2D location in a subsequent analysis of the same system can generally be identified with high confidence. Analogously, once an MT tag has been established for a particular biological system, a peptide eluting at the same normalized time and with the same mass in a subsequent analysis of the same system can generally be identified with high confidence. The confidence of the identification will depend strongly on the specificity provided by both the separation and the accuracy of the mass measurements.

20.3 ESSENTIAL COMPONENTS FOR EFFECTIVE LC-MS PROTEOMIC ANALYSIS

The technical foundation that enables the high-throughput AMT tag approach involves advanced separations combined with very accurate mass spectrometric measurements from FTICR and a supporting data analysis and management infrastructure. The use of FTICR presently provides a number of advantages over conventional MS platforms, including a dramatic reduction in the sample size needed and enhanced sensitivity for analysis of low-abundance proteins.[15–17] Mass spectra can be acquired with resolution in excess of 10^5 and with low to sub part-per-million (ppm) mass measurement accuracy (MMA).[18,19] These measurement qualities are important because they allow more complex mixtures to be characterized.

The following sections describe key supporting components of the FTICR-based analysis platform—electrospray ionization, high-efficiency LC separations, and downstream data processing methods—that are critical for optimum performance and that are being actively studied in our laboratory. Each component represents an independent link in the proteome chain of analysis since improvements in any one area eventually affect the results as a whole. There is little discussion of specific instrumentation advancements pertaining to the mass spectrometer itself, as this is beyond the scope of this chapter and would require significant discussion devoted to that topic alone. However, an overview of a specific MS technique that has led to dramatic increases in the dynamic range of detection for LC-FTICR analysis is provided.

20.3.1 Electrospray Ionization Efficiency

The application of electrospray ionization (ESI) to mass spectrometry by Fenn and coworkers in the 1980s provided a highly sought after component that allowed liquid separations to be directly coupled to MS for large protein analysis.[20] With the subsequent integration of ESI into the proteomic analysis platform, optimizing conditions to achieve the highest ESI efficiency is of critical importance. Depending upon analysis conditions, a challenging aspect of ESI is the possibility of "suppression" effects that occur due to the presence of solution matrix components or peptides eluting at the same point in the separation. Such suppression effects could make detecting the presence of some peptides difficult or determining the relative abundances of different peptides problematic. Since the extent of suppression is expected to be highly dependent upon the precise solution composition, even comparative measurements for the same peptide could be problematic. Therefore, it is best to conduct analyses under conditions in which ionization suppression effects are minimized.

The conditions under which ionization suppression occur are related to both analyte concentration and ESI volumetric flow rates and are relatively well understood.[21–23] Large, conventional flow rates result in greater compound-to-compound variation due to the increased analyte competition for the charge and exposure to the electrospray droplet surface. Converting the electrospray ionization from a conventional flow rate to a nanoflow regime (Figure 20.2)[24] produces smaller initial droplets that allow ions to form more rapidly and efficiently with less heating. Smaller droplets move toward the periphery of the electrospray plume due to their smaller inertia and higher charge density[22] (Figure 20.2, top); consequently, conventional electrospray is generally sampled off-axis of the MS inlet, where a majority of the ions are desolvated and the concentration of large problematic droplets or clusters is minimal. Depending on the flow rate, sample concentration, and surface activity, a large portion of the analyte (peptide sample) can form these charged clusters or residue particles

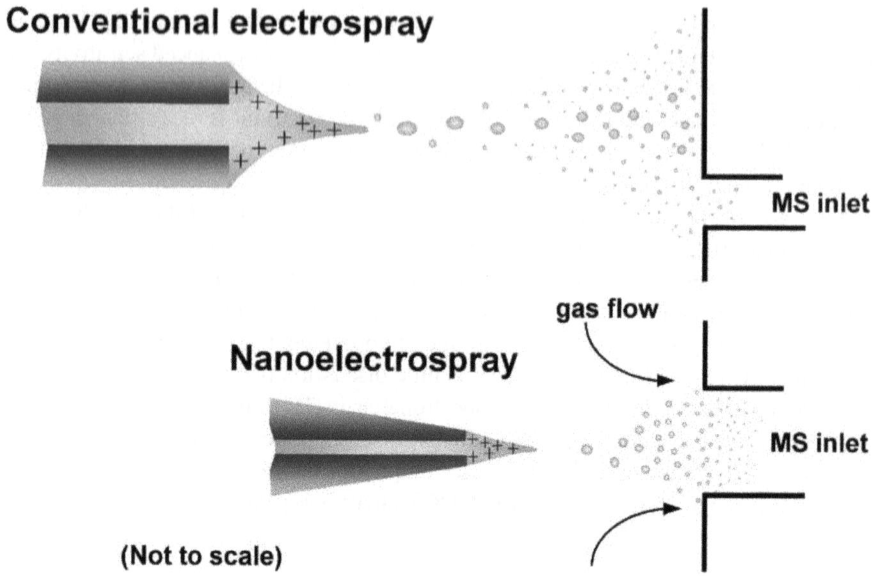

FIGURE 20.2 Normal flow rate electrospray (top) versus a nanoflow rate electrospray (bottom) that produces smaller droplets. By allowing closer proximity to the MS inlet, the lower flow rate electrospray affords more efficient ion introduction. Reproduced with permission from Ref.[24], copyright 2005 Am. Chem. Soc.

at conventional flow rates (μl/min). Also problematic is the limited ~1 cm spacing between the electrospray emitter and the MS inlet combined with the expansion of the electrospray plume, resulting in a reduction in MS inlet efficiency (Figure 20.2, top), as well as limiting sensitivity.

When smaller diameter capillaries are employed, higher sensitivity can be achieved when both the sample size and liquid flow rate are reduced.[25] For example, the use of 10 μm i.d. capillaries for CE-MS[26] provided low nl/min flow rates that resulted in a mass sensitive response due to the limited delivery of charge carrying species. The smaller electrospray droplet size enabled the emitter to be aligned on-axis and closer to the MS inlet, yielding a more efficient ion delivery and increase in sensitivity (Figure 20.2, bottom).[24,27]

With sufficiently low flow rates and concentrations delivered by the electrospray, ionization efficiency approaches 100%.[24] Even at reduced rates in the intermediate flow levels, ESI efficiencies will become more uniform,[24–28] and the efficiency of ion introduction through the MS inlet is enhanced. Additionally, the concentration where linear response ends can occur at a fairly sharp boundary.[28] Figure 20.3 shows ESI-FTICR peak intensities versus sample size for three abundant peptides obtained from an LC separation using a 30-μm-i.d. capillary.[29] For sample sizes below a given level (<100 ng for the example in Figure 20.3), MS peak intensities (or peak areas) are generally in the regime where signals increase linearly with sample size and provide the best quantitation.

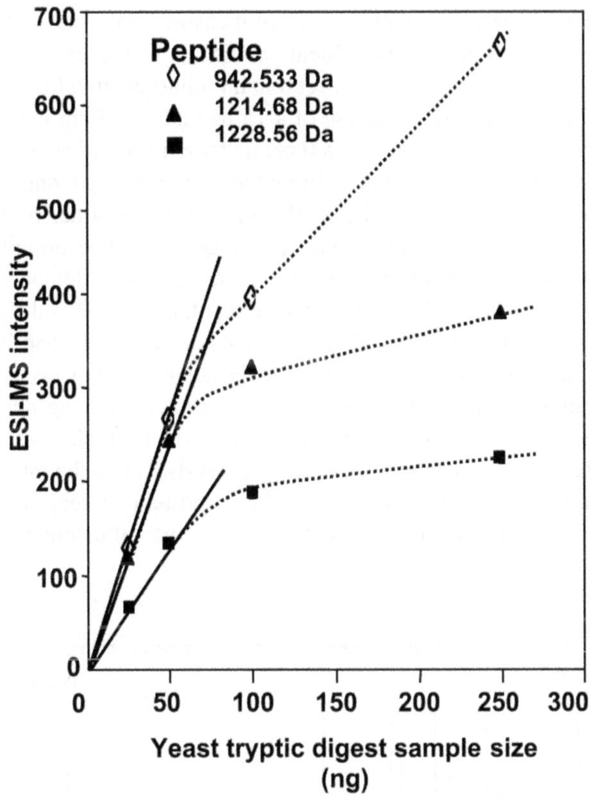

FIGURE 20.3 ESI-MS peak intensities versus the total proteome sample size for three of the more abundant peptides having the indicated molecular weights. The results were obtained using a 30-μm-i.d. packed capillary. Reproduced with permission from Ref.[30], copyright 2002 Am. Chem. Soc.

20.3.2 High-Efficiency LC Separations

With MS platforms such as FTICR that have a maximum ion trap capacity, the overall dynamic range achievable depends significantly on the resolution and peak capacity of the on-line separations used. The dynamic range achieved in any one FTICR spectrum is typically $\sim 10^3$, but the effective dynamic range achievable in proteome measurements with combined LC-FTICR can be improved to $>10^5$.[30] In addition, the use of high-efficiency, high-resolution separations to separate the abundant species from most of the low-abundant species enables not only better detection, but also quantitation.

Since smaller i.d. capillaries provide a linear response over a wider range of sample concentrations (Figure 20.3), a common approach, and one observed in our laboratory, has been toward decreasing LC capillary diameters (i.e., from 150 μm to 75 or 50 μm i.d. or even smaller). These smaller i.d. capillaries improve overall sensitivity and the practical dynamic range of measurements, particularly when the absolute sample size is constrained. Ideal conditions would use LC capillary columns that provide the lowest flow rates yet remain robust enough for high-throughput operations. As a demonstration, the use of a 15 μm i.d., ~85 cm long column at an optimal flow rate of ~20 nl/min at 10,000 psi increased ESI efficiency by ~100-fold over conventional 150 μm i.d. columns.[29] In another example, improving LC peak capacities from ~550 to ~1000 doubled the number of peptides identified by MS/MS.[31]

In another study, the sensitivity and the range of relative protein abundances measurable for small-sized complex proteomic samples was demonstrated using a 15-μm column and a tryptic digest of a mixture containing a 10^6:1 difference in protein abundances for two standards (75 femtomoles cytochrome *c* and 75 zeptomoles bovine serum albumin) and 5 ng of an $^{14}N/^{15}N$-labeled *Deinococcus radiodurans* lysate[30] (see Figure 20.4). The results show that proteins having approximately 6 orders of magnitude difference in relative abundances in a complex proteomic sample can be characterized from a single LC-FTICR analysis. While the ratios of MS peak intensities significantly deviated from the relative protein content during elution of the most abundant peptides, both this and previous work[29] illustrate that quantitative

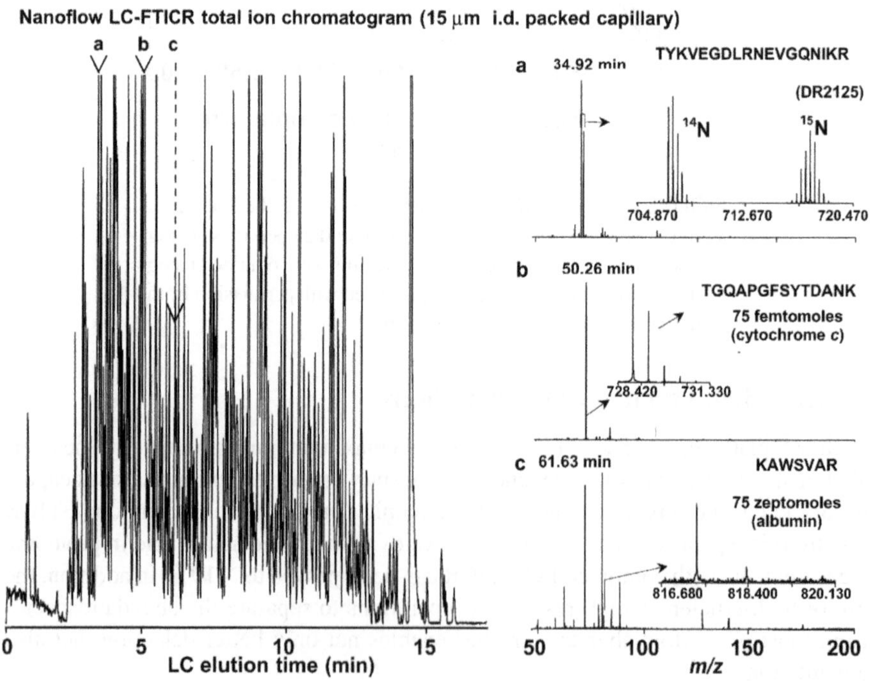

FIGURE 20.4 Examination of the analysis range of *D. radiodurans* relative protein abundances. The sample contained (a) 5 mg of $^{14}N/^{15}N$-labeled *D. radiodurans,* (b) 75 femtoles cytochrome *c*, and (c) 75 zeptomoles of bovine serum albumin tryptic digests. Reproduced with permission from Ref.[24], copyright 2004 Am. Chem. Soc.

analyses are achievable for other intermediate- and lower-abundance species. In general, our studies indicate that integrated peak intensities reflect relative abundances most precisely for lower-abundance species.

20.3.3 Expanded Dynamic Range for MS

For several years we recognized the potential of an approach that would expand the dynamic range of our MS instrumentation, but significant technological developments, as well as fast data-dependent computer control of the experiment, were required before the approach could be implemented. After overcoming several limitations, the dynamic range enhancement applied to MS (DREAMS)[32] was developed, which is based on the ejection of the most abundant ions in a mass spectrometer so as to provide more efficient use of the FTICR dynamic range for each spectrum.

To accomplish this task, we developed software that uses the peak intensities from an FTICR mass spectrum to calculate a set of frequencies that are then used to perform dipolar irradiation of ions in an external 2D quadrupole to remove the high intensity ions before the collection (or accumulation) step in the external quadrupole. The ions collected in this external quadrupole are then transferred to the FTICR ion trap. Because major ions are eliminated in this way, which normally would result in rapid filling of the external quadrupole ion accumulation device, longer ion accumulation times can be used to accumulate more of the low-abundance ions.

The end result of this process is that much greater sensitivity and an extended dynamic range are achieved. For demonstration, Figure 20.5 displays two partial chromatographic spectra obtained from one LC-FTICR analysis, using equal quantities of ^{15}N- and ^{14}N-labeled B16 mouse cells.[24] Two different ion chromatograms were reconstructed from this experiment, the first corresponding to the normal odd-numbered mass spectra and the second to the DREAMS even-numbered spectra in which the higher-abundant ions were ejected. Observed is the removal of the high-abundance ions, with a subsequence increase in detection of lower-abundant ions. The identification of additional high confidence $^{15}N/^{14}N$ pairs, previously unseen in the normal spectra, correlates to a large improvement in sensitivity and dynamic range that can be applied within a single LC-FTICR experiment. By implementing the DREAMS approach, we have typically seen proteomic identification rates increase by 35%.[33]

20.3.4 Data Processing Components

Large scale LC-MS proteomic analysis of biological samples generally produces vast quantities of data that can easily overwhelm researchers if steps are not taken to control the downstream dataflow. A single analysis often provides thousands of separate MS or MS/MS spectra that then need to be coupled to data analysis tools that can interpret and extract detected masses from the MS datasets and perform database searches to identify peptides on the basis of the detected masses. These complex multistage analyses also require experimental conditions and sample pedigree to be

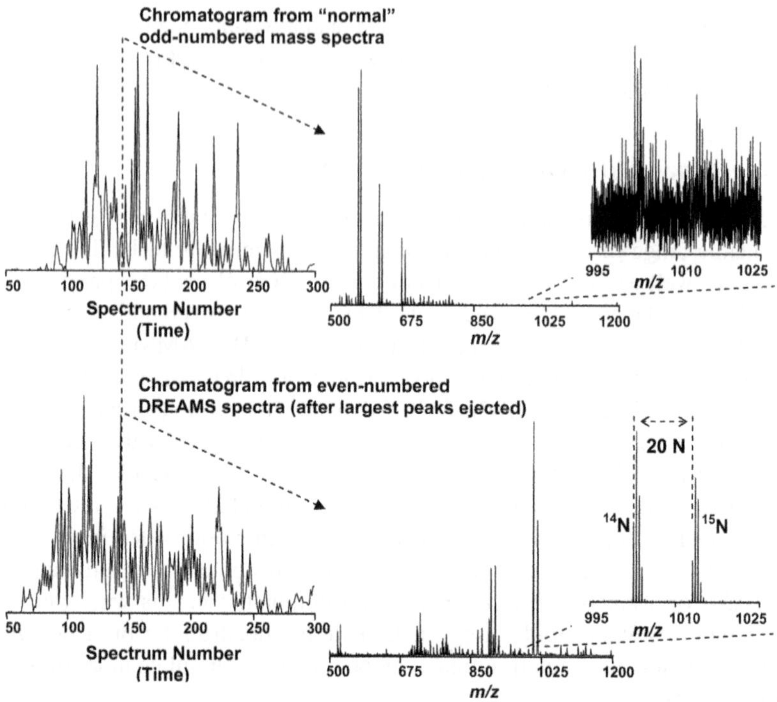

FIGURE 20.5 Partial chromatograms and examples of typical normal and DREAMS spectra from a capillary LC-FTICR analysis of peptides from a tryptic digest of a mixture of natural isotopic abundance and ^{15}N-labeled mouse B16 proteins. Top left: partial chromatogram reconstructed from the normal FTICR mass spectra. Bottom left: corresponding chromatogram from the DREAMS spectra for which high relative abundance species were ejected, allowing longer ion accumulation. The mass spectra (center) show the effective ejection of the major species in the top spectrum compared to the one shown on the bottom. The detail (right) shows a large gain in sensitivity and signal-to-noise ratio for a peptide pair providing a basis for quantitative comparison of protein abundances in the two cell cultures. Reproduced with permission from Ref.[24], copyright 2004 Am. Chem. Soc.

tracked and must have the flexibility to insert quality-control measures at several stages of the processing to ensure instrument performance and sample preparation quality.

To address these issues common to any LC-MS proteomic effort, our laboratory developed the Proteomics Research Information System and Management (PRISM) infrastructure[34] to store, track, and automate proteomic data analyses. PRISM is composed of distributed software components that operate cooperatively on several commercially available computer systems that communicate by means of standard network connections. The system not only collects data files directly from all mass spectrometers in our laboratory, but it also manages the storage and tracking of these data files and automates data processing to provide both intermediate results and final products.

The PRISM infrastructure encompasses two major subsystems: the Data Management System (DMS) and the Mass Tag System (MTS). DMS tracks both the analysis results files and maintains the background and data handling information of the analyzed sample (e.g., cell culture and sample processing conditions), experimental factors (e.g., fractionation or chromatographic separations used), and any MS-based parameters. The MTS produces, compiles, and maintains MT tag databases developed in the course of biological studies. Compiled data is segregated and maintained in experiment-specific MT tag databases that are used for the subsequent identification of peptides from FTICR analysis.

20.3.4.1 Data Reduction Processes

An important step in any proteomics data analysis pipeline is to reduce the large volumes of raw data to a usable amount for downstream analysis. Spectra from both ion trap LC-MS/MS and LC-FTICR analyses are involved in significant data-reduction steps. For example, a single normal 9.4 Tesla LC-FTICR experiment results in a 10 gigabyte raw data file, which is reduced to a table of detected masses and elution times approximately 10 megabytes in size, a reduction of three orders of magnitude. LC-MS/MS data reduction is less dramatic but still significant. A typical capillary LC-MS/MS ion trap raw dataset file is about 20 megabytes in size and generally yields a list of predicted peptides in the associated data file that is about 1 megabyte or smaller.

Data reduction for high-performance FTICR, and increasingly for time-of-flight (TOF) instrumentation, involves processing each spectrum to identify peaks contributing to the isotopic distributions of the detected species and then determining the corresponding neutral masses for each of the detections. A typical FTICR spectrum can contain thousands of isotopic distributions that, after processing to extract the peak information, will generate a table of detected masses and their corresponding intensities. This process of converting the isotopic distributions to tables of masses is referred to as mass transformation and deisotoping and is performed in our laboratory by an in-house software program ICR-2LS that utilizes an approach based on the THRASH algorithm.[35]

20.3.4.2 Peptide Identifications Using Accurate Mass and Time Measurements

With the AMT tag approach, an MT database is typically created by means of an extensive series of shotgun LC-MS/MS analyses, performed using a variety of sample fractionation or cell growth and treatment conditions. After peptides are identified from the MS/MS spectra using tools such as SEQUEST, we use an elution normalization algorithm to determine the corresponding LC elution time for each peptide so that all peptides are on the same elution time scale and any small run-to-run variations are corrected (normalized). This step requires that we use a rigorously controlled standard separation to provide high precision elution time information and reduce variation due to run-to-run variability in the LC separation process. The peptide identifications and normalized elution times are then saved for incorporation into the appropriate MT tag database.

The actual identification of peptides in LC-FTICR datasets is accomplished in our laboratory by our in-house developed software program VIPER, whose main objective is to connect these accurate mass and normalized elution time datasets with a previously established MT tag database for comparison and identification. The software performs MS data inspection, visualization, and analysis applications, allowing for rapid display and analysis of large datasets produced by both FTICR and TOF mass spectrometers. VIPER receives a text file created by the ICR-2LS that contains a list of molecular masses and their corresponding intensities and displays the result as a 2D plot with elution time (spectrum number) and molecular mass as the coordinate axes. Figure 20.6 provides an example of a 2D plot generated by VIPER.[36] The abundance of any isotopic distribution can be indicated by using a variable spot size, and color can be used to code abundance ratios and charge states of individual distributions.

Prior to displaying the results in a 2D plot, VIPER first combines the separately identified species (based upon similar molecular mass and elution time) into unique mass classes (UMCs). In separation terms, a UMC represents a single species that elutes as a single chromatographic peak. Each UMC has an accurate mass, central LC normalized elution time (NET), and abundance estimate, computed by summing the intensities of the MS peaks that comprise the UMC. Once characterized, a UMC can be searched against MT tag databases for peptides that match the accurate mass and NET information. After populating a 2D LC-FTICR peptide landscape with UMCs, VIPER allows one to look for groups of related data using data clustering algorithms that detect species having similar monoisotopic masses, elution times, intensities,

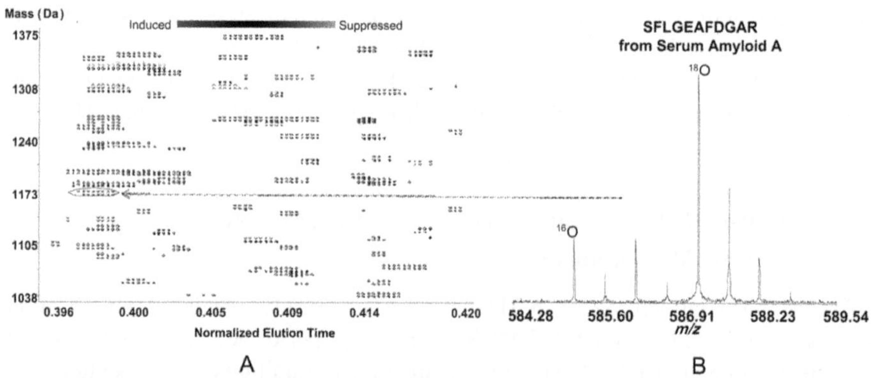

FIGURE 20.6 (A) A 2D plot generated by the software program VIPER showing $^{16}O/^{18}O$ peak intensity pairs using LC-FTICR technology; (B) Example peptide identification of an upregulated protein by comparing normal plasma and plasma from an LPS-administered subject identified by the AMT tag approach and ^{18}O labeling. Reproduced with permission from Ref.[36], copyright 2005 Future Drugs, Ltd.

and so on. VIPER's functionality, together with the UMC visualization and reporting capabilities, makes it an invaluable component in our proteomic data pipeline.

For complex proteomic systems, ambiguities often arise when trying to assign the correct identification to an observed accurate mass and elution time, even though stringent constraints (<5 ppm mass accuracy and ±5% elution time) have been applied. To address this issue, a statistically based algorithm that utilizes a least-squares method is employed to map the NET and mass of each UMC in an LC-FTICR dataset to the NETs and masses in the MT tag database. Figure 20.7 illustrates a situation in which several MT tags within the tolerance of a given UMC (most likely a peptide) result in an ambiguous assignment (i.e., to multiple possible peptides). In this figure, the mass accuracy defines the height, and the elution time reproducibility defines the width of the ellipses, that is, the acceptable error boundaries of the MT tags. In situations where several MT tags match a single feature, one can compute a probability of the most likely match based on the standardized squared distance between a given

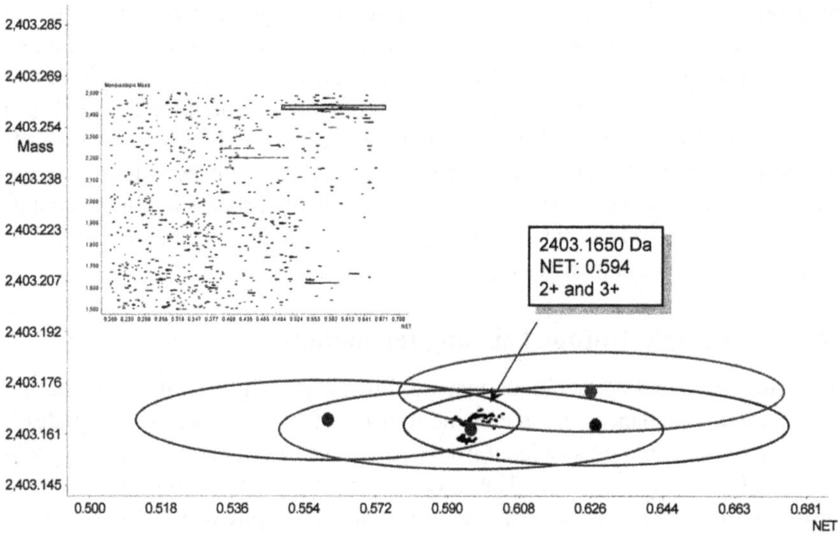

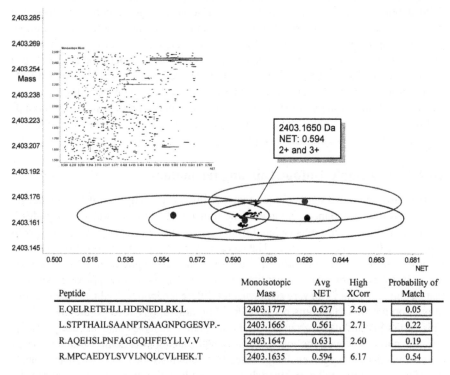

Peptide	Monoisotopic Mass	Avg NET	High XCorr	Probability of Match
E.QELRETEHLLHDENEDLRK.L	2403.1777	0.627	2.50	0.05
L.STPTHAILSAANPTSAAGNPGGESVP.-	2403.1665	0.561	2.71	0.22
R.AQEHSLPNFAGGQHFFEYLLV.V	2403.1647	0.631	2.60	0.19
R.MPCAEDYLSVVLNQLCVLHEK.T	2403.1635	0.594	6.17	0.54

FIGURE 20.7 The inset shows a 2D plot with normalized elution time (NET) on the horizontal axis and mass on the vertical axis. The region highlighted on the inset plot is illustrated by the larger plot. A single detected feature is present at 2403.1650 Da and 0.594 NET. Four MT tags match this feature with the tolerances dictated by the mass accuracy and elution time accuracy of the instrument. The probability of each match was computed using the standardized squared distance between the mass and elution time of the feature and mass and elution time of each MT tag. Reproduced with permission from Ref.[14], copyright 2005 Elsevier.

feature's mass and elution time and each MT tag's mass and elution time, as shown in the probability column in Figure 20.7. In this example, the fourth peptide is the most likely match because both its probability and its SEQUEST XCorr value are higher than those for the other peptides. A statistical foundation for such assignments that more broadly accounts for uncertainties is presently under development.

20.4 QUANTITATION STRATEGIES

Initial proteomic measurements have been mostly qualitative in nature, providing parts lists of proteins with ambiguous quantitative information and limited cotranslational and posttranslational modification data. While these more qualitative proteome measurements can be useful, there is a clear need for better quantitative measurements, which in turn requires a larger number of more accurate proteomic measurements to provide sufficient data for understanding even the simplest of biological systems.

Quantitative strategies often involve comparing protein abundances between two cellular populations that differ as a result of some change or perturbation to one of the populations. For comparative studies that employ stable-isotope labeling, the AMT tag approach can increase throughput and precision by directly comparing two proteomes in the same analysis; for example, comparing a perturbed system to a common reference proteome. A stable-isotope-labeled (e.g., ^{15}N or ^{18}O labeled) reference proteome provides an effective internal standard for many peptides.[37–39] A key advantage of this strategy is that variations due to sample processing (after mixing) and analysis are eliminated, which allows relative abundances to be determined to better than 10 to 20%, in many cases.[37,40]

20.4.1 ^{18}O Stable-Isotope Labeling Techniques

A stable-isotopic labeling technique that we found especially useful for comparing relative protein abundances is via enzyme transfer of ^{18}O from water to peptides.[41] With this technique, proteins isolated from two samples are separately digested with trypsin in either ^{16}O or ^{18}O water. The oxygen atom, either ^{16}O or ^{18}O, from water is incorporated into the newly formed C-terminus in each tryptic peptide, thus providing an isotope tag for relative quantitation. Initial work with this approach indicated that labeling efficiency can vary somewhat for different peptides, leading some peptides to incorporate two ^{18}O atoms per peptide while others only incorporate one ^{18}O.[42,43] This variability and its repercussions for quantitative analyses made its use problematic. Recently, however, the mechanism for ^{18}O transfer was demonstrated to be an enzyme-catalyzed oxygen exchange reaction[44] allowing much more consistent labeling of two ^{18}O atoms to be obtained during a postdigestion trypsin-catalyzed $^{16}O/^{18}O$ exchange reaction.[45]

The advantages of the enzymatic ^{18}O labeling approach include the ability to label all types of samples (e.g., tissues, cells, and biological fluids), the simplicity of the reaction and its specificity for the C-terminus of tryptic peptides, and the identical elution times for both light and heavy isotopic-coded peptides in a pair. The incorporation of two ^{18}O atoms provides a mass difference of 4 Da between the ^{16}O- and ^{18}O-labeled tryptic peptides and is most effective when using a high-resolution mass analyzer such as TOF or FTICR to effectively resolve the ^{16}O- and ^{18}O-labeled peptide pair peaks for quantitative determination of relative abundances.

Recently, our laboratory applied the trypsin-catalyzed $^{16}O/^{18}O$ labeling method and AMT tag approach to quantitative studies that looked at the relative changes in protein concentrations in treated (lipopolysaccaride) versus untreated human blood plasma samples.[46] Initial studies centered on optimization of human plasma analysis and the creation of a comprehensive MT database for the application of high mass accuracy studies.[47,48] Eventually multiple approaches were used, including extensive SCX fractionation, depletion of high-abundant proteins, and isolation of specific peptide subsets prior to LC-MS/MS analysis to obtain the necessary dynamic range in protein detection.[49] More than 600 LC-MS/MS analyses were performed to comprehensively identify peptides for the MT tag database. LC-FTICR analysis was then performed with $^{16}O/^{18}O$ labeled peptide mixtures from plasma. Figure 20.6 shows a partial 2D display of the results obtained from a single LC-FTICR analysis and an example of a peptide pair that illustrated the increase in protein concentration resulting from the lipopolysaccaride treatment.

With the increasing use of $^{16}O/^{18}O$ stable-isotope labeling, our laboratory found it beneficial to create isotope pair visualization software to help discern and identify correct isotopic peak pairs quickly and accurately. Figure 20.8, generated from an extension of the VIPER program, shows two different groups of UMCs that have been identified as an $^{16}O/^{18}O$ isotopic pair. Each group has been graphically displayed

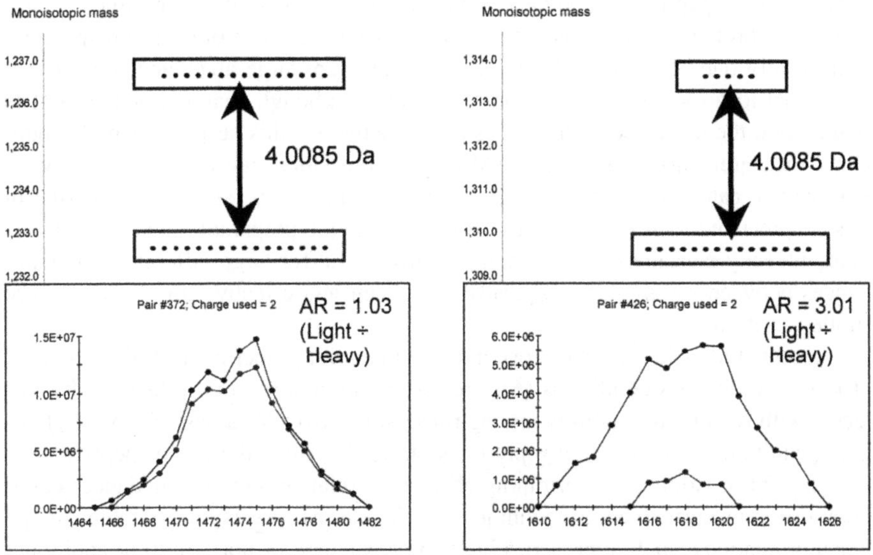

FIGURE 20.8 Graphic representation of two examples of $^{16}O/^{18}O$ isotopic pairs. A new addition to the software program VIPER allows you to graph potential isotopic pairs for a clearer visual picture for faster and more accurate verification of potential pairs. The spots above the charts correspond to the normal 2D display of UMCs by VIPER.

to show a more detailed description of the intensity information. One pair is identified as a near 1:1 ratio, and the second pair is identified as having an apparent abundance difference between the two samples that had been combined into one.

A new quantitative cysteine-peptide enrichment technology (QCET)[42] that enables high-throughput systematic identification and quantitation of proteins expressed in mammalian cells is also currently being applied in our laboratory. QCET combines quantitative $^{16}O/^{18}O$ labeling with specific capture and isolation of cysteine-containing peptides by means of a thiol-specific affinity resin. With QCET, proteins from two cell states or conditions are prepared and separately digested by trypsin under identical conditions, with tryptic peptides from both samples exclusively labeled with either ^{16}O or ^{18}O by immobilized-trypsin. The differentially labeled peptide samples are then combined, selectively enriched for cysteine content, and analyzed using the AMT tag approach for identification and quantification. This technology provides an alternative to the isotope-coded affinity tag approach (ICAT)[40] and can be readily applied to proteome-wide measurements of very small samples. Additional benefits include a reduction in sample complexity due to elimination of typically 80 to 90% of non-cysteine-containing peptides and improved peptide identification (since essentially all recovered peptides should contain a cysteine residue).

20.4.2 "Label-free" Quantitation Using Intensity Information

Comparative measurements based on isotopic labeling generally require that both versions of the peptide (labeled and unlabeled) be detected; however, large changes in relative protein abundances between two labeled samples often result in detection of only one of the peptides (e.g., when there are large abundance changes or low signal-to-noise levels for the measurements). Although quantitation approaches based upon the use of peak intensities are attractive for this purpose, peptide abundance measurements obtained using MS signal intensities may vary significantly for reasons that can include variations in ionization efficiencies (as discussed earlier in this chapter) and losses that occur during both sample preparation and separations. Peak intensity measurements are more readily useful for large differences in abundances between samples, but typically have been less effective for studying more subtle variations.

Recent studies in our laboratory indicate that with proper control of the sample process and analysis conditions (e.g., for electrospray ionization), data can be collected with high reproducibility among runs, which provides a basis for more effective quantitation. Other recent reports also have described the use of peptide peak intensity information for determining changes in relative protein abundances based upon different normalization techniques.[50,51] While less precise than stable-isotope labeling methods, such approaches have the advantage of not requiring additional processing steps to prepare samples with isotopic labels and are broadly applicable, regardless of sample type.

Label-free approaches can complement isotopic labeling approaches. Issues that dictate the applicability of utilizing this approach are primarily related to (1) the run-to-run reproducibility of proteome analyses, (2) the effectiveness of data normalization approaches, similar to those used for microarray data analysis, (3) the linearity of signal response as a function of protein concentration, and (4) other factors that can cause variation in response. We found that the run-to-run reproducibility of proteome analyses improved dramatically with implementation of a fully automated capillary LC-FTICR system.[52] Figure 20.9 shows the variation in intensities for peptides obtained from six replicate analyses of the same *Shewanella oneidensis* proteome sample. The average coefficient of variance is approximately 10% for the highest-abundance peptides and increases to about 40% for low-abundance peptides, illustrating the reproducibility obtainable using automated capillary LC-FTICR. These results support the feasibility of using quantitation approaches based on intensity data and are consistent with several other reports.

20.5 RECENT APPLICATIONS

The human blood plasma proteome, which is immensely complex and has an incredibly large range of relative protein concentrations, is an especially challenging system with regard to detectible dynamic range. More than 99% of the protein content in human plasma is due to only 22 proteins, with the most abundant protein, human serum albumin (HSA), representing at least half of the total content.[53] By utilizing a capillary LC separation system and extensive pre-MS separations coupled to an ion trap mass spectrometer, our laboratory was able to confidently identify >3,500 proteins in human blood plasma samples. The expanded sensitivity and dynamic range

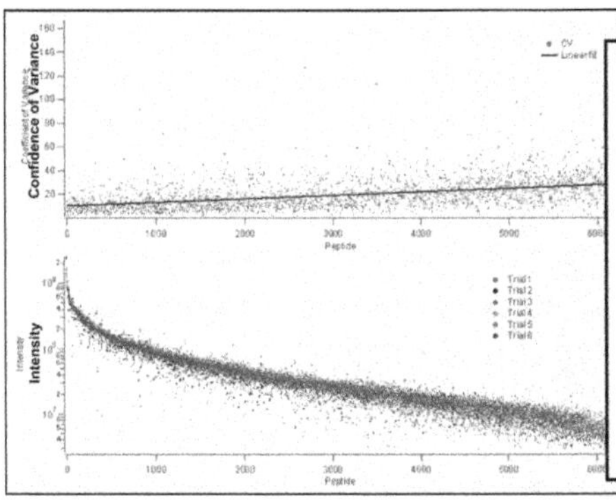

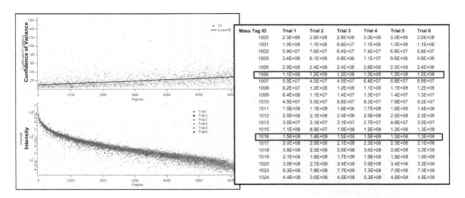

Mass Tag ID	Trial 1	Trial 2	Trial 3	Trial 4	Trial 5	Trial 6
1000	2.3E+08	2.6E+08	2.9E+08	3.0E+08	3.0E+08	3.0E+08
1001	1.0E+08	1.1E+08	9.8E+07	1.1E+08	1.0E+08	1.1E+08
1002	5.9E+07	7.5E+07	6.4E+07	7.4E+07	6.8E+07	6.8E+07
1003	3.4E+06	6.1E+06	6.8E+06	1.1E+07	9.8E+06	9.6E+06
1005	2.0E+08	2.4E+08	2.4E+08	2.6E+08	2.5E+08	2.4E+08
1006	1.1E+08	1.2E+08	1.2E+08	1.3E+08	1.2E+08	1.2E+08
1007	5.9E+07	4.0E+07	4.5E+07	6.4E+07	6.8E+07	6.8E+07
1008	9.2E+07	1.2E+08	1.2E+08	1.1E+08	1.1E+08	1.2E+08
1009	6.4E+06	1.1E+07	1.4E+07	1.3E+07	1.4E+07	1.3E+07
1010	4.5E+07	3.5E+07	6.6E+07	8.3E+07	7.9E+07	8.2E+07
1011	1.5E+08	1.1E+08	1.6E+08	1.7E+08	1.6E+08	1.4E+08
1012	2.5E+08	2.1E+08	2.4E+08	2.5E+08	2.5E+08	2.3E+08
1013	3.0E+07	3.1E+07	3.1E+07	2.7E+07	4.6E+07	3.5E+07
1015	1.1E+08	8.5E+07	1.0E+08	1.2E+08	1.2E+08	1.3E+08
1016	1.5E+09	1.4E+09	1.5E+09	1.5E+09	1.5E+09	1.3E+09
1017	2.0E+08	2.0E+08	2.1E+08	2.3E+08	2.3E+08	2.1E+08
1018	3.5E+08	2.3E+08	3.5E+08	3.6E+08	3.6E+08	3.3E+08
1019	2.1E+08	1.8E+08	1.7E+08	1.9E+08	1.9E+08	1.9E+08
1020	3.5E+08	2.7E+08	3.4E+08	3.6E+08	3.4E+08	3.2E+08
1023	6.3E+08	7.8E+08	7.7E+08	7.3E+08	7.0E+08	7.0E+08
1024	4.4E+08	3.0E+08	4.5E+08	5.3E+08	4.9E+08	4.8E+08

FIGURE 20.9 Left: Reproducibility of absolute abundance values for mass and time tags identified in six replicate capillary LC-FTICR analyses of a *S. oneidensis* tryptically digested proteome sample. The upper plot represents the coefficient of variance in intensity across the six replicates, and the lower plot compares the intensity values seen for each identified AMT tag across the replicates. Right: An excerpt from the raw intensity values is shown on the left. The two boxes illustrate the reproducibility of intensity values for a given AMT tag across the six replicates without normalization between the replicates. Reproduced with permission from Ref.[14], copyright 2005 Elsevier.

afforded by the separations, that is, about nine orders of magnitude in concentration (ranging from <30 pg/ml to ~30 mg/ml), allowed numerous lower-abundant species to be detected. Such a detectable dynamic range is needed for future endeavors aimed at identifying potential biomarkers for diagnostic purposes.

Another system currently under study at our laboratory is the human hepatocyte Huh-7.5 replicon system, which represents a highly permissive cell culture hepatitis C virus (HCV) replication model system. By using a combination of subcellular fractionation and multidimensional LC-MS/MS analysis, both Huh-7.5 (+) and (−) replicon samples were characterized to generate an MT tag database. More than 24,000 peptides were identified, which corresponded to >4,400 confidently identified Huh-7.5 proteins, and 7 of 10 HCV proteins.[54] Within the MS/MS data, significant protein abundance differences were observed between the (+) and (−) HCV samples and correlated to possible alterations in lipid metabolism that resulted from the presence of HCV proteins. The Huh-7.5 MT tag database was then used for preliminary high-sensitivity LC-FTICR analyses to identify proteins from small (<50 μg) human liver biopsy tissues. Greater than 1,500 proteins were identified from a single biopsy sample (see Figure 20.10), which represents the first significant proteomic coverage of such clinical samples. Future

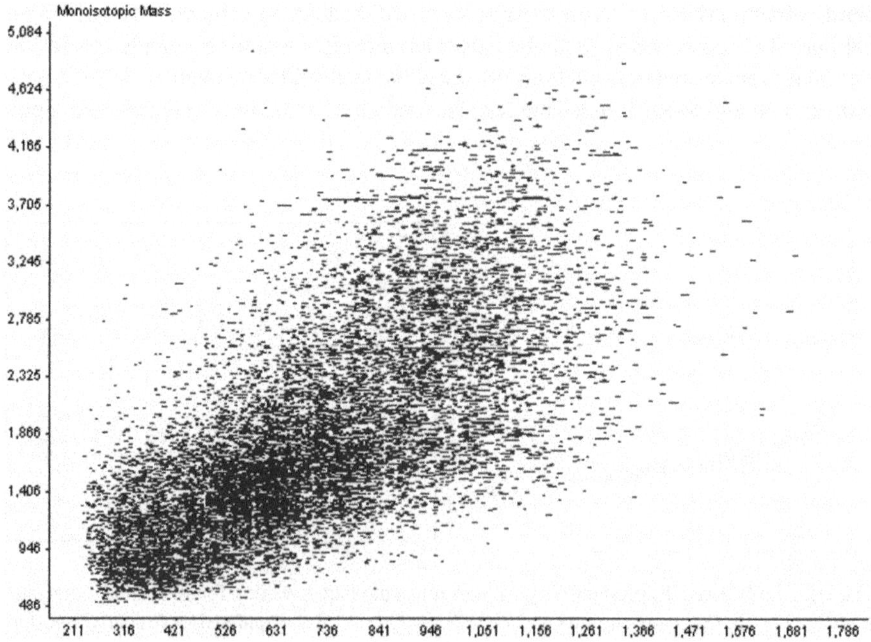

FIGURE 20.10 2D display of detected features from a single LC-FTICR-MS analysis of a tryptically digested liver biopsy sample. The display helps visualize possible peptides that are separated based upon elution time and molecular weight. Greater than 1,500 proteins were identified from correlating these peptides with the mass tag database generated from MS/MS analysis of the hepatocyte Huh-7.5 cell line.

work is now in progress to implement quantitative strategies for comparison of various pathologically differing biopsy samples.

In another study, the previously introduced QCET approach was initially applied to investigate the differential protein expression in human mammary epithelial cells (HMEC) following phorbol 12-myristate 13-acetate (PMA) treatment. An HMEC MT tag database was created by LC-MS/MS analysis following specific enrichment of cysteine-containing peptides from the tryptic digest of cellular lysates and fractionation by SCX chromatography. The MT tag database contained >6,000 identified peptides (covering >3,000 proteins). Subsequent LC-FTICR analysis to compare relative protein abundances between naïve and PMA-treated cells resulted in identification of 1,348 labeled peptide pairs from a single analysis (see Figure 20.11). Among these pairs, 935 were identified as AMT tags, which corresponded to 603

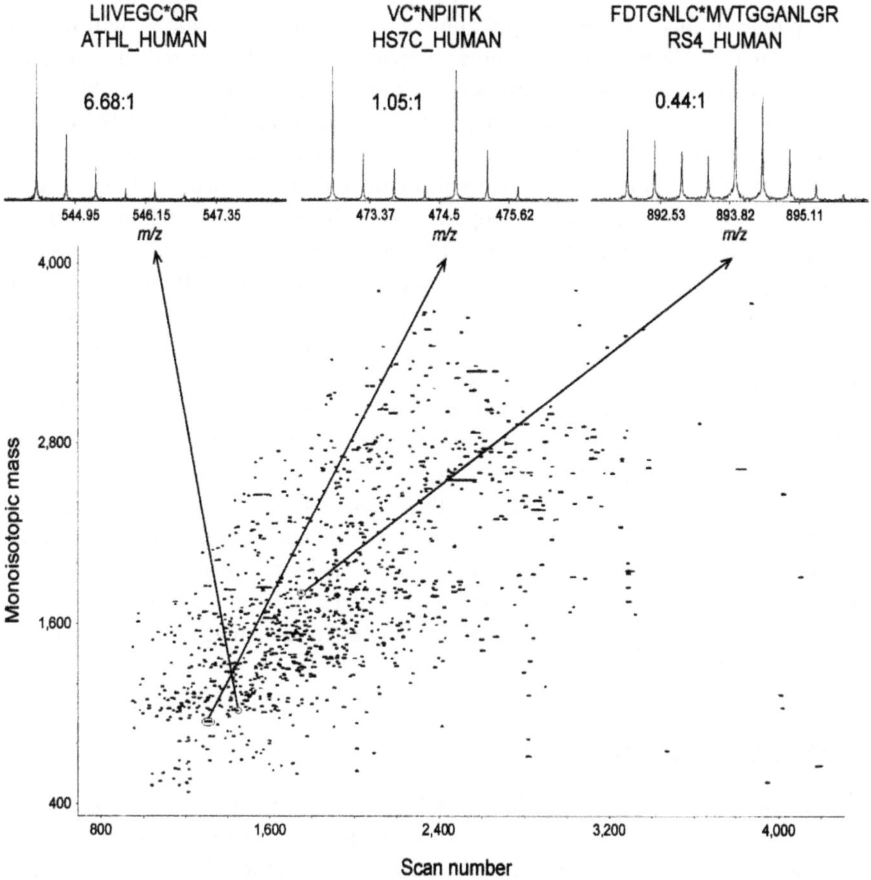

FIGURE 20.11 2D display of QCET isolated peptides from an HMEC sample resulting in 935 pairs identified as unique AMT tags. Insets show three examples of peptide pairs with their sequences, corresponding proteins, and the $^{16}O/^{18}O$ ratios. Reproduced with permission from Ref.[42], copyright 2004 Am. Chem. Soc.

proteins.[42] The reduction in overall sample complexity afforded by the use of this high-efficiency cysteine-containing peptide enrichment technique enables more effective application of the AMT tag approach for characterizing complex mammalian proteome samples. Although a disadvantage of the approach is that the number of peptides detected per protein is generally small, the confidence in protein identifications should still remain high due to the added constraint of identifying only cysteine-containing peptides.[42]

20.6 CONCLUSIONS

The technology platform and analysis pipeline described in this chapter for high-throughput global proteomic measurements has proven to be broadly effective for identifying proteins from both microbial and mammalian systems. The continued development of this high-throughput approach, and related variations, will involve targeting analysis to provide for a larger increase in separation speed (throughput). Also critical is the continued progression of supporting data analysis capabilities to allow effective analysis to extract biological information and extension of these capabilities to both modified peptides and top-down proteomics approaches.

Improvements in the field of proteomics will continue and will be driven largely by enhanced MS instrumentation, computational technologies, and quantitative methodologies. These improvements along with other advances are needed to address the diverse applications of the field. Improved electrospray ionization efficiency, achieved by decreasing the flow rate through the use of smaller i.d. capillaries, is expected to increase both the sensitivity of measurements and provide a basis for improved quantitation. High-resolution LC separations already have demonstrated large improvements in sensitivity, quantitation, and proteome coverage. Future efforts will seek to further improve the dynamic range of measurements, which in turn will lead to significantly expanded proteome coverage as a result of being able to better detect and identify low abundance species. As these capabilities continue to mature, they will push the limits of dynamic range detection, efficiency, and quantitation, while providing faster analyses with more reproducibility and resulting in a parallel increase in data production. Based on its performance to date, accurate mass LC-MS instrumentation promises to meet these future challenges for quantitative high-throughput proteomic studies.

ACKNOWLEDGMENTS

We are grateful to Penny Colton, Matthew Monroe, and Wei-jun Qian for assistance in preparing the materials. We thank the U.S. Department of Energy (DOE) Office of Biological and Environmental Research and the NIH National Center for Research Resources (Grant RR018522) for supporting portions of this research. We also thank the Environmental Molecular Sciences Laboratory, a national scientific user facility sponsored by DOE and located at Pacific Northwest National Laboratory (PNNL), for use of instrumentation. PNNL is operated by Battelle for the DOE under Contract No. DE-AC05–76RLO 1830.

REFERENCES

1. Wilkins, M.R., Williams, K.L., Appel, R.D., and Hochstrasser, D.F., Eds, *Proteome Research: New Frontiers in Functional Genomics*, Springer, Berlin, 1997.
2. Gygi, S.P. et al., Evaluation of two-dimensional gel electrophoresis-based proteome analysis technology, *Proc. Natl. Acad. Sci. USA*, 97, 9390–9395, 2000.
3. Liu, H., Lin, D., and Yates, J.R., III, Multidimensional separations for protein/peptide analysis in the post-genomic era, *BioTechniques* 32, 898–911, 2002.
4. Washburn, M.P., Wolters, D., and Yates, J.R., Large-scale analysis of the yeast proteome by multidimensional protein identification technology, *Nat. Biotechnol.*, 19, 242–247, 2001.
5. Wolters, D.A., Washburn, M.P., and Yates, J.R., An automated multidimensional protein identification technology for shotgun proteomics, *Anal. Chem.*, 73, 5683–5690, 2001.
6. Cox, K.A. et al., Quadrupole ion trap mass-spectrometry: Current applications and future-directions for peptide analysis, *Biol. Mass. Spectrom.*, 21, 226–241, 1992.
7. Huddleston, M.J., Bean, M.F., and Carr, S.A., Collisional fragmentation of glycopeptides by electrospray ionization LC MS and LC MS MS: Methods for selective detection of glycopeptides in protein digests, *Anal. Chem.*, 65, 877–884, 1993.
8. Jonscher, K. et al., Matrix-assisted laser desorption of peptides and proteins on a quadrupole ion trap mass-spectrometer, *Rapid. Commn. Mass Spectrom.*, 7, 20–26, 1993.
9. Loo, J.A., Edmonds, C.G., and Smith, R.D., Primary sequence information from intact proteins by electrospray ionization tandem mass spectrometry, *Science*, 248, 201–204, 1990.
10. Loo, J.A., Edmonds, C.G., and Smith, R.D., Tandem mass-spectrometry of very large molecules: Serum-albumin sequence information from multiply charged ions formed by electrospray ionization, *Anal. Chem.*, 63, 2488–2499, 1991.
11. Smith, R.D. et al., New developments in biochemical mass spectrometry: Electrospray ionization, *Anal. Chem.*, 62, 882–899, 1990.
12. Tomlinson, A.J. and Naylor, S., A strategy for sequencing peptides from dilute mixtures at the low femtomole level using membrane preconcentration-capillary electrophoresis-tandem mass spectrometry (mPC-CE-MS/MS), *J. Liquid Chromatogr.*, 18, 3591–3615, 1995.
13. Eng, J.K., McCormack, A.L., and Yates, J.R., An approach to correlate tandem mass spectral data of peptides with amino acid sequences in a protein database, *J. Am. Soc. Mass Spectrom.*, 5, 976–989, 1994.
14. Jacobs, J.M. et al., Ultra-sensitive, high throughput and quantitative proteomics measurements, *Int. J. Mass Spectrom.*, 240, 195–212, 2005.
15. Smith, R.D. et al., Rapid quantitative measurements of proteomes by Fourier transform ion cyclotron resonance mass spectrometry, *Electrophoresis*, 22, 1652–1668, 2001.
16. Pasa-Tolic, L. et al., Gene expression profiling using advanced mass spectrometric approaches, *J. Mass Spectrom.*, 37, 1185–1198, 2002.
17. Smith, R.D. et al., An accurate mass tag strategy for quantitative and high throughput proteome measurements, *Proteomics*, 2, 513–523, 2002.
18. Speir, J.P. et al., High-resolution tandem mass spectra of 37–67 kDa proteins, *J. Mass Spectrom.*, 30(1), 39–42, 1995.
19. Gorshkov, M.V. et al., Electrospray ionization-fourier transform ion cyclotron resonance mass spectrometry at 11.5 Tesla: Instrument design and initial results, *J. Am. Soc. Mass Spectrom.*, 9, 692–700, 1998.
20. Fenn, J.B. et al., Electrospray ionization-principles and practice, *Mass Spectrom. Rev.*, 9, 37–70, 1990.

21. Cech, N.B. and Enke, C.G., Practical implications of some recent studies in electrospray ionization fundamentals, *Mass Spectrom Rev.*, 20, 362–387, 2001.
22. Tang, K. and A. Gomez, On the structure of an electrostatic spray of monodisperse droplets, *Phys. Fluids*, 6, 2317–2332, 1994.
23. Bruins, A., Mass spectrometry with ion sources operating at atmospheric pressure, *Mass Spectrom. Rev.*, 10, 53–77, 1991.
24. Smith, R.D., Shen, Y., and Tang, K., Ultrasensitive and quantitative analyses from combined separations-mass spectrometry for the characterization of proteomes, *Acc. Chem. Res.*, 37, 269–278, 2004.
25. Wahl, J.H. et al., Attomole level capillary electrophoresis mass-spectrometric protein-analysis using 5-Mu-M-Id Capillaries, *Anal. Chem.*, 64, 3194–3196, 1992.
26. Smith, R.D. et al., Capillary electrophoresis/mass spectrometry, *Anal. Chem.*, 65, A574–A584, 1993.
27. Gale, D.C. and Smith, R.D., Small volume and low flow-rate electrospray ionization mass spectrometry of aqueous samples, *Rapid Commn. Mass Spectrom.*, 7, 1017–1021, 1993.
28. Schmidt, A., Karas, M., and Dülcks, T., Effect of different solution flow rates on analyte ion signals innano-ESI MS, or when does ESI turn into nano-ESI? *J. Amer. Soc. Mass Spectrom.*, 23, 492–500, 2003.
29. Shen, Y. et al., High-efficiency nanoscale liquid chromatography coupled on-line with mass spectrometry using nanoelectrospray ionization for proteomics, *Anal. Chem.*, 74, 4235–4249, 2002.
30. Shen, Y. et al., Nanoscale proteomics, *Anal. Bioanal. Chem.*, 378, 1037–1045, 2004.
31. Shen, Y. et al., High-efficiency on-line solid-phase extraction coupling to 15–150-μm i.d. column liquid chromatography for proteomic analysis, *Anal. Chem.*, 75, 3596–3605, 2003.
32. Belov, M.E. et al., Dynamic range expansion applied to mass spectrometry based on data-dependent selective ion ejection in capillary liquid chromatography fourier transform ion cyclotron resonance for enhanced proteome characterization, *Anal. Chem.*, 73, 5052–5060, 2001.
33. Pasa-Tolic, L. et al., Increased proteome coverage based upon high performance separations and DREAMS FTICR mass spectrometry, *J. Am. Soc. Mass Spectrom.*, 13, 954–963, 2002.
34. Kiebel, G.R. et al., Proteomics Research Information Storage and Management (PRISM) System, document PNNL-14567, http://ncrr.pnl.gov/software/ (last accessed April 2005).
35. Horn, D.M., Zubarev, R.A., and McLafferty, F.W., Automated reduction and interpretation of high resolution electrospray mass spectra of large molecules, *J. Am. Soc. Mass Spectrom.*, 11, 320–332, 2000.
36. Qian, W.J., Camp, D.G., and Smith, R.D., High throughput proteomics using fourier transform ion cyclotron resonance (FTICR) mass spectrometry, *Expert Rev. Proteomics*, 1, 89–97, 2004.
37. Oda, Y. et al., Accurate quantitation of protein expression and site-specific phosphorylation, *Proc. Natl. Acad. Sci. USA*, 96, 6591–6596, 1999.
38. Conrads, T.P. et al., Quantitative analysis of bacterial and mammalian proteomes using a combination of cysteine affinity tags and 15N-metabolic labeling, *Anal. Chem.*, 73, 2132–2139, 2001.
39. Pasa-Tolic, L. et al., High throughput proteome-wide precision measurements of protein expression using mass spectrometry, *J. Am. Chem. Soc.*, 121, 7949–7950, 1999.

40. Gygi, S.P. et al., Quantitative analysis of complex protein mixtures using isotope-coded affinity tags, *Nat. Biotechnol.*, 17, 994–999, 1999.
41. Yao, X. et al., Protelytic 18O labeling for comparative proteomics: Model studies with two serotypes of adenovirus, *Anal. Chem.*, 73, 2836–2842, 2001.
42. Liu, T. et al., High throughput comparative proteome analysis using a quantitative cysteinyl-peptide enrichment technology, *Anal. Chem.*, 76, 5345–5353, 2004.
43. Stewart, I., Thomson, T., and Figeys, D., 18-O labeling: A tool for proteomics, *Rapid Commn. Mass Spectrom.*, 15, 2456–2465, 2001.
44. Yao, X., Afonso, C., and Fenselau, C., Dissection of proteolytic 18O labeling: Endoprotease-catalyzed 16O-to-18O exchange of truncated peptide substrates, *J. Proteome Res.*, 2, 147–152, 2003.
45. Heller, M. et al., Trypsin catalyzed 16O-to-18O exchange for comparative proteomics: Tandem mass spectrometry comparison using MALDI-TOF, ESI-QTOF, and ESI-ion trap mass spectrometers, *J. Am. Soc. Mass Spectrom.*, 14, 704–718, 2003.
46. Qian, W.J., Monroe, M.E., Liu, T., Jacobs, J.M., Anderson, G.A., Shen, Y., Camp, D.G., II, Moore, R.J., Anderson, D.J., Zhang, R., Davis, R.W., Tompkins, R.G., and Smith, R.D., Quantitative proteome analysis of human plasma following *in vivo* lipopolysaccharide administration using $^{16}O/^{18}O$ labeling and the accurate mass and time tag approach, *Mol. Cell. Proteomics,* 4, 700–709, 2005.
47. Qian, W.J. et al., Comparitive proteome analyses of human plasma following lipopolysaccharide treatment using mass spectrometry, *Proteomics*, 5, 572–584, 2005.
48. Shen, Y. et al., High efficiency SCXLC/RPLC/MS/MS for high dynamic range characterization of the human plasma proteome, *Anal. Chem.*, 76, 1134–1144, 2004.
49. Jacobs, J.M., Diamond, D.L., Chan, E.Y., Gritsenko, M.A., Qian, W.J., Stastna, M., Baas, T., Camp, D.G., Carithers, R.L., Smith, R.D., and Katze, M.G., Proteome analysis of liver cells expressing a full-length hepatitis C virus (HCV) replicon and biopsies of post-transplanted liver from HCV-infected patients, *J. Virol.,* 12, 7558–7569, 2005.
50. Chelius, D. and Bondarenko, P.V., Quantitative profiling of proteins in complex mixtures using liquid chromatography and mass spectrometry, *J. Proteome Res.*, 1, 317–323, 2002.
51. Bondarenko, P.V., Chelius, D., and Shaler, T.A., Identification and relative quantitation of protein mixtures by enzymatic digestion followed by capillary reversed-phase liquid chromatography-tandem mass spectrometry, *Anal. Chem.*, 74, 4741–4749, 2002.
52. Belov, M.E. et al., An automated high performance capillary liquid chromatography-Fourier transform ion cyclotron resonance mass spectrometer for high-throughput proteomics, *J. Am. Soc. Mass. Spectrom.*, 15, 212–232, 2004.
53. Anderson, N.L. and N.G. Anderson, The human plasma proteome: History, character, and diagnostic prospects, *Mol. Cell. Proteomics*, 1, 845–867, 2002.
54. Jacobs, J.M., Adkins, J.N., Qian, W.J., Liu, T., Shen, Y., Camp D.G., and Smith, R.D., Utilizing human blood plasma for proteomic biomarker discovery, *J. Proteome Res.,* 4, 1073–1085, 2005.

21 Middle-Out Proteomics: Incorporating Multidimensional Protein Fractionation and Intact Protein Mass Analysis as Elements of a Proteomic Workflow

Scott J. Berger, Kevin M. Millea, Ira S. Krull, and Steven A. Cohen

CONTENTS

21.1 INTRODUCTION

Proteomic analyses have benefited enormously from advances in chromatography and mass spectrometry (MS), which together can provide a vast amount of data from the myriad of components present in typical samples. The daunting challenge in proteomics is to generate useful information from these data regarding the presence and abundance, either in absolute or relative terms, of the potentially thousands of proteins found in a biological system. With the recent shift from genomic to proteomic studies has come the realization that proteomics is far more complicated than the approximately 30,000 genes in the human genome would predict due to the multiplicity of protein structures stemming from a single gene.[1,2] Mammalian proteins are subject to a variety of posttranslational modifications, with over 200 different modifications being reported.[3] These modifications may be combinatorial as is often the case with glycosylation[4,5] and phosphorylation,[6–10] with a multiplicity of structure resulting from heterogeneity or incomplete occupation of one or more of the sites. Amino-terminal protein processing and acetylation produces additional structural variation from genome predicted protein sequences.[11–14] Each of these gene product modifications may have important physiological implications that result in altered biological activity, often as a normal consequence of cellular physiology, including mechanisms for regulating activity, interactions, cellular transport,[15–17] and protein lifetime.[18] Of interest to many proteomic researchers, such changes may also occur as a result of cellular insult or disease processes.[19–23]

Global proteomics studies are further complicated by the large dynamic range of protein concentrations present in a biological system. In a typical cell this range is 10^6, while in biofluids such as serum,[24,28] the dynamic range can be as high as 10.[11] This far exceeds the dynamic range accessible for any current MS detection technique, and hence limits the detectability of low concentration proteins in the presence of more highly abundant ones. In fact, detection of highly abundant proteins is often so ubiquitous that response from low abundance components is completely masked.

21.1.1 Proteomics Strategies

Proteomics analyses fall into two basic categories: methods that analyze the intact proteins (so-called top-down methods), and methods that directly resolve and analyze a digested sample (bottom-up methods). While these approaches each have their advantages and disadvantages, they can provide complementary information with regard to the identity and composition of components within a proteome.

Intact proteins contain valuable information that is obscured or lost upon enzymatic digestion. Closely related structures, such as those arising from transcriptional editing, cotranslational, or posttranslational modification are difficult to distinguish at the peptide level. Alternate RNA splicing may produce two or more proteins from the same RNA transcript, which are distinct, yet will contain most of the same peptides following digestion.[1,29,30] These modified forms can be functionally significant and play a role in a variety of biological processes such as signal transduction, regulation of enzymatic activity, cell growth, apoptosis, and protein localization. Differentiating between a modified and unmodified protein with even a single modification requires detection and characterization of the modified peptide that may be present at a small fraction of the concentration of peptides from the unmodified protein.[2,26,27] Other modifications (e.g., specific proteolytic processing events)[17,31–34] result in truncated structures yielding nearly identical sets of peptides after proteolysis. During typical bottom-up analyses, sequence coverage of identified proteins is low, often from only a few peptides, and the likelihood of identifying these truncated structures is low.

Top-down proteomics methods focus on the resolution of intact proteins for proteomic analysis. A one-dimensional sodium dodecyl sulfate–polyacrylamide gel electrophoresis (1D SDS-PAGE) gel can provide basic molecular weight information (typically $\pm 20\%$) that can be used in conjunction with in-gel digest data to help identify a protein.[35] Apparent masses significantly lower or higher than predicted may be evidence for gross structural changes from the predicted sequence, such as proteolytic processing or covalent associations, with itself, or other proteins.

Modern two-dimensional (2D) gels employ an immobilized pH gradient gel separation (allowing the determination of protein isoelectric point [pI]) in addition to the SDS-PAGE (molecular weight [MW]) separation. The added selectivity provided by the second separation dimension is sufficient to resolve many protein modifications that impact the charged nature of the protein. Such modifications include phosphorylation, glycosylation, lysine acylation, and deamidation. It is expected that the typical mammalian protein will be distributed over several spots. In more dramatic cases, highly phosphorylated and heterogeneously glycosylated proteins may appear as hundreds of spots (termed ladders or trains) across a narrow MW range.[26,36] Even with the expanded selectivity and resolution of a 2D gel, many spots are composed of more than one protein or modified form of a protein.

Identification for gel separated proteins is most commonly accomplished by in-gel digestion of stained protein spots, followed by matrix-assisted laser-desorption ionization (MALDI-MS)[35,37–39] or electrospray ionization tandem MS (ESI MS/MS)[40,41] analysis. The gel approach however does suffer from a number of shortcomings. Sensitivity and dynamic range are limited, good separation reproducibility is often more of an art than an exact science, and both the analysis and identification steps still require significant manual efforts.

More recently, high-resolution MS (e.g., hybrid quadropole-time of flight [QTOF], TOF-TOF, and Fourier transform ion cyclotron resonance [FTICR]) has been applied to tandem MS analysis of intact proteins.[42–48] High mass resolution, improved mass accuracy, and enhanced control of fragmentation conditions have permitted fragmentation of intact protein ions in the gas phase. These researchers have shown that sequence specific fragment ions are produced, and that these provide sufficient information to identify the proteins present. Features such as N- or C-terminal processing as well as the type and location of posttranslational modifications can also be identified using developmental software packages such as ProSight PTM.[49]

Despite these advances in top-down procedures using high-resolution MS, the technique is still in its nascent stage and has not been widely adopted for routine proteomic analysis. Expensive instrument acquisition costs, and the requirement for a high level of operator expertise are inhibitory for most laboratories. Moreover, the approach is not proven suitable for the analysis of complex mixtures typical in proteomics studies, necessitating extensive protein fractionation prior to analysis.

In the past few years, bottom-up approaches have engendered enormous interest, as the power of hyphenated liquid chromatography (LC)-MS/MS systems for peptide analysis has grown, and their applicability for identifying large numbers of proteins from complex peptide mixtures has been demonstrated. Also known as shotgun proteomics, the methods identify proteins by first digesting mixtures of proteins and sequencing the derived peptides, often by data-dependent acquisition of peptide MS/MS spectra (DDA). Most commonly, peptides in the digest mixture are separated by reversed-phase chromatography with on-line MS/MS detection,[50] although more recently MALDI MS/MS[51,52] has been applied to such mixtures following fraction collection or direct deposition of the LC effluent onto a target. With either MALDI or ESI MS/MS, the experimentally obtained spectra are then compared to a database of *in silico* generated spectra to provide identification based on various computerized scoring algorithms for the comparison.

Because of the tremendous complexity of a digested whole cell extract, and incomplete resolution in a single reversed-phase chromatographic step, Wolters et al.[53] developed a method they coined multidimensional protein identification technology (MudPIT) that incorporated two orthogonal chromatographic separations—an ion-exchange step followed by a reversed-phase step—linked to the ion-trap MS/MS analysis. The entire procedure including the 2D LC has been automated to allow perhaps hundreds of protein identifications to be made in a single analysis of a cellular extract digest.

Bottom-up analysis is not altogether foolproof, as current software tools for spectral interpretation and protein identification can produce a high number of false positive and negative protein identifications.[25,54] This error rate is compounded by the fact that most protein identifications from bottom-up techniques are produced from very few, or even one, detected peptides. The complementary nature of MALDI and ESI analysis methods may improve this situation,[52] but achieving comprehensive protein coverage (and characterization) from complex proteomic samples is still an unrealistic goal at this time.

21.1.2 Protein Fractionation

In recognition of the challenge posed by separating the entire mixture of peptides generated from global digestion of a cellular extract or a biofluid, many researchers have chosen to simplify the digest by fractionation at the protein level prior to the proteolytic step.[36,55–58] These approaches have two basic goals in mind. First, with fewer proteins present, the digest mixture will contain significantly fewer peptides than a global digest. Subsequent peptide analyses show better chromatographic resolution due to the reduced number of components present and improve the likelihood of successful protein identification by MS/MS. Also, by dividing the sample into discrete fractions, in comparison to the original sample, proteins are enriched relative to the total mass present in the fraction. This allows a greater amount of a single protein's peptides to be analyzed by the chromatographic step, thus increasing the sensitivity and overall dynamic range of the analysis.

Common fractionation modes include both gel electrophoresis and chromatographic protein separations. Liquid-phase electrophoretic separations, typically based on isoelectric focusing (IEF), have also been shown useful by several researchers.[59–62] Another increasingly popular method uses an initial separation by SDS-PAGE, followed by excision of the stained gel bands gel or simply cutting the gel into a preset number of slices, even without prior staining.[63,64] Gel bands are digested, and the subsequent peptides are analyzed by LC-MS. One consequence of this method is that identified proteins also can be characterized by the approximate molecular weight as calculated by their electrophoretic migration. Although not sufficient to confirm the exact sequence of an identified protein, this information can be useful in determining whether there has been significant protein processing.

Most applicable protein separation modes have been employed for chromatographic fractionation.[36,57,58,65–69] These include size-exclusion chromatography, both anion and cation exchange chromatography, reversed-phase chromatography, and affinity methods. Much recent attention has been focused on affinity methods for fractionating biofluids such as serum and plasma. Because these samples have a few highly abundant proteins such as albumin, transferrin, haptoglobin, and various immunoglobulins, less-abundant species can be difficult to identify and quantify in the presence of the overwhelming excess of the more common proteins. By selectively removing higher concentration species by affinity capture, their interference in detecting low concentration proteins is minimized.[58,70,71]

Several researchers have combined multiple protein separation techniques to increase the fractionation. Examples include coupling SDS-PAGE or liquid-phase IEF with reversed-phase LC,[72,73] ion-exchange followed by reversed-phase LC,[74] and free-flow electrophoresis (FFE) combined with reversed-phase LC.[75] Most of these approaches can still be considered bottom-up proteomics, as the ultimate protein identification relies solely on the information from MS and/or MS/MS analysis of complex peptide mixtures.

21.1.3 Combining Top-Down and Bottom-Up Proteomics

Combining intact protein analysis with peptide level identification forms the basis of a no-compromise approach for comprehensive proteomics. Several laboratories, including our own, have pioneered these approaches. Hayter et al.[76] combined gel-based peptide mass finger print (PMF) analysis with intact mass data gathered from intact protein infusion experiments, on a Q-TOF mass spectrometer, to analyze chicken muscle extracts. Hamler et al.[77] have used fractionation by liquid-phase IEF or chromatofocusing and analyzed collected fractions with a second protein separation step by reversed-phase chromatography. A portion of the LC effluent was analyzed by on-line ESI MS using an orthogonal ESI-TOF instrument, which provided intact MW for eluting proteins, with the bulk of the flow diverted to a fraction collector. Digested fractions were then analyzed by MALDI MS and PMF to identify proteins in the fractions. Finally, VerBerkmoes and colleagues combined the two methodologies using an FTICR MS platform to characterize over 850 proteins from *Shewanella oneidensis* at the peptide and protein level.[78]

More targeted approaches combining top-down and bottom methods have been reported by Ahn and Resing[79,80] for the analysis of ribosomal proteins and histones. The histone study was specifically aimed at deducing a variety of posttranslational modifications such as acetylation, methylation, and phosphorylation. Reversed-phase separations of the intact histones were monitored by triple-quadrupole MS that provided sufficient resolution and mass accuracy to tentatively identify the histone subtype and characterize multiple types of histone modification. Histone identification was confirmed by analysis of collected and subsequently digested fractions by LC-MS/MS.

Nemeth-Cawley et al.[42] used a Q-TOF instrument in a combined protein/peptide approach to study host-related protein impurities in a recombinant protein preparation. Although not a proteomics analysis per se, the use of amino acid sequence tags to search a genomic database and identify unknown proteins is essentially a proteomics-type study. Especially noteworthy was the ability to identify several protein impurities that indicated whether full length or clipped forms were present in particular SEC fractions.

As these studies indicate, the past several years has seen the development and integration of powerful tools for proteomics researchers. Improvements in the quality and automation of complex protein separations, combined with new capabilities for the MS analysis of proteins and peptides, are being applied to ever more complicated proteomics analyses. In the remainder of this chapter, we describe our middle-out approach, combining multidimensional protein LC and on-line protein ESI-TOF MS with fraction collection; followed by digestion and peptide level MS analysis, for the characterization of biological samples of varying complexity.

21.2 METHODS

21.2.1 Ribosomal Protein Analysis

Ribosomal proteins were purified and analyzed using procedures described in our earlier work.[74] Log-phase *Saccharomyces cerevisiae* were harvested, lysed, and subjected to differential centrifugation to enrich ribosomal protein particles. Acid extraction of those particles generated a protein sample (~50 μg) that was resolved in the first dimension of a 2D LC system (described later in the text and in Figure 21.3A) using a Shodex SP 420N 4.6 × 35 mm nonporous cation exchange column (obtained from Waters Corporation, Miford, Mass.) and a nine step (0 to 90% B; 10 % B/step) gradient at a flow rate of 400 μL/min. Eluent A was 50 mM methylamine, 6 M urea, 0.5 mM DTT, adjusted to pH 5.60 with acetic acid, containing 10% (v/v) acetonitrile. Eluent B was Eluent A plus 1 M NaCl. The individual elution steps were focused alternately onto one of two identical reversed-phase columns (Waters Symmetry 300 C_4, 3.5 μm, 2.1 × 50 mm), that were developed with a 20 to 50% B gradient over 18 minutes at 0.5 mL/min, where eluent A was 0.1% TFA in water and eluent B was 0.1% TFA in acetonitrile.

On-line intact protein MS analysis was performed on a Micromass™ LCT ESI-TOF MS (Waters Corporation, Milford, Mass.) equipped with an orthogonal electrospray interface and operated in the positive ion mode. Deconvoluted masses of sample components were determined by a combination of manual and automated application of a maximum entropy (MaxEnt1) deconvolution algorithm included in the MassLynx™ software package (Waters). MaxEnt spectra were generated from the entire acquired *m/z* range to produce deconvoluted neutral mass spectra in the mass range of 3,000 to 46,000 Da, using a bin size of 0.5 Da, a Gaussian damage model with a width of 0.75 Da, and a processing limit of 10 iterations or model convergence. LC/MS data was processed in an automated fashion by use of OpenLynx™ or FractionLynx Software, which automatically selected total ion chromatogram (TIC) peaks using the Peak ApexTrack™ function, combined spectra comprising the approximate full width of each peak, and generated deconvoluted spectra using the MaxEnt1 algorithm.

The collected ribosomal protein fractions, split from the SCX/RP/ESI-TOF MS experiment, were lyophilized and then dissolved in 50 mM ammonium bicarbonate containing 2 mM DTT and 0.1% RapiGest™ SF (Waters). Porcine trypsin (Promega, sequencing grade) was added at 1:50 (w/w) ratio and incubated for 1 hour at 37°C. PMF analyses of each of the digested fractions was performed on a Micromass M@LDI™ R TOF MS equipped with time lag focusing and operated in the positive ion reflector mode. One microliter of digest was applied on the MALDI target plate followed by an equal volume of α-cyano-4-hydroxycinnamic acid (10 mg/ml) in 49.95% methanol, 49.95% acetone and 0.1% TFA. Proteins were identified by the PeptideAuto PMF database-searching module, operating within ProteinLynx Global Server™ v1.1 searched against a nonredundant yeast database.

21.2.2 *Escherichia coli* Analysis

Stationary phase *E. coli* were resuspended in two volumes of 20 mM ammonium bicarbonate and lysed by three cycles of French press (15,000 psi). Cell debris was removed by centrifugation (30,000*g*, 30 minutes), and an S100 cytosolic fraction generated by ultracentrifugation (100,000*g*, 30 minutes, Ti90 rotor, Beckman). Cytosolic proteins were analyzed by a 2D strong anion exchange/reversed-phase (SAX/RP) LC system (described subsequently and in Figure 21.3B). *E. coli* cytosol (1.25 mg) was injected on the first-dimension BioSuite™ Q-PEEK (4.6 × 50 mm) SAX column and eluted with eight 50 mM NaCl steps, followed by two further salt steps up to 1 M NaCl (pH 8). Each of these steps was analyzed in the second-dimension reversed-phase separation using a 10 to 55% acetonitrile gradient containing 0.001% TFA/0.1% formic acid as a modifier. Thirty second fractions were collected during the first nine reversed-phase cycles. ESI-TOF MS analysis conditions were similar to those described above for ribosomal proteins, and FractionLynx software was used to control fraction collection and automatically process spectral data with reference to those collected fractions.

Fractions from the 2D separation were lyophilized, resuspended in 0.5% RapiGest™ SF/50 mM DTT solution in 50mM ammonium bicarbonate, alkylated with 200 mM iodoacetamide, and digested overnight with trypsin (0.01 mg/ml) at 37°C. RapiGest SF was degraded by addition of a 10% TFA and heating at 37°C. The sample was mixed

1:1 with matrix solution (5mg/mL α-cyanno hydroxycinnamic in 50/50 ethanol/ACN; 1% TFA), and 1.5μL was plated onto a stainless steel MALDI target. Digests were analyzed on a Micromass Q-TOF Ultima MALDI (Waters) instrument operating in the V reflecting mode, with a Glu-Fib peptide used as an external lockmass. An initial MALDI PMF spectra was acquired over an *m/z* range of 300–3000, and automated precursor ion selection for MS/MS was enabled for sequencing of peptides in decreasing order of abundance for up to five precursors per well. Spectra were processed with Protein Lynx Global Server and searched against the Swiss Protein database.[81]

21.2.3 AutoME Data Analysis

Automated maximum entropy (AutoME) displays of intact protein datasets were produced using a Visual Basic macro developed by Dr. Ignatius Kass (Waters) functioning within the MassLynx software. LC/MS datasests were time segmented by repeatedly summing a specified number of incremental scans and processing the summed spectra using the MaxEnt 1 deconvolution algorithm. Based on an absolute level of thresholding, a tab delimited data file for the various time segments (retention time, deconvoluted masses, intensity measurements) was created. 2D gel-like displays of this data were accomplished using the bubble plot feature in Microsoft Excel where the x axis was retention time, the y axis was mass, and bubble area represented intensity. Differential overlay plots (as shown later, in Figure 21.9) were generated by producing two bubble plots (solid bubble colors, black backgrounds) on separate layers in Adobe Photoshop and reducing the transparency of the uppermost layer to 50%.

21.3 RESULTS AND DISCUSSION

21.3.1 Overview

The concept of middle-out proteomic analysis relies on the synergy of combining fractionation at the intact protein level with acquisition of both intact mass (top-down) and peptide digest (bottom-up) data from the collected fractions. In our laboratory, we have utilized both single and multiple dimensions of fractionation in a workflow (Figure 21.1) that permits ESI MS of intact proteins, and concurrent collection of fractions for digest analysis. Applying 1D (reversed-phase) or 2D (typically ion-exchange / reversed-phase) chromatography for protein fractionation, roughly 10% of the effluent is diverted to ESI-TOF-MS analysis for intact protein mass determination, while the bulk of the flow is subjected to fraction collection. These fractions are digested and analyzed by a variety of peptide analysis techniques that provided the primary information used for protein identification. These identifications are related to the intact masses corresponding to that fraction, in many cases for confirmation of that identification, but in other cases to determine a deltamass value between the predicted and observed protein masses. The deltamass information can be used to infer protein modification state. Assigned modifications can be tested and confirmed using existing peptide MS/MS data, or through targeted analysis of the digested peptide fractions.

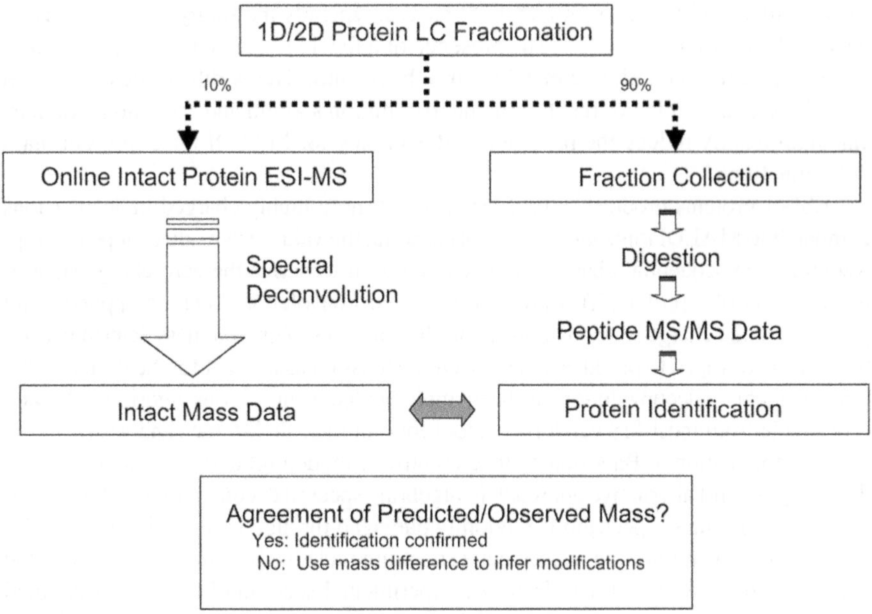

FIGURE 21.1 Workflow for middle-out proteomic analysis.

In the following sections of this chapter, we discuss the rationale for the design and implementation of this workflow and present illustrative examples of how data obtained from this workflow were interpreted.

21.3.2 From the Top Down

21.3.2.1 Mass Analysis of Intact Proteins

In the proteomics context, top down refers to analytical methodologies where samples are characterized at the intact protein level. As stated in the introduction, this level of analysis presents the opportunity to examine a protein *in situ* rather than assembling data about a protein by combining peptide information from a digested sample. Various workflows have been developed to apply the top-down approach for the analysis of complicated protein mixtures. In general, the types of methods fall into two categories: those that simply provide characteristic information about a protein (e.g., gel MW) to assist in identifying the protein or revealing modified forms of a protein, and those methods that by themselves may provide sufficient information to permit precise identification and determination of modification state. MS of intact proteins falls into the latter category, as the intact mass represents the entire history of that component from synthesis until analysis, including all of the processing, post-translational modifications, and sample handling induced changes to that protein.

Intact protein mass determinations can be made using both MALDI and ESI MS. MALDI protein analyses produce spectra typically dominated by peaks corresponding to molecular ions with a low number of charges, typically the singly, doubly, or triply charged forms of a protein.[82–84] The presence of a few low-charge state signals makes manual data interpretation somewhat straightforward. The width of these peaks in MALDI spectra, and poor resolution due to common sodium and potassium adducts, can significantly reduce the precision and accuracy for MALDI intact protein mass determinations.

ESI of proteins produces a broader series of more highly charged molecular ions compared to MALDI ionization. In simple spectra, the values from adjacent peaks (representing two sequential charge states) can be used to derive the zero charge state or neutral mass of a protein. This process, termed deconvolution,[85] can be applied to all pairs within the charge state envelope to produce a series of neutral mass determinations that can be averaged to produce a more accurate mass measurement for the protein. This type of algebraic deconvolution has been implemented using various algorithms[85–89] and is available from most MS vendors as a tool for biomolecule MS data analysis.

The application of Bayesian methods (entropic modeling) to this problem has produced a powerful alternative approach to algebraic spectral deconvolution, where a theoretical neutral mass spectrum is randomly damaged through multiple iterative cycles in an attempt to produce the neutral mass spectrum most likely to have given rise to the analyzed data.[90] When all data from the experimental spectrum have been accounted for by the model it is said that the model has converged on the most likely solution. It has been shown that, with proper implementation, this approach can produce quantitative measurements between runs in addition to qualitative characterization of components within an analysis.[91]

In our work, we have developed several automated workflows (Figure 21.2) for processing protein LC/MS datasets based on the MaxEnt entropic modeling spectral deconvolution approach pioneered by Skilling.[90–92] While the approaches we have undertaken will be generally applicable to laboratories working in this area, we have specifically developed these approaches using application tools found within the MassLynx suite of software.

Efficient LC separations of simple protein mixtures with limited dynamic range (e.g., many protein complexes) produce LC/MS analyses where peak information from the total ion chromatogram can be used to divide the analysis into discrete processing segments. In this case, an open access software tool was applied toward a workflow (Figure 21.2) that identifies the TIC peaks, sums mass spectra on both sides of the peak apex using an average peak width setting, applies MaxEnt1 deconvolution over a wide output mass range, and produces an interactive report file containing the resulting deconvoluted spectra within a viewer tool. Retention time and protein mass information is extracted from these report files for qualitative comparison with peptide MS information, using peak apex times and the range of retention times for each fraction as a cross reference.

As sample complexity and dynamic range of proteins within a sample increases, the TIC information becomes less relevant and other approaches can be employed to segment the data for analysis. For these samples we employ time-based fraction collection and data analysis, without regard to the underlying TIC trace. In one approach, time-based segmentation could be coordinated with physical fraction collection

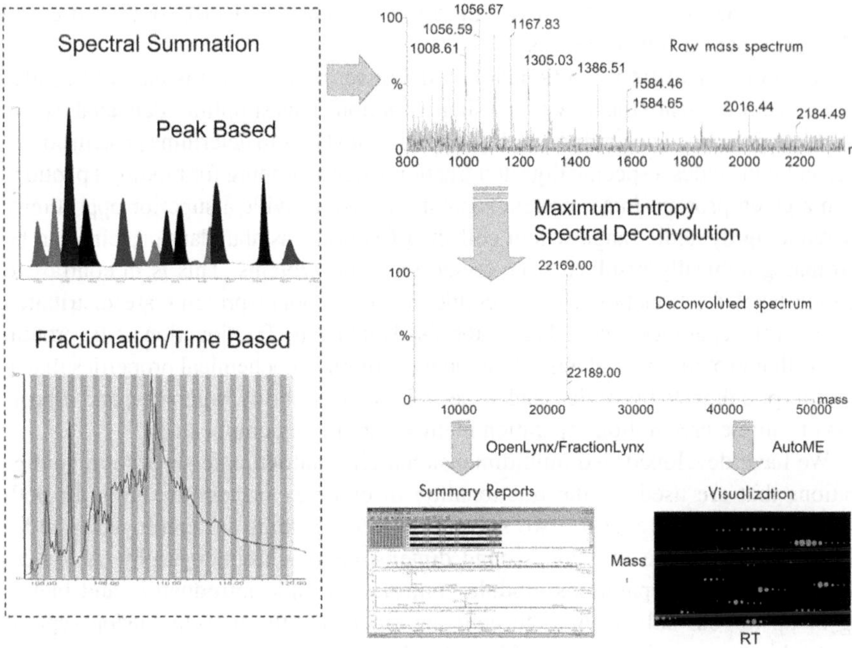

FIGURE 21.2 Workflows for analysis of intact protein LC/MS datasets.

through a software implementation (FractionLynx) that summed and deconvoluted spectra representing components present within a given fraction. This is the main approach that is used for qualitative characterization of a sample, as many proteins are split between two or more collected fractions, and direct relationships can be made with peptide data from corresponding digested fractions.

An alternative approach applies pure time-based segmentation to analyze a dataset without coordination with spectral or fractionation information. This permits spectral analysis of selectable resolution, which can be optimized to preserve chromatographic features of components within the analysis. This was accomplished using the AutoME macro to produce retention time versus mass versus intensity component lists in a format that could be imported into Microsoft Excel and other software tools for analysis and differential display of multiple samples. This data processing mode has been used extensively during methods development to address questions relating to chromatographic performance and reproducibility of the fractionation system. Similar tools have been developed by other groups to automate data analysis of intact protein LC/MS analyses for protein characterization,[89,93] and biomarker identification.[94]

21.3.2.2 Fractionation at the Intact Protein Level

In our application of the middle-out strategy, the 1D or 2D intact protein chromatographic separation serves two main purposes: a fraction of the flow is diverted for intact mass analysis, while the large majority of effluent is collected for digestion and subsequent peptide level analysis.

A major benefit from fractionation at the intact protein level is that all peptides for a given protein are found within a single fraction. This simplifies data analysis by eliminating the need to combine multiple peptide analyses to determine protein coverage, and it provides a specific digested fraction when searching for modified peptides from a given protein. Intact protein separations also provide a superior opportunity for reducing dynamic range within collected fractions, as abundant proteins can be chromatographically resolved from lesser abundant proteins. This is in contrast to peptide level fractionation where peptides from abundant proteins are distributed over the full separation space. The major disadvantage to fractionation at the protein level is that proteins exhibit a much broader set of physicochemical properties than a set of tryptic digested peptides, and issues relating to stability, solubility, and robustness of sample preparation/separation methods are more complicated.

We have developed two multidimensional chromatographic separation configurations that are used for the fractionation of complex protein mixtures. In both configurations (Figure 21.3), two solvent delivery systems are employed to independently service the first and second dimensions of chromatography. A Waters 2796 Alliance Bioseparations module controlled sample introduction and blending of up to four solvent lines to generate eluent for the first-dimension separation. A binary pumping system (1525μ) delivers the second-dimension eluent. In the first configuration (Figure 21.3A), a 10-port 2-position 2D column selection valve directs flow from the first-dimension column through one of two microbore reversed-phase columns used in the second-dimension separation, and on to waste. When the valve is switched, the alternate reversed-phase column is brought in-line with the first-dimension flow, the column formerly in-line with the first dimension is now placed in-line with the second-dimension eluent, and a gradient of increasing organic strength is applied. Since one of the reversed-phase columns is always in-line with the first-dimension column, all components are sampled by both dimensions, and the overall approach is termed a comprehensive 2D separation.

The second-dimension flow proceeds through the reversed-phase column and a second valve (desalting bypass) into a 9:1 split where the bulk of flow is directed to ultraviolet detection/fraction collection, and the rest into the ESI-TOF-MS interface. The desalting bypass valve is critical to on-line protein MS detection. Each time the 2D column selection valve is actuated, the column newly placed in-line with the MS detector contains a bolus of the IEX eluent. Failure to divert nonvolatile components away from the ESI source results in poor electrospray signal and potential blockage of the electrospray emitter. Each time a second-dimension column is switched in-line, the bypass valve is actuated and the reversed-phase gradient composition is maintained at initial conditions while the interfering substances are washed from the column. Middle-out analysis of yeast ribosomal proteins using a 2D (SCX/RP) separation clearly showed that washing steps of only seven column volumes produced spectra where the major identified components were proteins with sodium adducts, while washing with twice that amount produced deconvoluted spectra almost devoid of such adduct peaks.[74]

In a second comprehensive 2D LC configuration (Figure 21.3B), two reversed-phase trap columns replace the two second-dimension analytical columns, and a single

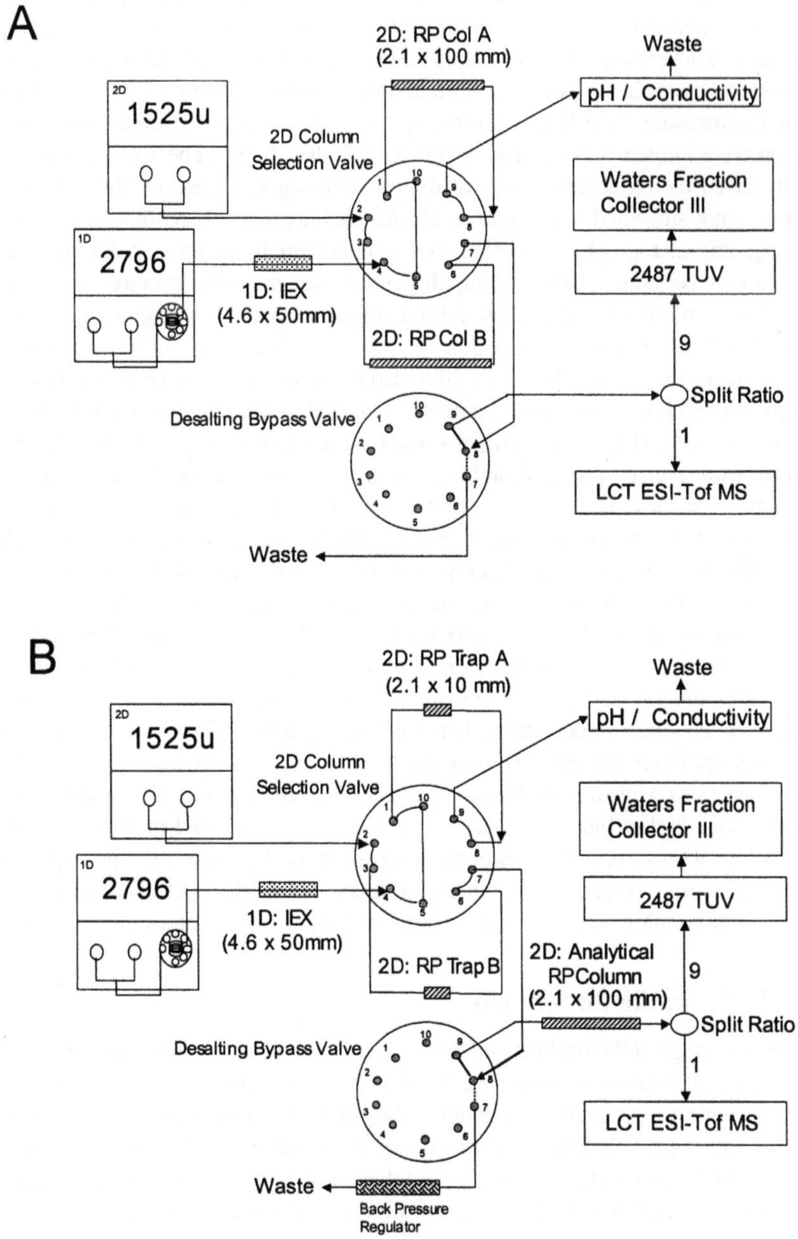

FIGURE 21.3 Comprehensive 2D LC/MS system diagrams representing (A) dual analytical; and (B) dual trap second-dimension column configurations.

reversed-phase analytical column is placed after the desalting bypass valve. To prevent rapid decompression of the system when the desalting valve is actuated, a pressure restrictor comparable to the backpressure of the analytical column is incorporated on the waste line. This configuration produces several advantageous changes to the behavior of the chromatographic system: (1) The first-dimension column no longer is subjected to the backpressure of a longer reversed-phase column, and first-dimension columns with pressure limited packings can be utilized; (2) desalting can be accomplished more rapidly given the reduced volumes of the trap cartridges; (3) the resolving analytical column is not subjected to deleterious eluents such as high pH buffers used in anion-exchange chromatography; and (4) analytes can potentially be refocused at the head of the resolving column, minimizing any dispersion encountered within the salt diversion valve. This configuration does present limitations on the ability to capture more poorly retained hydrophilic components (although this is not significant for most protein analytes), reduces the loading capacity of the second-dimension separation, and introduces additional complexity when developing methods and troubleshooting separations.

Most of our studies to date have involved the use of an ion-exchange first-dimension followed by a reversed-phase second dimension. The orthogonality of the two separation techniques provides high separation selectivity for the 2D separation, permitting more optimal use of the separation space. This selectivity has the additional benefit of separating proteins by their charge nature, enriching and resolving subsets of basic, neutral, and acidic proteins from one another. This likely reduces the magnitude of electrospray suppression effects that would be seen given separations based solely on protein hydrophobicity. Selection of ion-exchange media is limited by the requirements for column and sorbent pressure tolerance. The use of small diameter nonporous ion-exchange particles generates excellent resolution at system pressures as high as 3000 psi during multidimensional operation but exhibits more limited binding capacity than porous sorbents. Large pore (300 Angstrom or larger) sorbents based on methacrylate resin polymeric particles have higher binding capacities but are typically limited to operational pressures below 400 psi. For these lower-pressure columns, the use of the dual trap column configuration may be preferable, but can be made compatible with the dual analytical column configuration when the first-dimension flow is operated at lower flow rates.

21.3.3 From the Bottom Up

The second stage of the middle-out approach is the production and analysis of tryptic digests for fractions collected during the protein fractionation step. Efficient digestion of these fractions is critical to producing robust peptide datasets that can be combined with the intact protein data for total protein characterization. We have adopted the use of an MS compatible surfactant to aid digestion,[95–97] while others have identified additional modifiers and methodologies to improve digestion results.[98–100] Following digestion of the fractions, there are a variety of techniques that can be applied to peptide level (bottom-up) analyses within a middle-out workflow. The choice of analysis methodology is dependent on the overall complexity (number of components and dynamic range) of the sample and the number of fractions collected during the protein level separation.

As depicted in Figure 21.4, both MALDI and ESI MS approaches are capable of analyzing samples over a wide range of complexity. PMF techniques search patterns of peptide masses against theoretical digests of global, genomic, or subgenomic databases.[38,39] PMF approaches are capable of handling simple mixtures of proteins but suffer with mixtures containing multiple components at high dynamic range, as lower-abundance proteins may not have a sufficient number of detected peptides for confident identifications. PMF approaches also fall short for the analysis of very small proteins or some highly basic proteins that do not produce a sufficient number of peptides for an effective database search. With both ESI and MALDI PMF approaches, LC has been shown to provide somewhat superior performance, at the cost of increased fraction processing time.[101]

Approaches that utilize MS/MS techniques can develop protein identifications on spectra acquired for only a single peptide, if that peptide is unique to a given protein. The major limitation typically imposed by MS/MS techniques is that there is usually a limited opportunity to acquire comprehensive MS/MS spectra of all components before sample is exhausted (MALDI and nanospray), or more acutely when analysis is done in conjunction with an on-line separation (e.g., data-dependent LC-MS/MS). A second limitation of MS/MS techniques is the need to have sufficient precursor ion signal to generate detectable signal for many fragment ions. Again, this biases analyses toward more abundant components, but given the broad range of ionization properties of peptides from a given protein, this approach should be capable of identifying a greater number of low-abundance species in a mixture than the equivalent PMF analysis.

In our work to date, we have gravitated toward the use of MALDI PMF analysis for simple samples (e.g., complexes) and toward MALDI Q-TOF MS/MS analysis for analysis of fractions from more complex proteomic samples. This is primarily

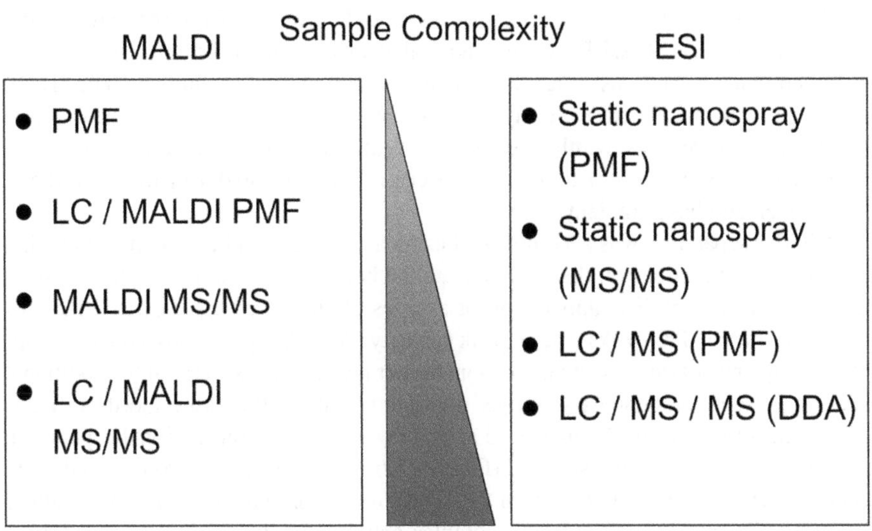

FIGURE 21.4 Typical MALDI and electrospray methodologies employed for the analysis of increasingly complicated peptide mixtures.

based on our ultimate desire to routinely collect a large number of fractions during a run, and automate rapid data acquisition, processing, and analysis of these fractions to the maximum extent possible. While it is true that other techniques can handle samples of greater complexity (and thus fewer fractions need be collected and analyzed), it has proven to our benefit to collect a greater number of fractions in order to minimize the combinations of protein identifications and intact masses that are matched using the middle-out methodology.

21.3.4 From the Middle Out

In this section, the middle-out methodology is illustrated with the analysis of a large protein complex (the yeast ribosome), and an extract of soluble proteins from *E. coli*. In both cases, more detailed examination of the resulting datasets can be obtained from work published elsewhere by our group,[74] or currently submitted for review in another publication (Millea, K.M., et al, unpublished results).

The yeast ribosome is a ribonuclear protein complex composed of 65 to 70 component proteins assembled into two major subunits, each organized on an RNA backbone.[102] Several of these proteins contain more than a single isoform, resulting in approximately 80 unique proteins that are expected within the complex. These proteins are roughly equimolar within the complex, typically smaller than the average-sized yeast protein, and many are among the most highly basic proteins in the cell. These characteristics, along with their high abundance and ease of purification, made the complex a useful tool for middle-out methods development.

Initial experiments focused on evaluating chromatographic systems for resolution of the components. In particular, the benefits and liabilities of multidimensional chromatography versus a single reversed-phase dimension were determined. Figure 21.5 displays the ESI-TOF-MS TICs obtained from a 1D reversed-phase separation and a comprehensive 2D (SCX/RP) separation (dual analysis column configuration) of ribosomal proteins. In each case, the total run time was limited to 200 minutes. The 1D RP separation was insufficient to resolve all ribosomal proteins, and only 53 peaks were observed. Average peak widths (of selected peaks with only one component) were measured at 30 seconds for a total peak capacity of roughly 240 during the utilized part of the reversed-phase gradient.

The 2D separation was accomplished using a nine-step gradient of KCl in the first dimension (dotted lines on the TIC indicate the beginning of the RP gradient associated with each step). The application of a series of shorter reversed-phase gradients produced sharper TIC peaks (useful peak capacity ~700, 107 peaks observed) that represent only one or two ribosomal components per peak. While several of the additional peaks represent components split over multiple salt steps, the vast majority of additional peaks result from the increased selectivity of the multidimensional separation. This improvement can be visualized (Figure 21.6) using the spectra from a single TIC peak (centered at ~87 minutes) from the 1D ribosomal protein analysis. The resulting spectrum (Figure 21.6, lower left) is complex and deconvolutes (inset) to reveal three ribosomal proteins. In the multidimensional analysis, Figure 21.6 shows that two of these components now resolve into distinct peaks (these components are now resolved

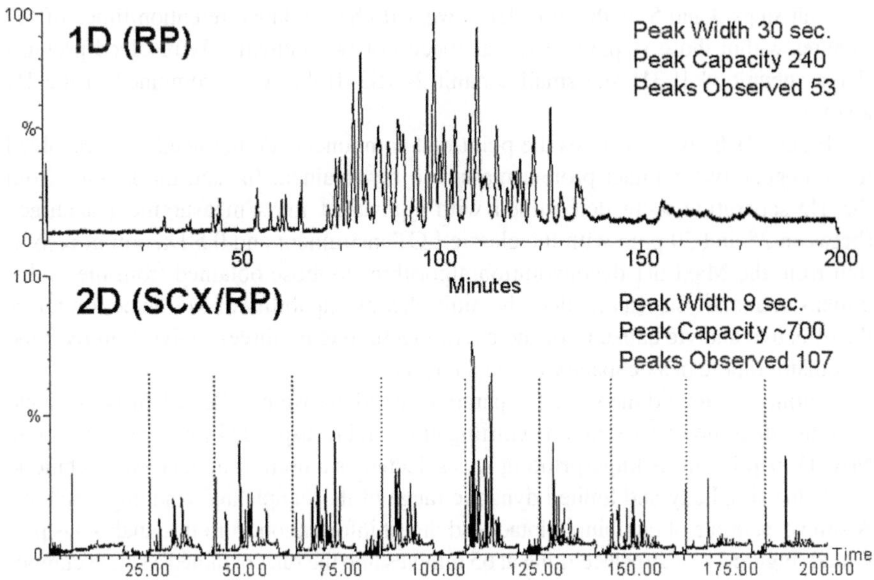

FIGURE 21.5 TIC comparison for yeast ribosomal proteins analyzed by 1D LC/MS (top), and 2D SCX/RP/MS (bottom)

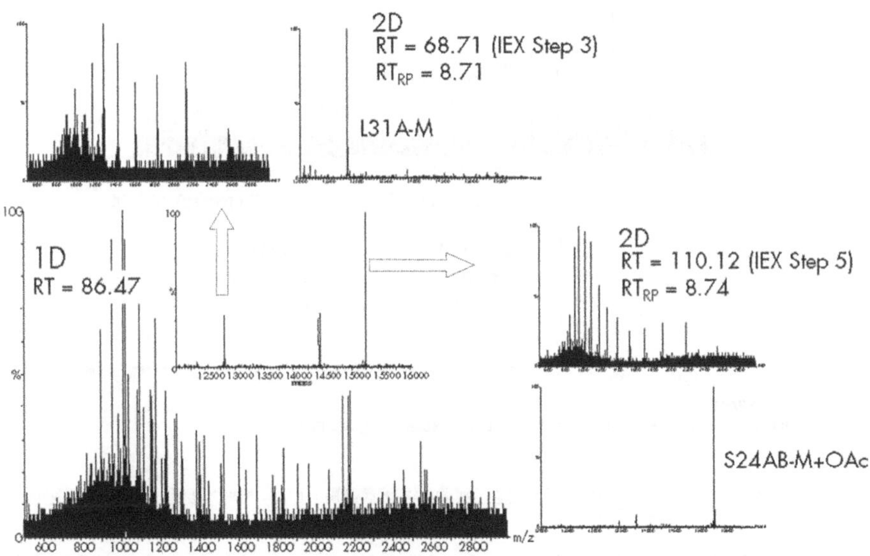

FIGURE 21.6 Example spectra showing how ribosomal protein components that cofractionate during 1D LC/MS analysis (lower left) are resolved with 2D LC/MS analysis (top, and right side spectra). This figure is modified from Liu et al.[74]

into salt steps 3 and 5, with equivalent reversed-phase relative retention times of 8.7 minutes within those steps) that reveal spectra of two individually resolved proteins (large subunit 31 [L31] and small subunit 24 [S24]) that were combined in the 1D analysis.

Figure 21.6 also illustrates the point that components do not need to be resolved to homogeneity for intact protein masses to be obtained. In fact, the masses from the 1D separation were determined with equivalent mass measurement accuracy (between 25 and 50 ppm with the classic LCT instrument and 0.5 Da output resolution from the MaxEnt1 deconvolution algorithm) to those obtained from the multidimensional analysis. In practice, the multiplexing capability of MS detection raises the effective analyte capacity of the overall technique by three- to five-fold over the chromatographic peak capacity of the analysis.

During the multidimensional separation, fractions were collected in two-minute intervals, dried down by vacuum centrifugation, and resuspended for tryptic digestion. MALDI peptide mass fingerprinting was selected as a bottom-up analysis technique due to the simplicity and limited dynamic range of the sample and resulting fractions. A simple example of combining intact and digest information from this analysis is presented (Figure 21.7). A peak eluting at 65 minutes into the run contained a single component of 14,103 Da (Figure 21.7, top). MALDI PMF data on the corresponding digested fraction identified a single isoform of ribosomal protein L26A with sequence coverage of 38%. In this case, the modification state determination is rather straightforward, as

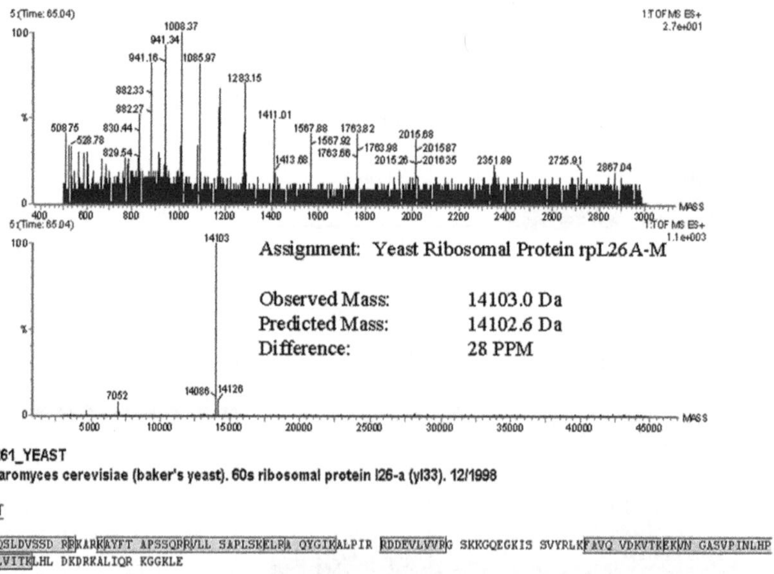

FIGURE 21.7 Combination of intact mass (*m/z* and deconvoluted spectra, top) and peptide mass fingerprint data (bottom) used to characterize ribosomal protein L26A from a 2D LC/MS analysis, combined with MALDI PMF analysis of a single collected fraction. PMF identified peptides are shaded.

simple cotranslational processing, removal of the N-terminal methionine, accounted for the observed mass. Within the peptide data, this N-terminal processing was confirmed by the observation of a N-terminal tryptic peptide with a single missed cleavage. The lack of acetylation on the N-terminal alanine is noteworthy as the terminus falls within the canonical substrates capable of modification by methionine processing and cotranslational N-acetylation.[11–14]

The most universal set of protein modifications, from bacteria to humans, is cotranslational N-terminal processing.[11–14,103,104] As a protein is synthesized on the ribosome, the initiating methionine residue may be cleaved by a methionine aminopeptidase revealing the penultimate amino acid on the N-terminus. Experiments have shown distinctive N-terminal sequence preferences for this enzymatic activity. Slightly later in translation, N-termini that are small and uncharged can be acted upon by N-terminal acetyltransferases to form a peptide linkage between the protein N-terminus and a two carbon acyl group. This combination of processing events results in four likely potential masses for any predicted protein sequence.

In prokaryotes and simple eukaryotes these four potential masses should be sufficient to identify the majority of proteins, while in higher organisms fewer proteins will be amenable to such direct assignments. In practice, it has been found that annotated sequence databases such as the Swiss Protein Database[81] and the deltamass list (http://www.abrf.org/index.cfm/dm.home) of known protein modifications and masses can provide additional help with securing protein assignments based on intact mass. In highly modified proteins, the combinatorial nature of the additional modifications may preclude such direct assignments based on intact mass information alone.

Data from a more complex fraction collected from this same 2D analysis (Figure 21.8) shows a collected fraction that encompassed four TIC peaks (Figure 21.8A). The spectra were summed over the entire collected fraction and deconvoluted to produce a spectrum (Figure 21.8B) containing four apparent peaks that resolved into six total components upon closer analysis. Intact masses for five of these components were consistent with the methionine processed forms of ribosomal proteins L38, L33 (two isoforms), and L31 (two isoforms); and one peak (mass 20,307 Da) that could not be assigned to any yeast protein, even considering the possibility of cotranslational processing. MALDI PMF data (Figure 21.8C) confirmed the four tentative identifications and was able to identify the single peptide that distinguishes the two isoforms of L31 and L33 (shown by arrows in the sequences presented in Figure 21.8C). In these cases, the N-terminal peptide was not identified and the modification information would be lost without the intact protein data. PMF data was not able to confirm the identification of L38 due to the lack of detectable peptides from this small (9 kD) and highly basic (pI ~10.9) protein. However, almost 60% sequence coverage was obtained for another ribosomal protein (L20) not identified directly by intact mass. Searching the observed mass against this sequence with the BioLynx application identified a fragment of L20 lacking the three N-terminal amino acids. This truncated form had not been known before our original report[74] and has been seen with multiple ribosomal protein preparations. Such proteolytic processing is virtually undetectable using peptide analysis methods alone and can play a vital role in regulation of protein function and structure.[31,32,34,105]

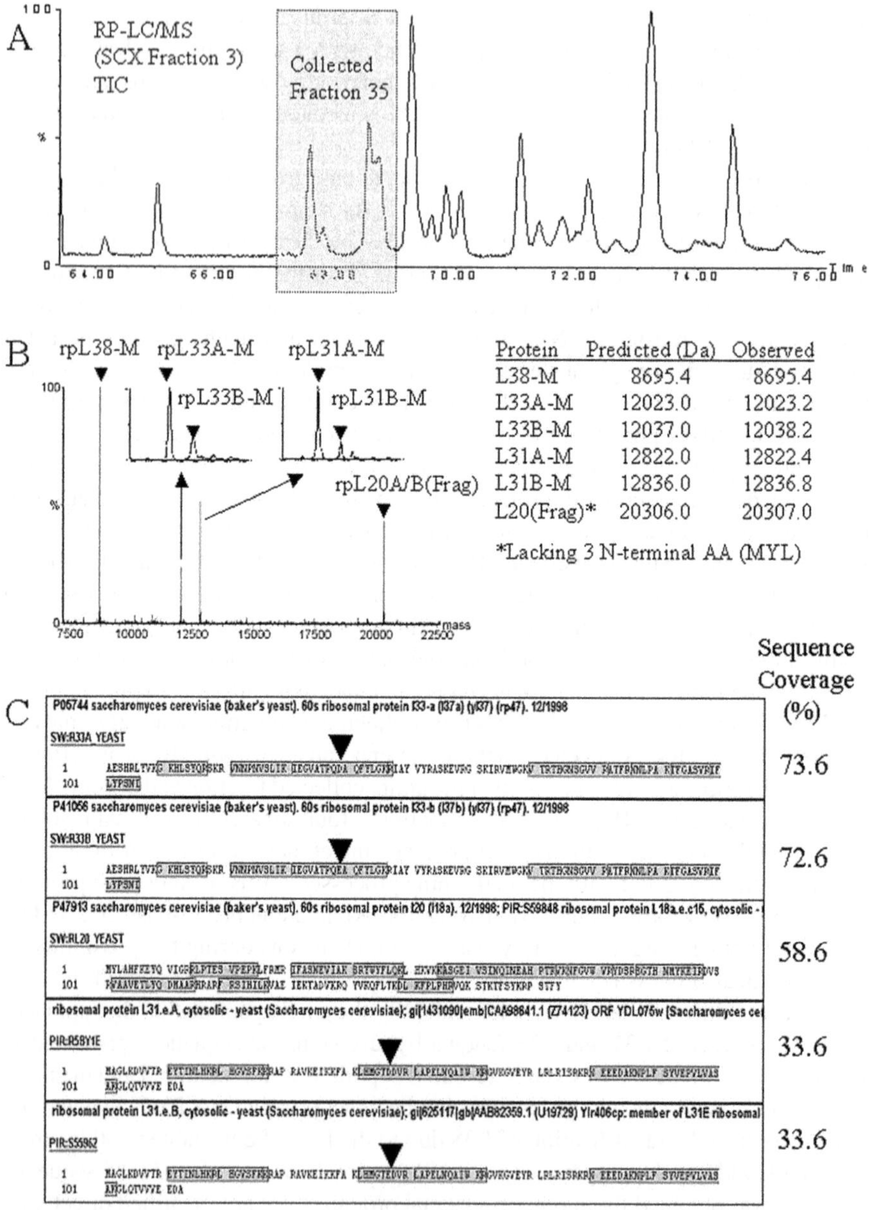

FIGURE 21.8 Combination of intact mass and peptide mass fingerprint data from a complicated 2D LC/MS fraction. Multiple ribosomal proteins were identified by intact mass and MALDI PMF analysis of fraction 35. (A) The TIC information for the fraction; (B) resulting deconvoluted mass spectra of this region (expanded in insert); and (C) PMF data generated from this fraction are displayed.

As was indicated above, the potential analytical capacity of our current approach (700 chromatographic peak capacity multiplied by five-component MS capacity) is on the order of several thousand proteins, whereas the peak capacity of a single reversed-phase separation with MS would be approximately 1200. This was sufficient for ribosomal protein analysis but would be insufficient for a more complicated mixture such as a whole cell cytosolic preparation. Thus the ribosomal study, while eminently suitable for methods development, only showed modest benefit in transitioning from a 1D to 2D LC system, where roughly equivalent detection (~85%) of the known ribosomal protein subunits was obtained using the 1D and 2D fractionation approaches. In an attempt to challenge the analytical capability of the 2D system, we next attempted to analyze samples of much greater complexity and dynamic range.

Cytosol from *E. coli* was used to study the overall reproducibility of the multidimensional protein fractionation and the utility of applying a peptide analysis workflow incorporating MALDI MS/MS data acquisition. Unlike ribosomal analyses, where ESI-TOF MS sensitivity limited the maximum protein mass loaded for the protein fractionation, rather than the column capacity, the cytosolic protein samples are loaded up to the proactical limits of the first-dimension column. The intended effect of utilizing the full system loading capacity for these experiments was to produce an analysis where the maximum number of components (up to 3000 predicted cytosolic proteins) are above MS and MS/MS detection thresholds. Several chromatographic configurations were tested, and the combination of porous strong anion exchange and reversed-phase 2D step chromatography in the dual trap column configuration (Figure 21.3B) was adopted. In this case, trap columns were especially useful in minimizing exposure of the reversed-phase packing to the high pH environment encountered when coupling to anion exchange chromatography. Secondarily, the trap column configuration permitted use of a 4.6 mm porous ion-exchange column, having increased loading capacities up to several mg, but without the robust pressure tolerance of a nonporous column.

Reproducibility studies of a 2D Step (AEX/RP/ESI-TOF-MS) fractionation of *E. coli* cytosol were conducted as described in the Methods section of this chapter. The TIC traces from three consecutive runs are shown in Figure 21.9 (top). The overall TIC patterns, and the region expanded from these TICs (Figure 21.9, middle), show general conservation of gross features between the runs, but microheterogeneities when comparing specific peaks across the runs. An overlay plot (retention time versus mass versus intensity [spot size]) derived from AutoME analysis of the fourth salt step (Figure 21.9, bottom) shows reproducibility of chromatographic elution for components identified from two of the three cytosol runs. In this display, components arising from only the first run are displayed in green, those from the second run in red, and those overlapping the two runs in an orange color. As can be seen from Figure 21.9, the vast majority of components replicate between runs with good conservation of component intensities and elution pattern. The use of chromatographic additives and revised methodologies (e.g., use of a linear gradient in the first dimension) are currently being evaluated as possibilities for enhancing fractionation robustness and quantitative capabilities of the overall technique.[106]

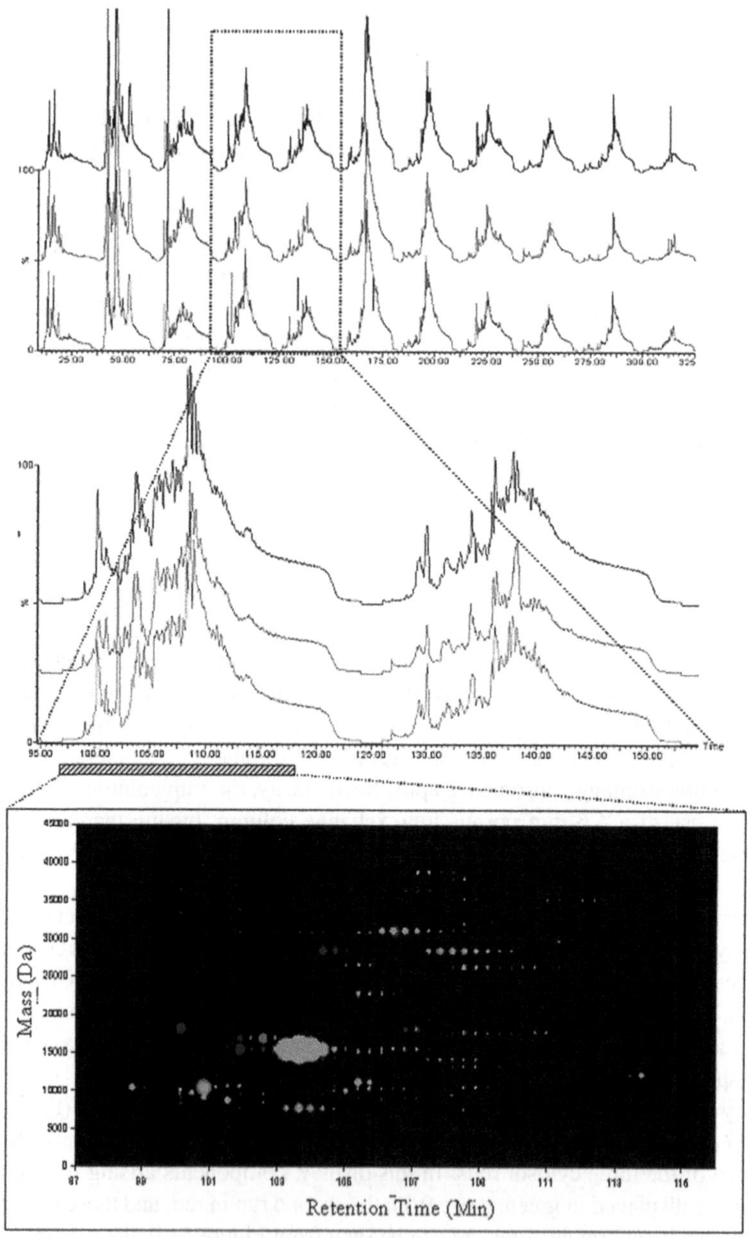

FIGURE 21.9 Reproducibility of three 2D SAX/RP/MS analyses of *E. coli* is demonstrated using the overall TIC data (top), expanded TIC of SAX steps three and four (middle), and the AutoME overlay display (bottom) of step four from two of the three runs. A detailed color view of the plots (viewed on the Web at www. gives a clear indication of mass versus retention time versus intensity (spot size). In the colored plot, component profiles from the first run are displayed in green, the second run in red, and overlapping component distributions in orange.

The use of MALDI MS/MS for characterizing digests of collected fractions has provided analytical capabilities comparable to those shown for PMF analysis of the yeast ribosomal proteins. One of the *E. coli* runs shown in Figure 21.9 was collected into 4 × 96 (384 total) fractions. ESI-TOF-MS (Figure 21.10, lower left) and MALDI Q-Tof MS/MS (Figure 21.10, top) analyses of one of these fractions are shown. The FractionLynx report for this fraction indicated two major components (13002 Da and 18161 Da) were present within this fraction. The MS/MS data were used to identify the YIFE_ECOLI protein (Figure 21.10, top left) with 24 PMF peptide matches (97% sequence coverage) and four peptides sequenced by MS/MS (middle panel indicated in red). The MS/MS spectrum for an indicated peptide (red triangle) is displayed in the lower panel. The corresponding intact mass is consistent with removal of the initiating methionine.

The second major component in this fraction was identified as OSMY_ECOLI with 17 PMF matches (72% sequence coverage) and two MS/MS identifications (Figure 21.10, top right, blue highlighted peptides). Combinations of N-terminal processing were not consistent with the observed mass, which was significantly lower

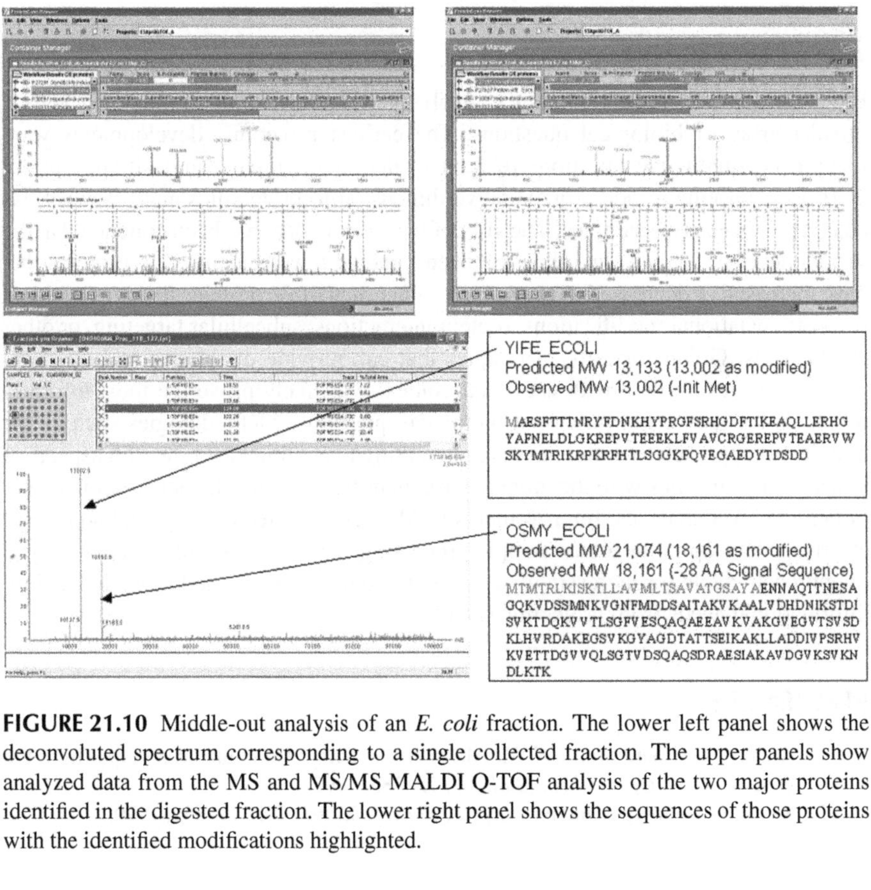

FIGURE 21.10 Middle-out analysis of an *E. coli* fraction. The lower left panel shows the deconvoluted spectrum corresponding to a single collected fraction. The upper panels show analyzed data from the MS and MS/MS MALDI Q-TOF analysis of the two major proteins identified in the digested fraction. The lower right panel shows the sequences of those proteins with the identified modifications highlighted.

than the predicted mass of that protein. The observed mass was searched against the protein sequence, and results were consistent with the observed form of the protein lacking the N-terminal 28 amino acids. A search of the Swiss-Prot database reveals this protein is localized to the *E. coli* periplasmic space and contains a 28 amino acid targeting sequence that is proteolytically processed upon delivery to the compartment. Thus, even with the more complicated sample, a processing event not easily identifiable by intact protein or peptide-based proteomic methodologies alone was readily determined using the combination of both methodologies.

Current and future efforts with the middle-out proteomic methodology are oriented toward defining and improving the comprehensiveness and robustness of the approach. In particular, identifying conditions for stable extraction and chromatographic resolution of membrane proteins will significantly impact the types of biological questions and systems we can address. While the comprehensiveness of the technique has been established with analysis of a complicated protein complex, there are still fundamental questions of analytical penetration into samples of much higher dynamic range. Technological innovations in MS will certainly expand these capabilities, but the importance of developing robust and reproducible sample processing and bioseparations should not be undervalued.

21.4 PERSPECTIVE

Proteomics as a science continues to evolve and attempt to answer an increasingly broader array of biological questions. The earliest proteomic developments were focused on qualitative questions relating to defining sample composition. Current work is attempting to add a quantitative dimension to proteomics and ask what has changed between two samples. The bulk of the work to date has been conducted at the peptide level of analysis and thus has been targeted toward identifying proteins that have increased or decreased in abundance, rather than those proteins that are regulated by posttranslational modifications, protein interactions, subcellular targeting, or other mechanisms. Going forward, techniques that can provide qualitative and quantitative capabilities in the broadest cellular context will likely prove the most informative tools for answering important biological questions. Methodologies such as the middle-out approach to proteomics are attempting to combine the powerful aspects of intact protein analysis with the more precise identification and characterization power of peptide level analyses. Improvements to MS detection technology, and additional efforts to develop more robust sample processing workflows, should make this type of approach viable for increasingly complex biological systems and more demanding informational needs from proteomic samples.

REFERENCES

1. Roberts, G.C. and Smith, C.W., Alternative splicing: Combinatorial output from the genome, *Curr. Opin. Chem. Biol.*, 6, 375–383, 2002.
2. Klose, J., Genotypes and phenotypes, *Electrophoresis* 20, 643–652, 1999.

3. Kuster, B. and Mann, M., Identifying proteins and post-translational modifications by mass spectrometry, *Curr. Opin. Struct. Biol.*, 8, 393–400, 1998.
4. Haltiwanger, R.S. and Lowe, J.B., Role of glycosylation in development, *Annu. Rev. Biochem.*, 73, 491–537, 2004.
5. Schulenberg, B., Beechem, J.M., and Patton, W.F., Mapping glycosylation changes related to cancer using the Multiplexed Proteomics technology: A protein differential display approach, *J. Chromatogr. B Analyt. Technol. Biomed. Life Sci.*, 793, 127–139, 2003.
6. Ahn, N.G. and Resing, K.A., Toward the phosphoproteome, *Nat. Biotechnol.*, 19, 317–318, 2001.
7. Kalume, D.E., Molina, H., and Pandey, A., Tackling the phosphoproteome: Tools and strategies, *Curr. Opin. Chem. Biol.*, 7, 64–69, 2003.
8. Ray, R. and Haystead, T.A., Phosphoproteome analysis in yeast, *Methods Enzymol.*, 366, 95–103, 2003.
9. Resing, K.A. and Ahn, N.G., Protein phosphorylation analysis by electrospray ionization-mass spectrometry, *Methods Enzymol.*, 283, 29–44, 1997.
10. Oda, Y., Nagasu, T., and Chait, B.T., Enrichment analysis of phosphorylated proteins as a tool for probing the phosphoproteome, *Nat. Biotechnol.*, 19, 379–382, 2001.
11. Huang, S., Elliot, R.C., Liu, P.-S., Koduri, R.K., Weickmann, J.L., Lee, J.-H., Blair, L.C., Ghosh-Dastidar, P., Bradshaw, R.A., Bryan, K.M., Einarson, B., Kendall, R.L., Kolacz, K.H., and Saito, K., Specificity of cotranslational amino-terminal processing in yeast, *Biochemistry*, 26, 8242–8246, 1987.
12. Driessen, H.P.C., de Jong, W.W., Tesser, G.I., and Bloemendal, H., The mechanism of N-terminal acetylation of proteins, *CRC Crit. Rev. Biochem.*, 18, 281–325, 1985.
13. Moershell, R.P., Hosokawa, Y., Tsunasawa, S., and Sherman, F., The specificities of yeast methionine aminopeptidase and acetylation of amino-terminal methionine in vivo. *J. Biol. Chem.*, 265, 19,638–19,643, 1990.
14. Tsunasawa, S., Stewart, J.W., and Sherman, F., Amino-terminal processing of mutant forms of yeast iso-1-cytochrome c, *J. Biol. Chem.*, 260, 5382–5391, 1985.
15. Mostov, K.E. and Cardone, M.H., Regulation of protein traffic in polarized epithelial cells, *Bioessays*, 17, 129–138, 1995.
16. Blenis, J. and Resh, M.D., Subcellular localization specified by protein acylation and phosphorylation, *Curr. Opin. Cell. Biol.*, 5, 984–989, 1993.
17. Lively, M.O., Signal peptidases in protein biosynthesis and intracellular transport, *Curr. Opin. Cell. Biol.*, 1, 1188–1193, 1989.
18. Varshavsky, A., The N-end rule pathway of protein degradation, *Genes Cells*, 2, 13–28, 1997.
19. Celis, J.E., Kruhoffer, M., Gromova, I., Frederiksen, C., Ostergaard, M., Thykjaer, T., Gromov, P., Yu, J., Palsdottir, H., Magnusson, N., and Orntoft, T.F., Gene expression profiling: Monitoring transcription and translation products using DNA microarrays and proteomics, *FEBS Lett.*, 480, 2–16, 2000.
20. Chambers, G., Lawrie, L., Cash, P., and Murray, G.I., Proteomics: A new approach to the study of disease, *J. Pathol.*, 192, 280–288, 2000.
21. Corthals, G.L. and Nelson, P.S., Large-scale proteomics and its future impact on medicine, *Pharmacogenomics J.*, 1, 15–19, 2001.
22. Rosenblatt, K.P., Bryant-Greenwood, P., Killian, J.K., Mehta, A., Geho, D., Espina, V., Petricoin, E.F., III, and Liotta, L.A., Serum proteomics in cancer diagnosis and management, *Annu. Rev. Med.*, 55, 97–112, 2004.

23. Marko-Varga, G. and Fehniger, T.E., Proteomics and disease: The challenges for technology and discovery, *J. Proteome Res.*, 3, 167–178, 2004.
24. Corthals, G.L., Wasinger, V.C., Hochstrasser, D.F., and Sanchez, J.C., The dynamic range of protein expression: A challenge for proteomic research, *Electrophoresis*, 21, 1104–1115, 2000.
25. Patterson, S.D., How much of the proteome do we see with discovery based proteomics, and how much do we need to see? *Curr. Proteomics*, 1, 3–12, 2004.
26. Wilkins, M.R., Sanchez, J.C., Williams, K.L., and Hochstrasser, D.F., Current challenges and future applications for protein maps and post-translational vector maps in proteome projects, *Electrophoresis*, 17, 830–838, 1996.
27. Gygi, S.P., Corthals, G.L., Zhang, Y., Rochon, Y., and Aebersold, R., Evaluation of two-dimensional gel electrophoresis-based proteome analysis technology, *Proc. Natl. Acad. Sci. U S A*, 97, 9390–9395, 2000.
28. Anderson, N.L. and Anderson, N.G., The human plasma proteome: history, character, and diagnostic prospects, *Mol. Cell Proteomics*, 1, 845–867, 2002.
29. Lopez, A.J., Alternative splicing of pre-mRNA: Developmental consequences and mechanisms of regulation, *Annu. Rev. Genet.*, 32, 279–305, 1998.
30. Black, D.L., Mechanisms of alternative pre-messenger RNA splicing, *Annu. Rev. Biochem.*, 72, 291–336, 2003.
31. Ehrman, M., and Clausen, T., Proteolysis as a regulatory mechanism, *Annu. Rev. Genet.*, 38, 709–724, 2004.
32. Gottesman, S., Proteolysis in bacterial regulatory circuits, *Annu. Rev. Cell Dev. Biol.*, 19, 565–587, 2003.
33. Link, A.J., Robison, K., and Church, G.M., Comparing the predicted and observed properties of proteins encoded in the genome of *Escherichia coli* K-12, *Electrophoresis*, 18, 1259–1313, 1997.
34. Randall, L.L., Josefsson, L.G., and Hardy, S.J., Processing of exported proteins in *Escherichia coli*, *Biochem. Soc. Trans.*, 8, 413–415, 1980.
35. Mann, M., Hojrup, P., and Roepstorff, P., Use of mass spectrometric molecular weight information to identify proteins in sequence databases, *Biol. Mass Spectrom.*, 22, 338–345, 1993.
36. Pieper, R., Gatlin, C.L., Makusky, A.J., Russo, P.S., Schatz, C.R., Miller, S.S., Su, Q., McGrath, A.M., Estock, M.A., Parmar, P.P., Zhao, M., Huang, S.T., Zhou, J., Wang, F., Esquer-Blasco, R., Anderson, N.L., Taylor, J., and Steiner, S., The human serum proteome: Display of nearly 3700 chromatographically separated protein spots on two-dimensional electrophoresis gels and identification of 325 distinct proteins, *Proteomics*, 3, 1345–1364, 2003.
37. Henzel, W.J., Billeci, T.M., Stults, J.T., Wong, S.C., Grimley, C., and Watanabe, C., Identifying proteins from two-dimensional gels by molecular mass searching of peptide fragments in protein sequence databases, *Proc. Natl. Acad. Sci. U S A*, 90, 5011–5015, 1993.
38. James, P., Quadroni, M., Carafoli, E., and Gonnet, G., Protein identification by mass profile fingerprinting, *Biochem. Biophys. Res. Commn.*, 195, 58–64, 1993.
39. Pappin, D.J., Hojrup, P., and Bleasby, A.J., Rapid identification of proteins by peptide-mass fingerprinting, *Curr. Biol.*, 3, 327–332, 1993.
40. Eng, J.K., McCormack, A.L., and Yates, J.R., III, An approach to correlate tandem mass spectral data of peptides with amino acid sequences in a protein database, *J. Am. Soc. Mass Spectrom.*, 5, 976–989, 1994.
41. Mann, M. and Wilm, M., Error-tolerant identification of peptides in sequence databases by peptide sequence tags, *Anal. Chem.*, 66, 4390–4399, 1994.

42. Nemeth-Cawley, J.F., Tangarone, B.S., and Rouse, J.C., "Top Down" characterization is a complementary technique to peptide sequencing for identifying protein species in complex mixtures, *J. Proteome Res.*, 2(5), 495–505, 2003.
43. Kjeldsen, F., Haselmann, K.F., Budnik, B.A., Sorensen, E.S., and Zubarev, R.A., Complete characterization of posttranslational modification sites in the bovine milk protein PP3 by tandem mass spectrometry with electron capture dissociation as the last stage, *Anal. Chem.*, 75, 2355–2361, 2003.
44. Suckau, D. and Resemann, A., T3-sequencing: Targeted characterization of the N- and C-termini of undigested proteins by mass spectrometry, *Anal. Chem.*, 75, 5817–5824, 2003.
45. Reid, G.E. and McLuckey, S.A., 'Top down' protein characterization via tandem mass spectrometry, *J. Mass Spectrom.*, 37, 663–675, 2002.
46. Kelleher, N.L., Taylor, S.V., Grannis, D., Kinsland, C., Chiu, H.J., Begley, T.P., and McLafferty, F.W., Efficient sequence analysis of the six gene products (7–74 kDa) from the *Escherichia coli* thiamin biosynthetic operon by tandem high-resolution mass spectrometry, *Protein Sci.*, 7, 1796–1801, 1998.
47. Meng, F., Cargile, B.J., Patrie, S.M., Johnson, J.R., McLoughlin, S.M., and Kelleher, N.L., Processing complex mixtures of intact proteins for direct analysis by mass spectrometry, *Anal. Chem.*, 74, 2923–2929, 2002.
48. Meng, F., Forbes, A.J., Miller, L.M., and Kelleher, N.L., Detection and localization of protein modifications by high resolution tandem mass spectrometry, *Mass Spectrom. Rev.*, 24, 126–134, 2005.
49. LeDuc, R.D., Taylor, G.K., Kim, Y.B., Januszyk, T.E., Bynum, L.H., Sola, J.V., Garavelli, J.S., and Kelleher, N.L., ProSight PTM: An integrated environment for protein identification and characterization by top-down mass spectrometry, *Nucleic Acids Res.*, 32 (Web Server issue), W340–345, 2004.
50. Link, A.J., Eng, J., Schieltz, D.M., Carmack, E., Mize, G.J., Morris, D.R., Garvik, B.M., Yates, J.R., III, Direct analysis of protein complexes using mass spectrometry, *Nat. Biotechnol.*, 17, 676–682, 1999.
51. Rejtar, T., Hu, P., Juhasz, P., Campbell, J.M., Vestal, M.L., Preisler, J., and Karger, B.L., Off-line coupling of high-resolution capillary electrophoresis to MALDI-TOF and TOF/TOF MS, *J. Proteome Res.*, 1, 171–179, 2002.
52. Stapels, M.D. and Barofsky, D.F., Complementary use of MALDI and ESI for the HPLC-MS/MS analysis of DNA-binding proteins, *Anal. Chem.*, 76, 5423–5430, 2004.
53. Wolters, D.A., Washburn, M.P., Yates, J.R., III, An automated multidimensional protein identification technology for shotgun proteomics, *Anal. Chem.*, 73, 5683–5690, 2001.
54. Resing, K.A., Meyer-Arendt, K., Mendoza, A.M., Aveline-Wolf, L.D., Jonscher, K.R., Pierce, K.G., Old, W.M., Cheung, H.T., Russell, S., Wattawa, J.L., Goehle, G.R., Knight, R.D., and Ahn, N.G., Improving reproducibility and sensitivity in identifying human proteins by shotgun proteomics, *Anal. Chem.*, 76, 3556–3568, 2004.
55. Righetti, P.G., Castagna, A., Herbert, B., Reymond, F., and Rossier, J.S., Prefractionation techniques in proteome analysis, *Proteomics*, 3, 1397–1407, 2003.
56. Millea, K.M. and Krull, I.S., Subproteomics in analytical chemistry: Chromatrographic fractionation techniques in the characterization of proteins and peptides, *J. Liquid Chromatogr. Rel. Technol.*, 26, 2195–2224, 2003.
57. Wang, H. and Hanash, S., Multi-dimensional liquid phase based separations in proteomics, *J. Chromatogr. B Analyt. Technol. Biomed. Life Sci.*, 787, 11–18, 2003.

58. Wang, H., and Hanash, S., Intact-protein based sample preparation strategies for proteome analysis in combination with mass spectrometry, *Mass Spectrom. Rev.*, 24, 413–426, 2005.
59. Weber, G. and Bocek, P., Recent developments in preparative free flow isoelectric focusing, *Electrophoresis*, 19, 1649–1653, 1998.
60. Tan, A., Pashkova, A., Zang, L., Foret, F., and Karger, B.L., A miniaturized multi-chamber solution isoelectric focusing device for separation of protein digests, *Electrophoresis*, 23, 3599–3607, 2002.
61. Zuo, X. and Speicher, D.W., Microscale solution isoelectrofocusing: A sample pre-fractionation method for comprehensive proteome analysis, *Methods Mol. Biol.*, 244, 361–375, 2004.
62. Zuo, X., Echan, L., Hembach, P., Tang, H.Y., Speicher, K.D., Santoli, D., and Speicher, D.W., Towards global analysis of mammalian proteomes using sample prefractionation prior to narrow pH range two-dimensional gels and using one-dimensional gels for insoluble and large proteins, *Electrophoresis*, 22, 1603–1615, 2001.
63. Gavin, A.C., Bosche, M., Krause, R., Grandi, P., Marzioch, M., Bauer, A., Schultz, J., Rick, J.M., Michon, A.M., Cruciat, C.M., Remor, M., Hofert, C., Schelder, M., Brajenovic, M., Ruffner, H., Merino, A., Klein, K., Hudak, M., Dickson, D., Rudi, T., Gnau, V., Bauch, A., Bastuck, S., Huhse, B., Leutwein, C., Heurtier, M.A., Copley, R.R., Edelmann, A., Querfurth, E., Rybin, V., Drewes, G., Raida, M., Bouwmeester, T., Bork, P., Seraphin, B., Kuster, B., Neubauer, G., and Superti-Furga, G., Functional organization of the yeast proteome by systematic analysis of protein complexes, *Nature*, 415, 141–147, 2002.
64. Schirle, M., Heurtier, M.A., and Kuster, B., Profiling core proteomes of human cell lines by one-dimensional PAGE and liquid chromatography-tandem mass spectrometry, *Mol. Cell Proteomics*, 2, 1297–1305, 2003.
65. Badock, V., Steinhusen, U., Bommert, K., and Otto, A., Prefractionation of protein samples for proteome analysis using reversed-phase high-performance liquid chromatography, *Electrophoresis*, 22, 2856–2864, 2001.
66. Bauer, A. and Kuster, B., Affinity purification-mass spectrometry: Powerful tools for the characterization of protein complexes, *Eur. J. Biochem.*, 270, 570–578, 2003.
67. Shefcheck, K., Yao, X., and Fenselau, C., Fractionation of cytosolic proteins on an immobilized heparin column, *Anal. Chem.*, 75, 1691–1698, 2003.
68. Wissing, J., Heim, S., Flohe, L., Bilitewski, U., and Frank, R., Enrichment of hydrophobic proteins via Triton X-114 phase partitioning and hydroxyapatite column chromatography for mass spectrometry, *Electrophoresis*, 21, 2589–2593, 2000.
69. Butt, A., Davison, M.D., Smith, G.J., Young, J.A., Gaskell, S.J., Oliver, S.G., and Beynon, R.J., Chromatographic separations as a prelude to two-dimensional electrophoresis in proteomics analysis, *Proteomics*, 1, 42–53, 2001.
70. Pieper, R., Su, Q., Gatlin, C.L., Huang, S.T., Anderson, N.L., and Steiner, S., Multi-component immunoaffinity subtraction chromatography: An innovative step towards a comprehensive survey of the human plasma proteome, *Proteomics*, 3, 422–432, 2003.
71. Hinerfeld, D., Innamorati, D., Pirro, J., and Tam, S.W., Serum/plasma depletion with chicken immunoglobulin y antibodies for proteomic analysis from multiple mammalian species, *J. Biomolec. Technol.*, 15, 184–190, 2004.
72. Wall, D.B., Kachman, M.T., Gong, S.S., Parus, S.J., Long, M.W., and Lubman, D.M., Isoelectric focusing nonporous silica reversed-phase high-performance liquid chromatography/electrospray ionization time-of-flight mass spectrometry: A three-dimensional

liquid-phase protein separation method as applied to the human erythroleukemia cell-line, *Rapid Commn. Mass Spectrom.*, 15, 1649–1661, 2001.

73. Wall, D.B., Kachman, M.T., Gong, S., Hinderer, R., Parus, S., Misek, D.E., Hanash, S.M., and Lubman, D.M., Isoelectric focusing nonporous RP HPLC: A two-dimensional liquid-phase separation method for mapping of cellular proteins with identification using MALDI-TOF mass spectrometry, *Anal. Chem.*, 72, 1099–1111, 2000.
74. Liu, H., Berger, S.J., Chakraborty, A.B., Plumb, R.S., and Cohen, S.A., Multidimensional chromatography coupled to electrospray ionization time-of-flight mass spectrometry as an alternative to two-dimensional gels for the identification and analysis of complex mixtures of intact proteins, *J. Chromatogr. B Analyt. Technol. Biomed Life Sci.*, 782, 267–289, 2002.
75. Moritz, R.L., Ji, H., Schutz, F., Connolly, L.M., Kapp, E.A., Speed, T.P., and Simpson, R.J., A proteome strategy for fractionating proteins and peptides using continuous free-flow electrophoresis coupled off-line to reversed-phase high-performance liquid chromatography, *Anal. Chem.*, 76, 4811–4824, 2004.
76. Hayter, J.R., Robertson, D.H., Gaskell, S.J., and Beynon, R.J., Proteome analysis of intact proteins in complex mixtures, *Mol. Cell Proteomics*, 2, 85–95, 2003.
77. Hamler, R.L., Zhu, K., Buchanan, N.S., Kreunin, P., Kachman, M.T., Miller, F.R., and Lubman, D.M., A two-dimensional liquid-phase separation method coupled with mass spectrometry for proteomic studies of breast cancer and biomarker identification, *Proteomics*, 4, 562–577, 2004.
78. VerBerkmoes, N.C., Bundy, J.L., Hauser, L., Asano, K.G., Razumovskaya, J., Larimer, F., Hettich, R.L., and Stephenson, J.L., Jr., Integrating "top-down" and "bottom-up" mass spectrometric approaches for proteomic analysis of *Shewanella oneidensis*, *J. Proteome Res.*, 1, 239–252, 2002.
79. Louie, D.F., Resing, K.A., Lewis, T.S., and Ahn, N.G., Mass spectrometric analysis of 40 S ribosomal proteins from Rat-1 fibroblasts, *J. Biol. Chem.*, 271, 28,189–28,198, 1996.
80. Galasinski, S.C., Resing, K.A., and Ahn, N.G., Protein mass analysis of histones, *Methods*, 31, 3–11, 2003.
81. Boeckmann, B., Bairoch, A., Apweiler, R., Blatter, M.C., Estreicher, A., Gasteiger, E., Martin, M.J., Michoud, K., O'Donovan, C., Phan, I., Pilbout, S., and Schneider, M., The SWISS-PROT protein knowledgebase and its supplement TrEMBL in 2003, *Nucleic Acids Res.*, 31, 365–370, 2003.
82. Karas, M. and Hillenkamp, F. Laser desorption ionization of proteins with molecular masses exceeding 10,000 daltons, *Anal. Chem.*, 60, 2299–2301, 1988.
83. Tanaka, K., Waki, H., Ido, Y., S., A., and Yoshida, Y., Protein and polymer analysis up to m/z 100,000 by laser ionization time-of-flight mass spectrometry, *Rapid Commn. Mass Spectrom.*, 2, 151–153, 1988.
84. Hillenkamp, F., Karas, M., Beavis, R.C., and Chait, B.T., Matrix-assisted laser desorption/ionization mass spectrometry of biopolymers, *Anal. Chem.*, 63, 1193A–1203A, 1991.
85. Fenn, J.B., Mann, M., Meng, C.K., Wong, S.F., and Whitehouse, C.M., Electrospray ionization for mass spectrometry of large biomolecules, *Science*, 246, 64–71, 1989.
86. Horn, D.M., Zubarev, R.A., and McLafferty, F.W., Automated reduction and interpretation of high resolution electrospray mass spectra of large molecules, *J. Am. Soc. Mass Spectrom.*, 11, 320–332, 2000.
87. Zhang, Z. and Marshall, A.G., A universal algorithm for fast and automated charge state deconvolution of electrospray mass-to-charge ratio spectra, *J. Am. Soc. Mass Spectrom.*, 9, 225–233, 1998.

88. Zheng, H., Ojha, P.C., McClean, S., Black, N.D., Hughes, J.G., and Shaw, C., Heuristic charge assignment for deconvolution of electrospray ionization mass spectra, *Rapid Commn. Mass Spectrom.*, 17, 429–436, 2003.
89. Pearcy, J.O. and Lee, T.D., MoWeD, a computer program to rapidly deconvolute low resolution electrospray liquid chromatography/mass spectrometry runs to determine component molecular weights, *J. Am. Soc. Mass Spectrom.*, 12, 599–606, 2001.
90. Skilling, J., *Classic Maximum Entropy*, Kluwer Academic, Norwell, Mass., 1989, pp. 45–52.
91. Ferrige, A.G., Seddon, M.J., Green, B.N., Jarvis, S.A., and Skilling, J., Disentangling electrospray spectra with maximum entropy, *Rapid Commn. Mass Spectrom.*, 6, 707–711, 1992.
92. Ferrige, A.G., Seddon, M.J., Skilling, J., and Ordsmith, N., The application of MaxEnt to high resolution mass spectrometry, *Rapid Commn. Mass Spectrom.*, 6, 765–770, 1992.
93. Lee, S.W., Berger, S.J., Martinovic, S., Pasa-Tolic, L., Anderson, G.A., Shen, Y., Zhao, R., and Smith, R.D., Direct mass spectrometric analysis of intact proteins of the yeast large ribosomal subunit using capillary LC/FTICR, *Proc. Natl. Acad. Sci. U S A*, 99, 5942–5947, 2002.
94. Williams, T.L., Leopold, P., and Musser, S., Automated postprocessing of electrospray LC/MS data for profiling protein expression in bacteria, *Anal. Chem.*, 74, 5807–5813, 2002.
95. Arnold, R.J., Hrncirova, P., Annaiah, K., and Novotny, M.V., Fast proteolytic digestion coupled with organelle enrichment for proteomic analysis of rat liver, *J. Proteome Res.*, 3, 653–657, 2004.
96. Suder, P., Bierczynska, A., Konig, S., and Silberring, J., Acid-labile surfactant assists in-solution digestion of proteins resistant to enzymatic attack, *Rapid Commn. Mass Spectrom.*, 18, 822–824, 2004.
97. Yu, Y.Q., Gilar, M., Lee, P.J., Bouvier, E.S., and Gebler, J.C., Enzyme-friendly, mass spectrometry-compatible surfactant for in-solution enzymatic digestion of proteins, *Anal. Chem.*, 75, 6023–6028, 2003.
98. Hixson, K.K., Rodriguez, N., Camp, D.G., 2nd, Strittmatter, E.F., Lipton, M.S., and Smith, R.D., Evaluation of enzymatic digestion and liquid chromatography-mass spectrometry peptide mapping of the integral membrane protein bacteriorhodopsin, *Electrophoresis*, 23, 3224–3232, 2002.
99. Li, N., Shaw, A.R., Zhang, N., Mak, A., and Li, L., Lipid raft proteomics: Analysis of in-solution digest of sodium dodecyl sulfate-solubilized lipid raft proteins by liquid chromatography-matrix-assisted laser desorption/ionization tandem mass spectrometry, *Proteomics*, 4, 3156–3166, 2004.
100. Russell, W.K., Park, Z.Y., and Russell, D.H., Proteolysis in mixed organic-aqueous solvent systems: Applications for peptide mass mapping using mass spectrometry, *Anal. Chem.*, 73, 2682–2685, 2001.
101. Wall, D.B., Berger, S.J., Finch, J.W., Cohen, S.A., Richardson, K., Chapman, R., Drabble, D., Brown, J., and Gostick, D., Continuous sample deposition from reversed-phase liquid chromatography to tracks on a matrix-assisted laser desorption/ionization precoated target for the analysis of protein digests, *Electrophoresis*, 23, 3193–3204, 2002.
102. Planta, R.J. and Mager, W.H., The list of cytoplasmic ribosomal proteins of *Saccharomyces cerevisiae*, *Yeast*, 14, 471–477, 1998.
103. Arnold, R.J. and Reilly, J.P., Observation of *Escherichia coli* ribosomal proteins and their posttranslational modifications by mass spectrometry, *Anal. Biochem.*, 269, 105–112, 1999.

104. Arnold, R.J., Polevoda, B., Reilly, J.P., and Sherman, F., The action of N-terminal acetyltransferases on yeast ribosomal proteins, *J. Biol. Chem.*, 274, 37,035–37,040, 1999.
105. Neurath, H. and Walsh, K.A., Role of proteolytic enzymes in biological regulation (a review), *Proc Natl Acad Sci U S A*, *73*, 3825–3832, 1976.
106. Millea, K.M., Kass, I.J., Cohen, S.A., Krull, I.S., Gebler, J.C., and Berger, S.J., Evaluation of multidimensional (ion-exchange/reversed-phase) protein separations using linear and step gradients in the first dimension, *J. Chromatogr. A.*, 1079, 287–298, 2005.

22 Polymeric Monolithic Capillary Columns in Proteomics

Alexander R. Ivanov

CONTENTS

22.1 INTRODUCTION AND OVERVIEW

Monolithic stationary phases are separation media for liquid-phase chromatography and provide an alternative to stationary phases based on microparticles. First attempts to make porous monolithic separation media and overcome limitations of granulate stationary phases including relatively large interparticular void volume and slow diffusional mass transfer within the pores of microbeads were made almost 40 years ago.[1–3] Although the first attempts of making the monolithic columns were not very successful, another wave of interest to new approaches toward monoliths came in the late 1980s and grew in the 1990s. Commercialization of several high-efficiency separation products based on monolithic media occurred in the late 1990s and early 2000s.[4–9] The monolithic chromatographic bed consists of a single-piece rigid porous structure determining a network of macropores and micropores without any interstitial void volume.[10,11] Because of the absence of interparticular voids, all the mobile phase is forced to flow through the porous media, which leads to a substantial enhancement of the mass transfer. Such convective mass transport results in a significant increase in the separation efficiency and the speed of separation, especially for high molecular weight molecules, in contrast to slow diffusional mass transfer mechanism in the pores of granular stationary phases.[12–14]

Monolithic columns can be divided into two main groups: polymeric monoliths and silica-based sol-gel monoliths. In this chapter, we describe some applications of polymeric monolithic columns in proteomic research. Polymeric rigid monoliths molded *in situ* were introduced by Svec et al. in 1992.[11] Polymeric monolithic columns are rigid organic polymer-based porous rods that can be prepared from polyacrylates,[11] polymethacrylates,[5,15] polyacrylamides,[16] and polystyrenes.[4,14,17] The porous structure is resulted from phase separation, which occurs during the free radical polymerization when appropriate amounts of inert porogenic solvents are present.[4,18] Typically, the mold (tube, capillary, channel, etc.) is filled with a mixture of monomer(s), cross-linker, porogenic solvent(s), and free radical initiator. Then, the mold is sealed, and polymerization occurs at carefully controlled conditions.[18,19] The seals then are replaced with fittings if necessary. The columns are flushed from remaining components of polymerization mixture with the stream of compressed gas and (or) suitable solvents and grafted with chemicals, altering chromatographic surface with specific functional groups if necessary. Then, the monoliths are ready to use.

Several approaches for synthesis of polymeric monoliths combining well-controlled porosity with appropriate surface chemistry in which the reaction mixture is placed in a mold, followed by ultraviolet (UV)- or thermally-induced *in situ* polymerization, have been published.[4,5,14,17,20–23] The chemistry of the column and surface of separation media is controlled by the choice of monomer, so columns can be produced to target specific separation applications. Also, the surface functionalities can be altered by heat- or UV-induced chemical grafting.[21–23] UV-transparent molds (typically polytetrafluoroethylene [PTFE] coated capillaries) are usually used for the light-induced polymerization and grafting.[22,23] Fast polymerization and potential for synthesis of monoliths with different functionalities in the same mold are among the most important advantages of UV-initiated polymerization over its thermally induced alternative.

The monolithic material can be shaped differently for specific applications. First monolithic columns were commercialized in the 1990s in the form of short disks. Typically, the monolith is prepared in a mold and cut into disks several millimeters thick. A disk is mounted in a special cartridge with adaptors for mobile-phase flow. The short flow path and low pressure are the advantages of the discs.[18,19] Thus, a number of cartridges can be mounted in the system, allowing separation based on column functionalities. Unlike disks, monolithic polymer columns are polymerized *in situ* in a fused silica, plastic, or stainless steel tube. Monolithic polymeric separation media was also employed in annular and conical shapes.[18] Companies such as BIA Separations (Ljubljana, Slovenia), LC Packings / Dionex (Amsterdam, The Netherlands), ISCO (Lincoln, Nebr.), Sepragen (San Leonardo, Calif.) and BioRad (Richmond, Calif.)[19] are current manufacturers of monliths in different formats.

Monolithic polymeric stationary phases are being actively investigated as an alternative to particulate columns in a variety of applications in capillary nanoflow liquid chromatography (LC),[4,24–26] capillary electrochromatography (CEC),[5,21,27–29] immunoaffinity chromatography,[30,31] chiral chromatography,[32] ion-exchange chromatography,[33] solid-phase extraction,[34,35] and microchip separations.[23,36,37] The summarized advantages of such phases include simplicity of preparation, avoidance of frits and column packing procedures, robustness, high porosity and permeability, absence of silanols causing unwanted secondary interactions with analytes, and high chemical stability over a wide pH range. The monolithic media have properties of fast mass transfer and low pressure resistance, enabling rapid separations. The monoliths can be shaped for specific applications. The surface of the monolith can be chemically modified to perform chromatographic separations in different modes.

The era of proteomics has made a nano and capillary format of liquid-phase separation methods extremely popular mainly due to their compatibility with mass spectrometry (MS). The concept of monolithic separation media is particularly favorable for the fabrication of capillary columns.[28,37–39] Nanoflow LC, using monolithic columns, offers the advantages of high resolution, high mass sensitivity, and low sample and mobile phase consumption.[4,39] Monolithic nanocapillary columns can be a good alternative to microparticle-packed columns because of their relative ease of manufacture, without the need of high pressure, and their high-performance characteristics. High efficiencies of the narrow bore capillary columns are found, due to decreased flow dispersion and a homogeneous packing bed structure. Highly sensitive detection is also achievable when using ultra narrow-bore liquid chromatography–electrospray ionization–mass spectrometry (LC-ESI-MS) due to the fact that ESI is a primarily concentration-sensitive technique over a wide range of flow rates. Relative to packed columns, porous monolithic nanocapillary columns also result in potentially less clogging of the ESI tip and do not require frits[39] due to covalent immobilization of the monolith to the wall of a capillary, high chemical stability, and absence either of loose particles in the column bed or frits.

Polymeric monolithic materials offer excellent biocompatibility, are mechanically stable, stable over a wide pH range, and can also be cleaned, without damage, with caustic mobile phases. Described characteristics of monoliths suggest that nanoflow monolithic columns are highly promising for proteomics and other applications. Increasing dependence in proteomics on reversed-phase liquid chromatography (RPLC) for the separation of complex peptide mixtures also should be noted. The current need in proteomics for even higher resolution, sensitivity, and throughput is a reason for the great interest of the research community in advancing RPLC of peptides using monoliths.[40]

The focus of this chapter is to present several applications of monolithic polymeric capillary columns in proteomics and provide examples of pressure- and electric-field driven and two-dimensional (2D) chromatography using monolithic capillary columns coupled to electrospray MS in proteomic applications.

22.2 HIGH-EFFICIENCY PEPTIDE ANALYSIS ON MONOLITHIC MULTIMODE CAPILLARY COLUMNS

The rapid acceleration of research in the area of proteomics requires the development of new, efficient, robust, and high-throughput separation technologies. Capillary electrochromatography (CEC) was brought into a focus as an alternative separation technique due to its vastly enhanced efficiency and peak capacity compared with traditional liquid-phase separation techniques. Electrochromatography was introduced in 1974 by Pretorius et. al[41] and rapidly developed in the late 1990s and early 2000s. The major innovation in electrochromatography is the complete or at least a partial replacement of the pressure-driven flow of the mobile phase typical for LC with an electroosmotic flow.[42] This electroosmotic flow is generated by applying a high voltage across the column. The combination of high efficiencies given by electrically driven flow with the peptide and protein structure identification of MS makes a coupling of these two methods a powerful bioanalytical tool for the proteomic research.

Recently, we presented high-efficiency peptide analysis using multimode pressure-assisted CEC/CE (pCEC/pCE) monolithic polymeric columns.[21] We demonstrated the separation of model peptide mixtures and protein digests by isocratic and gradient elution under an applied electric field with UV and ESI-MS detection. Capillary multipurpose columns were prepared in silanized fused-silica capillaries of 50, 75, and 100 μm i.d. by thermally induced *in situ* copolymerization of methacrylic monomers (methyl methacrylate and glycidyl methacrylate), with ethylene glycol dimethacrylate as a cross-linking agent in the presence of *n*-propanol, formamide as inert porogens, and azobisisobutyronitrile as the initiator. The subsequent covalent attachment of *N*-ethylbutylamine provided a positively charged chromatographic surface common for anion-exchange capillary LC columns. The resulting columns were examined by scanning electron microscopy (SEM) to visualize the structure of the monolith and its uniformity, as well as the bonding of the polymeric support to the silanized capillary wall.

Monolithic columns were termed as multipurpose or multimode columns because they showed mixed modes of separation mechanisms under different conditions. Anion-exchange separation ability in the LC mode can be determined by the cationic chromatographic surface of the monolith functionalized with *N*-ethylbutylamine. At acidic pHs and high voltage applied across the column, we observed that the monolithic stationary phase provided conditions for predominantly capillary electrophoretic migration of peptides along with limited chromatographic retention (CE mode), as shown earlier.[27] Since the peptides and the surface of the porous monolith are both positively charged, the peptides are repelled from the chromatographic surface; nevertheless, a moderate level of solvophobic interaction[43,44] may result in weak chromatographic retention. On the other hand, at basic pHs and electric field across the column, enhanced chromatographic retention of peptides on monolithic capillary column made CEC mechanisms of migration responsible for separation. With increasing pH, the positive charge of the peptides decreases so that their electrophoretic mobility and overall migration rate also decrease. In alkaline media, electroosmotic flow (EOF) is still toward the anode but the flow rate is lower than in acidic media because of decreased protonation of the tertiary amino functionalities on the chromatographic surface. At alkaline conditions (CEC mode), peptides are generally negatively charged and migrate in the same direction as the EOF, with stronger chromatographic retention on the less protonated surface of the monolith.

The influence of mobile phase organic content and ionic strength on separation in positively charged monoliths was also examined. Increasing ionic strength of the mobile phase caused decrease in peptide migration rates due to the decrease in EOF. The monolith has both anion-exchange and hydrophobic (solvophobic) functionalities on its chromatographic surface. Generally, the elution strength of the mobile phase in ion-exchange chromatography depends primarily on its ionic strength, so the retention of the analytes decrease with increasing ionic strength. Because this was not observed with the described monolith, the major migration mechanism is suggested to be electrophoretic.[21] We found that the migration time for peptides increased with increasing acetonitrile content in the mobile phase for small hydrophilic peptides. Larger peptides behaved similarly to proteins due to the greater chromatographic retention, that is, k' values for proteins decreased with increasing acetonitrile concentration. Variations in separation efficiency and retention at different mobile phase contents suggest also the use of solvent gradients in pCEC/pCE as an additional means for optimizing separations on these monolithic columns. For separations of complex peptide mixtures, such as tryptic digests of proteins, organic solvent gradient elution conditions typically provided better peak shapes and less band broadening for hydrophobic peptides than isocratic elution.[21]

In mixed mode separations such as pCEC/pCE, the flow rate and electrophoretic migration can be independently tuned with applied pressure and voltage, providing an additional degree of freedom for optimization of separation performance. Moreover, as it was discussed previously, while slow diffusive mass-transfer kinetics are often the major limiting factor in speed and efficiency of separations on granulate stationary phases,[15,45] monoliths have greatly improved convective mass-transfer properties because the entire eluent stream is forced to flow through the large pores of the separation media.[5,12] The concomitant use of pressure assistance increases the mobile phase flow rate to take advantage of the enhanced mass transfer of the monolithic stationary phases. The combination of the electrophoretic flat flow profile and the pressure-driven parabolic flow profile results in less peak broadening than pressure-driven flow alone.[21,28,46,47]

The columns demonstrated the chemical and mechanical stability of the chromatographic surface and polymeric architecture of the monolith, which led to a good reproducibility and durability. The monoliths could withstand prolonged exposure to different mobile phases at different pH values (pH 2.2–10) for months.[21]

Taking advantage of electroosmotic flow, high separation efficiency for peptides (exceeding 300,000 plates per meter) has been achieved using monolithic capillary columns in both modes due to the relatively flat flow profile with selection of appropriate pore sizes (Figure 22.1). Flow rates compatible with ESI allowed direct interfacing to MS in order to provide structural information in the MS and tandem MS (MS/MS) modes. Coupling to MS requires separation buffers of low conductivity. It was observed that the electrospray was more robust when volatile low conductive acidic buffers prepared from acids with a low ionization constant (e.g., acetic acid) were used than with ammonium acetate buffer or buffers prepared from acids with high ionization constants (e.g. TFA, formic acid). Low buffer conductivity also preserves the electric field strength and reduces Joule heating. The high efficiency of the column, its favorable mass transfer properties, manipulation of the electric field, and gradient conditions allowed separation within 5 to 7 minutes.[21] Figure 22.2 shows an example of protein identification using data-dependent MS/MS scanning by high-throughput pCEC/pCE-ESI-MS. The analysis of bovine serum albumin (BSA) tryptic digest at the level of 50 fmol followed by database search gave correct protein identification with sequence coverage of 26% and a SEQUEST score of 218 using Bioworks 3.0 (ThermoFinnigan) with a search against the whole nonredundant protein database.[21] Comparison of results for MS/MS protein identification at separation of the BSA digest within 5 and 20 minutes showed that a four-fold increase of the separation time yields only 3 to 5% increase in sequence coverage for the identified protein. The aforementioned results show potential for pressure-assisted CEC/CE-ESI-MS using monolithic capillary columns in separation of complex peptide mixtures.

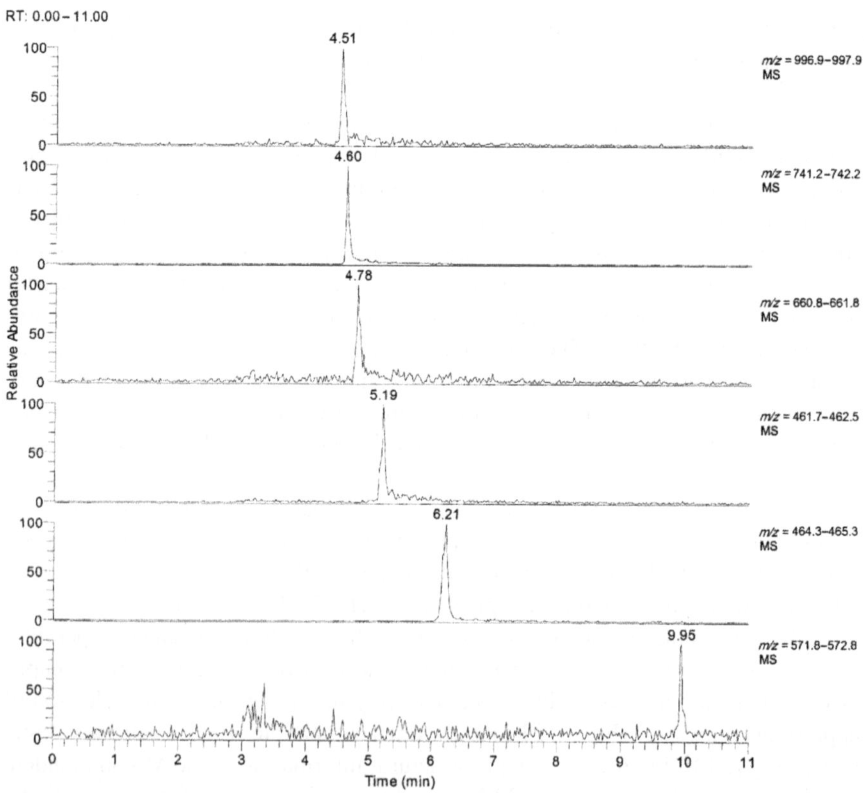

FIGURE 22.1 Extracted ion electrochromatograms for selected peptide ions of BSA digest (50 fmole; injection volume, 10 nL) separated under isocratic elution conditions. Column: 100 μm i.d., 365 μm o.d., 19 cm; flow rate, 120–130 nl/min; eluent 25% HPLC solvent B (v/v). Solvents: A, 0.1% (v/v) acetic acid in water; B, 0.1% (v/v) acetic acid in acetonitrile; applied voltage: 10 kV on the inlet, +2.3 kV on the outlet; electric field strength: 647 V/cm. Efficiencies for peaks from top to bottom: 237,000 pl/meter, 386,000 pl/meter, 185,000 pl/meter, 218,000 pl/meter, 176,000 pl/meter, 802,000 pl/meter. Average efficiency: 334,000 plates/meter. From Ivanov et al., 2003.[21]

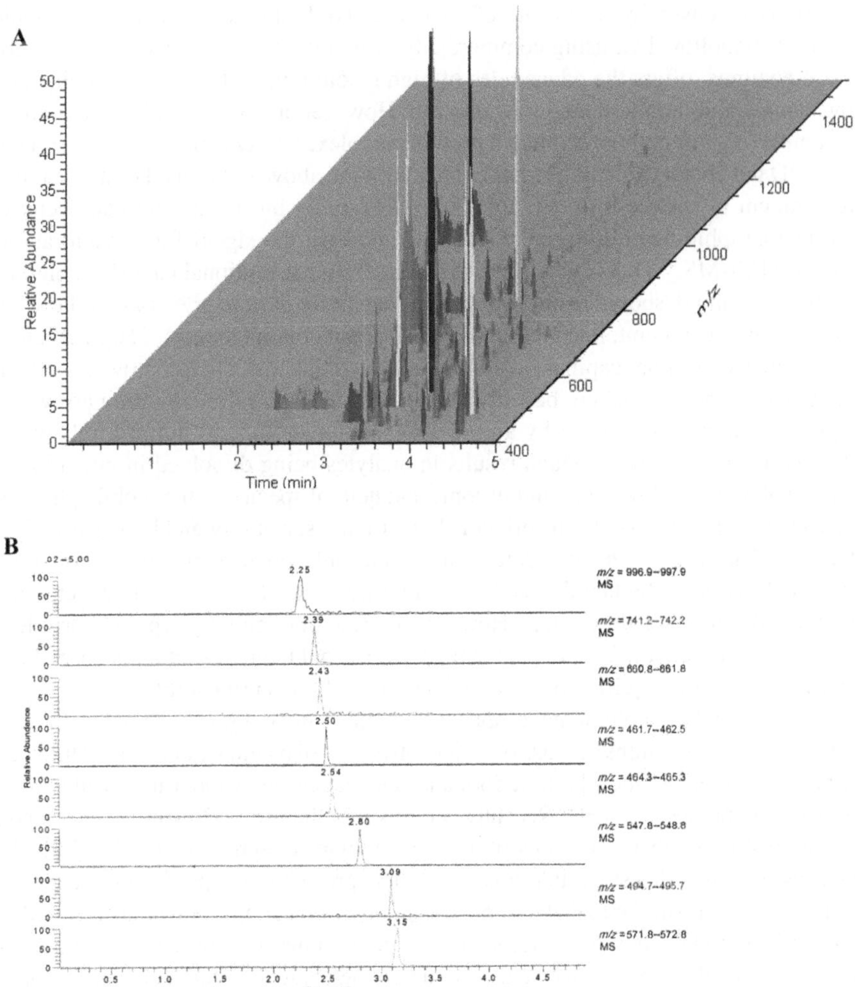

FIGURE 22.2 High-throughput separation of 50 fmole of BSA tryptic digest. Monolithic column: 50 μm i.d., 365 μm o.d., 22.5 cm. Solvents: A, 0.1% acetic acid in water; B,– 0.1% acetic acid in acetonitrile. Gradient: 0 min, 5% B; 10 min, 30% B; flow rate: 180–185 nl/min. Applied voltage: an inlet of the column –15 kV, outlet + 2 kV; electric field strength: 755 V/cm. (A) 3D overlay CEC/CE-MS electrochromatogram; (B) extracted ion electrochromatograms for some selected peptides. Modified from Ivanov et al., 2003.[21]

22.3 HIGH-SENSITIVITY ESI-MS ANALYSIS OF PROTEIN TRYPTIC DIGESTS USING ULTRA LOW I.D. POLYSTYRENE-DIVINYLBENZENE MONOLITHIC CAPILLARY COLUMNS

The high mass sensitivity identification of large numbers of peptides from protein digests is one of the major goals of proteomics. A main characteristic of nanocolumns is high mass sensitivity, as a result of the decreased dilution of the chromatographic band.[45,48] Nanoflow LC, using commercially available 75 and 100 μm i.d. reversed-phase columns, offers the advantages of high resolution, high mass sensitivity, and low sample and mobile phase consumption. However, analysis of a limited amount of sample (e.g., immunoprecipitated protein complexes, laser capture microdissected cells, 2D gel spots) can still be challenging with the above columns. For a fixed limited amount of sample injected, columns with smaller inner diameter can decrease chromatographic band dilution[45,48] and thus increase the signal for concentration-sensitive ESI-MS.[49] Theoretically, downscaling from conventional nano-LC columns of 75 to 20 μm i.d. should result in a gain in sensitivity of $(d_1/d_2)^2 \approx 14$ (for the same injected sample amount, linear column velocity, and column length).[39] High efficiencies of the narrow bore capillary columns are found due to decreased flow dispersion and a homogeneous packing bed structure. The bulk liquid flow in ultra narrow i.d. capillary columns is reduced by an order of magnitude (15–50 nl/min), relative to 75–100 μm i.d. columns, which results in analytes being dissolved in much lower eluent volume with lower amount of ionic and neutral species of the mobile phase in the chromatographic band. The effect is higher mass sensitivity and higher ESI ability.[45,48–54] Thus, highly sensitive detection is achievable when using ultra narrow-bore LC-ESI-MS due to the fact that ESI is a primarily concentration-sensitive technique over a wide range of flow rate.[54] However, narrow-bore columns (particularly less than 50 μm i.d.) are difficult to pack with conventional microparticles because of the very high pressure required to overcome the low column permeability.[49,55,56]

Ultra low i.d. monolithic nanocapillary columns can be a good alternative to microparticle packed columns because of their relative ease of manufacture, without the need of high pressure, and their high-performance characteristics without use of ultra high pressure LC pumps.[4,12,14,17,25,39] Recently, we reported the use of 20 μm i.d. polymeric polystyrene-divinylbenzene monolithic nanocapillary columns for the LC-ESI-MS analysis of tryptic digest peptide mixtures.[39] The procedure for producing the monolithic column was similar to that described previously for cationic ones and consisted of (*a*) covalent modification of the fused silica capillary inner wall by vinyltrimethoxysilane; (*b*) filling the capillary with a degassed polymerization mixture containing monomer (styrene), cross-linker (divinylbenzene), and inert porogens (tetrahydrofuran and *n*-octanol); (*c*) thermally induced *in situ* polymerization in the presence of initiator in a capillary pressurized from both ends to form a macroporous, rigid, and uniform structure covalently attached to the capillary inner surface; and (*d*) flushing out the porogens and remaining polymerization mixture. Polymerization and cross-linking led to a phase separation to create permanent channels in the rigid polymeric material (Figure 22.3). Covalent immobilization helped reduce shrinkage of the monolith during the polymerization[4] and eliminated the need for retaining frits. Pressurization during polymerization[57] reduced monolith structure irregularities by both minimizing shrinkage[57,58] and bubble formation with porous monoliths in the 20 μm i.d. column. Typically, 40 to 50 psi applied from both ends of the capillary filled with polymerization mixture helped to achieve a uniform continuous monolithic structure.

Besides the factors improving the sensitivity described above, ultra low i.d. monolithic nanocapillary columns operating at a flow rate of 20 to 50 nl/min may also take

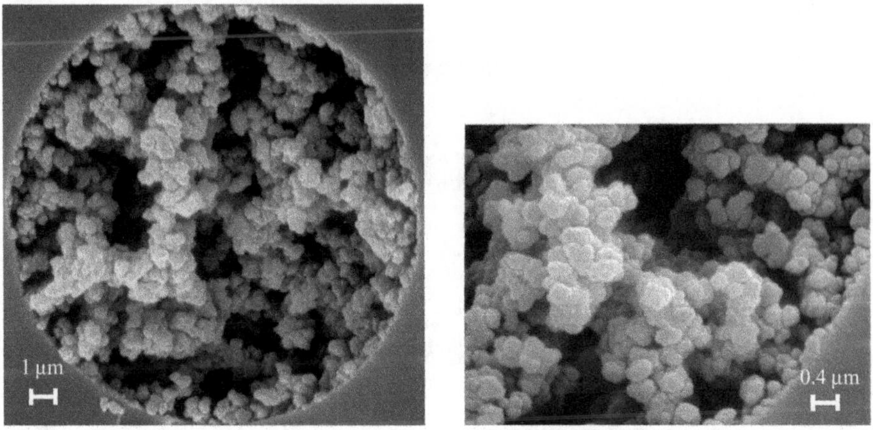

A B

FIGURE 22.3 Scanning electron micrographs of the polystyrene-divinylbenzene monolithic packing in a 1 cm section of fused-silica capillaries of 20 μm i.d. The images correspond to (A) 5000×; and (B) 12,500T magnification. From Ivanov et al., 2003.[39]

advantage of nanospray ionization to increase MS sensitivity.[54] Stable ESI conditions at ultra low flow rate can potentially result in the more efficient transfer of ions into the entrance of the mass spectrometer.[48–54] The effectiveness of the approach was demonstrated in protein identification at 1 and 10 attomole amounts of tryptic digest of a protein mixture (Figure 22.4). Typically, a mass sensitivity of 5 to 10 amol was observed for tryptic peptides in the MS mode at a signal-to-noise ration ≥5 but was very dependent on peptide primary structure, as was shown before.[59,60] High mass sensitivity in the MS and MS/MS modes using an ion-trap MS was found, a factor of up to 20-fold improvement over 75 μm i.d. nanocolumns. The number of peptides identified in the MS and MS/MS modes at this level using the 75 μm i.d. column was comparable to that at 5 to 10 amol level of injection on 20 μm i.d. monolithic column.[39]

The poly (styrene-divinylbenzene) (PS-DVB) monolithic columns with pseudobeads smaller than 1 μm (Figure 22.3) demonstrated high efficiency, over 100,000 plates/meter (calculated from extracted ion chromatograms) for peptides using a 10 cm long column, with typical peak widths at half height of 5 of 15 seconds, as well as reasonable backpressure of 100 to 300 bars. A wide linear dynamic range (~4 orders of magnitude, 100 fg to 1000 pg) was achieved, and good run-to-run and column-to-column reproducibility of isocratic and gradient elution separations were found.[39] The columns could be flushed with solvents in either direction up to at least 2000 psi pressure without any damage to the monolithic structure according to our experience, and the columns could withstand prolonged exposure to mobile phases from pH 2.0 to pH 11.5 for at least several months and at least 1000 injections per column. Clogged columns can be reused after cutting a short piece (~1 mm) from the clogged end or after grinding the end of the column by a diamond sand lapper.

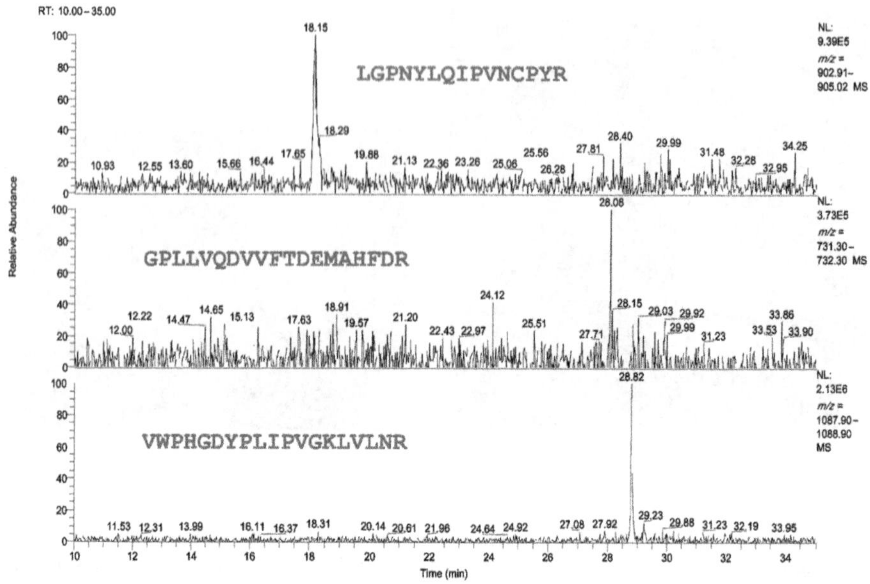

A

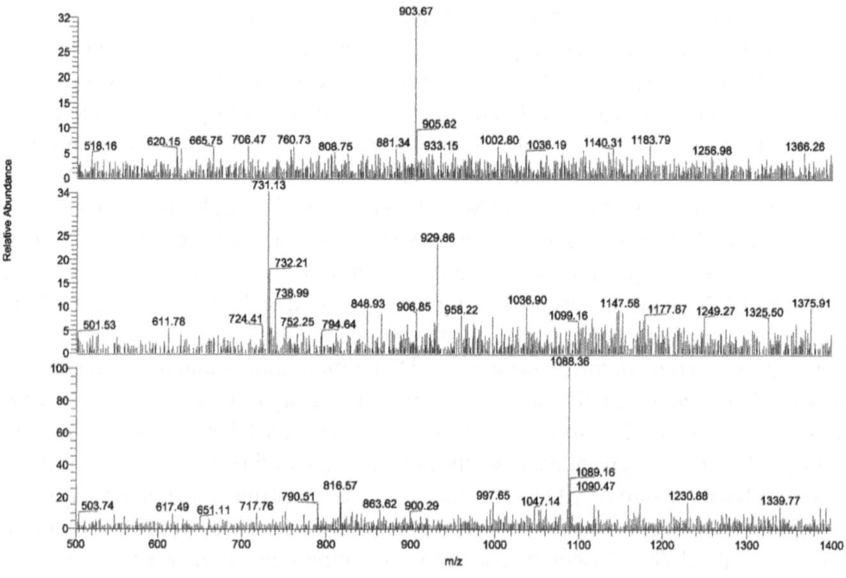

B

FIGURE 22.4 Gradient nano-LC-ESI-MS of a tryptic digest of bovine catalase on the monolithic column (10 amol of digest injected on the column). (A) Extracted ion chromatograms for selected peptides; (B) MS spectra at the peak maximum of the same peptides. From Ivanov et al., 2003.[39]

To demonstrate the potential of the ultra low i.d. monoliths in the high-sensitivity proteomic analysis, a tryptic digest of protein extract of a breast ductal carcinoma tissue section (10^3 cell equivalent) was also analyzed allowing identification of tens of most abundant proteins. The above results open the possibility to utilize such narrow bore monolithic columns for analysis of tissue samples with a limited number of cells and samples difficult to access. The column characteristics suggest that ultra narrow-bore monolithic columns are highly promising for proteomics and other applications.

22.4 POLYSTYRENE MONOLITHIC CAPILLARY COLUMNS IN ON-LINE 2D LC-ESI-MS PROTEOMIC ANALYSIS

Proteomic samples can contain several thousands of proteins resulting in millions of peptides after enzymatic digestion. Bottom-up shotgun LC/MS proteomics strategy requires high resolving power of liquid separation at the peptide level. Various separation methods utilized in proteomics prior to MS include one-dimensional (1D) and 2D polyacrylamide gel electrophoresis (PAGE), 1D nano-LC, multidimensional protein identification technology (MudPIT), isoelectric focusing (IEF), and SPE, among others. Each method has its own limitations, which makes further development of separation methodologies in proteomics necessary. Multidimensional LC coupled to MS has become the most powerful alternative to 2D PAGE technology because of its higher sensitivity, speed, flexibility, robustness, and ease of automation.[61,62] Recently, we reported the use of the polymeric monolithic capillary columns for the analysis of complex proteomic samples with 2-dimensional liquid chromatography electrospray ionization trap mass spectrometry (2D LC-ESI-IT-MS).[63] 2D separation enhanced the protein identification rate approximately three-fold in comparison to single dimension separation, as expected. We demonstrated preliminary data for a comparison of two on-line 2D LC-MS approaches: 2D LC-MS with the salt plug injections vs. comprehensive (dual gradient) 2D LC-MS. In first method, digests of cell lysates were injected on a strong cation exchange (SCX) column and eluted by the KCl salt plugs of constantly increasing salt concentration from 0 to 500 mM. Then, they were trapped on a reversed-phase precolumn, desalted, and, finally, eluted from the trapping column and separated on monolithic nano-LC column at gradient elution conditions. In dual gradient experiments, a salt gradient from 0 up to 500 mM of KCl instead of the salt plug injections and two identical trapping columns instead of one were used. Polystyrene monolithic capillary columns were prepared in silanized fused-silica capillaries of 75 μm I.D. by thermally induced *in situ* polymerization as described above. *Schizosaccharomyces pombe* cell lysates (~0.2 to 0.4 mg of total starting protein amount) were subjected to either on-line 2D LC-MS/MS analysis with the salt plug injections (Figure 22.5) or to comprehensive (dual-gradient) on-line 2D LC-MS/MS analysis followed by the database search. The method demonstrated that the monolithic columns are feasible to use in various proteomic applications with wide dynamic range and high complexity of a sample. The method showed high efficiency in the pressure-driven elution mode for the second dimension (over 100,000 plates/meter), ease of manufacture in comparison to particulate columns of similar

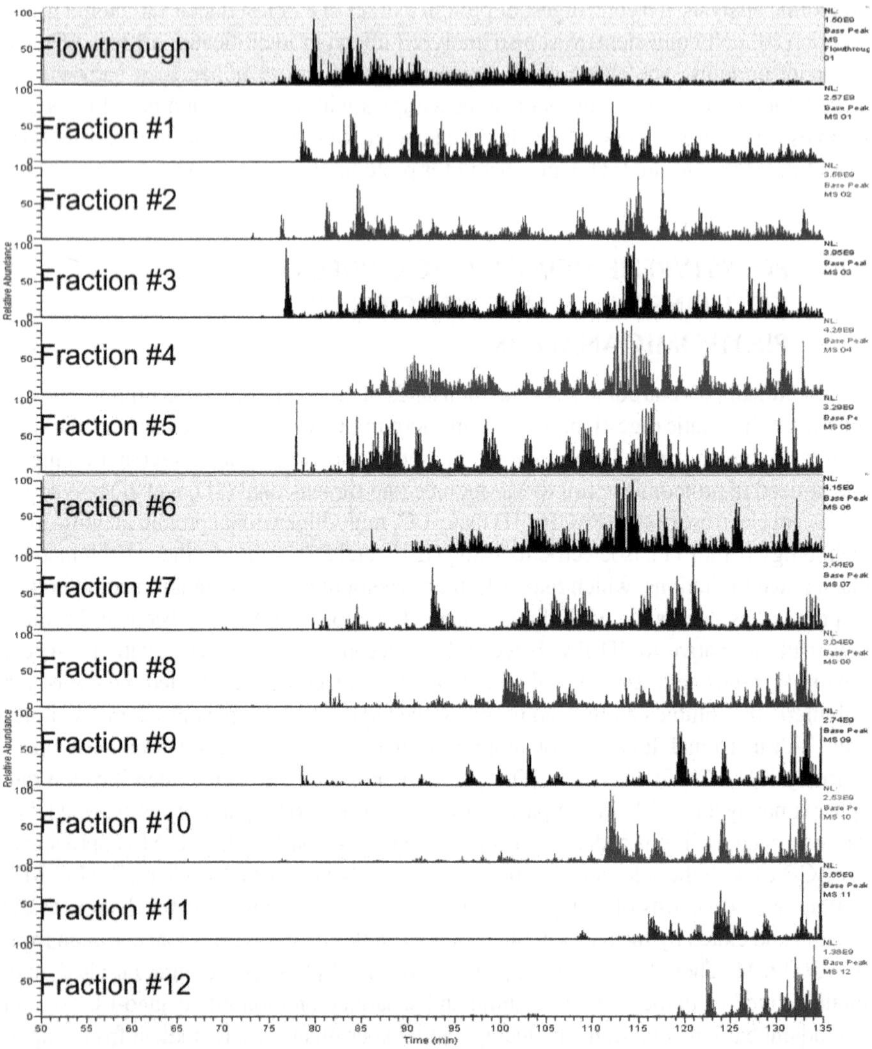

FIGURE 22.5 Base ion chromatograms for reversed-phase separations of 13 salt plug injection fractions on 75 μm i.d. monolithic column in on-line 2D LC-ESI-MS analysis of *S. pombe* lysate. PS-DVB monolithic column: 75 μm i.d., 365 μm o.d., 15 cm; trapping column: 500 μm i.d. × 10 mm, packed with Styros 2R, 5 μm (OraChrom, Mass.); SCX column: 300 μm i.d. × 1.5 cm, packed with POROS BioX-SCX, 5 μm (LC Packings, Calif.). Solvents for reversed-phase gradient: (A) 2% acetonitrile, 0.1% formic acid, pH 3.0; (B) 85% acetonitrile, 5% isopropanol, 0.1% formic acid, pH 3.0. Reversed phase gradient: 0–45% in 90 minutes, 45–85% in 10 minutes. Solvents for IEX salt plugs were prepared from; (C) 5 mM KH2PO4, 5% acetonitrile, 0.05% formic acid, pH 3.0; (D) 500 mM KCl, 5 mM KH2PO4, 5% acetonitrile, 0.05% formic acid, pH 3.0.

dimensions, good column lifetime (up to 1000 injections per column) and reproducibility, and ease of a column cleanup under harsh elution conditions, helping to avoid cross-contamination between different samples injected. The value of the approach was shown in dual-gradient elution and salt plug injection 2D-LC-ESI-MS analyses of *S. pombe* cell lysate protein digests. Both methods were compared in their resolving power and impact on protein identification rate. Thus, dual-gradient 2D LC-MS gave about 10 to 20% more protein identifications than the alternative 2D LC-MS technique with the salt plug injections at the same geometry of the columns used in both experiments, same number of fractions eluted from the first separation dimension (SCX), and similar elution conditions (KCl salt concentration increment). An impact of an SCX column length (e.g., column efficiency and capacity) on protein identification rate in the salt plug injection 2D LC-MS mode was also briefly compared. The use of 15 cm long 300 μm i.d. SCX column instead of 1.5 cm × 300 μm i.d. gave a 10 to 35% gain in protein identification with more evenly distributed eluted peptides among the SCX fractions. Moreover, we showed that repetitive multiple analyses of the same sample greatly improves protein identification rate (up to ~40% for two repetitive runs versus a single one), which was expected and in agreement with recently published reports.[64,65]

22.5 CONCLUSION

Polymeric monolithic capillary columns have become popular as alternatives to granulate packed columns due to the simplicity of their preparation and handling, advantageous flow transfer properties and separation performance at acceptable backpressure, diversity of their chemistries and functionalities of their chromatographic surfaces, and the elimination of need for retaining frits. In contrast to packed columns, polymeric monoliths demonstrate outstanding chemical and mechanical stability as a result of their highly cross-linked chemical structure and covalent attachment to the inner wall of the column. A few examples of peptide separation on monolithic columns coupled to ESI-MS shown above demonstrate the potential of the monoliths for various proteomic applications.

ACKNOWLEDGMENTS

The author would like to thank Li Zang, Michael Schmidt, and Dr. Tomas Rejtar for their help and fruitful discussions in the preparation of this manuscript. Special thanks to Dr. Barry L. Karger (Northeastern University, Barnett Institute, Boston) and Dr. Dieter A. Wolf (Harvard School of Public Health, Boston) for their strong support.

REFERENCES

1. Kubin, M., Spacek, P., and Chromecek, R., Gel permeation chromatography on porous poly(ethylene glycol methacrylate), *Collect. Czech. Chem. Commn.*, 32, 3881–3888, 1967.

2. Hileman, F.D., Sievers, R.E., Hess, G.G., and Ross, W.D., In situ preparation and evaluation of open pore polyurethane chromatographic columns, *Anal. Chem.*, 45, 1126–1130, 1973.
3. Ross, W.D. and Jefferson, R.T., In situ-formed open-pore polyurethane as chromatography supports, *J. Chrom. Sci.*, 8, 386–389, 1970.
4. Premstaller, A., Oberacher, H., Walcher, W., Timperio, A.M., Zolla, L., Chervet, J.P., Cavusoglu, N., van Dorsselaer, A., and Huber, C.G., High-performance liquid chromatography-electrospray ionization mass spectrometry using monolithic capillary columns for proteomic studies, *Anal. Chem.*, 73, 2390–2396, 2001.
5. Peters, E.C., Petro, M., Svec, F., and Fréchet, J.M., Molded rigid polymer monoliths as separation media for capillary electrochromatography, *Anal. Chem.*, 69, 3646–3649, 1997.
6. Tennikov, M.B., Gazdina N.V., Tennikova, T.B., and Svec, F., Effect of porous structure of macroporous polymer supports on resolution in high-performance membrane chromatography of proteins, *J. Chromatogr. A*, 798, 55–64, 1998.
7. Branovic, K., Latner, G., Barut, M., Strancar, A., Josic, D., and Buchacher, A., Very fast analysis of impurities in immunoglobulin concentrates using conjoint liquid chromatography on short monolithic disks, *J. Immunol. Methods*, 271, 47–58, 2002.
8. Ishizuka, N., Minakuchi, H., Nakanishi, K., Soga, N., Nagayama, H., Hosoya, K., and Tanaka, N., Performance of a monolithic silica column in a capillary under pressure-driven and electrodriven conditions, *Anal. Chem.*, 72, 1275–1280, 2000.
9. Belenkii, B.G., Podkladenko, A.M., Kurenbin, O.I., Mal'tsev, V.G., Nasledov, D.G., and Trushin, S.A., Peculiarities of zone migration and band broadening in gradient reversed-phase high-performance liquid chromatography of proteins with respect to membrane chromatography, *J. Chromatogr.*, 645, 1, 1993.
10. Hjerten, S., Liao, J.-L., and Zhang, R., High-performance liquid chromatography on continuous polymer beds, *J. Chromatogr. A*, 473, 273–275, 1989.
11. Svec, F., Fréchet, J.M.J., Continuous rods of macroporous polymer as high-performance liquid chromatography separation media, *Anal. Chem.*, 64, 820–822, 1992.
12. Meyers, J.J. and Liapis, A.I., Network modeling of the convective flow and diffusion of molecules adsorbing in monoliths and in porous particles packed in a chromatographic column, *J. Chromatogr A.*, 852, 3–23, 1999.
13. Liapis, A.I., McCoy, M.A., Perfusion chromatography: Effect of micropore diffusion on column performance in systems utilizing perfusive adsorbent particles with a bidisperse porous structure, *J. Chromatogr. A*, 660, 85–96,1994.
14. Petro, M., Svec, F., and Fréchet, J.M.J., Molded continuous poly(styrene-co-divinylbenzene) rod as a separation medium for the very fast separation of polymers: Comparison of the chromatographic properties of the monolithic rod with columns packed with porous and non-porous beads in high-performance liquid chromatography of polystyrenes, *J. Chromatogr. A.*, 752, 59–66, 1996.
15. Svec, F., and Fréchet, J.M.J., Modified poly(glycidyl methacrylate-co-ethylene dimethacrylate) continuous rod columns for preparative-scale ion-exchange chromatography of proteins, *J. Chromatogr. A.*, 702, 89–95, 1995.
16. Hjertén, S., Mohammad, J., and Nakazato, K., Improvement in flow properties and pH stability of compressed, continuous polymer beds for high-performance liquid chromatography, *J. Chromatogr. A.*, 646, 121–128, 1993.
17. Moore, R.E., Licklider, L., Schumann, D., and Lee, T.D., A microscale electrospray interface incorporating a monolithic, poly(styrene-divinylbenzene) support for on-line

liquid chromatography/tandem mass spectrometry analysis of peptides and proteins, *Anal. Chem.*, 70, 4879–4884, 1998.
18. Svec, F., Tennikova, T.B., and Deyl, Z., Eds., *Monolithic Materials: Preparation, Properties and Applications*, Elsevier, Amsterdam, 2003, 13 pp.
19. Svec. F., Porous monoliths: Emerging stationary phases for HPLC and related methods, *LCGC*, 18–21, 2004.
20. Rohr, T., Hilder, E.F., Donovan, J.J., Svec, F., and Fréchet, J.M.J., Photografting and the control of surface and inner volume chemistry in three-dimensional porous polymer monoliths, *Macromolecules*, 36, 1677–1684, 2003.
21. Ivanov, A.R., Horvath, C., and Karger, B.L., High-efficiency peptide analysis on monolithic multimode capillary columns: Pressure-assisted capillary electrochromatography/capillary electrophoresis coupled to UV and electrospray ionization-mass spectrometry, *Electrophoresis*, 24, 3663–3673, 2003.
22. Viklund, C., Svec, F., Fréchet, J.M.J., and Irgum, K., Monolithic, "molded," porous materials with high flow characteristics for separation, catalysis, or solid phase chemistry: Control of porous properties during polymerization, *Chem. Mater.*, 8, 744–750, 1996.
23. Ngola, S.M., Fintschenko, Y., Choi, W.Y., and Shepodd, T.J., Conduct-as-cast polymer monoliths as separation media for capillary electrochromatography, *Anal. Chem.*, 73, 849–856, 2001.
24. Premstaller, A., Oberacher, H., and Huber, C.G., High-performance liquid chromatography-electrospray ionization mass spectrometry of single- and double-stranded nucleic acids using monolithic capillary columns, *Anal. Chem.*, 72, 4386–4393, 2000.
25. Huang, X., Zhang, S., Schultz, G.A., and Henion J., Surface-alkylated polystyrene monolithic columns for peptide analysis in capillary liquid chromatography-electrospray ionization mass spectrometry, *Anal. Chem.*, 74, 2336–2344, 2002.
26. Xie, S., Allington, R.W., Svec, F., and Fréchet, J.M., Rapid reversed-phase separation of proteins and peptides using optimized "moulded" monolithic poly(styrene–*co*-divinylbenzene) columns, *J. Chromatogr. A.*, 865, 169–174, 1999.
27. Zhang, S., Huang, X., Zhang, J., and Horváth, C., Capillary electrochromatography of proteins on an anion-exchanger column, *J. Chromatogr. A.*, 887, 465–477, 2000.
28. Gusev, I., Huang, X., and Horváth, C., Capillary columns with in situ formed porous monolithic packing for micro high-performance liquid chromatography and capillary electrochromatography, *J. Chromatogr. A.*, 855, 273–90, 1999.
29. Lämmerhofer, M., Svec, F., Fréchet, J.M., and Lindner, W., Capillary electrochromatography in anion-exchange and normal-phase mode using monolithic stationary phases, *J. Chromatogr. A.*, 925, 265–277, 2001.
30. Schuster, M., Wasserbauer, E., Neubauer, A., and Jungbauer, A., Short cut of protein purification by integration of cell-disrupture and affinity extraction, *Bioseparation*, 9, 259–268, 2000.
31. Hahn, R., Podgornik, A., Merhar, M., Schalaun, E., Jungbauer, A., Affinity monoliths generated by in situ polymerization of the ligand. *Anal. Chem.*, 73, 5126–5132, 2001.
32. Lämmerhofer, M., Peters, E.C., Yu, C., Svec, F., and Fréchet, J.M., Chiral monolithic columns for enantioselective capillary electrochromatography prepared by copolymerization of a monomer with quinidine functionality. I. Optimization of polymerization conditions, porous properties, and chemistry of the stationary phase, *Anal. Chem.*, 72, 4614–4622, 2000.

33. Sykora, D., Svec, F., and Fréchet, J.M., Separation of oligonucleotides on novel monolithic columns with ion-exchange functional surfaces, *J. Chromatogr. A*. 852, 297–304, 1999.
34. Yu, C., Davey, M.H., Svec, F., and Fréchet, J.M., Monolithic porous polymer for on-chip solid-phase extraction and preconcentration prepared by photoinitiated in situ polymerization within a microfluidic device, *Anal. Chem.*, 73, 5088–5096, 2001.
35. Tan, A., Benetton, S., and Henion, J.D., Chip-based solid-phase extraction pretreatment for direct electrospray mass spectrometry analysis using an array of monolithic columns in a polymeric substrate, *Anal Chem.*, 75, 5504–5511, 2003.
36. Throckmorton, D.J., Shepodd, T.J., and Singh, A.K. Electrochromatography in microchips: Reversed-phase separation of peptides and amino acids using photopatterned rigid polymer monoliths, *Anal. Chem.*, 74, 784–789, 2002.
37. Ericson, C., Holm, J., Ericson, T., and Hjertén, S., Electroosmosis- and pressure-driven chromatography in chips using continuous beds, *Anal. Chem.*, 72, 81–87, 2000.
38. Oberacher, H., Premstaller, A., and Huber, C.G., Characterization of some physical and chromatographic properties of monolithic poly(styrene-co-divinylbenzene) columns, *J. Chromatogr. A.*, 1030, 201–208, 2004.
39. Ivanov, A.R., Zang, L., and Karger, B.L., Low-attomole electrospray ionization MS and MS/MS analysis of protein tryptic digests using 20-microm-i.d. polystyrene-divinylbenzene monolithic capillary columns. *Anal Chem.*, 75, 5306–5316, 2003.
40. Xiong, L., Zhang, R., and Regnier F.E., Potential of silica monolithic columns in peptide separations, *J. Chromatogr. A*, 1030, 187–194, 2004.
41. Pretorius, V., Hopkins, B.J., and Schieke, J.D., Electroosmosis—a new concept for high-speed liquid chromatography, *J. Chromatogr.* 99, 23–30, 1974.
42. Hilder, E.F., Svec, F., and Fréchet, J.M., Development and application of polymeric monolithic stationary phases for capillary electrochromatography, *J. Chromatogr. A*, 1044, 3–22, 2004.
43. Horváth, C. and Melander, W., Liquid chromatography with hydrocarbonaceous bonded phases: Theory and practice of reversed phase chromatography, *J. Chromatogr. Sci.*, 15, 393–404, 1977.
44. Vailaya, A. and Horváth, C., Solvophobic theory and normalized free energies of nonpolar substances in reversed phase chromatography, *J. Phys.Chem. B*, 101, 5875–5888, 1997.
45. Unger, K.K., Packings and stationary phases in chromatographic techniques, Unger, K.K., Ed., Marcel Dekker, New York, 1990.
46. Adam, T. and Unger, K.K., Comparative study of capillary electroendosmotic chromatography and electrically assisted gradient nano-liquid chromatography for the separation of peptides, *J. Chromatogr. A.*, 894, 241–251, 2000.
47. Huang, P., Jin, X., Chen, Y., Srinivasan, J.R., and Lubman, D.M., Use of a mixed-mode packing and voltage tuning for peptide mixture separation in pressurized capillary electrochromatography with an ion trap storage/reflectron time-of-flight mass spectrometer detector, *Anal. Chem.*, 71, 1786–1791, 1999.
48. Haskins, W.E., Wang, Z., Watson, C.J., Rostand, R.R., Witowski, S.R., Powell, D.H., and Kennedy, R.T., Capillary LC-MS^2 at the attomole level for monitoring and discovering endogenous peptides in microdialysis samples collected in vivo, *Anal. Chem.*, 73, 5005–5014, 2001.
49. Shen, Y., Zhao, R., Berger, S.J., Anderson, G.A., Rodriguez, N., and Smith, R.D., High-efficiency nanoscale liquid chromatography coupled on-line with mass spectrometry using nanoelectrospray ionization for proteomics, *Anal. Chem.*, 74, 4235–4249, 2002.

50. Wilm, M. and Mann M., Theoretical electrospray and Taylor-Cone theory: Dole's beam of macromolecules at last? *Int. J. Mass Spectrom. Ion Proc.*, 136, 167–180, 1994.
51. Gale, D.C. and Smith, R.D., Small volume and low flow-rate electrospray ionization mass spectrometry of aqueous samples, *RCMS* 7, 1017–1021, 1993.
52. Fenn, J.B., Ion formation from charged droplets: Roles of geometry, energy and time, *J. Am. Soc. Mass Spectrom.*, 4, 524–535, 1993.
53. Kebarle, P. and Tang, L., From ions in solution to ions in the gas phase, *Anal. Chem.*, 65, 972A–986A, 1993.
54. Emmett, M.R., Caprioli, R.M., Micro-electrospray mass spectrometry: Ultra-high-sensitivity analysis of peptides and proteins. *J. Am. Soc. Mass Spectrom.*, 5, 605–613, 1994.
55. Wu, N., Lippert, J.A., Lee, M.L., Practical aspects of ultrahigh pressure capillary liquid chromatography, *J. Chromatogr A.*, 911, 1–12, 2001.
56. MacNair, J.E., Patel, K.D., and Jorgenson, J.W., Ultrahigh-pressure reversed-phase capillary liquid chromatography: Isocratic and gradient elution using columns packed with 1.0-micron particles, *Anal. Chem.*, 71, 700–708, 1999.
57. Bente, P.F., III and Myerson, J., Method of Preventing Shrinkage Defects in Electrophoretic Gel Columns, U.S. Patent 4,810,456, Mar. 7, 1989.
58. Huber. C., Oberacher, H., and Premstaller, A., Method and Apparatus for Separating Polynucleotids Using Monolithic Columns, U.S. Patent Application 0088753 A1, Jul.11, 2002.
59. Cech, N.B., Krone, J.R., and Enke, C.G., Predicting electrospray response from chromatographic retention time, *Anal Chem.*, 73, 208–213, 2001.
60. Cech, N.B. and Enke, C.G., Relating electrospray ionization response to nonpolar character of small peptides, *Anal. Chem.*, 72, 2717–2723, 2000.
61. Link, A.J., Eng, J., Schieltz, D.M., Carmack, E., Mize, G.J., Morris, D.R., Garvik, B.M., and Yates, J.R., III, Direct analysis of protein complexes using mass spectrometry, *Nat. Biotechnol.*, 17, 676–82, 1999.
62. Washburn, M.P., Wolters, D., and Yates, J.R., III, Large-scale analysis of the yeast proteome by multidimensional protein identification technology, *Nat. Biotechnol.*, 19, 242–247, 2001.
63. Ivanov, A.R., Monolithic Capillary Columns in On-Line 2D-LC ESI-MS Proteomic Analysis, paper presented at the 52nd ASMS Conference on Mass Spectrometry and Allied Topics, Nashville, TN, May 23–27, 2004.
64. Shen, Y., Jacobs, J.M., Camp, D.G., II, Fang, R., Moore, R.J., Smith, R.D., Xiao, W., Davis, R.W., and Tompkins, R.G., Ultra-high-efficiency strong cation exchange LC/RPLC/MS/MS for high dynamic range characterization of the human plasma proteome, *Anal. Chem.*, 476, 1134–1144, 2004.
65. Durr, E., Yu, J., Krasinska, K.M., Carver, L.A., Yates, J.R. III, Testa, J.E., Oh, P., and Schnitzer, J.E., Direct proteomic mapping of the lung microvascular endothelial cell surface in vivo and in cell culture, *Nat. Biotechnol.*, 22, 985–992, 2004.

Part V

Related Techniques

23 Proteins Staining in Polyacrylamide Gels*

Gary B. Smejkal

CONTENTS

23.1 INTRODUCTION

For nearly five decades, polyacrylamide gel electrophoresis (PAGE) has endured as an essential method for macromolecular separations. Obligatory preamble aside, the potential of PAGE has not been fully realized. It remains in its formative years, with further development still ongoing.

Consider that immobilized pH gradients (IPGs) are presently capable of separating proteins differing in charge by 0.002 isoelectric point (pI) units such that thousands of proteins can theoretically be resolved over the pH 3–10 range. This number of hypothetical plates is nearly squared when a second dimension of electrophoresis is introduced, such that the number of proteins that can be arrayed by two-dimensional gel electrophoresis (2DE) extends into the hundreds of thousands, even by the most conservative estimates. Experimentally, however, the resolving capacity of 2DE is usually less than 1000 proteins. This is largely attributable to the broad concentration range in which proteins are expressed, spanning possibly 12 orders of magnitude. While isoelectric fractionation, a divide and conquer strategy for isolating very narrow

* Portions of this review were excerpted from an Editorial titled "The Coomassie Chronicles" by G.B. Smejkal, *Expert Rev. Proteomics,* 1, 381–387, 2005. Used with permission from Future Drugs Limited, London, U.K., copyright 2005.

segments of a proteome for individual analyses, and affinity depletion of interfering proteins will eventually lead to the exploration of the subliminal layers of proteomes, this must coincide with the development of new detection methods of higher sensitivity and improved compatibility with downstream procedures.

It is incorrect to assume that most proteins contain acidic, basic, and hydrophobic residues in numbers that are directly proportional to polypeptide chain length, and therefore, that staining based on electrostatic or hydrophobic interaction will correlate with protein mass. To the contrary, proteins are diversified in amino acid composition such that the selectivity of a stain for specific residues renders attempts to quantify one protein relative to another inaccurate. The idealized stain would bind proteins nonspecifically without influence of amino acid composition, relative hydrophobicity, or the capacity to bind sodium dodecylsulfate (SDS), and its interaction would necessarily be based on a denominator common to all residues.

23.2 ORGANIC DYES

Fazekas De St. Groth et al.[1] first used Coomassie Brilliant Blue (CBB) R-250, a wool dye borrowed from the textile industry, for staining proteins separated electrophoretically in cellulose acetate. Logically, this dye would soon be applied to polyacrylamide gel staining[2], as would its dimethylated derivative G-250.[3] The Coomassie dyes, named in commemoration of the British occupation of the Ashanti capital Kumasi in Ghana, have remained in use for over 40 years. Members of the first Electrophoresis Societies often referred to themselves as the Bluefingers, and some German practitioners were mistaken for counterfeiters because of the residual blue stain on their hands.

The single most important advance in CBB staining would come from Neuhoff et al.[4] who thoroughly investigated the colloidal properties of these dyes and demonstrated sensitivity approaching silver staining. Exploiting the relatively low solubility of CBB G-250, suspensions are produced where the dye exists predominantly in micellar form too large to penetrate most gel matrices. These colloids are presumably stabilized by the interaction of H^+ ions with SO_3^- groups aligned at the surface of the colloid. This mechanism is analogous to the lowering of the critical micelle concentration (CMC) and increase in the molar concentration of micelles of SDS that coincides with increasing counterion concentration. Background free staining occurs by localized solvation of the dye into protein zones. Suspensions of CBB G-250 and R-250 are unsuitable for staining proteins in agarose gels since these colloids are small enough to penetrate the macroporous gel structure where they remained trapped, resulting in unacceptably high backgrounds that can require days to remove. Alternatively, Coomassie Violet R-150 forms extremely large unstable colloids impenetrable to agarose gels.[5,6]

The Coomassie dyes are disulfonated triphenylmethane dyes that bind proteins primarily through electrostatic interaction with basic amino acid residues in their protonated state, but also through hydrophobic interaction with hydrophobic residues and sodium SDS micelles. CBB staining adheres to the Lambert-Beer Law and is linear at protein concentrations over at least three orders of magnitude, as indicated by its use in the Bradford protein assay. However, different equations derive from

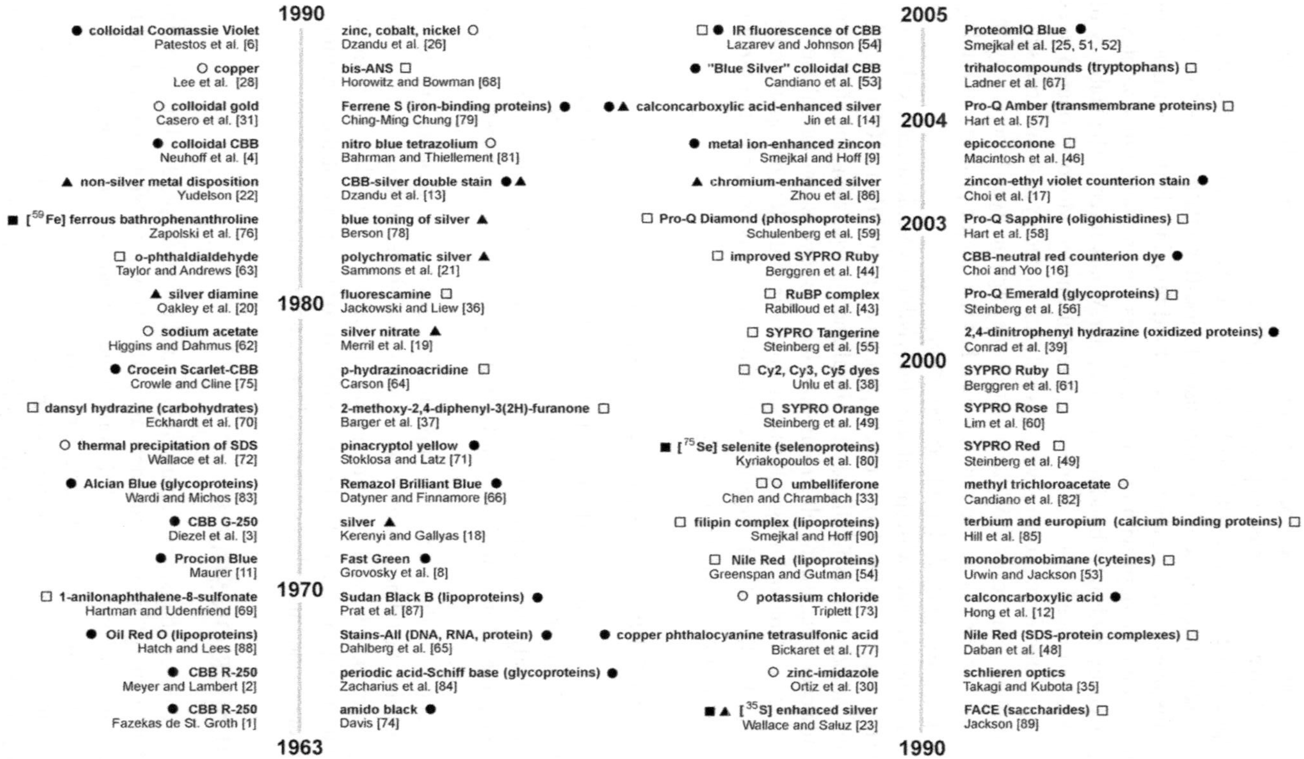

FIGURE 23.1 Timeline of the major developments in polyacrylamide gel staining over the past 42 years. Stains were categorized as visible dyes (●), fluorescent stains (□), negative stains (○), silver stains (▲), or radioactive (■) methods. The most important, but not necessarily the first, publication is cited. The specificity of some stains is indicated in parentheses.

linear regression of absorbance vs. mass when different proteins are used to calibrate such assays. Differential binding of CBB is demonstrative of selectivity based on amino acid composition, and at least one group has reported that CBB staining may overestimate relative protein mass in polyacrylamide gels.[7]

Other triphenylmethane dyes such as Fast Green were also explored, and despite their relative insensitivity, these were found useful for differentiating very basic proteins such as histones.[8] Most of the organic dyes bind proteins via the interaction of one or more sulfonate groups with protonated amines, and the resulting complex is further stabilized by hydrogen bonding and Van der Waals forces. The stability of these complexes is related to the number of dye sulfonates. Monosulfonated dyes are rapidly dissociated from protein,[9] whereas polysulfonated dyes form more stable protein-dye-protein complexes.[10] While most of the organic dyes function by this mechanism, an exception is Procion blue, which has the unique property of binding covalently to proteins via its dichlorotriazinyl group.[11]

While it has not gained widespread use, the visible dye calconcarboxylic acid has been used to stain proteins during electrophoresis. The addition of 0.01% of this reagent to the cathodic buffer facilitates the simultaneous separation and staining of proteins.[12] However, because the electrophoretic influx of dye is constant, these gels must be destained to remove backgrounds. Sensitivity is in the low nanogram range. Like CBB,[13] the binding of calconcarboxylic acid to proteins promotes the binding of silver ions and enhances silver staining.[14]

23.3 STAINING PROTEINS WITH TANNINS

Interestingly, Syed and Sayeed[93] stained proteins in polyacrylamide gels with henna, an extract from the leaves of *Lawsonia inermis* used for dying hair and on the Indian subcontinent for body ornamentation. Staining was reportedly more sensitive than CBB for some proteins and, unlike the sulfonated dyes, was effective over the entire pH 2–12 range.

Staining occurs via the interaction of plant tannins with both proteins and carbohydrates. Arginine, histidine, lysine, and tryptophan residues of proteins bind electrostatically to carboxylic and phenolic groups of tannins. This class of pigments chelate certain metal ions at a pH above the pKa of one or more of the substituent phenolic groups,[94] and this induces a hypsochromic shift in staining.

23.4 COUNTERION STAINS

The use of counterion stains, which pairs anionic and cationic dyes that bind proteins by two different charge interactions, appears particularly promising. When used in combination, anionic dyes such as CBB form large ion pair complexes with cationic dyes. Only a small proportion of each dye remains in free form. This lessens the selectivity of the individual dyes, which is otherwise biased by amino acid composition. Whereas CBB or other anion binds positively charged amino groups, the cation simultaneously binds to negatively charged carboxyl groups. Counterion stains using CBB and Bismark Brown R[15] or neutral red[16] as the cation are described. (In addition

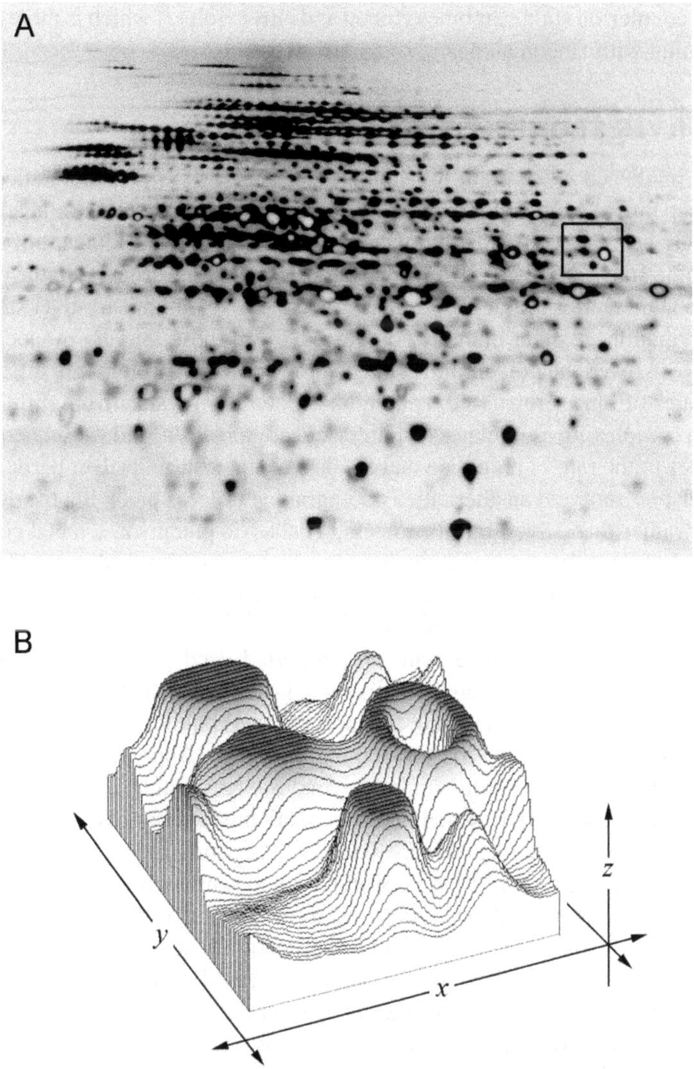

FIGURE 23.2 The propensity of silver staining toward negative staining of proteins. (A) Two-dimensional gel of *Escherichia coli* cytosolic proteins (20 micrograms total protein) stained with glutaraldehyde-free silver stain; (B) three-dimensional surface plot of integrated spot densities for the region outlined in (A) illustrating how negative staining skews quantitation. The crater at the right is the negatively stained protein spot. Pixel density is plotted on the *z* axis. The depression results in an artificially lowered value of integrated spot density for that protein. The other three major spots are of saturated density rendering quantitation unreliable, even on a relative basis. Silver-stained gel provided by Myra Robinson of Proteome Systems.

to electrostactic interactions, neutral red incorporates into SDS micelles.) The most effective counterion stain combines zincon and ethyl violet,[17] which is more sensitive than staining with zincon alone.[9]

23.5 SILVER STAINING

The first published silver stain by Kerenyi and Gallyas[18] was an emulation of earlier histochemical stains and enabled the visualization of proteins in agarose gels. Later, Merril et al.[19] would revolutionize staining with the development of an acidic silver nitrate stain for polyacrylamide gels that extended sensitivity into the picogram range. Soon thereafter, Oakley et al.[20] described the first ammonical silver stain based on the formation of silver diamine complexes. From these initially monochromatic stains have evolved the polychromic silver stains[21] in which proteins are differentially stained hues of blue, brown, red, and yellow. Coloration results from differences in silver ion complexing as it relates to amino acid composition, and subsequently, from differences in the rate of reduction and the size of the resulting silver particles.

Yudelson[22] reported an alternative photographic process involving the deposition of nickel rather than silver. In this process, a catalytic precursor, a metal coordinate compound of the platinum series, is first adsorbed to proteins in the polyacrylamide gel. Reduction of the adsorbed metal transforms the protein bands into catalytic sites for the reduction of nickel to its metallic form.

Perhaps the most sensitive stain ever reported used [^{35}S] thiourea for toning of silver stained polyacrylamide gels. Adapted directly from the autoradiographic method used by astrophysicists to enhance photographs of faint distant galaxies, sensitivity was extended into the low femtogram range. Appropriately, Wallace and Saluz[23] would title their *Nature* article "Beyond Silver Staining."

Merril and Goldman[24] reported that silver staining is linear only at very low protein concentrations, spanning over the 0.02 to 2.0 ng/mm^2 range. (This would more appropriately be expressed in terms of mm^3, since sensitivity is also a function of gel thickness. Protein bands and spots are three-dimensional entities.) The slopes obtained for eight standard proteins had a coefficient of variation of 0.78. Moreover, silver staining correlates poorly with other stains. Smejkal et al.[25] demonstrated that for 944 different bacterial proteins, the correlation coefficient of silver staining with SYPRO Ruby and CBB staining was 0.48 and 0.66, respectively.

Silver ions bind proteins by electrostatic interaction with COO^- groups of Asp and Glu or by complexing with the imidazole, SH, SCH_3, and NH_3 groups of His, Cys, Met, and Lys, respectively. Thermodynamics favor the reduction of free silver ions in the gel over the reduction of silver complexed with protein, thus there is predisposition toward negative staining of proteins. While silver impregnation may be cross-sectional, the reduction of silver occurs only at the gel surface where the unbound silver ion concentration is significantly lowered by washing the gel prior to development. Hence, reproducible silver staining of proteins and moderation of background staining relies on the precise control over these washing steps; prolonged washing lowers sensitivity while insufficient washing results in high background development.

Most importantly, silver stains use aldehydes that covalently modify proteins rendering most of these methods incompatible with mass spectrometry. The removal of glutaraldehyde improves compatibility with mass spectrometry but coincides with decreased sensitivity, and formaldehyde is still necessary for the reduction of silver ions. Further, the removal of metallic silver prior to mass spectrometry requires the use of strong oxidizing agents.

23.6 NEGATIVE STAINS

From the use of metal salts have evolved a number of negative stains in which transparent protein zones are formed in contrast to a developing background. (The first silver stains were negative stains due to the faster rate at which free silver ions in the polyacrylamide gel are reduced.) Other metals such as copper, zinc, nickel, and cobalt are not reduced, but instead form insoluble complexes with components of the polyacrylamide gel other than the separated proteins.[26] The negative image develops within minutes and staining is completely reversible. This is in contrast to CBB staining, which binds proteins so tenaciously that it may not be completely removed from trypsin digested peptides, as evidenced from spectra showing mass shifts corresponding to multiples of CBB molecular mass.[27] Since proteins are not precipitated in the gel, trypsin digestion and the recovery of peptides is more effective.

Divalent transition metals react with carbonates and hydroxides produced during electrophoresis to form complexes that are apparently stabilized by Tris and SDS, but are dissociated by glycine, dithiothreitol (DTT), or ethylenediaminetetraacetic acid (EDTA).[28] In the zinc-imidazole method, SDS at concentrations below its CMC stabilizes zinc-imidazolate complexes in the surrounding gel, whereas the localized high concentration of SDS associated with proteins prevents the formation of these complexes.[29,30]

Casero et al.[31] described a negative gold stain with sensitivity reportedly equal to silver. (Such claims are rather ambiguous, since they usually emanate from comparisons with a particular staining method, not necessarily the most sensitive one.) While the underlying mechanisms of gold staining remain obscure, it is conjectured that mixed gold-detergent micelles are formed within the polyacrylamide matrix, where they remain trapped. For this reason, negative gold staining is ineffectual in agarose gels where macroporosity impedes the entrapment of nascent micelles.

Gombocz and Cortez[32] exploited the intrinsic fluorescence of carrier ampholytes used in isoelectric focusing (IEF) for protein shadowing. Fluorescent scanning of these gels during electrophoresis revealed the localized quenching of carrier ampholyte fluorescence by the focused protein zones, which registered as inverted peaks. Similarly, Chen and Chrambach[33] described gels fortified with the uncharged fluorescent compound umbelliferone and the quenching of fluorescence by unlabeled proteins.

23.7 SCHLIEREN OPTICS

In 1966, Allen and Moore[34] first used schlieren optics to monitor the electrophoresis of unstained proteins in polyacrylamide gels. Sensitivity was in the low microgram range. This approach was revisited two decades later by Takagi et al.[35] for real-time visualization of ionic boundaries in SDS PAGE. Using improved schlieren optics, bands corresponding to 100–300 ng of protein were detected.

23.8 FLUORESCENT STAINS

Numerous very sensitive methods for the fluorescent labeling of proteins prior to electrophoresis have been described. Both fluorescamine and 2-methoxy-2,4-diphenyl-3(2H)-furanone (MDPF) enable sensitivity in the low nanogram range,[36,37] and because these compounds and their hydrolysis products do not fluoresce, the detection of low levels of protein is not impaired by background fluorescence. The nonfluorescent reagent MDPF reacts with lysine amino groups to form covalent fluorescent adducts that are more photostable than their fluorescamine counterparts. Significant shifts in protein mass and isoelectric point are induced which negate IEF and complicate analysis of mass spectra. To overcome the obstacles presented by covalent modification in a manner that is permissive of IEF, proteins have otherwise been labeled with the cyanine dyes Cy2, Cy3, and Cy5 in a limiting reaction in which only 1% to 2% of lysines are modified.[38] Another approach enabling IEF is to label the proteins following focusing, but prior to second-dimension PAGE. Conrad et al.[39] described DNP derivatization of protein carbonyls in IPG strips during the equilibration process that precedes PAGE.

Yamamoto et al.[40] detected picomolar quantities of proteins in polyacrylamide gels based on the labeling of protein sulfhydryls with the fluorescent thiol *N*-(7-dimethylamino-4-methylcoumarinyl) maleimide (DACM). Later, the poor correlative of thiol-reactive fluorophores with polypeptide mass was very clearly demonstrated by Berggren et al.[41] using monobromobimane for detection of the subunits of *Escherichia coli* ATP synthase complex. The application of these reagents is limited since they stain only a specific residue such as cysteine, and consequently, they are insensitive for the detection of polypeptides that do not contain the reactive residue in sufficient numbers.

SYPRO Ruby has traditionally been used to bridge the gap between CBB and silver staining sensitivity. SYPRO Ruby has a broader linear dynamic range than silver, enables better sequence coverage, and in some instances, stains proteins that are poorly stained with silver. For example, SYPRO Ruby stains human epidermal keratin with 20 times greater sensitivity than acidic silver nitrate stain and with eight times greater sensitivity than alkaline silver diamine stain.[42] Rabilloud et al.[43] demonstrated a ruthenium II Tris bathrophenanthroline disulfonate complex with sensitivity greater than the original SYPRO Ruby, but this was later countered with a report of an improved formulation of SYPRO Ruby.[44] The staining mechanism of sulfonated lanthanide chelates is primarily through electrostatic interaction. The ruthenium ion coordinates three bathrophenanthroline disulfonates resulting in a hexasulfonated complex that avidly binds proteins in their protonated state. A correlative of 0.83 was found between SYPRO Ruby and CBB staining.[25]

Bell and Karuso[45] recently isolated a compound from the fungus *Epicoccum nigrum* useful as a fluorescent stain for polyacrylamide gels. Distributed by Amersham

Biosciences under the trade name Deep Purple, it is reportedly several times more sensitive than SYPRO Ruby.[46] The reactive component of the Deep Purple stain is epicocconone, a nonfluorescent azophilone that becomes fluorescent when it interacts with protein.[47] The shift to fluorescence is induced by the reaction of epicocconone with lysine amines. However, more epicocconone apparently binds to the proteins than can be converted intrinsically, and subsequently, gels must be incubated in dilute ammonia to fully convert the azophilone to its fluorescent form. As a consequence, residual epicoccone in the gel is also converted in the process, which generates background fluorescence and decreases the signal-to-noise ratio. Another disadvantage of the stain is its photoinstability.[25] The primary mechanism of epicocconone binding is through hydrophobic interaction with lypophilic tail of SDS bound to protein, similar to the

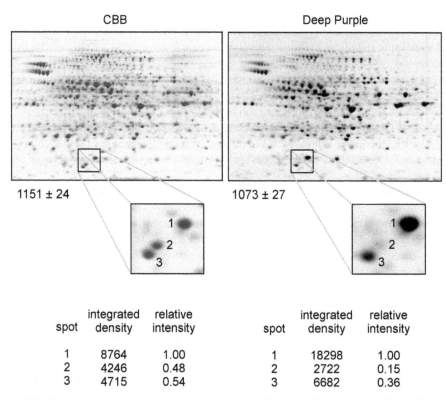

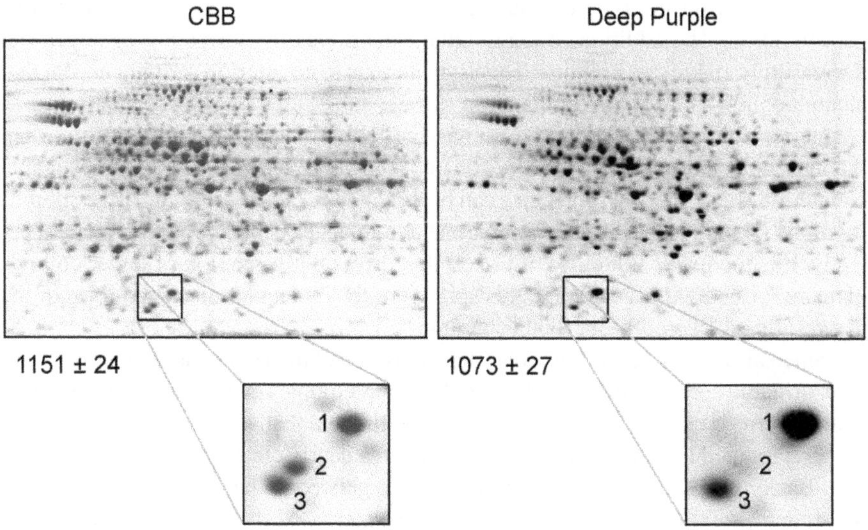

spot	integrated density	relative intensity
1	8764	1.00
2	4246	0.48
3	4715	0.54

spot	integrated density	relative intensity
1	18298	1.00
2	2722	0.15
3	6682	0.36

FIGURE 23.3 Colloidal CBB staining according to Smejkal et al.[51,52] compared to Deep Purple fluorescent stain. The number of spots detected in triplicate gels is shown. CBB detected 7.3% more protein spots than Deep Purple. Insets show the insensitivity in which Deep Purple detects the second, middle protein spot shown in the enlarged regions of the gel. The integrated density of each spot from which estimates of relative protein amount were derived for each stain are tabulated. First-dimension IEF was performed on IPGs pH 4–7 (left to right, acidic to basic proteins). Second-dimension SDS PAGE was performed on 6–15% polyacrylamide gradient gels.

mechanisms of Nile Red[48] and the SYPRO Red and Orange and dyes.[49] Staining based strictly on this interaction with SDS makes the incorrect assumption that all proteins bind SDS to a similar degree. To the contrary, glycoproteins bind significantly less SDS than other proteins,[50] and therefore, would be stained less efficiently with epicocconone. Hence, Deep Purple will be insensitive for some proteins (Figure 23.3, enlarged regions). A correlation coefficient of 0.84 was estimated between Deep Purple and CBB staining.[25]

23.9 RECENT DEVELOPMENTS IN COLLOIDAL COOMASSIE CHEMISTRY

The colloidal properties of CBB, or other ionic dyes for that matter, can be manipulated by altering the type of counterion and its concentration relative to the dye concentration. The addition of ammonium salts drives the formation of larger, but decreasingly stable dye colloids. This is analogous to counterion staining except that an ammonium salt, usually ammonium sulfate, substitutes for the secondary dye as counterion for stabilization of the primary colloid. Some stabilization occurs when the intermolecular repulsion of disulfonates is lowered in the presence of a suitable counterion, such that dye molecules can be arranged in closer proximity. The solubility of the G and R forms of the Coomassie dye in water is 50 and 70 mg/ml, respectively. It is the lower solubility of G-250 that drives the formation of larger colloids than are formed with R-250, and this shift from the molecular dispersed form of the dye results in lowered background staining.

Smejkal et al.[51] described the controlled transition of molecular dispersions of G 250 to an unstable colloid which nearly completely precipitates over a 10- to 12-hour time course. Termed equilibration staining, it was estimated that 99.1% ± 0.4% of the dye was precipitated leaving the gel to equilibrate in a nearly clear residual staining solution. Background staining was minimal and the gels could be scanned directly from the staining solution. Background staining was increased if the equilibrated gel was transferred to methanol, acetic acid, or even water since the dye resolubilized into the surrounding media and diffused randomly back into the gel. To counter this effect, the stained gels were stored in a dilute ammonium sulfate solution. Distributed by Proteome Systems under the tradename ProteomIQ Blue, sensitivity is comparable to SYPRO Ruby and Deep Purple fluorescent stains.[52]

The Blue Silver stain described by Candiano et al.[53] was more sensitive than any of the colloids previously described by Neuhoff et al.[4] Increased binding of CBB was attributed to increasing the phosphoric acid concentration, hence lowering the pH, to enable more comprehensive protonation of Asp and Glu residues. Unlike the name implies, it was not as sensitive as silver staining.

Recently, Lazarev and Johnson[54] integrated CBB fluorescence in gels scanned with an infrared laser scanner (LiCor, Nebraska). CBB fluoresces maximally near 680 nm excitation wavelength and an increase in sensitivity approaching two orders of magnitude was reported (Figure 23.4). As a fluorescent stain, CBB is at least twice as sensitive as SYPRO Ruby.

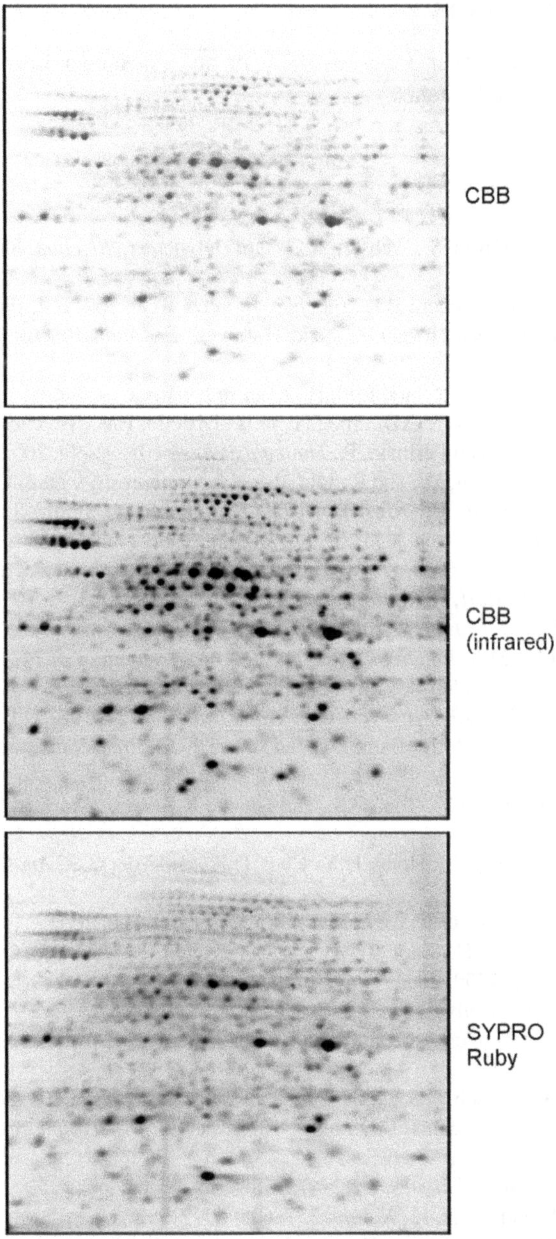

FIGURE 23.4 Two-dimensional gel of *Escherichia coli* cytosol stained with CBB and visualized by white light transillumination (top). The same gel scanned with a LiCor Odyssey infrared laser at excitation wavelength 680 nm (middle) compared to a duplicate gel stained with SYPRO Ruby fluorescent stain (bottom). First-dimension IEF was performed on IPGs pH 4–7 (left to right, acidic to basic proteins). Second-dimension SDS PAGE was performed on 6–15% polyacrylamide gradient gels. Images provided by Alexander Lazarev and Peter Johnson.

ACKNOWLEDGMENTS

I would like to thank Peter Johnson of LiCor and Alexander Lazarev and Myra Robinson of Proteome Systems.

REFERENCES

1. Fazekas, D.S., Groth, S., Webster, R.G., and Datyner, A., *Biochim. Biophys. Acta*, 71, 377–385, 1963.
2. Meyer, T.S. and Lambert, B.L., *Biochim. Biophys. Acta*, 107, 144, 1965.
3. Diezel, W., Koppenschlager, G., and Hofmann, E., *Anal. Biochem.*, 48, 617–620, 1972.
4. Neuhoff, V., Stamm, R., and Eibl, H., *Electrophoresis*, 6, 427–448, 1985.
5. Shainoff, J.R., Ratnoff, O.D., Smejkal, G.B., DiBello, P.M., Welches, W.R., Lill, H., Mitkevich, O.V., and Periman, P., *Thrombosis Res.*, 101, 91–99, 2001.
6. Patestos, N.P., Fauth, M., and Radola, B.J., *Electrophoresis*, 9, 488–496, 1988.
7. Zhu, H., Hargrove, M., Xie, Q., Nozaki, Y., Linse, K., Smith, S., Olson, J., and Riggs, A., *J. Biol. Chem.*, 271, 29,999–30,006, 1996.
8. Grovosky, M.A., Carlson, K., Rosenbaum, J.L., *Anal. Biochem.*, 35, 351–359, 1970.
9. Smejkal, G.B., Hoff, H.F., *Biotechniques*, 34, 486–488, 2003.
10. Righetti, P.G., *J. Chromatography*, 157, 243–251, 1978.
11. Maurer, H.R., *Disc Electrophoresis and Related Techniques of Polyacrylamide Gel Electrophoresis*, Walter de Gruyer, Berlin, 1971, pp. 73–76.
12. Hong, H.Y., Choi, J.K., and Yoo, G.S., *Anal. Biochem.*, 214, 96–99, 1993.
13. Dzandu, J.K., Deh, M.E., Barrett, D., and Wise, G.E., *Proc. Natl. Acad. Sci. USA*, 81, 1733–1737, 1984.
14. Jin, L.T., Hwang, S.Y., Yoo, G.S., and Choi, J.K., *Electrophoresis*, 25, 2486–2493, 2004.
15. Choi, J.K., Yoon, S.H., Hong, H.Y., Choi, D.K., and Yoo, G.S., *Anal. Biochem.*, 236, 82–84, 1996.
16. Choi, J.K. and Yoo, G.S., *Arch. Pharm. Res.*, 25, 704–708, 2002.
17. Choi, J.K., Tak, K.H., Jin, L.T., Hwang, S.Y., Kwon, T.I., and Yoo, G.S., *Electrophoresis*, 23, 4053–4059, 2002.
18. Kerenyi, L. and Gallyas, F., *Clin. Chem. Acta*, 38, 465–457, 1972.
19. Merril, C.R., Switzer, R.C., and Van Keuren, M.L., *Proc. Natl. Acad. Sci. USA*, 76, 4335–4340, 1979.
20. Oakley, B.R., Kirsch, D.R., and Morris, N.R., *Anal. Biochem.*, 105, 361–363, 1980.
21. Sammons, D.W., Adams, L.D., and Nishizawa, E.E., *Electrophoresis*, 2, 135–140, 1981.
22. Yudelson, J., *Biotechniques*, 2, 42–47, 1984.
23. Wallace, A. and Saluz, H., *Nature*, 357, 608–609, 1992.
24. Merril, C.R. and Goldman, D., *Two-Dimensional Gel Electrophoresis of Proteins*, Academic Press, Orlando, 1984, pp. 93–109.
25. Smejkal, G.B., Robinson, M.H., and Lazarev, A., *Electrophoresis*, 25, 2511–2519, 2004.
26. Dzandu, J.K., Johnson, J.F., Wise, G.E., *Anal. Biochem.*, 174, 157–167, 1988.
27. Righetti, P.G., Stoyanov, A.V., Zhukov, M.Y., *The Proteome Revisited: Theory and Practice of All Relevant Electrophoretic Steps*, Elsevier, Amsterdam, pp. 313–314.

28. Lee, C., Levin, A., and Branton, D., *Anal. Biochem.*, 166, 308–312, 1987.
29. Castellanos-Serra, L. and Hardy, E., *Electrophoresis*, 22, 864–873, 2001.
30. Ortiz, M.L., Calero, M., Fernandez-Patron, C., Patron, C.F., Castellanos, L., and Mendez, E., *FEBS Lett.*, 296, 300–304, 1992.
31. Casero, P., Battista de Campo, G., and Righetti, P.G., *Electrophoresis*, 6, 367–372, 1985.
32. Gombocz, E. and Cortez, E., *Electrophoresis*, 20, 1365–1372, 1999.
33. Chen, N. and Chrambach, A., *Anal. Biochem.*, 242, 64–67, 1996.
34. Allen, R.C. and Moore, D.J., *Anal. Biochem.*, 16, 457–465, 1966.
35. Takagi, T. and Kubota, H., *Electrophoresis*, 11, 361–366, 1990.
36. Jackowski, G. and Liew, C.C., *Anal. Biochem.*, 102, 321–325, 1980.
37. Barger, B.O., White, R.C., Pace, J.L., Kemper, D.L., and Ragland, W.L., *Anal. Biochem.*, 70, 327–335, 1976.
38. Unlu, M., Morgan, M.E., and Minden, J.S., *Electrophoresis*, 18, 2071–2077, 1997.
39. Conrad, C.C., Choi, J., Malakowsky, C.A., Talent, J.M., Dai, R., Marshall, P., and Gracy, R.W., *Proteomics*, 1, 829–834, 2001.
40. Yamamoto, K., Sekine, T., and Kananoka, V., *Anal Biochem.*, 79, 83–94, 1977.
41. Berggren, K.N., Chernokalskaya, E., Lopez, M.F., Beecham, J.M., Patton, W.F., *Proteomics*, 1, 54–65, 2001.
42. Berggren, K., Chernokalskaya, E., Steinberg, T.H., Kemper, C., Lopez, M.F., Diwu, Z., Haugland, R.P., Patton, W.F., *Electrophoresis*, 21, 2509–2521, 2000.
43. Rabilloud, T., Strub, J.M., Luche, S., Van Dorsselaer, A., Lunardi, J., *Proteomics*, 1, 699–704, 2001.
44. Berggren, K.N., Schulenberg, B., Lopez, M.F., Bogdanova, A., Smejkal, G., Wang, A., Patton, W.F., *Proteomics*, 2, 486–498, 2002.
45. Bell, P.J. and Karuso, P., *J. Am. Chem. Soc.*, 125, 9304–9305, 2003.
46. Mackintosh, J.A., Choi, H.Y., Bae, S.H., Veal, D.A., Bell, P.J., Ferrari, B.C., Van Dyk, D.D., Verrills, N.M., Paik, Y.K., and Karuso, P., *Proteomics*, 3, 2273–2288, 2003.
47. Ferrari, B.C., Attfield, P.V., Veal, D.A., Bell, P.J., *J. Microbiol. Methods*, 52, 133–135, 2003.
48. Daban, J.R., Bartolome, S., and Samso, M., *Anal. Biochem.*, 199, 169–174, 1991.
49. Steinberg, T.H., Jones, L.J., Haugland, R.P., Singer, V.L., *Anal. Biochem.*, 239, 223–237, 1996.
50. Segrest, J.P., Jackson, R.L., Andrews, E.P., Marchesi, V.T., *Biochem. Biophys. Res. Commn.*, 44, 390–395, 1971.
51. Smejkal, G.B., *Mol. Cel. Proteomics*, 2, 851, 2003.
52. Smejkal, G.B., Lazarev, A., Robinson, M.H., and McCarthy, J.T., *PharmaGenomics*, 4, 14–15, 2004.
53. Candiano, G., Bruschi, M., Musante, L., Santucci, L., Ghiggeri, G.M., Carnemolla, B., Orecchia, P., Zardi, L., and Righetti, P.G., *Electrophoresis*, 25, 1327–1333, 2004.
54. Smejkal, G.B., *Expert Rev. Proteomics*, 1, 381–387, 2005.
55. Urwin, V.E., Jackson, P., *Anal. Biochem.*, 209, 57–62, 1993.
56. Greenspan, P. and Gutman, R.L., *Electrophoresis*, 14, 65–68, 1993.
57. Steinberg, T.H., Lauber, W.M., Berggren, K., Kemper, C., Yue, S., and Patton, W.F., *Electrophoresis*, 21, 497–508, 2000.
58. Steinberg, T.H., Pretty On Top, K., Berggren, K.N., Kemper, C., Jones, L., Diwu, Z., Haugland, R.P., and Patton, W.F., *Proteomics* 1, 841–855, 2001.
59. Hart, C., Schulenberg, B., and Patton, W.F., *Electrophoresis*, 25, 2486–2493, 2004.

60. Hart, C., Schulenberg, B., Diwu, Z., Leung, W.Y., and Patton, W.F., *Electrophoresis*, 24, 599–610, 2003.
61. Schulenberg, B., Aggeler, R., Beechem, J.M., Capaldi, R.A., Patton, W.F., *J. Biol. Chem.*, 278, 27,251–27,255, 2003.
62. Lim, M.J., Patton, W.F., Lopez, M.F., Spofford, K.H., Shojaee, N., Shepro, D., *Anal. Biochem.*, 245, 184–195. 1997.
63. Berggren, K., Steinberg, T.H., Lauber, W.M., Carroll, J.A., Lopez, M.F., Chernokalskaya, E., Zieske, L., Diwu, Z., Haugland, R.P., and Patton, W.F., *Anal. Biochem.*, 276, 129–143, 1999.
64. Higgins, R.C. and Dahmus, M.E., *Anal. Biochem.*, 93, 257–260, 1979.
65. Taylor, M.D. and Andrews, A.T., *Electrophoresis*, 2, 76–81, 1981.
66. Carson, S.D., *Anal. Biochem.*, 78, 428, 1977.
67. Dahlberg, A.E., Dingman, C.W., and Peacock, A.C., *J. Mol. Biol.*, 4, 139, 1969.
68. Datyner, A. and Finnamore, E., *Anal. Biochem.*, 52, 45–55, 1973.
69. Ladner, C.L., Yang, J., Turner, R.J., Edwards, R.A., *Anal. Biochem.*, 326, 13–20, 2004.
70. Horowitz, P.M. and Bowman, S., *Anal. Biochem.*, 165, 430–434, 1987.
71. Hartman, B.K. and Udenfriend, S.A., *Anal. Biochem.*, 30, 391–394, 1969.
72. Eckhardt, A.E., Hayes, C.E., and Goldstein, I.J., *Anal. Biochem.*, 73, 192–197, 1976.
73. Stoklosa, J.T. and Latz, H.W., *Biochem. Biophys. Res. Commn.*, 60, 590–596, 1974.
74. Wallace, R.W., Yu, P.H., Dieckert, J.P., and Dieckert, J.W., *Anal. Biochem.*, 61, 86–92, 1974.
75. Triplett, B.A., *Biotechniques*, 13, 352–354, 1992.
76. Davis, B.J., *Ann. N.Y. Acad. Sci.*, 121, 404, 1964.
77. Crowle, A.J. and Cline, C.J., *J. Immunol. Meth.*, 17, 379–381, 1977.
78. Zapolski, E.J., Gersten, D.M., and Ledley, R.S., *Anal. Biochem.*, 123, 230–234, 1982.
79. Bickar, D., Reid, P.D., *Anal. Biochem.*, 203, 109–115, 1992.
80. Berson, G., *Anal. Biochem.*, 134, 230–234, 1983.
81. Ching-Ming Chung, M., *Anal. Biochem.*, 40, 498–502, 1985.
82. Kyriakopoulos, A., Hammel, C., Gessner, H., and Behne, D., *Am. Biotech. Lab.*, 4, 22, 1996.
83. Bahrman, N. and Thiellement, H., *Electrophoresis*, 6, 357–358, 1985.
84. Candiano, G., Porotto, M., Lanciotti, M., and Ghiggeri, G.M., *Anal. Biochem.*, 243, 245–248, 1996.
85. Wardi, A.H. and Michos, G.A., *Anal. Biochem.*, 49, 6007–609, 1972.
86. Zacharius, R.M., Zell, T.E., Morrison, J.H., and Woodlock, J.J., *Anal. Biochem.*, 30, 148, 1969.
87. Hill, I.E., Hogue, C.W.V., Clark, I.D., MacManus, J.P., and Szabo, A.G., *Anal. Biochem.*, 216, 439–443, 1994.
88. Zhou, Z.D., Liu, W.Y., and Li, M.Q., *Biotechnol. Lett.*, 21, 1801–1804, 2003.
89. Prat, J.P., Lamy, J.N, Weill, J.D., *Bull. Soc. Chim. Biol.*, 51, 1357, 1969.
90. Hatch, F.T., Lees, R.S., *Adv. Lipid Res.*, 6, 1, 1968.
91. Jackson, P., *Biochem J.*, 270, 705–713, 1990.
92. Smejkal, G.B. and Hoff, H.F., *Electrophoresis*, 15, 922–925, 1994.
93. Ali, R. and Sayeed, A.S., *Electrophoresis*, 11, 343–344, 1990.
94. McDonald, M., Mila, I., and Sealbert, A.J., *Agric Food Chem.*, 44, 599, 1996.

24 Multiplexed Proteomics: Fluorescent Detection of Proteins, Glycoproteins, and Phosphoproteins in Two-Dimensional (2D) Gels

Birte Aggeler-Schulenberg

CONTENTS

24.1 INTRODUCTION

The detection of posttranslational modifications of proteins is becoming increasingly important as it is being recognized to play a key role in cell signaling. The two most widely studied phenomena are glycosylation and phosphorylation of proteins.

Glycosylation can be divided into O-linked (attached to serine or threonine residues through an hydroxyl group) or N-linked (attached to asparagines through a β-amide bond) oligosaccharides.[1,2] Changes in N-linked branching structures of the carbohydrate moieties have been associated with the malignant transformation of cells[3–6] as well as cell–cell adhesion. A variety of methods have been developed both for gel electrophoresis and for pure mass spectrometry approaches (reviewed in[7]) to identify the glycoproteins as well as their carbohydrate content.

Protein phosphorylation has been studied extensively using a broad range of methods from immobilized metal affinity columns (IMAC) in conjunction with mass spectrometry analysis, radiography of incorporated ^{32}P (or ^{33}P) to immunoblotting against anti-phosphoserine, threonine, or tyrosine residues. Reversible phosphorylation plays a critical role in biological regulation (as in the Krebs cycle or cell signaling) as well as in the activation of oncogenes.[8–10]

In order to ease the detection of either glycosylated or phosphorylated proteins in polyacrylamide gels or microarrays, we developed the Multiplexed Proteomics® platform consisting of fluorescent stains that are capable of specifically detecting carbohydrate (Pro-Q® Emerald 300 glycoprotein gel stain) or phosphate groups (Pro-Q Diamond phosphoprotein gel stain) after sodium dodecyl sulfate–polyacrylamide gel electrophoresis (SDS-PAGE) or 2D gel electrophoresis.[11,12] Pro-Q Emerald glycoprotein stain is based on a periodic acid and Schiff Base chemistry that leads to a covalent bond of the dye to the carbohydrate moiety of the protein. Since the dye does not interact with the amino acids themselves, trypsin digest fingerprinting of proteins is still possible, especially in conjunction with an endoglycosidase treatment (like PNGase F). Pro-Q Diamond stain, on the other hand, is a noncovalent stain and therefore does not interfere with mass spectrometry either. Both stains can be sequentially stained for total protein detection using SYPRO® Ruby protein gel stain, which allows for differential display analysis[13,14] and relative quantitation of expression levels as well as subsequent mass spectrometric identification of proteins. If the posttranslational protein stains are followed by a total protein detection method like SYPRO Ruby, the degree of shift in phosphorylation or glycosylation due to the versus normal cells can be determined.[15]

The three dyes employed in the Multiplexed Proteomics platform span a wide range of emission wavelengths. The Pro-Q Emerald 300 dye has a ultraviolet (UV) excitation maximum and cannot be used on laser-based gel scanners, whereas the Pro-Q Diamond dye is optimally excited with a 532 nm laser. SYPRO Ruby protein gel stain is very versatile in that both UV light as well as 473 or 488 nm lasers can be used for excitation. This chapter is designed to help both new and experienced users of fluorescent stains to understand the basic principles as well as some pitfalls of using the Multiplexed Proteomics platform on 2D gels.

24.2 MATERIALS

24.2.1 2D Gel Staining

Pro-Q Emerald 300 glycoprotein gel stain, Pro-Q Diamond phosphoprotein gel stain, and SYPRO Ruby protein gel stain were obtained from Molecular Probes, Inc. (Eugene, Oreg.).

Human plasma and dimethylformamide (DMF) or dimethyl sulfoxide (DMSO) were obtained from SIGMA Chemical Company (St. Louis). IPG strips used for this study were purchased from Amersham Life Sciences (Piscataway, N.J.). The volumes used for fixation and washing large format 2D gels are typically 500 ml/gel. For the staining step, a minimum of 350 ml/gel (see protocol below) is required (larger volumes of 400 to 500 ml might be advisable, depending on the container size used). For smaller gels, a volume of stain that is roughly ten times the volume of the gel should be employed.

24.2.2 Glycoprotein Detection Using Pro-Q Emerald 300 Glycoprotein Gel Stain

The buffers and solutions needed for glycoprotein detection included in the Multiplexed Proteomics glycoprotein gel stain kit are as follows:

1. Pro-Q Emerald 300 reagent, component A, enough for 1l of total staining solution, which can be used to stain up to three large 2D gels (20 × 20 cm).
2. Pro-Q Emerald 300 staining buffer (component B), 1l.
3. Periodic acid, 10 g (component C).
4. SYPRO Ruby protein gel stain, 1l.

In addition, the following solutions should be made up fresh before use:

1. Fixation solution: 50% methanol, 10% acetic acid in dH_2O.
2. Wash solution: 3% acetic acid in dH_2O.
3. Carbohydrate oxidation solution: 1% periodic acid in 3% acetic acid. Add 1l3% acetic acid to component C and mix until completely dissolved. This solution is stable for several months.
4. Pro-Q Emerald staining solution: Add 24 ml of DMF to the Pro-Q Emerald 300 reagent (component A) and mix until completely dissolved. If DMSO is used instead of DMF, the gel background might increase. Unused solution should be stored in the freezer (≤–20°C) and is stable for several months. Each large format 2D gel is stained with 350 ml (absolute minimum in order to obtain high quality staining and good sensitivity) of diluted Pro-Q Emerald 300 dye.
5. Wash solution for SYPRO Ruby protein gel stain: 10% methanol (or ethanol), 7% acetic acid.

24.2.3 Phosphoprotein Detection in 2D Gels

The following solutions have to be made up fresh before use:

1. Fixation solution: 50% methanol, 10% acetic acid in dH_2O
2. Destain solution: 20% acetonitrile, 50 mM sodium acetate pH 4.0 (can be stored for several weeks at room temperature). Enough solution for

three washes should be prepared ahead of time. It is helpful to have a 1 M stock of sodium acetate at hand (pH 4) that can be diluted to 50 mM final concentration in 20% acetonitrile.

24.2.4 Total Protein Detection Using SYPRO Ruby Protein Gel Stain

For washing the gels after staining, enough of a 10% methanol (or ethanol), 7% acetic acid solution should be made up fresh for two to three washes.

24.2.5 Instrumentation for Imaging Fluorescent Stains and Data Analysis

In order to image all members of the Multiplexed Proteomics technology family, both a UV light based charge-coupling device (CCD) camera system and a laser-based gel scanner would be ideal. The following instruments can be used for imaging the fluorescent stains.

1. Pro-Q Emerald 300 dye: The LumiImager (Roche Biochemicals, Mannheim, Germany) or any standard UV light based imaging system with a 300 nm excitation is suitable. As an emission filter the 520 nm bandpass in the LumiImager or a Wratten #9 photographic filter are well suited.
2. Pro-Q Diamond dye: For optimal imaging results, a 532 nm excitation wavelength as obtained by a frequency doubled Nd-YAG laser is recommended. In case of the BioRad FX laser based imaging system, a 555 longpass filter is recommended for the emission, whereas the 580 nm longpass filter is optimally suited when using the FUJI FLA 3000. Optimal filters on other laser-based imaging systems have to be determined empirically.
3. SYPRO Ruby dye: This is the most versatile dye of the family with respect to suitable excitation wavelengths. SYPRO Ruby dye can be imaged on either a UV light based systems just like the Pro-Q Emerald 300 dye using the same emission filters or a laser-based imaging system. Optimal excitation using laser scanners is obtained with either a 473 nm (FUJI FLA 3000) or a 488 nm laser (BioRad FX). To match the emission wavelength, the same filters as used for Pro-Q Diamond dye are best suited.
4. Data analysis software: The number of good 2D analysis software packages has increased in the last five years, and the pros and cons cannot be discussed in this chapter. Some of the commercially available software packages include: Delta 2-D (Decodon, Greifswald Germany), Progenesis (Nonlinear dynamics, Newcastle upon Tyne, UK), Z3 (Compugen, TelAviv, Israel), and PDQuest (BioRad Laboratories, Hercules, Calif.). Virtual overlays for visual inspection can also be performed in Adobe Photoshop, but it is a bit more tricky than using the specialized software.

24.3 METHODS

Multiplexed Proteomics technology staining can be performed by sequential staining for one posttranslational modification combined with total protein poststain as well as by combining both posttranslational specific stains sequentially followed again by a total protein detection. It is important to image one stain before continuing to the next because the Pro-Q Diamond dye signal is removed during the second fixation. Doing the glycoprotein and phosphoprotein specific stains on separate gels is definitely the most reliable way, if quantitative data are the goal, as well as obtaining maximum sensitivity and specificity. However, it is possible to stain a single 2D gel first for glycoproteins, followed by an overnight fixation, before continuing on to the glycoprotein and total protein detection. This approach can lead to an increase in unspecific staining in the glycoprotein detection, as well as a reduction in signal intensity and loss in sensitivity in the Pro-Q Emerald 300 dye staining. The fixation in between the phosphoprotein stain and the glycoprotein stain was found to be extremely important in obtaining reliable results. For the untrained user, it might be helpful to stain one set of gels for each posttranslational modification separate at first as a control, before doing a sequential three color stain on a single gel. Simultaneous imaging of any of the stains is not possible.

24.3.1 Detection of Phosphoproteins

When considering the protein load for 2D gels, it is advisable to start with a load appropriate for a Colloidal Coomassie stain, which is similar to the sensitivity of Pro-Q Diamond phosphoprotein gel stain for singly phosphorylated proteins. We routinely load 150 to 250 μg of protein of a cell lysate for large format 2D gel (4–7 or 3–10 IPG strips). After running the 2D polyacrylamide gels, it is important to fix the gels in 50% methanol, 10% acetic acid overnight. All steps are conducted with gentle agitation (50 rpm). Complete removal of SDS ensures specific staining especially if a lower grade SDS is used. In some cases a second fix for 1 hour might be needed. The next day the fixative is washed out with three 15-minute washes in dH_20. The washes can be extended for up to 30 minutes each without negative influence on the final results. The gels are then stained for 1.5 to 2 hours in Pro-Q Diamond dye solution followed by three 30-minute destainings. Before imaging, it is important to wash the gels in dH_2O to prevent corrosion of the instruments due to vapors. Gels are then imaged using a 532 nm laser excitation and a 580 nm longpass filter (FUJI FLA 3000) or a 555 nm longpass filter (BioRad FX). If the background is still high or uneven, perform another destaining step for 30 minutes and repeat the water wash before imaging the gel again. If a UV light source is used for excitation, the sensitivity can be up to ten-fold lower, which has to be adjusted for with the protein load on the gel. It is helpful to have a positive control on the gel either as a 1D marker lane or a known phosphoprotein in the lysate that can be seen easily in order to adjust the grayscale correctly. If the gel background is brought up to extreme levels, low amounts of unspecifically labeled proteins can appear as very dark spots. Since the stain is of a noncovalent nature, proteins can still be identified after staining by trypsin digest and mass spectrometry methods.

24.3.2 Detection of Glycoproteins

Gels are fixed in 50% methanol, 10% acetic acid overnight like for Pro-Q Diamond dye staining. The next day, the gels are washed three times 15 minutes in 3% acetic acid, followed by an oxidation step for 1 to 2 hours. The periodic acid is removed by three to four 20-minute washes in 3% acetic acid. This step is very important to ensure good labeling with the Pro-Q Emerald dye. The staining step is performed for 2.5 to 3 hours in the diluted Pro-Q Emerald dye using the provided staining buffer. To remove the background staining of the gels, two to three 20-minutes of washing in 3% acetic acid is performed, before imaging. Again the gels could be imaged after two washes and again after three wash steps if necessary.

For gel imaging a UV light source is used for excitation in combination with a 520 nm band pass filter or a Wratten #9 longpass filter. Even though the stain binds covalently to carbohydrates, proteins can still be identified by mass spectrometry, especially, if the carbohydrate moiety is cleaved of before trypsin digestion using PNGaseF, for example, for N-linked carbohydrates.

24.3.3 Detection of Total Protein

After obtaining satisfactory images, each of the stains described in the previous sections can be directly followed by SYPRO Ruby protein gel stain (or any other total protein stain) without further fixation. It is advisable to stain the gels overnight to ensure the brightest signal. The next day the background signal has to be reduced by two to three 20-minute washes in 10% methanol, 7% acetic acid followed by another 15 minutes in dH_2O before imaging the gels.

24.3.4 Three-Color Staining of Gels

In some cases it might be desirable to stain a sample for glycol as well as phosphoprotein content followed by total protein staining. In those cases, the gel first has to be stained with the Pro-Q Diamond dye as described in section 24.3.1 followed by the complete protocol described in 24.3.2 including the fixation step, which is very important to be repeated. This ensures better specificity. The SYPRO Ruby dye staining can be performed straight after the Pro-Q Emerald 300 dye without another fixation. The brightness of the Pro-Q Emerald 300 dye signal is compromised to some extent if used in a three-color staining setup (see Figure 24.1 B, C) Human plasma is a very nice test case due to the presence of reasonable amounts of glycosylated and phosphorylated proteins.

24.4 CONCLUSIONS

A variety of methods for posttranslational modification identifications have evolved over the past five years, including mass spectrometry methods, Western blot based analysis with improved antibodies, as well as affinity-based purifications methods. The Multiplexed Proteomics technology platform is a good addition in regard to sensitivity, specificity, and ease of use. Through mixing and matching of the three different protein

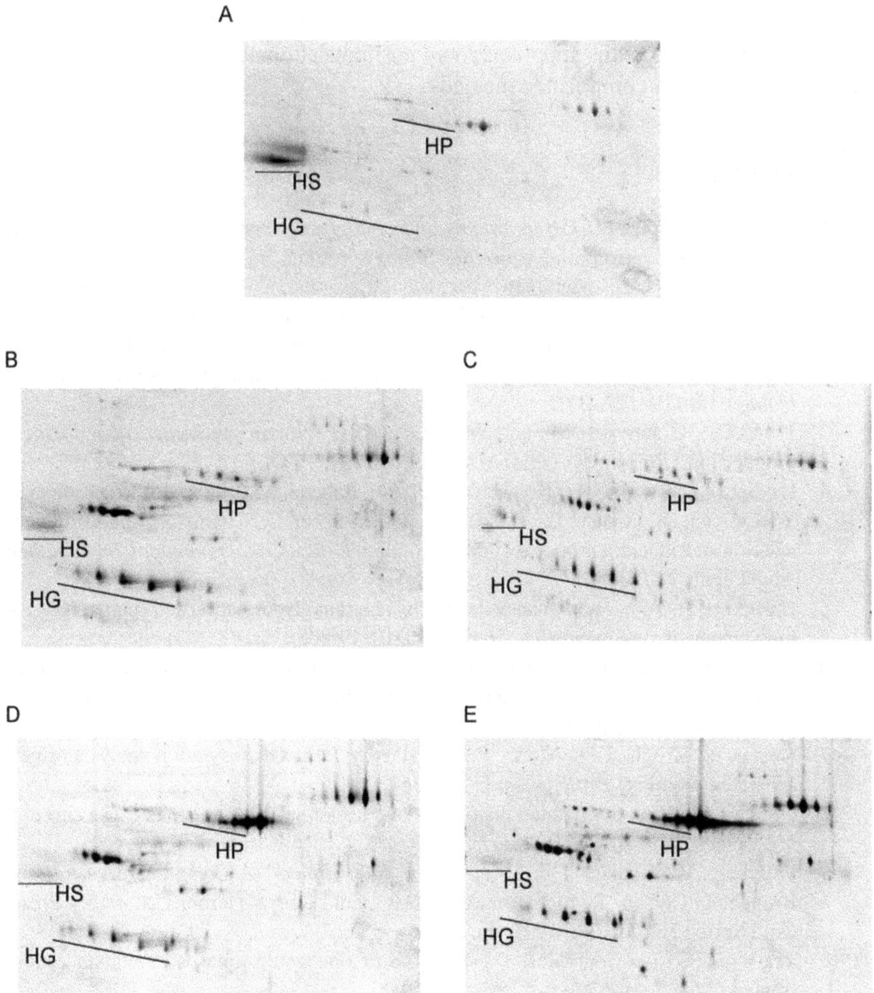

FIGURE 24.1 Human plasma after albumin depletion was separated on a 2D gel using 4–7 IPG strips and a 10% SDS-PAGE gel. (A) Gel stained with Pro-Q Diamond dye; (B) gel stained only with Pro-Q Emerald 300 dye; (C) gel stained with Pro-Q Emerald 300 dye after Pro-Q Diamond dye; (D) SYPRO Ruby dye staining of gel shown in B; (E): SYPRO Ruby dye staining of the gel shown in C. HG, haptoglobin β chain; HS, α2 HS glycoprotein; HP, hemopexin. Images for Pro-Q Emerald dye were obtained on the LumiImager (Roche Biochemicals). All other images were obtained using the BioRad FX (see Methods for details).

stains involved, this technology is flexible and adaptable to different questions. It can also be used as the front end of any mass spectrometry based analysis by streamlining the number of proteins that have a potentially interesting posttranslational modification, rather than using a shotgun approach. It can also be used in conjunction with Western blot analysis (e.g., the identification of the global glycoprotein pattern first, followed by specific lectin Western blots or in the case of phosphoprotein identification the Pro-Q Diamond dye can be paired with antibody detection of phosphor-Ser, -Thr, or -Tyr residues). Using this approach, more posttranslationally modified proteins can be identified with less complicated methods.

REFERENCES

1. Beeley, J., Chapter 1 in *Glycoprotein and Proteoglycan Techniques: Laboratory Techniques in Biochemistry and Molecular Biology*, Volume 16, Burdon, R. and van Knippenberg, P., Eds., Elsevier, New York, 1987, pp. 5–28.
2. Sasai, K., Ikeda, Y., Fuji, T., Tsuda, T., and Taniguchi, N., UDP-GlcNAc concentration is an important factor in the biosynthesis of beta 1,6-branched oligosaccharides: regulation based on the kinetic properties of N-acetylglucosaminyltransferase V, *Glycobiology*, 12, 119–127, 2002.
3. Dennis, J., Granovsky, M., and Warren, C., Glycoprotein glycosylation and cancer progression, *Biochim Biophys. Acta*, 1473, 21–34, 1999.
4. Laidler, P., Litynska, A., Hoja-Lukowicz, D., Labedz, M., Przybylo, M., Ciolczyk-Wierzbicka, D., Pochec, E., Trebacz, E., and Kremser, E., Characterization of glycosylation and adherent properties of melanoma cell lines, *Cancer Immunol. Immunother.* Aug 2 epub. 1–7, 2005.
5. Hakemori, S., Tumor malignancy defined by aberrant glycosylation and sphingo(glyco)-lipid metabolism, *Cancer Res.*, 56, 5309–5318, 1996.
6. Kobata, A. and Amano, J., Altered glycosylation of proteins produced by malignant cells, and application for the diagnosis and immunotherapy of tumours, *Immunol. Cell Biol.,* 83, 429–439, 2005.
7. Küster, B., Krogh, T. N., Mørtz, E., and Harvey, D. J., Glycosylation analysis of gel-separated proteins, *Proteomics*, 1, 350–361, 2001.
8. Grander, D., How do mutated oncogenes and tumor suppressor genes cause cancer?, *Med. Oncol.*, 15, 20–26, 1998.
9. Weinberg, R., Chapter 1 in *Origins of Human Cancer: A Comprehensive Review*, Brugge, J., Curran, T., McCormick, F., Eds., Cold Spring Harbor Laboratory Press, Cold Spring Harbor, NY, 1991, pp. 1–16.
10. Tsatsanis, C. and Spandilos, D., The role of oncogenic kinases in human cancer (review), *Int. J. Mol. Med.*, 5, 583–590, 2000.
11. Steinberg, T., Pretty On Top, K., Berggren, K., Kemper, C., Jones, L., Diwu, Z., Rapid and simple single nanogram detection of glycoproteins in polyacrylamide gels and on electroblots, Haugland, R., and Patton, W., *Proteomics*, 1, 841–855, 2001.
12. Steinberg, T., Agnew, B., Gee, K., Leung, W., Goodman, T., Schulenberg, B., Hendrickson, J., Beechem, J., Haugland, R., and Patton, W., Global quantitative phosphoprotein analysis using Multiplexed Proteomics technology, *Proteomics*, 3, 1128–1144, 2003.
13. Nishihara, J.C. and Champion, K.M., Quantitative evaluation of proteins in one- and two-dimensional polyacrylamide gels using a fluorescent stain, *Electrophoresis*, 23, 2203–2215, 2002.
14. Antelmann, H., Darmon, E., Noone, D., Veening, J.W., Westers, H., Bron, S., Kuipers, O. P., Devine, K.M., Hecker, M., and van Dijl, J. M., The extracellular proteome of *Bacillus subtilis* under secretion stress conditions, *Mol. Microbiol.*, 49, 143–156, 2003.
15. Schulenberg, B., Beechem, J., and Patton, W., Mapping glycosylation changes related to cancer using the Multiplexed Proteomics technology: a protein differential display approach, *J. Chromatogr. B, Analyt. Technol. Biomed. Life Sci.,* 793, 127–139, 2003.

25 Glyoxyl Agarose and Its Composite Gels: Advantageous Alternatives to Polyacrylamide Gel Electrophoresis for Very Large and Small Proteins and Peptides and All Sizes in Between

John R. Shainoff

CONTENTS

25.1 INTRODUCTION

Large proteins do not penetrate gels for polyacrylamide gel electrophoresis (PAGE), and small proteins and peptides cannot be fixed by blotting for detection or measurement. Even molecular weight markers do not transfer reliably, but their transfer is inversely dependent on molecular size.[1] We were driven to develop an alternative to

PAGE and Western blotting because fibrinogen and its multimeric derivatives, the proteins of principal interest to us, do not blot-transfer at all. By example, studies using Western blotting failed to detect all but degraded forms of fibrin(ogen), while our studies using direct immunoprobing showed that degraded fibrinogen comprise a miniscule percentage of the total (Figure 25.1) and intact fibrinogen and its cross-linked polymers comprise nearly all of the deposits.[2] The intact fibrinogen and its oligomers remain in the poyacrylamide gels like highly insoluble, precipitated blocks of coal as they separate from the SDS. Our solution was to develop an agarose-based derivative with acetaldehyde substituent groups that interact too weakly with amino groups to interfere with electrophoresis but could covalently link the protein and peptide to the gel matrix as alkyl amine adducts by simply immersing the gel in buffer containing cyanoborohydride to catalyze reductive amination of the aldehyde substituent groups.[3,4] Further, the gels could be used for multiple immunoprobing for tricolor staining of fibrinogen α, β, and γ chains[5] and enhancing sensitivity because retained antibody can be fixed in-place for multiple processing without loss. The advantages are illustrated by their utility for detecting numerous fibrin(ogen) derivatives not known before[2,6,7,8] and detection of small peptides such as heme-octapeptide that cannot be fixed by conventional fixatives.[4] Even the usual molecular weight standards have variable blot-transfer characteristics, varying inversely with molecular size.[1] However,

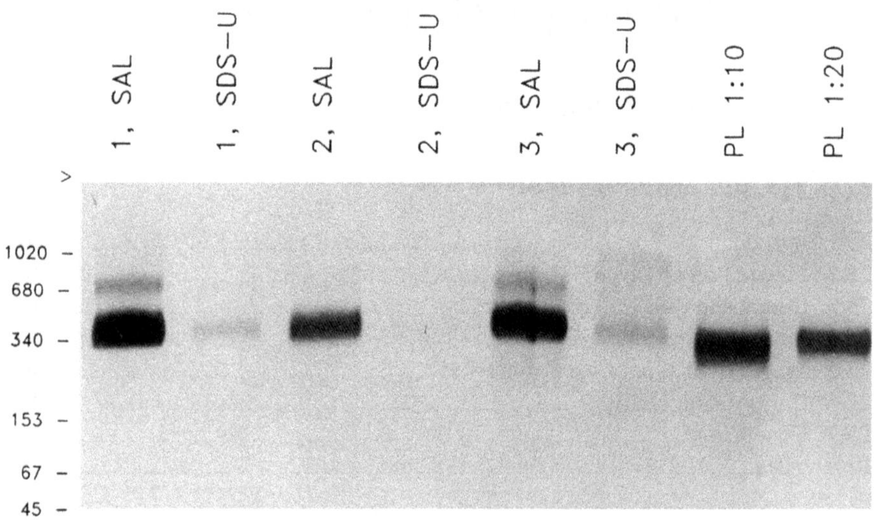

FIGURE 25.1 Glyoxyl agarose electropherograms of saline (SAL) and SDS-urea (SDS-U) extracts of three atherosclerotic aortic intimas probed directly with anti-fibrinogen antibody showing only intact fibrinogen and oligomers, and virtually no degraded forms. The two lanes at the right (designated PL) are of a normal plasma sample run at indicated dilutions. Prior studies using Western blotting uncovered only degraded forms from intimal extracts because intact fibrinogen does not blot transfer.[2] Likewise, reduced gels failed to reveal degraded polypeptide chains of fibrinogen, only intact chains and cross-linked high molecular weight oligomers, which also do not blot-transfer well at all.

direct probing of thick gels is much more time consuming and requires more attention to detail than the much simpler probing of thin blotting membranes. Yet, blotting fails to reveal many proteins. We have devised a simplified approach to direct immunoprobing by converting the thick electrophoresis gels to thin membranes that can be probed as easily and more certainly than blotting membranes.

25.2 PREPARATION OF GLYOXYL AGAROSE

Glyoxyl agarose is best prepared by reaction of agarose gels, rather than powders or molten solutions, with glycidol and oxidation of resultant glyceryl agarose, as described.[3] A simple method involving derivatization of agarose powder was described,[4] but this approach yields a low melting form that can partially dissolve in SDS solutions. The gels used for derivatizing have protected hydroxyls involved in aggregation of the agarose, whereas those hydroxyls are left unprotected in the powders and in molten agarose. Importantly, the glycidol should have low polymer content (viscosity below 4 centipoise at 20°C), obtained either from fresh lots from the manufacturer or purified by vacuum distillation and stored at −70°C. The derivatized agarose can be stored at 4°C for years in 0.1 mM HCl. For gel preparation, the glyoxyl agarose should be mixed with twice as much unmodified agarose prior to melting.

25.3 COMPOSITE GELS

Agarose gels, by themselves, do not adequately sieve proteins of moderate molecular size. To enhance sieving we constructed composites with degradable (removable) polyacrylamide fillers, as described.[4,7] These fillers are constructed using dihydroxyethylene-bis-acrylamide (DHEBA) as a cross-linker because it is easily hydrolyzed at alkaline pH. We also add 1-allyloxy-2,3-proparnediol to the polymerization mixture because it helps produce small polyacrylamide fragments that are easily washed out of the glyoxyl agarose matrix with its retained (immobilized) proteins or peptides.[4] Upon fixing the protein or peptide and removal of the PAG filler, the gels are open to direct probing with antibody.

25.4 SIMPLIFIED IMMUNOPROBING WITH COMPRESSED GELS

The problem with direct immunoprobing is that gels are thick (nominally 1.5 mm), and because of dissociability of antigen and antibody, complexes require specialized washout procedures, best achieved by precise suctioning buffer through the gels. That imposes a burden, and except where necessary for fibrinogen and von Willebrand factor[7,9] the approach has been ignored. To remove that obstacle, we devised a method for constructing gels that could be compressed from 1.8 to 0.3 mm, which enabled washout of unbound antibody in 20 seconds and made direct immunoprobing as simple as Western blotting.[10] However, this simplified approach needs prefabricated gels to become adopted because few laboratories prepare their own gels and want gels that can be run in a standard format for comparing results with international databases, regardless of the deficiencies of the gels.

25.5 MASS PRODUCTION OF THE COMPOSITES

The simplest approach, in our opinion, would be to photopolymerize the acrylamide filler in the composite gels using methylene blue as chromophore,[11] which was shown to be superior to persulfate-catalyzed polymerization in both acid and basic buffers. Further, by using a moving illumination shield, it should be possible to create gradient gels that can separate and fix proteins from $10^{<1}$ to 10^8 kDa without worrying about fixation or blot-transfer characteristics.

25.6 CURRENT AVAILABILITY

Glyoxyl agarose was initially available as NuFix from FMC Bioproducts (now Cambrex); however, they discontinued production because their method of preparation yielded a low-melting form, synthesized in molten form, which proved no better than other immobilizing media despite its low cost for beaded affinity gels. Glyoxyl agarose is currently available from Hispanagar (Madrid, Spain) distributed in the U.S. by Global Imports (Tampa, Fla., www.Abtbeads.com).

25.7 CONCLUSION

What do you do when PAGE or Western blotting fails your research efforts because your proteins do not enter the gels or blot transfer efficiently? Consider glyoxyl agarose and its composites, which are permeable to anything and also eliminate, by direct probing, the uncertainties of Western blotting. Glyoxyl agarose and its composites can also be probed repetitively for either enhanced sensitivity or multicolor discrimination.[4,10]

REFERENCES

1. Smejkal, G.B. and Gallagher, S., Determination of semidry protein transfer efficiency with transverse gradient gel electrophoresis, *Biotechniques*, 16, 196–199, 1994.
2. Valenzuela, R., Shainoff, J.R., DiBello, P.M., Urbanic, D.A., Anderson, J.M., Matsueda, G.R., and Kudryk, B.J., Immunoelectrophoretic and immunohistochemical characterizations of fibrinogen derivatives in atherosclerotic aortic intimas and vascular prosthesis pseudo-intimas, *Am. J. Pathology*, 141, 861–880, 1992.
3. Shainoff, J.R., Zonal immobilization of proteins, *Biochem. Biophys. Res. Commn.*, 95, 690–695, 1980.
4. Shainoff, J.R., Electrophoresis and direct immunoprobing on glyoxyl agarose and polyacrylamide composites, *Adv. Electrophoresis*, 6, 61–177, 1993.
5. Shainoff, J.R. and Urbanic, D.A., Multicolour immunostaining of fibrinogen polypeptide chains for identification of their derivatives in electrophoregrams, *Blood Coagulation Fibrinolysis*, 1, 479–484, 1990.

6. Shainoff, J.R., Smejkal, G.B., DiBello, P.M., Mitkevich, O.V., Levy, P.J., Lill, H., and Dempfle, C.E., Isolation and characterization of the fibrin-intermediate arising from cleavage of one fibrinopeptide A from fibrinogen, *J. Biol. Chem.*, 271, 24,129–24,137, 1996.
7. Shainoff, J.R., Urbanic, D.A., and DiBello, P.M., Immunoelectrophoretic characterizations of the cross-linking of fibrinogen and fibrin by plasma- and tissue-transglutaminase. Identification of a rapid mode of hybrid α/γ-chain cross-linking that is promoted by the γ-chain cross-linking, *J. Biol. Chem.*, 266, 6429–6437, 1991.
8. Shainoff, J.R., Valenzuela, R., Urbanic, D.A., DiBello, P.M., Lucas, F.V., and Graor, R., Fibrinogen Aα- and γ-chain dimers as potential differential indicators of atherosclerotic and thrombotic vascular disease, *Blood Coagulation Fibrinolysis*, 1, 499–503, 1990.
9. Hoyer, L.W. and Shainoff, J.R., Factor VIII-related protein circulates in normal human plasma as high molecular weight multimers, *Blood*, 6, 1056–1059, 1980.
10. Smejkal, G.B., Shainoff, J.R., and Yakubenko, A.V., Direct chemiluminescent immunodetection of proteins in agarose gels, *Electrophoresis*, 23, 979–984, 2002.
11. Caglio, S., Chiari, M., and Righetti, P.G., Gel polymerization in detergents: Conversion efficiency of methylene blue vs. persulfate catalysis, as investigated by capillary zone electrophoresis, *Electrophoresis*, 15, 209–214, 1994.

26 Nuclear Magnetic Resonance–Driven Chemical Proteomics: The Functional and Mechanistic Complement to Proteomics

Phani Kumar Pullela and Daniel S. Sem

CONTENTS

26.1 INTRODUCTION

Typically in biochemistry, protein purification is followed by functional studies to characterize a purified protein. Likewise, in the emerging field of proteomics, separation of the many proteins in the proteome will need to be followed with functional characterizations, but now on a much larger scale since more proteins are involved. To this end, the field of chemical proteomics[1,2] has emerged as a subdiscipline of proteomics, devoted to the study of protein–ligand interactions in a proteome-wide or systems-based manner. We define systems-based strategies broadly to include not only the simultaneous study of proteins related by metabolic or regulatory pathway,[3] but also those defined by binding site shape and chemical make-up (pharmacophore).[4–6] Thus, proteins are grouped by the network of interactions that occur either directly between them or indirectly through the ligands they bind. Studies can be of either (*a*) mixtures of proteins (either in cells or cell lysates) analyzed as a pool, or (*b*) purified proteins analyzed in parallel.

While chemical proteomic strategies were initially developed using methods such as affinity- or activity-based profiling,[7,8] nuclear magnetic resonance (NMR) methods are increasingly being used to broadly explore protein function.[9,10] We explore these methods in this chapter. Strategies include (*a*) NMR-guided design of gene-family focused libraries of chemical proteomic probes using NMR SOLVE;[4,11,12] (*b*) affinity fingerprinting with Cofactor fingerprinting with saturation transfer difference (CF STD) NMR;[9,10,13] (*c*) NMR-constrained docking studies;[12,14] and (*d*) methods for assaying protein–ligand interactions *in vivo* (*in cell* NMR[15] and magnetic resonance imaging[13]). Studies of protein–ligand interactions across gene families are discussed as a means for understanding the basic biochemical roles of proteins and the design of drugs and chemical genomic probes and inhibitors to determine function *in vivo*. With regard to drug and inhibitor design, proteins and protein families are presented as targets and antitargets in terms of whether desired or undesired protein–ligand interactions occur in the biological milieu.

26.2 BACKGROUND

26.2.1 Chemical Proteomics as Systems-Based Characterization of Protein Function

The field of chemical genomics (or chemogenomics) relies on chemical probes to explore biological function in the context of a whole organism—effectively using chemicals to destroy the function of a protein target, thereby generating the chemical equivalent of a genetic knockout.[16] To the extent that the underlying biology of an organism is determined by the biochemical functions of all its proteins (its proteome), the strategy of chemical genomics is to probe the biological function by chemically knocking out one protein at a time and studying the biological consequences of each knockout. Likewise, chemical proteomics (or chemoproteomics) relies on chemical probes to address protein function at the level of the protein itself, rather than the whole organism. It does this in a systems-based or parallel manner within an organism, organ, or organelle, or on panels of purified proteins. Emphasis is on protein systems related by metabolic or regulatory pathways or by binding-site pharmacophore. Networks of proteins related by binding site tend to bind the same ligands or chemical probes and are called pharmacofamilies.[4–6] The field of chemical proteomics was defined by Jeffery and Bogyo as being focused on the method or mechanism through which the structure, function, and role of proteins in different biological systems can be studied using chemical probes.[1] These are typically affinity-based probes used to label proteins with some detectable label such as a fluorescent label that can be detected in the context of a protein gel (Figure 26.1). Examples of such probes have been used to label active site thiol nucleophiles of proteases.[17] Isotope coded affinity tags (ICATs) are another widely used chemical proteomic strategy, used to label proteins either in the solid or solution phase and then later identified using mass spectrometry.[18] Such global labeling has been used to detect changes across a proteome in response to some effector such as a drug. Solution-phase studies of protein function using chemical probes are too numerous to review here, but include most prominently mass spectrometry,[19] ultraviolet spectroscopy,[20] and NMR.[21] The new challenge that proteomics presents to the field of biochemistry is how to perform such studies either on pools of proteins or in broadly executed parallel studies of many related proteins. This chapter focuses on some current uses and future potential of NMR methods for such studies in chemical proteomics.

26.2.2 NMR-Based Characterization of Protein Structure

NMR spectroscopy offers the flexibility of studying either the chemical probe or the protein receptor, or both. NMR methods have been used widely to determine protein structures, dating back four decades,[21] and NMR as a tool for such studies was the topic of the shared 2002 Nobel Prize in Chemistry. The first structures of biological macromolecules determined by NMR spectroscopy were based solely on [^{1}H,^{1}H] correlation experiments for proteins up to a size of about 10 kDa. However, for larger proteins, the increased resonance overlap made structure determination increasingly difficult. The introduction of isotopic labeling of proteins with the ^{15}N and ^{13}C NMR active nuclei has now extended the molecular weight range to beyond 20 kDa, by reducing resonance overlap through separation of the peaks along one or more heteronuclear frequency dimensions.[22] Suitable protein samples can be expressed in bacteria that are grown on minimal media that contains ^{15}N-labeled ammonium salts as the sole nitrogen source and ^{13}C-labeled glucose as the sole carbon source. For larger proteins, deuteration is necessary, which can be achieved by growing bacteria in D_2O. An additional problem with even larger proteins (>20 kDa) is they display broad lines due to slower tumbling rates and efficient transverse relaxation.

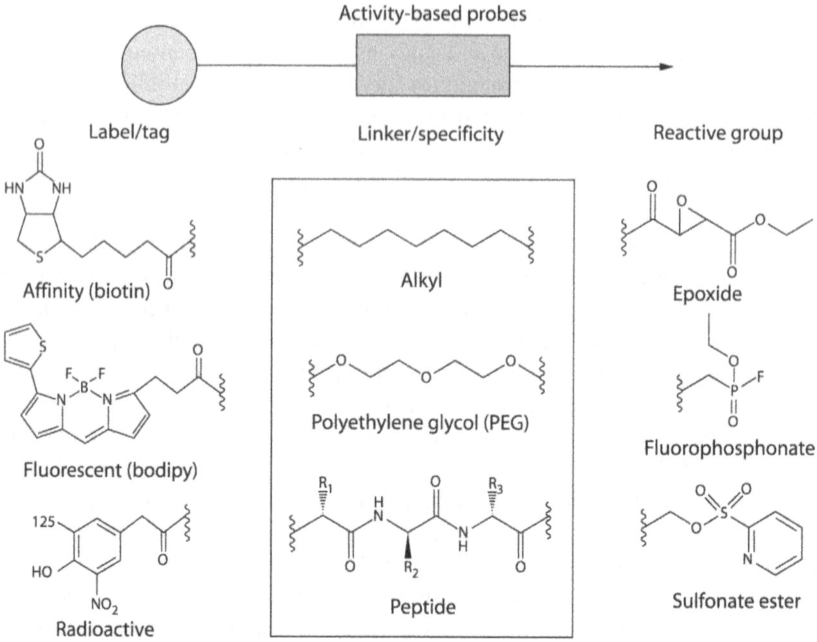

FIGURE 26.1 Examples of chemical probes used in chemical proteomic studies to label proteins with detectable groups. Electrophilic groups in the right column are attacked by active site nucleophiles, yielding proteins that are tagged with a detectable group. Reproduced with permission from.[1]

Transverse relaxation optimized spectroscopy (TROSY)[23] has now enabled the recording of high-resolution NMR spectra of macromolecules and multimolecular assemblies with good resolution for masses up to 100 kDa. The diminishing molecular weight constraints and the ability to study amorphous, nanocrystalline or microcrystalline materials (as opposed to crystalline solids for X-ray diffraction) has also led to the use of solid state NMR as an additional tool for the study of biological systems, especially membrane-bound proteins.[24]

NMR structures now constitute 15 to 20% of those submitted to the Protein Data Bank (PDB), and the method is making significant contributions in the field of structural proteomics.[25,26] Since there is a direct relationship between protein structure and protein function, it is felt that function can be better addressed in a broad manner if we have access to the structural complement of the proteome, or at least for a representative subset of proteins in the proteome. NMR along with x-ray crystallography are central in this effort,[25] and NMR is especially useful for smaller proteins that are not easily crystallized. But NMR structural studies extend beyond full characterization of the three-dimensional (3D) structure of a protein. Recently, there has been much interest in using NMR as a tool to guide the modular design of protein inhibitors for drug discovery[26] and to probe the function and mechanism of proteins.[4] Such studies are discussed later in the chapter, while here we turn next to studies of the dynamic or mobile nature of protein structures.

26.2.3 Protein Dynamics: From Snapshots to Movies

Although NMR is playing a vital role in structural proteomics efforts, its main contribution may be in a related effort. In the actual proteome as it exists in the cell, proteins are not static structures but rather are constantly in motion—and NMR is able to characterize protein motions on various timescales. Proteins are constantly in motion, with entire domains and regions of secondary structure and even sidechains moving, to accommodate binding events as well as changes in structure needed for the protein to accomplish its function as a receptor, enzyme, ion channel, molecular motor, and so forth. Dynamic motion is vital to protein function, which depends on alterations in 3D structure in response to specific molecular interactions with other proteins or with small molecule ligands—the focus of chemical proteomic studies. Multidimensional NMR methods, combined with isotope labeling, can provide access to dynamic information for virtually every atom in a protein.[22] Typical experiments involve measurements of the relaxation rates of ^{15}N nuclear spin and of the ^{1}H–^{15}N cross relaxation rate, the latter being measured via the steady state ^{15}N{^{1}H} nuclear Overhauser effect (NOE).[22] ^{15}N relaxation studies of proteins are typically performed at protein concentrations approaching the millimolar range. This is a considerable drawback, as it is not always possible to obtain labeled protein in sufficient quantities or to be able to create protein solutions that are concentrated enough for the required measurements. However, there has recently been a revolutionary improvement in sensitivity, made possible with cryoprobe technology. The receiver coil is cooled to near liquid helium temperatures (~25 K), thus decreasing electronic thermal noise in the receiver and resulting in increased signal-to-noise ratio without affecting temperature in the sample chamber. The sensitivity enhancement available with cryoprobe technology is an impressive 3.1- to 3.3-fold, compared with the 2-fold improvement obtained in going from 500 to 800 MHz.

26.2.4 NMR-Based Fragment Assembly: Modular Inhibitor Design in NMR Legoland™

NMR-based modular approaches to drug design began in the mid-1990s with the introduction of the structure activity relationship (SAR) by NMR technique,[26] although the general modular approach to inhibitor design predates this and was elaborated on extensively in the field of enzymology by Jencks et al.[27] SAR by NMR is based on the use of protein chemical shift changes to screen for low-affinity ligands, followed by structural characterization to determine which ligands are proximal to each other, then these are chemically joined through a linker. Thus, structural information is used to direct a linked fragment approach to enhancing binding affinities. Other NMR-based fragment screening methods have been reported in recent years, such as SHAPES.[28] In SHAPES, the goal is to assess the binding of a fairly small but diverse library of low molecular weight scaffolds to a drug target using NMR techniques that detect ligand. An important aspect of the SHAPES strategy is that the fragments being screened for binding are carefully preselected to be druglike building blocks that represent a basis set of fragments that were chosen from an analysis of preexisting drugs. This can be done by choosing fragments that are easily joined, to address ease of synthesis earlier in the process. To facilitate methods such as SHAPES, various ligand-screening methods have been reported that rely on one-dimensional NMR experiments to detect transfer of magnetization between the protein and the ligand, as an indicator of binding. Among these methods, saturation transfer difference (STD),[13] and water ligand observation by gradient spectroscopy (WaterLOGSY)[29] are widely used, due to their high sensitivity and ease of implementation. NMR has also emerged as a tool to address the needs of functional genomics and proteomics by extending these fragment assembly strategies into the realm of combinatorial chemistry, with application to families of proteins. To enable the focusing of combinatorial libraries of chemical probes to a particular gene family, nuclear magnetic resonance–structurally oriented library valency engineering (NMR SOLVE)[4,11,12] was developed and is discussed further in section 26.3.

26.2.5 The Emerging Role of NMR Spectroscopy in Chemical Proteomics

The field of chemical proteomics is about studying protein–ligand interactions on a large scale, and NMR is emerging as a powerful method for such studies. NMR plays a role not just in designing chemical proteomic probes, as just discussed, but also in the biophysical characterization of protein–ligand interactions. An important asset of NMR spectroscopy in chemical proteomics is its ability to detect ligand binding for tight binding ligands as well as compounds with lower affinities. This unique property makes NMR-based studies an extremely flexible tool for both modular inhibitor and drug design and subsequent drug discovery (screening).[9] Methods for detecting binding fall into two main categories: those that monitor NMR signals from the protein, such as chemical shift mapping, and those that monitor the ligand signal. The latter exploit either magnetization transfer from protein to ligand or the large differences in the rates of rotational and translational motions of a small molecule in the free-state relative to when it is bound to a macromolecule. The utility of NMR in chemical proteomics therefore includes studies of protein structure and dynamics as well as design of inhibitors that can be used as functional probes or as drug leads. NMR can also be used to help define a protein's function by using cofactor fingerprinting methods described subsequently, which identify the major cofactors required for function and therefore provide a starting point in defining the chemical mechanism for an enzyme. These *in vitro* NMR studies of biochemical function can even be extended to chemical proteomic studies *in vivo* by probing protein ligand interactions within cells using *in cell* NMR[15] or within the context of a whole organism using molecular imaging.[13,30] The next section discusses experimental details for a few of these chemical proteomic applications that make central use of NMR methods.

26.3 CHEMICAL PROTEOMIC APPLICATIONS OF NMR *IN VITRO*

26.3.1 A NMR-Guided Design of Gene-Family-Focused Libraries of Chemical Proteomic Probes Using NMR SOLVE

The NMR SOLVE method exploits the fact that large families of cofactor-dependent proteins have adjacent binding sites, of which one is conserved throughout the family since it binds the cofactor. An advantage of the method is that it derives structural information in a very pragmatic way by using selective isotope labeling to observe and assign only a small number of key protons in a binding site and mapping the binding sites with chemical shift perturbations by cofactor reference ligands.[4,11,12] It can thus derive structural information in the absence of complete NMR assignments and avoid the complexity of NMR spectra of uniformly isotope-labeled large proteins. NMR SOLVE finds greatest utility in chemical proteomics, in the sense that it can be used to design chemical probes that are useful across a whole family of proteins related by a common cofactor binding site, such as oxidoreductases. The NMR SOLVE strategy therefore differs from SAR by NMR in that it allows the development of a library of inhibitors for a family of proteins. This kind of parallel production of inhibitors across a gene family, such as oxidoreductases, will have applications in chemogenomic and functional genomic efforts to define protein functions, as well as in drug design.

NMR SOLVE was recently used to design biligand libraries off of a privileged scaffold for the oxidoreductases.[4] Oxidoreductases comprise two adjacent binding sites: the NAD(P)H cofactor binding site and the substrate-binding site. The binding site for the nicotinamide ring of the cofactor is always close to the substrate site, since the nicotinamide ring is involved in a hydride transfer reaction with the substrate. After assigning cross-peaks for the cofactor and substrate binding sites, the next step is to appropriately orient or dock a cofactor mimic with NOE and chemical shift perturbation data so that a linker can be placed on it next to the substrate site. This linker will serve as an attachment point for expanding a combinatorial library, directed from the privileged scaffold into the substrate site. Fragments to be attached here are a diverse library of druglike fragments, and because of the chelate effect, attachment of even a weak binding fragment can lead to high-affinity binding. This type of joining of two weak binding fragments has been proposed to provide a contribution as large as 45 entropy units to the binding energy, corresponding to an increase in affinity of 10^8 fold.[4,27] The realization of this 10^8-fold increase in affinity is under ideal thermodynamic conditions and will be difficult to obtain in a practical sense. Factors that can decrease the magnitude of this effect include flexibility of linkers, nonoptimal placement of ligands, and molecular repulsions between binding site residues and ligands or linkers. Still, this combinatorial strategy for fragment linkage is an efficient way of focusing a library, since even an imperfect linkage can produce large affinity boosts. Figure 26.2A shows a cofactor mimic of NADH docked into the binding site of the enzyme dihydrodipicolinate reductase (DHPR), overlaid

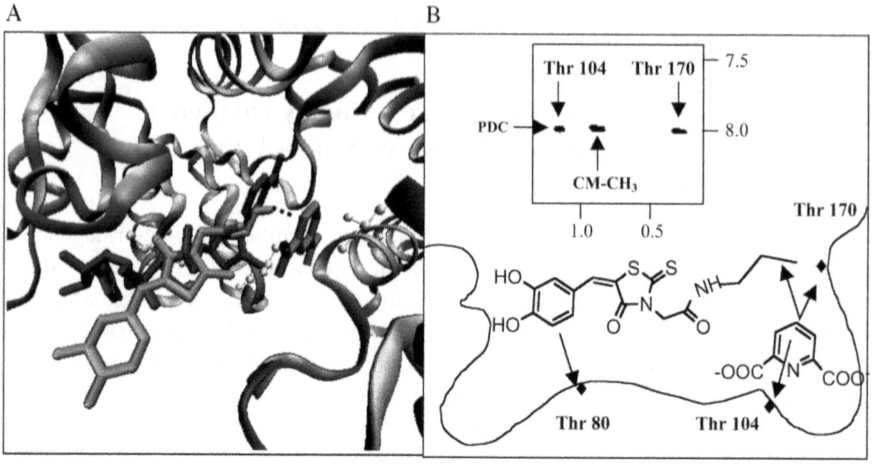

FIGURE 26.2 NMR SOLVE–based design of biligand libraries. (A) The binding site region of *E. coli* DHPR, with reference ligands NADH and 2,6-pyridine dicarboxylate (PDC) shown adjacent to each other. The NADH cofactor mimic (CM) was computationally docked using MOE (Chemical Computing Group) software using the MMFF94 forcefield and coordinates for DHPR (1arz), and overlaid on NADH. The threonines present in the binding site are rendered as balls and sticks. Proximity of PDC to the methyl group on the CM's propyl amide is indicated with a dashed line; (B) NMR SOLVE experiments provided the data needed to orient the cofactor mimic (CM) into the binding site, relative to PDC. The NOESY spectrum of the ternary complex shows that the methyl terminus of the propylamide functionality is positioned close to the substrate-binding site, suggesting this as the location for expanding a combinatorial library. Adapted from.[4]

on NADH to show relative binding mode. The structures of cofactor mimic-1 and biligand inhibitors are shown in Table 26.1, along with dissociation constants for binding to three oxidoreductases. The propylamide derivative of this cofactor mimic shows a clear NOE (Figure 26.2B) between its terminal methyl group and 2,6-pyridinedicarboxylate (PDC, reference substrate antagonist). Based on this NOE to PDC, the end of the propylamide functionality can be structurally placed next to the substrate binding site. Furthermore, since PDC shows the same NOEs to protein methyl groups as it had when NADH was bound, it seems likely that the DHPR structure has not changed dramatically from that of the native ternary complex. It should be noted that the NMR SOLVE experiments would have suggested the same library expansion point in the absence of any protein structural information, since cross-peaks for threonine residues were assigned based on proximity to reference ligands (NADH and PDC). The last step of this systems-based approach to inhibitor design is the expansion of a biligand library on this portion of the linker. After construction and screening of such a biligand library, potent and selective inhibitors were identified for three different oxidoreductases, with three different fragments attached to this linker (Table 26.1). It is also evident from Table 26.1 that the cofactor mimic binds weakly to the three enzymes with K_d values in 25 to 100 μM range, making it a good

TABLE 26.1 Chemical Structures of the Cofactor Mimic (CM) and Biligand Inhibitors Containing the CM Skeleton

Structure	DHPR	DOXPR	LDH
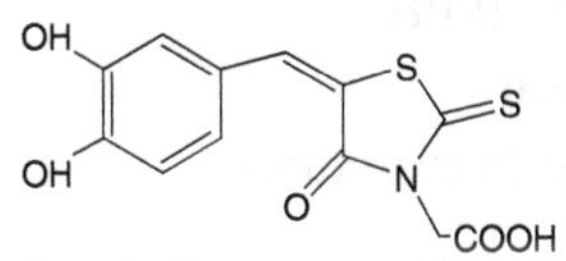 CM	26 μM	>50 μM	55 μM
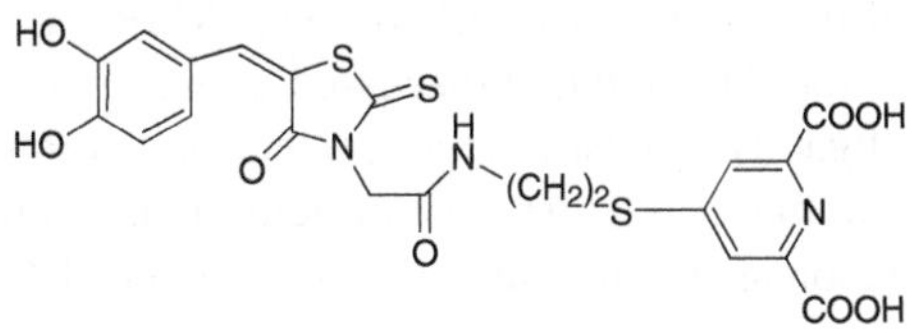	100 nM	7.9 μM	620 nM
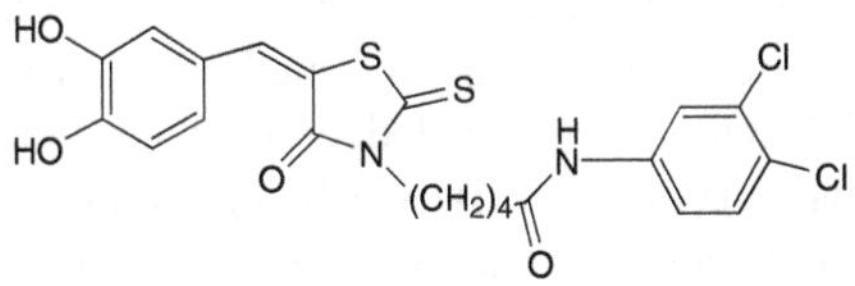	>50 μM	10 μM	42 nM
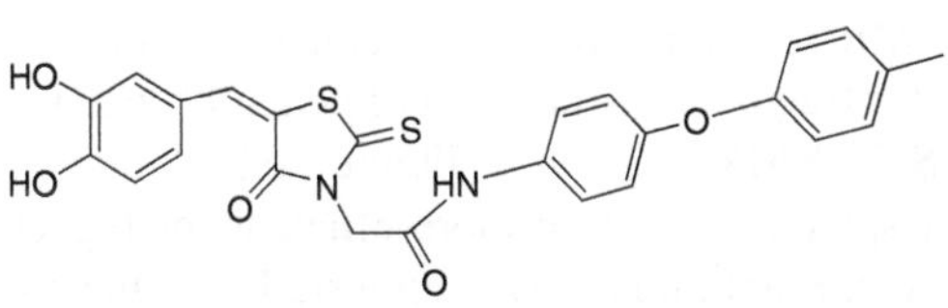	>25 μM	202 nM	12 μM

Designed using NMR SOLVE. Dissociation constants are given for three oxidoreductases. LDH, lactate dehydrogenase; DHPR, dihydrodipicolinate reductase; DOXPR, 1-deoxy-D-xylulose-5-phosphate reductoisomerase. Adapted from.[4]

privileged scaffold. Hence, using NMR SOLVE, a biligand library can be built with an appropriately chosen cofactor mimic and well-placed linkers, which will produce nM inhibitors for any given enzyme family with adjacent binding sites. Such inhibitors can be used for chemical proteomic or chemogenomic studies of proteins in the oxidoreductase gene family, so the value of NMR SOLVE is in the efficient construction of such chemical probes.

26.3.2 Cofactor Fingerprinting with STD NMR: Addressing the Needs of Functional Proteomics by Identifying Binding Preferences

Functional proteomics efforts are devoted to assigning functions to the proteins identified in genomics and subsequent proteomics projects. In a recent review article, Stockman and Dalvit proposed a means by which NMR screening could be used in functional genomics efforts, by screening a basis set of 200 small molecule ligands to identify those that bind to proteins of unknown function.[9] An ideal NMR-based screening tool for such studies might be the STD method discussed above. STD is a methodology for screening a pool of ligands against a target protein in a combinatorial fashion.[31] It is based on transfer of magnetization from protein to the bound ligand. Saturation of a single protein resonance can result in rapid spread of the saturation over the entire protein and then to ligand, via spin diffusion, thereby identifying which ligands in a pool bind to the protein. It identifies the tightest binding ligands, as long as they are in fast exchange on the NMR timescale.

Recently, we have shown[10] that STD-NMR screening could be used to measure the competitive binding of cofactors relative to one another, thereby enabling one to assess the specificity of a protein for these cofactors. In this method, cofactor fingerprinting with STD NMR (CF STD NMR), a prior knowledge of protein function is not required, since binding rather than enzymatic activity is being monitored. In general, most enzymes use cofactors in the chemical reactions that they catalyze, and these cofactors bind at the active site. Therefore, the binding preferences for certain cofactors can give significant insights into the function of a particular enzyme. Furthermore, since most cofactors bind with K_d values greater than 1 μM, cofactor binding will be in fast exchange on the NMR timescale so STD NMR screening should work well. For CF STD NMR screening, one could either use pools of intact cofactors, or even pools of cofactor fragments. A basis set of potential cofactor fragments that could be used in CF STD NMR is given in Figure 26.3. Although screening against the smaller set of fragments might be more efficient for higher throughput functional genomics efforts, our preliminary validation studies with proteins of known function suggest that it is safer to screen with full size cofactors, since affinity for fragments is sometimes so weak that binding is not observed in STD NMR experiments.[10]

We have applied the CF STD NMR strategy to determine the cofactor preference for a protein suspected of binding cyclic nucleotides, based on a bioinformatics analysis.[32,33] A set of 13 cofactors and 8 cofactor fragments were studied using CF STD NMR with this protein. In a typical CF STD NMR experiment, a difference

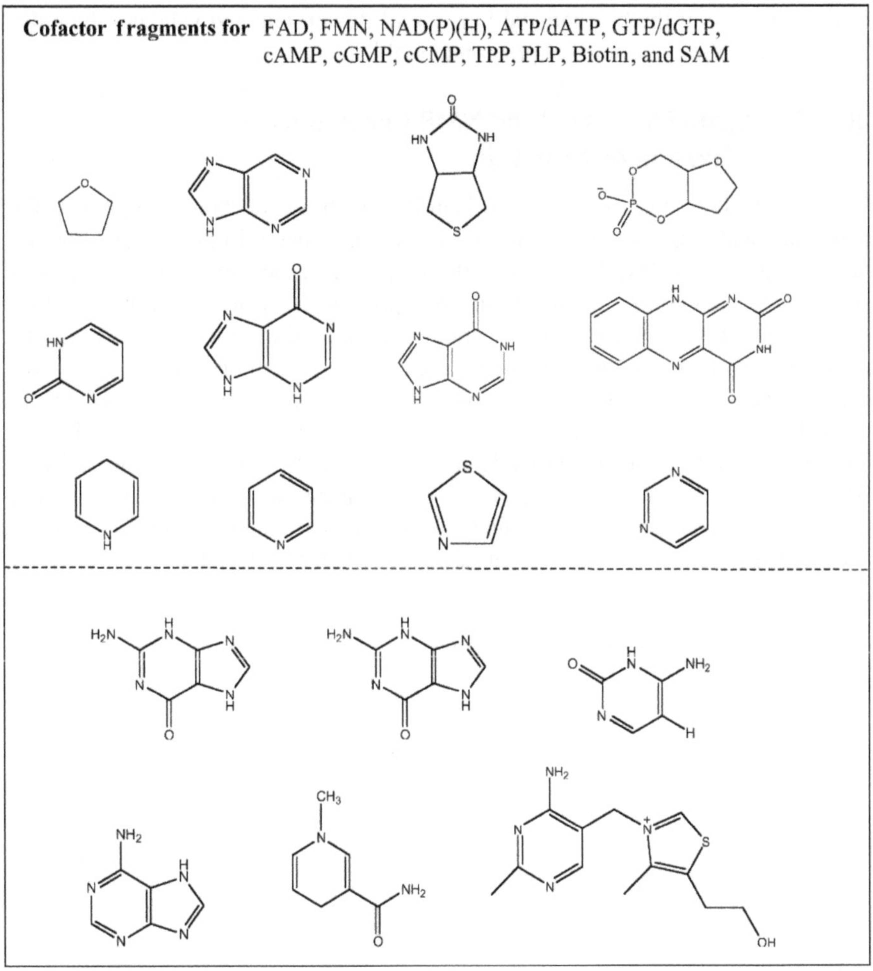

Cofactor fragments for FAD, FMN, NAD(P)(H), ATP/dATP, GTP/dGTP,
cAMP, cGMP, cCMP, TPP, PLP, Biotin, and SAM

FIGURE 26.3 Heterocyclic ring fragments that comprise the commonly used cofactors in biochemistry. Fragments in the top panel were obtained using the fragmentation tool contained in the Pipeline Pilot software package (SciTegic, Inc., San Diego). In the bottom panel, some of these fragments were decorated with additional functional groups that are likely to be important for protein binding. These fragments serve as a basis set of compounds for affinity fingerprinting proteins of unknown function.

spectrum is generated with and without preirradiation of proton resonances for protein atoms in ^{1}H NMR spectrum, in a region where cofactor protons do not resonate. As most cofactors have proton resonances above 3 ppm, we choose to irradiate at 1 to 1.2 ppm, a region usually occupied by methyl groups of amino acids such as Ala, Val, Thr, Leu, and Ile. Figure 26.4 shows the CF-STD spectrum, where the cofactors cCMP (peaks 7, 10, 13) and 5′ AMP (peaks 1, 4, 9) are observed to bind to the protein. This CF STD NMR study therefore provides a starting point for further characterization of the biochemical function of this protein and validates the utility of CF STD NMR as a chemical proteomic tool.[10]

26.3.3 Ligand Docking Using NMR Constraints: T_1 Relaxation or NOE Data

While CF STD NMR can be used to identify which ligands bind to a protein, further functional insights can be obtained if there is structural information about the protein–ligand complex. Although NMR is a powerful tool for determining protein structures (see section 26.3.2) and is playing a prominent role in structural proteomics efforts,[25] it can also provide the structural data needed to guide the docking of a ligand into a protein structure. Thus, it serves the chemical proteomic role of providing information about protein–ligand interactions in a manner that lends itself to high throughput and potentially highly parallel studies of many proteins. The distance data used to guide the docking of a ligand can come from either T_1 relaxation data or NOE data. With regard to the former, one can measure the distance between a paramagnetic center such as the iron atom of a heme cofactor and the protons on a bound substrate or inhibitor. Roberts[14] has used T_1 relaxation effects induced by the iron of

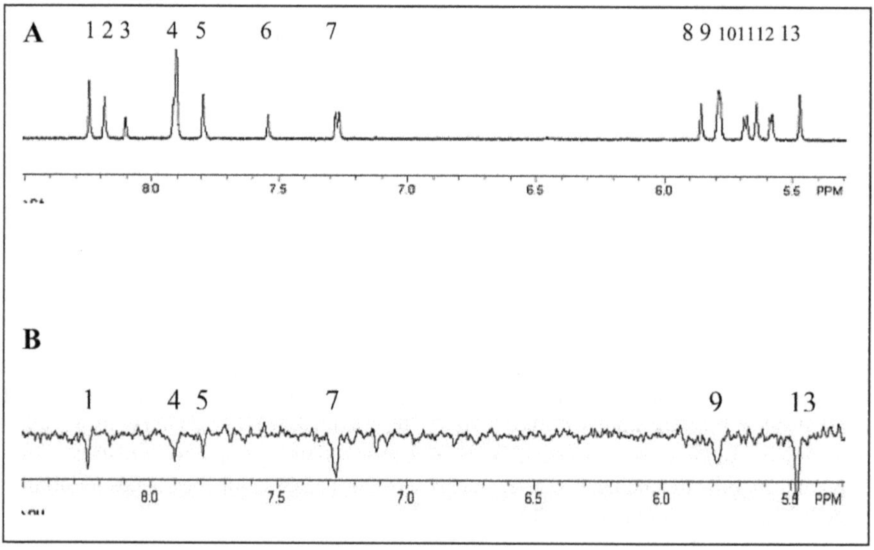

FIGURE 26.4 Probing protein function with CF-STD NMR. (A) 1D proton NMR spectrum of a sample containing RSP2 and a mixture of six cofactors (ATP, GTP, cAMP, cGMP, cCMP, and 5′ AMP). Based on individual proton spectra of cofactors, the peaks were assigned as follows: 5′ AMP (1, 4, 9), ATP (2, 4, 9), cAMP (4, 8), GTP (5, 12), cGMP (6, 11), and cCMP (7, 10, 13); peak 3 corresponds to an impurity present in cCMP. The concentration of RSP2 was 10 μM, and each cofactor was 1 mM; (B) the CF-STD NMR spectrum recorded in the presence of 4 mM Mg^{2+}, with irradiation at 1.21 ppm. The NMR buffer used was 20 mM sodium phosphate, 200 mM NaCl, 100% D_2O, pH 7.4. Spectra obtained at 298 K and 600 MHz.

the heme in cytochrome P450D6 to dock the drug codeine into its binding site (Figure 26.5A). Such studies could be done in a fairly high throughput manner to generate large databases of cheminformatic information that could be used to describe the binding preferences for families of proteins like the cytochromes P450, as we have recently proposed.[33] Such data would complement an affinity fingerprint[34] by providing a complementary structural fingerprint that could be used as a 3D quantitative structure activity relationship (QSAR) descriptor to guide the design of molecules for or against binding to a certain protein.[33]

In addition to T_1-based strategies, one can measure NOEs between a ligand and a protein to generate distance constraints to guide the docking process. This can be done either using a crystal or NMR structure of the protein without ligand bound, or even a homology modeled protein, as the docking target. A first step is to assign some key protons in the protein binding site, either using traditional sequential assignment methods[22] or cofactor mapping and chemical shift perturbation methods.[12,35] As a demonstration of the latter strategy, we have assigned methyl protons in the binding site of DHPR and measured NOEs between these protons and protons on a furoic acid mimic of the NADH cofactor, in order to guide the docking of this ligand into the binding site (Figure 26.5D). The presence of an NOE between two protons indicates that they are within 5 angstroms, so pairs of protons on ligand and protein are therefore constrained in this way as part of a docking calculation.

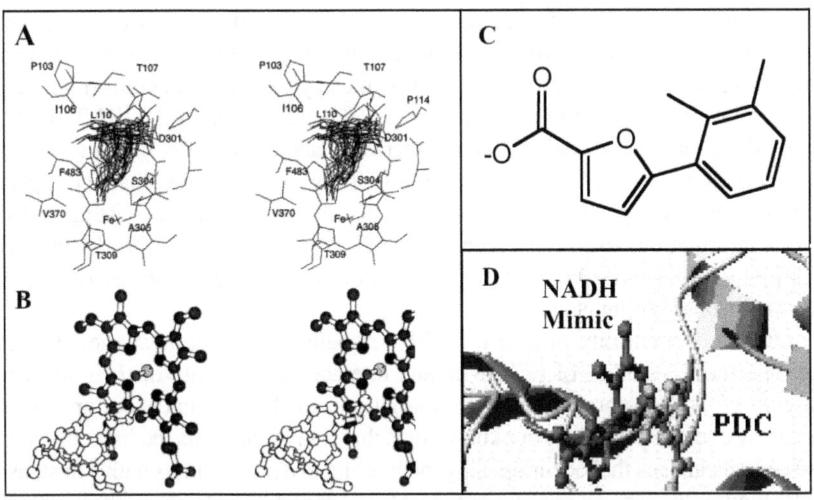

FIGURE 26.5 NMR-based docking studies. Docking of codeine into the binding site of cytochrome P450 2D6, using T_1 relaxation data for codeine protons. (A) Overlay of multiple orientations of codeine, all consistent with the T_1 data, as well as; (B) the most representative codeine structure (open circles) relative to the heme (closed circles). Reproduced with permission from;[14] (C) a furoic acid NADH cofactor mimic was docked into the binding site of; (D) DHPR, based on NOEs measured between protein methyl groups and protons on the ligand. Adapted from.[12]

26.4 CHEMICAL PROTEOMIC APPLICATIONS OF NMR *IN VIVO*

26.4.1 *In Cell* NMR: Chemical Proteomics in the Cellular Milieu

NMR is the only method that can provide high-resolution structures of biological macromolecules in solution under near physiological, solution conditions.[21] Still, there is concern that *in vitro* studies in solution do not fully reproduce the native environment inside a cell, and that this could on occasion have functional implications for the protein being studied.[36] This is because the buffer conditions are not selected for their closest match to the natural environment of the protein but instead to optimize experimental parameters such as solubility and sensitivity or to minimize NMR buffer signals that could interfere with the spectrum. The protein being studied may be sensitive to interactions that occur in the cell, but are not being properly reproduced in the NMR tube. Such effects could range from the presence of coactivator proteins, unrecognized cofactors, or prosthetic groups including metal ions, the physical environment in the cell (pH, ionic strength, viscosity, water activity), and interactions with structural elements in the cell, such as cytoskeletal components or membranes. Recent developments in NMR technology have led researchers and clinicians to study protein–ligand interactions in living systems, giving rise to new fields of study such as *in vivo* or *in cell* NMR[15,37,38,39] and molecular imaging.[16,30]

In cell NMR is a tool for monitoring changes in protein structure that occur upon interactions with other cellular components such as ligands, coactivators, and prosthetic groups, in the context of the cellular environment (Figure 26.6). In one elegant demonstration of how the cellular environment can yield a protein conformation that differs from that observed *in vitro*, Serber et al.[15] have measured the [^{1}H,^{15}N]-HSQC *in cell* NMR spectrum of calmoduline selectively labeled with ^{15}N on all lysine residues, measured in living *Escherichia coli* cells. They observed more than the expected eight cross-peaks expected for the eight lysines in calmodulin, suggesting that different conformations exist for calmodulin in *E. coli* relative to the NMR tube. This work also demonstrates the utility of using selectively labeled amino acids as NMR structural probes within cells.

In terms of the utility of *in cell* NMR for chemical proteomic studies, Hubbard et al.[39] performed studies of protein–ligand interactions. The aim of their study was to use *in cell* NMR to probe *in vivo* conformation and changes that occur upon ligand binding by cheY in *E. coli*. To our knowledge, this study represents the first observation of structural changes that occur upon ligand binding to a protein in its native host, using NMR. Figure 26.7 shows regions of the [^{1}H,^{15}N]-HSQC spectrum of cheY, obtained after addition of a ligand to growing *E. coli* cells, compared with those obtained after addition of the compound *in vitro*. Comparison of the [^{1}H,^{15}N]-HSQC spectra for *in vivo* versus *in vitro* binding suggests that the mode of binding of the ligand to cheY is similar in both environments, thereby providing important information that suggests the ligand has quite similar interactions both *in vitro* and *in vivo*. Also this observation suggested that the ligand is able to penetrate through the cell membrane

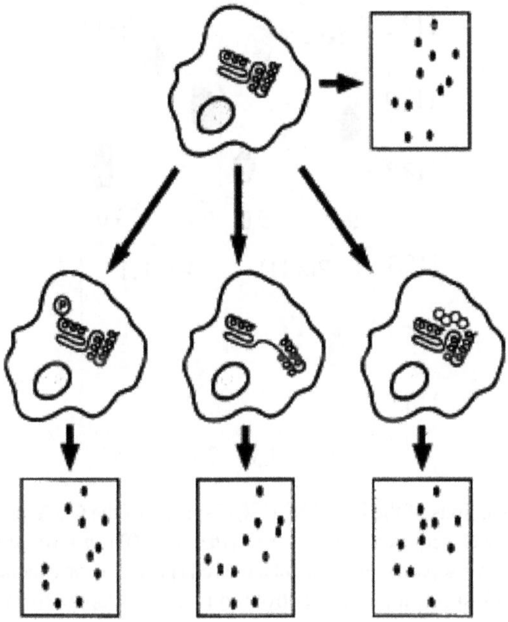

FIGURE 26.6 Schematic representation of *in cell* NMR changes due to possible posttranslational (left), conformational (middle), or binding (right) mechanisms, which induce changes in 2D NMR spectra relative to an unperturbed state (top). Reproduced with permission from.[36]

and selectively bind to the cheY protein. These results show that *in cell* NMR also has the potential for answering questions about the bioavailability of both drug molecules, as well as chemical proteomic probes, within a living biological system.

26.4.2 Molecular Imaging: Magnetic Resonance Contrast Agents as Chemical Proteomic Probes in Whole Organisms

Molecular imaging is simply imaging at the molecular level, where proteins interact with ligands. Molecular imaging methods are likely to have profound impact on both basic research and advanced drug discovery efforts because they allow the probing of protein–ligand interaction within the context of a whole organism. Several examples are described in this section from the chemical proteomics perspective, where imaging is serving to report on protein–ligand binding events. The image is formed on the basis of the ^{1}H NMR signal from the protons of bulk water, as is typically the case in traditional magnetic resonance imaging (MRI). The signal intensity is a function of the water concentration and the relaxation times (T_1 and T_2) but is often quite

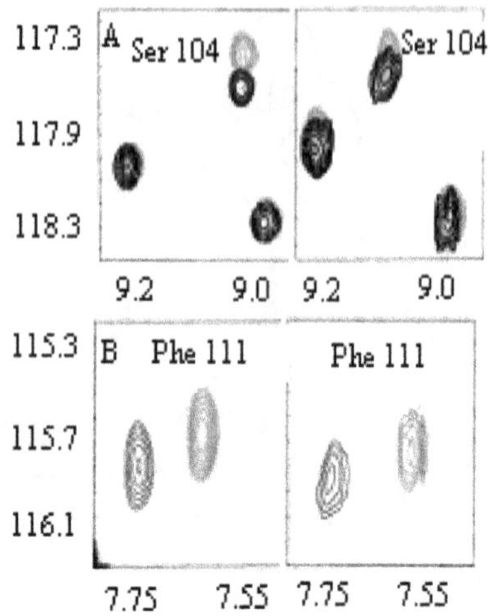

FIGURE 26.7 Two regions of the [^{1}H-^{15}N] HSQC spectra of the CheY protein in the presence and absence of a ligand, acquired under (A) *in vitro*; and (B) *in vivo* conditions. The figure shows the overlay of two separate experiments, the red spectrum corresponding to the presence of ligand and the black spectrum to the control (no ligand). Adapted from.[39]

low.[40] Therefore, in order to increase the intrinsic contrast observed in an MRI scan, a paramagnetic contrast agent is often administered before performing the MRI scan.[41] One widely used class of contrast agents are gadolinium chelates. Gd^{3+} is ideal since it is paramagnetic and has a very strong magnetic moment and causes significant relaxation of water protons—an effect called high relaxivity.[41] The relaxation effect only occurs significantly for water molecules that bind in the first coordination sphere of the metal, but since they exchange fast with bulk water the effect is amplified. Although general use of gadolinium chelates provides useful anatomical information in MRI scans, the innovation that molecular imaging offers is to somehow trigger relaxivity increases in response to some biological event. Such events have included changes in metabolism, pH, pO_2, protein expression levels, and Ca^{2+}.[13]

For purposes of *in vivo* chemical proteomics, the most interesting changes monitored with molecular imaging have involved measurement of enzyme activity or direct monitoring of protein–ligand interactions. With regard to the latter, the binding to a macromolecule substantially slows molecular rotation of a Gd^{3+} complex, resulting in an additional increase in the relaxivity and therefore tissue contrast, a phenomenon know as receptor-induced magnetization enhancement (RIME).[42] Though this enhancement is significant, the RIME approach alone cannot produce images with adequate sensitivity since most protein targets under study exist at nanomolar concentrations. To address this sensitivity issue, Meade et al.[13] and Louie et al.[43] introduced enzyme activation of contrast agents. In this approach, the enzyme substrate or chelating agent occupy all nine coordination sites, inhibiting water access to the paramagnetic ion, as shown in Figure 26.8. The substrate that caps the Gd^{3+} ion is a sugar that can

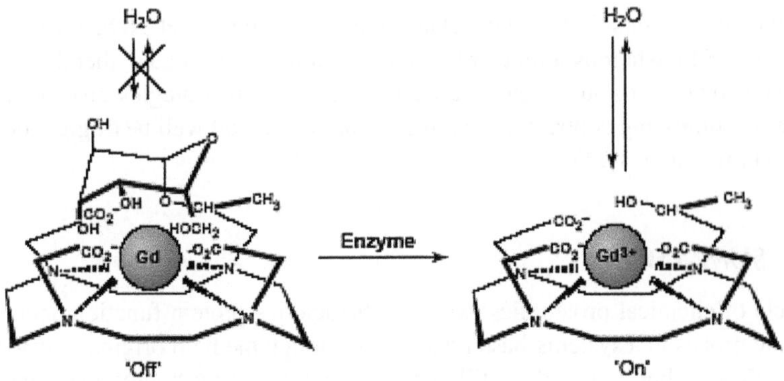

FIGURE 26.8 Enzyme-activated magnetic resonance contrast agents. The Gd-tetraazamacrocycle complex, having a galactopyranosyl ring at the apex position, creates the off state, since the inner sphere cannot interact with the bulk water. Upon cleavage of the sugar residue by β-galactosidase, the complex switches to an on state, where the Gd^{3+} becomes more accessible to surrounding water, allowing exchange between bulk water and the inner coordination sphere, thereby increasing the relaxivity effect on bulk water, since Gd^{3+} is paramagnetic. Reproduced with permission from.[13]

be cleaved by the enzyme β-galactosidase. When this happens the cap is removed, thereby freeing up coordination sites on the Gd^{3+} so that water can exchange. In this manner, it was possible to monitor *in vivo* levels of β-galactosidase enzyme, since enzymatic activation of the contrast agent yielded a high relaxivity state producing significant signal (T_1) contrast in images of *Xenopus laevis*.[43] In a related approach Nivorozhkin et al.,[42] created a contrast agent containing lysine masking groups that could be removed *in vivo* by carboxypeptidase B.[42] When this happened, the contrast agent was then able to bind to human serum albumin (HSA). When bound to HSA, the molecular weight increase caused a large increase in the rotational correlation time (τ_R) of the Gd^{3+} chelate complex, thereby increasing relaxivity. Therefore, while the first example of enzyme activation yielded increased relaxivity through increased water exchange by freeing up coordination sites on Gd^{3+}, this example relies on an increase in size of the complex to increase relaxivity. Thus, the general strategy of enzyme activation can be used in various ways to increase contrast in imaging studies.

While the above examples provide information on the presence of *in vivo* enzyme activities, they only provide measurements of enzyme–ligand interactions indirectly by effectively measuring enzyme-catalyzed increases in relaxivity. Direct protein–ligand interactions were observed in one case by linking the Gd^{3+} chelate to a peptide ligand for Gal20,[44,45] and relying on the increase in relaxivity afforded by increases in τ_R, as discussed above with binding to HSA. Strategies such as this hold great promise for the *in vivo* characterization of protein–ligand interactions, thus extending the reach of NMR-based chemical proteomic studies to whole organisms.

One final note on sensitivity issues relates to the use of chemical exchange saturation transfer (CEST), whereby amide or hydroxyl protons are irradiated, thereby transferring magnetization to bulk water via exchange.[46–49] Such strategies provide another avenue for improving contrast in imaging studies and could well be of great utility in chemical proteomic studies.

26.5 SUMMARY

The field of chemical proteomics exists to characterize protein function using small molecule probes in a systems-based manner. Although the field originally made little use of NMR technology and is still currently dominated by fluorescence and mass spectrometric methods, NMR techniques are playing an increasingly important role both *in vitro* and *in vivo*. *In vitro* studies of protein–ligand interactions focus on highly parallel approaches to the structural characterization of protein–ligand complexes, based on T_1- or NOE-derived distance constraints that can guide docking calculations. Such strategies are a nice complement to initial binding studies using STD NMR, which can provide early clues as to protein function. Although these methods are powerful ways to characterize protein function, they attack the functional proteomics problem one isolated protein at a time. For this reason, some of the most exciting NMR applications in chemical proteomics involve studies of protein–ligand interactions *in vivo*, as with *in cell* NMR or molecular imaging. The latter is perhaps in the earliest stages of development but holds the greatest promise since there are tremendous benefits to being able to study protein–ligand interactions in the context of the whole proteome, in the context of a living organism. This may be where NMR methods have their greatest impact in chemical proteomics, if it is possible to overcome the sensitivity issues associated with the low concentrations of receptors and enzymes in cells. And an efficient source of small molecule probes for such *in vivo* studies could be from the systems-based design of inhibitors that systematically target different proteins in a gene family, as with NMR SOLVE. These can then be used for complementary chemical proteomic studies of protein–ligand interactions and chemical genetic studies of phenotypic changes, thereby allowing whole-organism structure-function studies to be made by correlating phenotype with *in vivo* protein–ligand interactions.

REFERENCES

1. Jeffery, D. and Bogyo, M., Chemical proteomics and its application to drug discovery, *Curr. Opinion Biotechnol.*, 11, 602–609, 2000.
2. Adam, G.C., Sorensen, E.J., Cravatt, G.F., Chemical strategies for functional proteomic, *Mol. Cell. Proteomics*, 1, 781–790, 2002.
3. Weston, A.D. and Hood, L. Systems biology, proteomics, and the future of health care: Toward predictive, preventative, and personalized medicine, *J. Proteomics Res.*, 3, 179–196, 2004.
4. Sem, D.S., Bertolaet, B., Baker, B., Chang, E., Costache, A., Coutts, S., Dong, Q., Hansen, M., Hong, V., Huang, X., Jack, R.M., Kho, R., Lang, H., Meininger, D.,

Pellecchia, M., Pierre, F., Villar, H., and Yu, L., Systems-based design of bi-ligand inhibitors of oxidoreductases: Filling the chemical proteomic toolbox, *Chem. Biol.*, 11, 185–194, 2004.
5. Kho, R., Baker, B.L., Newman, J.V., Jack, R.M., Sem, D.S., Villar, H.O., and Hansen, M.R., A path from primary protein sequence to ligand recognition, *Proteins*, 50, 589–599, 2003.
6. Kho R., Newman, J.V., Jack, R.M., Villar, H.O., and Hansen, M.R., Genome-wide profile of oxidoreductases in viruses, prokaryotes, and eukaryotes, *J. Proteome Res.*, 2, 626–632, 2003.
7. Campbell, D.A. and Szardenings, A.K., Functional profiling of the proteome with affinity labels, *Curr. Opin. Chem. Biol.*, 7, 296–303, 2003.
8. Greenbaum, D.C., Arnold, W.D., Lu, F., Hayrapetian, L., Baruch, A., Krumrine, J., Toba, S., Chehade, K., Bromme, D., Kuntz, I.D., and Bogyo, M., Small molecule affinity fingerprinting: A tool for enzyme family subclassification, target identification, and inhibitor design, *Chem. Biol.*, 9, 1085–1094, 2002.
9. Stockman, B.J. and Dalvit, C., NMR screening techniques in drug discovery and drug design, *Prog. Mag. Res. Spec.*, 41, 187–231, 2002.
10. Yao, H. and Sem, D.S., Cofactor fingerprinting with STD-NMR to characterize proteins of unknown function: identification of a rare cCMP cofactor preference, *FEBS Letters*, 579, 661–666, 2005.
11. Sem, D.S., Yu, L., Coutts, S.M., and Jack, R., An object-oriented approach to drug design enabled by NMR SOLVE, the first real-time structural tool for characterizing protein-ligand interactions, *J. Cell. Biochem.*, 37, S99, 2001.
12. Pellecchia, M., Meininger, D., Dong, Q., Chang, E., Jack, R., and Sem, D. S., NMR-based structural characterization of large protein-ligand interactions, *J. Biomol. NMR*, 22, 165, 2002.
13. Meade, T.J., Taylor, A.K., and Bull, S.R., New magnetic resonance contrast agents as biochemical reporters, *Curr. Opinion Neurobiol.*, 13, 597–602, 2003.
14. Modi, S., Paine, M.J., Sutcliffe, M.J., Lian, L.-Y., Primrose, W.U., Wolf, C.R., and Roberts, G.C.K., A model for human cytochrome P450 2D6 based on homology modeling and NMR studies of substrate binding, *Biochemistry*, 35, 4540–4550, 1996.
15. Serber, Z., Ledwidge, R., Miller, S. M., and Dotsch, V., Evaluation of parameters critical to observing proteins inside living *Escherichia coli* by in-cell NMR spectroscopy, *J. Am. Chem. Soc.*, 123, 8895–8901, 2001.
16. Bredel, M. and Jacoby, E., Chemogenomics: An emerging strategy for rapid target and drug discovery, *Nat. Rev. Genet.*, 5, 262–275, 2004.
17. Greenbaum, D., Medzihradszky, K.F., Burlingame, A., and Bogyo, M., Epoxide electrophiles as activity-dependent cysteine protease profiling and discovery tools, *Chem. Biol.*, 7, 569–581, 2000.
18. Gygi, S.P., Rist, B., Gerber, S.A., Turecek, F., Gelb, M.H., and Aebersold, R., Quantitative analysis of complex protein mixtures using isotope-coded affinity tags, *Nat. Biotechnol.*, 17, 994–999, 1999.
19. Naylor, S. and Kumar, R., Emerging role of mass spectrometry in structural and functional proteomics, *Adv. Protein Chem.*, 65, 217–248, 2003.
20. Minks, C., Huber, R., Moroder, L., and Budisa, N., Noninvasive tracing of recombinant proteins with "fluorophenyl-fingers," *Anal. Biochem.*, 284, 29–34, 2000.
21. Wuthrich, K., *NMR of Proteins and Nucleic Acids*, Wiley, New York, 1986.
22. Cavanagh, J., Fairbrother, W.J., Palmer, A.G., III, and Skelton, N.J., *Protein NMR Spectroscopy*, Academic Press, San Diego, 1996.

23. Pervushin, K., Reik, R., Wider, G., and Wuthrich, K., Attenuated T2 relaxation by mutual cancellation of dipole-dipole coupling and chemical shift anisotropy indicates an avenue to NMR structures of very large biological macromolecules in solution, *Proc. Natl. Acad. Sci. USA*, 94, 12366, 1997.
24. Marassi, F.M. and Opella, S.J., NMR structural studies of membrane proteins, *Curr. Opin. Struct. Biol.*, 8, 640–648, 1998.
25. Staunton, D., Owen, J., Campbell, I.D., NMR and structural genomics, *Acc. Chem. Res.*, 36, 207–214, 2003.
26. Shuker, S. B., Hajduk, P. J., Meadows, R. P., and Fesik, S. W., Discovering high-affinity ligands for proteins: SAR by NMR, *Science*, 274, 1531, 1996.
27. Page, M.I. and Jencks, W.P., Entropic contributions to rate accelerations in enzymic and intramolecular reactions and the chelate effect, *Proc. Natl. Acad. Sci. USA*, 68, 1678–1683, 1971.
28. Fezjo, J., Lepre, C. A., Peng, J. W., Bemis, G. W., Ajay, Murcko, M. A., and Moore, J. M., The SHAPES strategy: An NMR-based approach for lead generation in drug discovery, *Chem. Biol.*, 6, 755, 1999.
29. Dalvit, C., Pevarello, P., Tato, M., Veronesi, M., Vulpetti, A., and Sundstrom, M., Identification of compounds with binding affinity to proteins via magnetization transfer from bulk water, *J. Biomol. NMR*, 18, 65–68, 2000.
30. Herschman, H. R. Molecular imaging: Looking at problems, seeing solutions, *Science*, 302, 605–608, 2003.
31. Mayer, M. and Meyer, B., Characterization of ligand binding by saturation transfer difference NMR spectroscopy, *Angew. Chem. Int. Ed. Engl.*, 38, 1784–1788, 1999.
32. Yang, P., Yang, C., and Sale, W.S., Flagellar radial spoke protein 2 is a calmodulin binding protein required for motility in *Chlamydomonas reinhardtii*, *Eukaryotic Cell*, 3, 72–81, 2004.
33. Yao, H. and Sem, D.S. Cofactor fingerprinting with STD NMR to characterize proteins of unknown function: Identification of a rare cCMP cofactor preference. *FEBS Letters*, 579, 661–666, 2005.
34. Kauvar, L.M., Higgins, D.L., Villar, H.O., Sportsman, J.R., Engqvist-Goldstein, A., Bukar, R., Bauer, K.E., Dilley, H., and Rocke, D.M., Predicting ligand binding to proteins by affinity fingerprinting, *Chem. Biol.*, 2, 107–118, 1995.
35. Medek, A., Hajduk, P.J., Mack, J., and Fesik, S.W., The use of differential chemical shifts for determining the binding site location and orientation of protein-bound ligands, *J. Am. Chem. Soc.*, 122, 1241–1242, 2000.
36. Serber, Z. and Dotsch, V., In-cell NMR spectroscopy, *Biochemistry*, 48, 14,317–14,323, 2001.
37. Serber, Z., Keatinge-Clay, A. T., Ledwidge, R., Kelly, A. E., Miller, S. M., and Dotsch, V., High-resolution macromolecular NMR spectroscopy inside living cells, *J. Am. Chem. Soc.*, 123, 2446–2447, 2001.
38. Wieruszeski, J.-M., Bohin, A., Bohin, J.-P., and Lippens, G., *In vivo* detection of the cyclic osmoregulated periplasmic glucan of *Ralstonia solanacearum* by high-resolution magic angle spinning NMR, *J. Magn. Reson.*, 151, 118–123, 2001.
39. Hubbard, J. A., Maclachlan, L. K., King, G. W., Jones, J. J., and Fosberry, A. P., Nuclear magnetic resonance spectroscopy reveals the functional state of the signaling protein CheY *in vivo* in *Escherichia coli*, *Mol. Microbiol.*, 49, 1191–1200, 2003.
40. Louie, A. Y., Duimstra, J. A., and Meade, T. J., in *Brain mapping: The methods*, 2nd ed., Toga, W. A., Mazziotta, C. J., Eds., Elsevier Science, San Diego, 2002.

41. Toth, E., Helm, L., and Merbach, A.E., Relaxivity of gadolinium (III) complexes: Theory and mechanism, in *The Chemistry of Contrast Agents in Medical Magnetic Resonance Imaging*, Toth E. and Merbach, A.E., Eds., John Wiley and Sons, New York, 2001, pp. 45–120.
42. Nivorozhkin, A.L., Kolodziej, A.F., Caravan, P., Greenfield, M.T., Lauffer, R.B., and McMurray, T.J., Enzyme-activated Gd^{3+} magnetic resonance imaging contrast agents with a prominent receptor-induced magnetization enhancement, *Angew. Chem. Int. Ed. Engl.*, 40, 2903–2906, 2001.
43. Louie, A.Y., Huber, M.M., Ahrens, E.T., Rothbacher, U., Moats, R., Jacobs, R.E., Fraser, S.E., and Meade, T.J., In vivo visualization of gene expression using magnetic resonance imaging, *Nat. Biotechnol.*, 18, 321–325, 2000.
44. Aime, S., Fasano, M., and Terreno, E., Lanthanide(III) chelates for NMR biomedical applications, *Chem. Soc. Rev.*, 27, 19–29, 1998.
45. Ogawa, S., Tank, D.W., Menon, R., Ellerman, J.M., Kim, S.G., Merkle, H., and Ugurbil, K., Intrinsic signal changes accompanying sensory stimulation: functional brain mapping with magnetic resonance imaging, *Proc. Natl. Acad. Sci. USA*, 89, 5951–5955, 1992.
46. Aime, S., Barge, A., Delli Castelli, D., Fedeli, F. , Mortillaro, A., Nielson Flemming, U., and Terreno, E., Paramagnetic lanthanide(III) complexes as pH-sensitive chemical exchange saturation transfer (CEST) contrast agents for MRI applications, *Magn. Reson. Med.*, 47, 639–648, 2002.
47. Zhang, S., Michaudet, L., Burgess, S., and Sherry, S.D., The amide protons of an ytterbium dota tetraamide complex act as efficient antennae for transfer of magnetization to water, *Angew. Chem. Int. Ed. Engl.*, 41, 1919–1921, 2002.
48. Zhang, S., Winter, P., Wu, K., and Sherry, A.D., A novel Europium(III)-based MRI contrast agent, *J. Am. Chem. Soc.*, *123*, 1517–1518, 2001.
49. Aime, S., Delli Costelli, D., Fedeli, F., and Terreno, E., A paramagnetic MRI-CREST agent responsive to lactate concentration, *J. Am. Chem. Soc.*, 124, 9364–9365, 2002.

27 Electrophoretic Nuclear Magnetic Resonance in Proteomics: Toward High-Throughput Structural Characterization of Biological Signaling Processes

Qiuhong He and Xiangjin Song

CONTENTS

27.1 INTRODUCTION

The postgenomic research quickly expands to characterize the human proteome, including the posttranscriptional modifications of all human proteins. Many clinical proteomic projects are launched to search for new biomarkers of human diseases and drug targets for personalized therapies for the future years of molecular medicine. Successful proteomic analysis of human serum and urine samples have shown feasibility to reveal distinctive protein profiles of healthy human subjects and cancer patients.[2]

The proteomic samples are analyzed with effective separation techniques such as the surface-enhanced laser desorption and ionization time-of-flight mass spectroscopy (SELDI-TOF MS),[3,4] and two-dimensional gel electrophoresis.[5] The disease protein profiles were identified from cell extracts employing laser capture microdissection to catch cells from tissue specimens.[6] Certain biological signaling networks were mapped to depict a delicate balance of protein–protein, protein–RNA, and protein–DNA interactions.[1,7–9] An example signaling pathway is displayed here for the tumor necrosis factor TNF-α that actives the transcriptional factor NF-κB (Figure 27.1).[1,10] Such a proteomic map may serve as a guide to design effective anti-cancer therapies.[11,12]

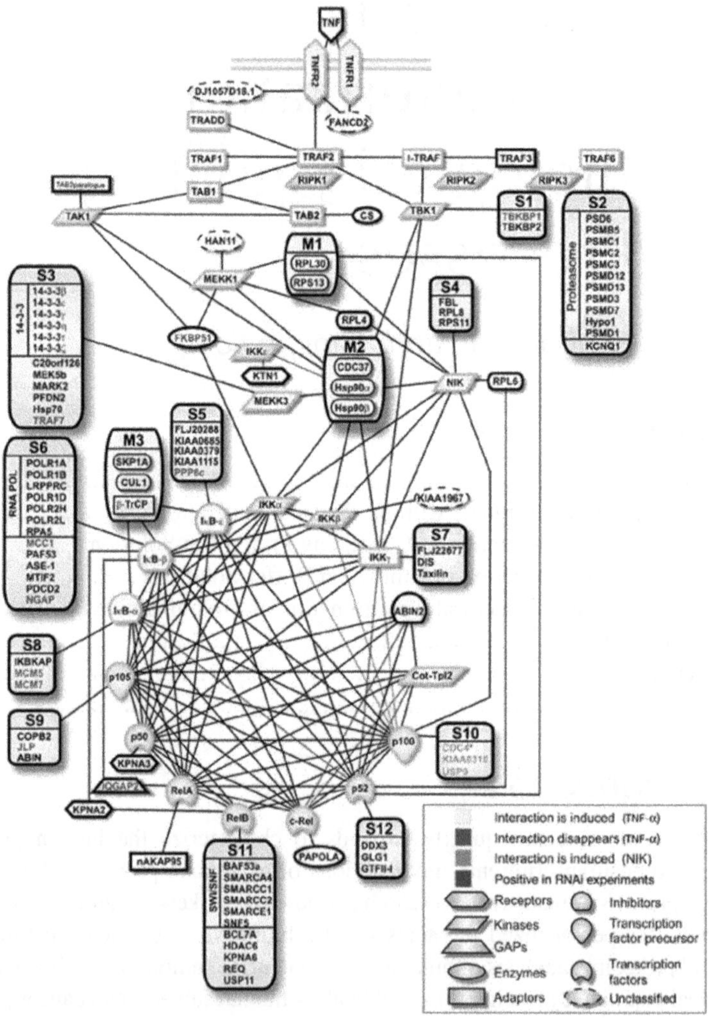

FIGURE 27.1 The TNF-α/NF-κB signaling pathway visualized as a network. From Ref.[1], with permission.

The proteomic endeavor continues the research of the pregenomic era using high-resolution x-ray crystallography, nuclear magnetic resonance (NMR) spectroscopy, and other techniques such as atomic force microscopy to study protein families and genetic evolutions across species by examining relationship between protein structures and functions.[13–16] Nuclear magnetic resonance is perhaps the most suitable method to study biological signaling events and protein interactions in aqueous solution[17] and *in vivo* in human and animals.[18–20] To map the conformational changes of the interacting proteins, however, we encounter paramount technical challenges due to overlapping NMR signals from different proteins coexisting in solution. In this chapter, we present an emerging technology of electrophoretic NMR (ENMR) for structural and clinical proteomic research. The multidimensional NMR spectra of coexisting proteins can be sorted out without physical separation of the protein components.[21,22] In principle, the ENMR techniques can be used to identify the chemical structure for each protein detected from body fluid and cell extracts in the electrophoretic flow dimension of an ENMR spectrum. Therefore, the ENMR proteomic profiles may be potentially used for early diagnosis of human diseases. In addition, structural visualization of protein interactions *in situ* is possible by the three-dimensional (3D) conformational mapping at the protein reaction interfaces (or so-called active pockets). Thus, structural characterization of biological signaling events by ENMR is possible in the presence of the entire proteomic assembly of biomacromolecules.

27.2 SEPARATING PROTEIN SIGNALS WITHOUT PHYSICAL SEPARATION

Multidimensional electrophoretic NMR (nD-ENMR) was initially designed to study 3D structures of coexisting proteins and protein conformations in solution without physical separation of the protein components. By applying a DC electric pulse, the NMR signals of the proteins can be resolved in a new dimension of electrophoretic flow at different resonant frequencies proportional to their electrophoretic migration rates (Figure 27.2).[21] In a 3D-ENMR experiment, for example, the overlapping two-dimensional (2D) NMR spectra of different molecules can be resolved into different subspectral planes as if they were independently obtained from pure protein solutions. Each subspectrum of the component protein gives the NMR parameters of chemical shifts, spin J-coupling constants, or dipolar coupling parameters for sequential and stereospecific structural assignments.

The feasibility of separating protein signals in the electrophoretic flow dimension by ENMR was demonstrated using a protein mixture of bovine serum albumin (BSA) and ubiquitin (Figure 27.3A). The data were acquired from a modified stimulated echo sequence (Figure 27.3B), which was composed of three 90° RF pulses, two spatial-encoding gradients (g_1, δ_1), and a spoiler (crusher) gradient (g_2, δ_2) in the storage period of the z-magnetization. An incrementing DC electric field (E_{dc}, Δ) was inserted in this period between the two spatial-encoding gradients to drive the electrophoretic migration of proteins. In the z-storage period (Δ), the T_1 process dominates the spin

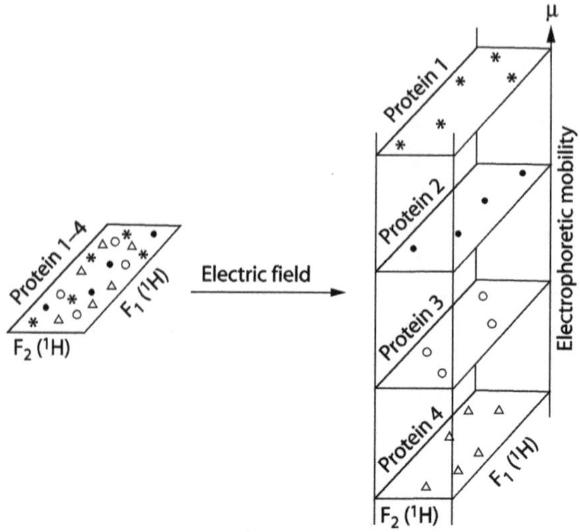

FIGURE 27.2 Two-dimensional NMR spectra of four different proteins (represented by ♦, •, *, and Δ, respectively), sorted by their electrophoretic mobilities (μ).[21] (with permission)

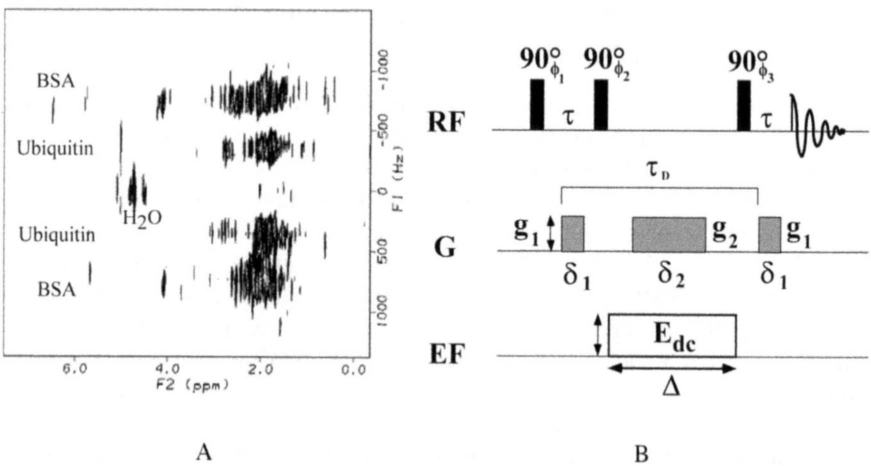

FIGURE 27.3 (A) A contour plot of the 2D stimulated echo ENMR spectrum acquired from 0.15 mM BSA and 1 mM ubiquitin mixed with 26 mM ethylenediamine in D_2O. The solution conductivity $\kappa = 0.692$ mS·cm^{-1} and pH = 10.37. $E_{dc} = 0$ to 38.45 V cm^{-1} with 21 incrementations, $\Delta = 0.4085$ s. $\tau = 2.535$ ms, $\tau_D = 0.4111$ s, $T_R = 1.5$ s, NS = 128, $g_1 = g_2 = 48.72$ Gauss cm^{-1}, $\delta_1 = 2$ ms, and $\delta_2 = 8$ ms. (B) The stimulated-echo ENMR pulse sequence. From Ref.[21], with permission.

relaxation, permitting a long period of electrophoretic flow of the proteins. The electrophoretic flow introduces a cosine factor that modulates the amplitude of the stimulated echo as a function of electric field amplitude (E_{dc}) and duration (t_1), the gradient strength ($K = \gamma g\delta$), and electrophoretic mobility (μ) of the molecule,

$$M\left(K,\Delta,E_{dc}\right)=\frac{M(0)}{2}\exp\left[-DK^2\left(\tau_D-\frac{\delta}{3}\right)-\frac{2\tau}{T_2}-\frac{\Delta}{T_1}\right]\cos\left[\left(KE_{dc}t_1\right)\mu\right] \quad (27.1)$$

where $M(0)$ is the initial magnetization after the first 90° pulse, γ is the gyromagnetic ratio, D is the diffusion coefficient, τ_D is the diffusion delay between the two spatial-encoding gradients, and τ is the delay between the first two 90° pulses, during which spin T_2 relaxation dominates. U-shaped electrophoretic cells were used (Figure 27.4A), with a cross-sectional area A = 3.758 mm^2. Two platinum electrodes were inserted into the protein solution in the U-tube arms. The ENMR experiment was carried out on a Bruker AM 500 NMR spectrometer, using a commercial 5 mm NMR probe. To eliminate electroosmotic effect, the inner U-tube glass surface was coated with (γ-glycidoxypropyl) trimethoxysilane and methylcellulose.[26]

This experiment proves the feasibility of ENMR applications in studying protein mixtures and its potential to provide unique proteomic profiles from human

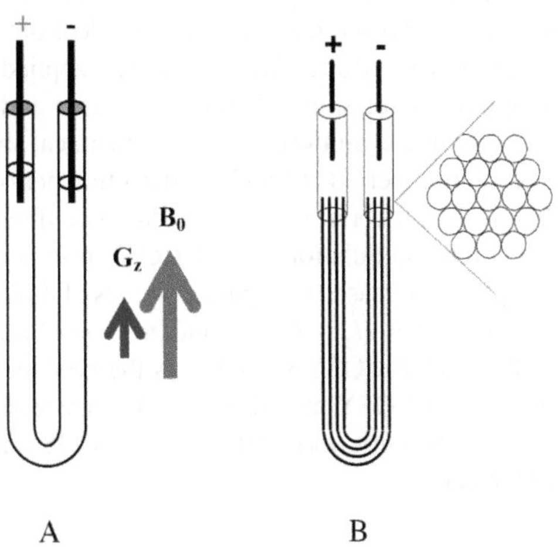

FIGURE 27.4 U-shaped electrophoretic cells for vertical bore magnet. The two Pt-electrodes are vented to the atmosphere at the top of the U-tube arms to avoid gas bubbles. Different variations such as concentric ENMR tubes are used to give phase sensitive mode ENMR spectra in the flow dimension;[22,23] (B) The CA-ENMR sample cell was constructed from parallel capillaries to block the convective current and increase electric field for biological samples of high ionic strength.[24] A horizontal CA-ENMR configuration has also been proposed to maximize the effective electric field applied to the sample.[25]

serum, urine, and cell extract samples. The 1D- or nD-NMR spectra of the component proteins are displayed in the conventional NMR dimensions to identify the chemical nature of the protein components. Since no protein ionization would occur (as in the SELDI-TOF MS experiments) and no protein separation is required (as in the 2D gel electrophoresis), the ENMR proteomic profiles would be generated from native protein conformations, including the spectral features of protein interactions to probe the biological signaling pathways. We have recently demonstrated that the high-resolution ENMR separation in the flow dimension can be achieved using the maximum entropy method for signal processing (unpublished results), or the new ENMR devices using the high-voltage electric field generator (He *et al.*, unpublished results). A low-voltage electric field device would also produce the high-resolution proteomic profiles using the horizontal ENMR cell chambers and the constant-time ENMR pulse schemes to remove the spectral artifacts from electric current induced magnetic field gradient.[25]

27.3 STRUCTURAL CHARACTERIZATION OF COEXISTING PROTEINS IN SOLUTION

A 3D ENMR experiment is necessary to give the two-dimensional correlation spectra of different protein components in solution that can be used for simultaneous sequential and stereospecific assignments of small proteins (<20 kD) coexisting in solution. To study large proteins, multinuclear multidimensional NMR sequences can be employed and modified accordingly for the ENMR experiments.[28] The first 3D ENMR experiment was demonstrated using the 3D electrophoretic COSY pulse sequence (Figure 27.5).[28] A DC electric field pulse was applied between a pair of pulsed magnetic field gradients for spin coherence selection and spatial labeling of the moving molecules. The time evolution of the spin chemical shift correlations and J-coupling during the evolution period (t_1) and the detection period (t_2) generated the 2D COSY type of chemical shift correlations in the first two dimensions. In the flow dimension, the electrophoretic modulation of each COSY resonance was produced as the electric field was stepwise increased to separate signals of different molecules. For a weakly coupled two-spin system ($I_1 = I_2 = ½$ and $|\omega_1 - \omega_2| \geq |2\pi J|$), the final spin density matrix ($\boldsymbol{\sigma}$) of the 3D EP-COSY sequence is the product of the spin density matrix of the conventional 2D COSY experiment and a cosine factor, $\cos(K\mu E_{dc}\Delta)$, which describes the electrophoretic modulation of the COSY resonances using a U-shaped ENMR sample cell:

$$\boldsymbol{\sigma}(t_1,t_2)=\frac{i}{4}\left\{\begin{array}{l} I_1^- \exp\left[i\omega_1 t_2 - i\omega_1(t_1+\Delta)\right]\cos(\pi J t_2)\cos\left[\pi J(t_1+\Delta)\right] \\ \quad +\frac{1}{2} I_2^- \exp\left[i\omega_2 t_2 - i\omega_1(t_1+\Delta)\right]\sin(\pi J t_2)\sin\left[\pi J(t_1+\Delta)\right] \end{array}\right\} \\ \times\exp\left[-i(\phi_1 - 2\phi_2)\right]\exp\left[-DK^2\Delta - \frac{t_1+\Delta}{T_2}\right]\cos(K\mu E_{dc}\Delta) \tag{27.2}$$

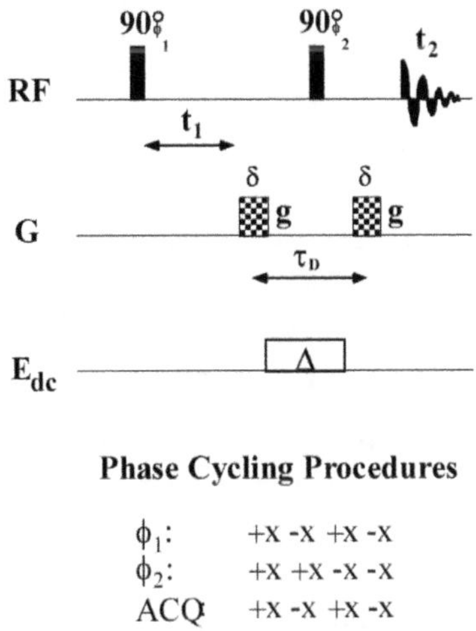

FIGURE 27.5 Pulse sequence for 3D electrophoretic mobility ordered COSY. E_{dc} and Δ are the amplitude and duration of the applied electric field pulse.[28] (with permission)

The COSY resonances of molecules $\{i, i = 1, 2, 3...\}$ migrating at different electrophoretic mobilities were distinguished and displayed at different frequencies $\{V_i\}$ in the third dimension of electrophoretic flow velocity, where $v_i = \pm(KI_\delta\Delta/2\pi\kappa A)\mu_i$ and I_δ is the increment of the applied DC electric current. The experiment was carried out using a solution mixture containing L-aspartic acid and 148 mM 4,9-dioxa-1,12-dodecanediamine in D_2O on a Bruker 500 NMR spectrometer equipped with an actively shielded *z*-magnetic field gradient. Capillary array ENMR (CA-ENMR) sample cells (Figure 27.4B) were uncoated after cleaning the capillaries with 1 M HCl, deionized water, and 1 M NaOH. The migration rates of L-aspartic acid and 4,9-dioxa-1,12-dodecanediamine were enhanced by the solution electroosmotic flow in the capillaries.[29]

In the 3D EP-COSY experiment, chemical shifts and J-coupling constants can be measured for the two molecules in the separated COSY planes (Figure 27.6). Similarly, the 3D EP-COSY can sort out the COSY subspectra of individual molecules in the solution mixture for simultaneous structural assignments of proteins. We have also developed a 3D electrophoretic HSQC (EP-HSQC) scheme to obtain the pure-absorption mode HSQC spectra of different molecules (Zhang, Li, and He, unpublished results). In principle, the multidimensional NMR spectra of proteins and protein conformations can be obtained simultaneously to characterize protein conformational changes and protein interactions.

27.4 HEAT-INDUCED CONVECTION IN BIOLOGICAL BUFFER SOLUTIONS

The heat-induced convection due to electric power deposition in an ENMR experiment may hamper the ENMR studies of protein samples in biological buffer solutions

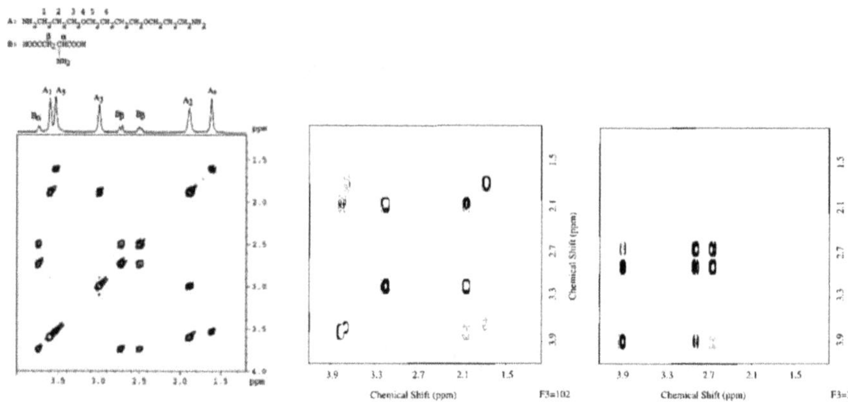

FIGURE 27.6 (A) A control 2D COSY spectrum from the mixture of L-aspartic acid and 4, 9-dioxa-1, 12-dodecanediamine using conventional 2D COSY method; (B) & (C) the two 2D COSY planes from the 3D Electrophoretic COSY matrix displaying the COSY spectrum of each component molecule.[28] (with permission)

of high salt concentrations. When electric conductivity (κ) is high, the electric field ($E_{dc} = I_e/\kappa A$) may become too low to detect the cosinusoidal electrophoretic signal modulations. Generation of sufficient electric field in high salt buffer solutions of proteins requires relatively large electric current, I_e, and causes unavoidable sample heating. In an ENMR experiment, fast heat removal at the edge of the sample tube by cooling air (or cooling fluid) produces a temperature gradient across the sample, which in turn generates a fluid density gradient to cause bulk solution convection. It is difficult, if not impossible, to measure the electrophoretic flow superimposed on an irregular convective flow. To solve the convection problem, we have developed the capillary array electrophoretic NMR (CA-ENMR)[24] and the convection compensated ENMR (CC-ENMR)[30] techniques. In CA-ENMR, the sample chambers were made of capillary bundles to break the convective current loops and the electrical eddy current. In addition, the reduced cross-sectional area of the sample chamber increases the strength of electric field. In an experimental demonstration, the reduced bulk convection and increased electric field were observed in high salt conditions, using a sample solution of 1 mM protein lysozyme in 50 mM NaH_2PO_4/D_2O (Figure 27.7).

An alternative CC-ENMR method was developed to sensitize the electrophoretic motion in the presence of certain amount of bulk convective flow (Figure 27.8).[30] This was accomplished by gradient moment nulling and switching polarity of the applied DC electric field. In the previous ENMR experiments, the spin-echo formation requires that the zeroth gradient moment, m_0, be zero, whereas the electrophoretic flow measurement requires that the first gradient moment, m_1, to be nonzero. The values of the higher-order terms of the gradient moments are nonzero in the presence

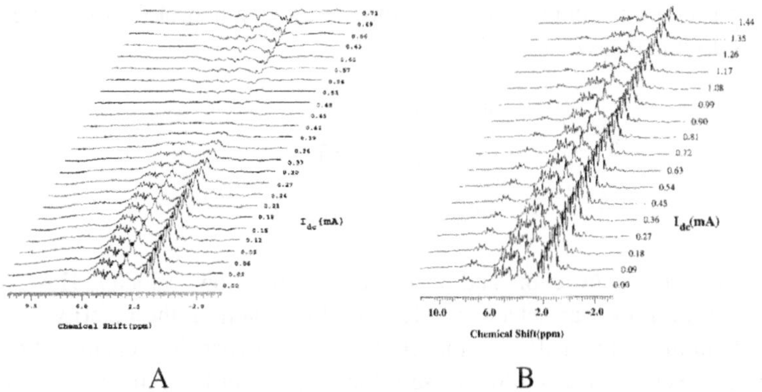

FIGURE 27.7 Electrophoretic interferograms of 1 mM lysozyme in a D_2O solution of 50 mM NaH_2PO_4. Data were obtained at 25°C with (A) a 12-bundle capillary array U-tube (I.D. = 250 μm) that gives electrophoretic signal oscillation; (B) as a comparison, the same sample in a conventional glass U-tube did not receive sufficient electric field to generate the electrophoretic signal modulations. Sample conductivity is 4.21 mS • cm^{-1}. pH = 5.89.[24] (with permission)

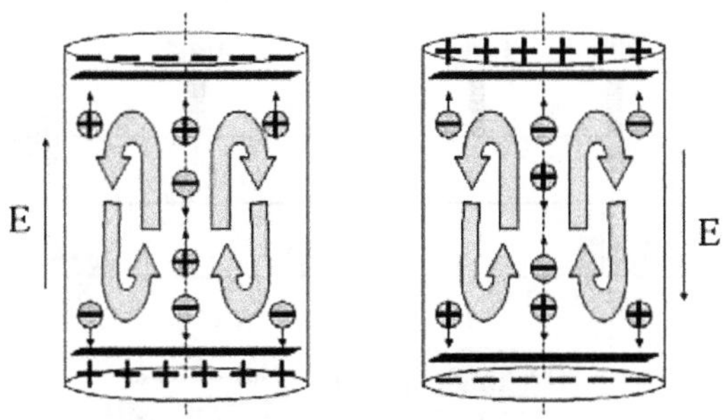

FIGURE 27.8 Switching the polarity of electric field altered the electrophoretic flow direction (solid arrows); however, the procedure exerted no influence on the convective flow circulation (curled arrows).[30] (with permission)

of convection (and zero otherwise).[30] Therefore, these ENMR sequences were prone to the heat-induced convection. To reduce the convection artifacts in the ENMR spectra, we have constructed a CC-ENMR sequence by connecting back-to-back the two spin-echo ENMR sequences (Figure 27.9). Because the coherence order, $p(t)$, had opposite signs in the two halves of the CC-ENMR sequence, the gradient moments, $m_n = \int_0^t p(t)G(t)t^n dt$, changed signs in the middle of the experiment. As a result, all order terms of the gradient moment were zero: $m_k = 0$, where $k = 0, 1, 2, 3\ldots$. The convection-induced spin phase accumulation in the first half of the experiment ($\phi^c_{k,a}$) was refocused in the second half ($\phi^c_{k,a}$):

$$\begin{aligned}\phi^c_{k,a} &= \gamma v^c_0 m^c_k \\ \phi^c_{k,b} &= \gamma v^c_0 \left(-m^c_k\right) \\ \phi^c_{k,\text{tot}} &= \phi^c_{k,a} + \phi^c_{k,b} = 0\end{aligned} \tag{27.3}$$

where c denotes convection and v^c_0 the convective flow velocity. However, the electrophoretic flow was selectively detected by switching the polarity of the DC electric field in the two halves of the CC-ENMR sequence. The change of the ionic migrating direction canceled the phase-refocusing effect from the gradient moment nulling. Consequently, the phase accumulation of the transverse magnetization from the electrophoretic flow was maintained:

$$\begin{aligned}\phi^E_{1,a} &= \gamma v^E_0 m^E_1 \\ \phi^E_{1,b} &= \gamma\left(-v^E_0\right)\left(-m^E_1\right) \\ \phi^E_{1,\text{tot}} &= \phi^E_{1,a} + \phi^E_{1,b} = 2\gamma v^E_0 m^E_1\end{aligned} \tag{27.4}$$

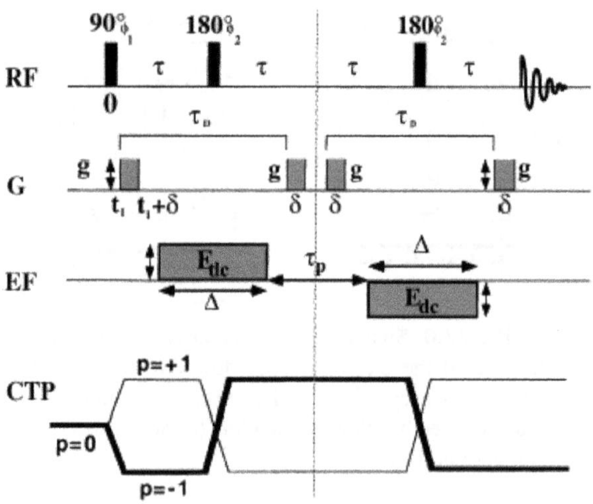

Phase cycling Procedures:

ϕ_1: +x -x -x +x +y -y -y +y
ϕ_2: +y -y -y +y +x -x -x +x
ACQ: +x -x -x +x +y -y -y +y

FIGURE 27.9 The CC-ENMR pulse sequence to remove the convection artifacts.[30] (with permission) where E designates electrophoretic flow at the velocity, v_0^E. The polarity change of the electric field did not affect the heat-induced convection, which originated from temperature gradient and gravitational forces, and the gradient moment nulling remained effective to remove convection artifacts.

The effectiveness of CC-ENMR for convection compensation was assessed with a high-salt solution containing 100 mM L-aspartic acid and a 100 mM 4,9-dioxa-1,12-dodecanediamine mixed in D_2O. As the electric field increased, convection artifacts grew progressively more pronounced in the control spin-echo ENMR experiment, as seen from the phase distortions of the water signal and the signal decays of the ionic species (Figure 27.10A). These convection artifacts were clearly reduced in the CC-ENMR spectrum (Figure 27.10B), with detectable consinusoidal electrophoretic signal oscillations of L-aspartic acid and 4,9-dioxa-1,12-dodecanediamine.

27.5 HIGH-THROUGHPUT STRUCTURAL DETERMINATION OF PROTEIN ACTIVE POCKETS

Since NMR is a major tool for high-resolution structural determination of proteins, ENMR has an unique advantage for proteomic studies—not only can a group of proteins be identified as disease markers (as in SELDI-TOF-MS and 2D electrophoresis methods), but also the identified proteins can be characterized for their chemical identification with detailed structural information. In addition, ENMR experiments can be designed to visualize protein conformational changes during protein interactions in the presence of the other proteomic molecules. By applying a DC electric field, multicomponent protein interactions can be studied using established intermolecular NOE-based screening methods for reactive residues in polypeptide chains, or other

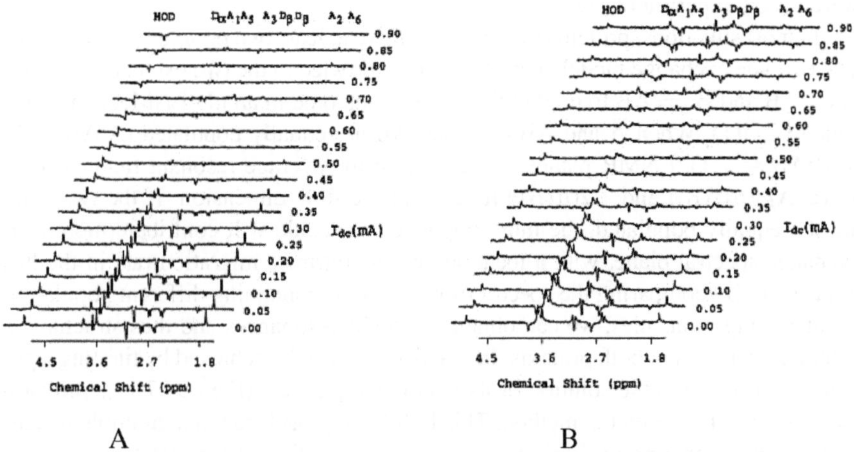

FIGURE 27.10 (A) the control proton spin-echo ENMR spectrum; and (B) The proton CC-ENMR spectrum with reduced convection artifacts. From Ref.[30], with permission.

structural mapping methods based on altered chemical shifts, dynamic parameters, and residual dipole coupling patterns.[17] For example, ENMR can detect drug-binding affinities to different proteins in the same solution by comparing the interacting spectral patterns of the molecules.

In this section, we focus our discussion on exchange ENMR (Ex-ENMR) for high-throughput structural mapping of the active reaction interfaces of proteins. In the fast-exchange limit, the signals from reactive residues can be obtained in the exchange ENMR spectral planes at the average migration rates of the interacting proteins in the flow dimension. In principle, Ex-ENMR can solve structures of the active pockets without determining full structures of the proteins. Since an increasing number of protein structures are deposited into protein banks, we will eventually have a full scaffold structure available from the entire human proteome. However, studying protein interactions and reaction kinetics still needs a high-throughput approach for structural characterization of protein reactivities to understand biological signaling pathways in normal and pathological conditions.

Our mathematical model of Ex-ENMR was derived based on a hypothetical case of a two-site protein exchange: A(in protein A) $\underset{k_{BA}}{\overset{k_{AB}}{\rightleftharpoons}}$ B(in protein B), where k_{AB} and k_{BA} are the chemical exchanging rates, assuming the reactive protein A and protein B have different electrophoretic mobilities, v_a and v_b, respectively. When the chemical exchange rate is much faster than the difference of the two protein migration rates ($k_{AB} = k_{BA} \gg 1/2|v_a - v_b|$), the signals from the interacting residuals would appear at the average migration rates at $v_{exchange} = 1/2(v_a + v_b)$ in the flow dimension, assuming the two states are equally populated. In the slow exchange limit when $k_{AB} = k_{BA} \ll 1/2|v_a - v_b|$, the signals of the interacting proteins can be distinguished at v_a or v_b in the electrophoretic flow dimension (unpublished results). In practice, many protein reactions are in fast-exchange limit in the ENMR time scale (ca. 50–1000 milliseconds). Therefore, the exchanging spin resonances would be extracted into separate 2D ENMR planes in a 3D ENMR spectrum, if such case were tested experimentally.

In most situations, protein interactions involve at least two proteins A + B $\rightleftharpoons$ AB. The two-site exchange model needs to be applied twice to the two exchange reactions A $\rightleftharpoons$ AB and B $\rightleftharpoons$ AB to predict the spectral outcome in an nD-ENMR experiment (Figure 27.11). When A and AB or B and AB are equally populated (A:AB = 1:1 or B:AB = 1:1), ENMR planes will appear at the average resonant frequencies of $1/2(\mu(A)+\mu(AB))$ and $1/2(\mu(B)+\mu(AB))$ in the flow dimension. If the two states are not equally populated, the more populated molecule will contribute more to its exchange spectral plane, with a location near its intrinsic migration rate in the flow dimension. By comparing the spectra from solutions containing different populations of interacting molecules, we can obtain the NMR resonances and thus structures of reaction interfaces in both proteins. In practice, this can be achieved by titrating a protein (or a drug) into the solution of its interaction partners (Figure 27.12), similar to the chemical shift mapping method. This ENMR approach can map multiple reactive proteins in a signaling network without selective labeling of the interested proteins. Of course, molecular diffusion and spin relaxation also affect the spectral outcome of the Ex-ENMR exchanging planes.[31]

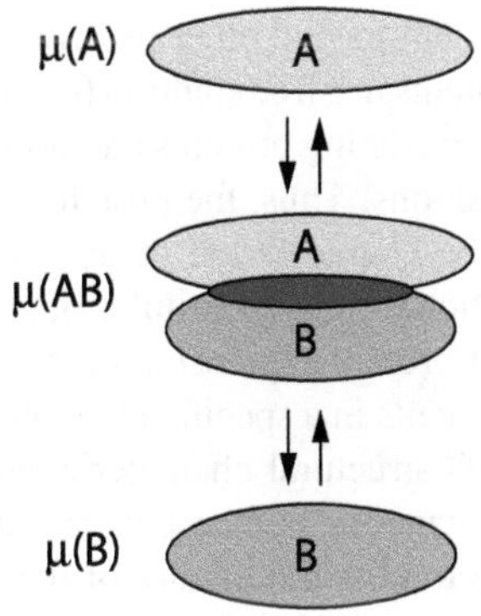

FIGURE 27.11 Schematics of two interacting proteins A and B undergoing chemical exchange processes, where μ(A), μ(B), and μ(AB) are the electrophoretic migration rates of free protein A, free protein B, and protein complex AB, respectively.

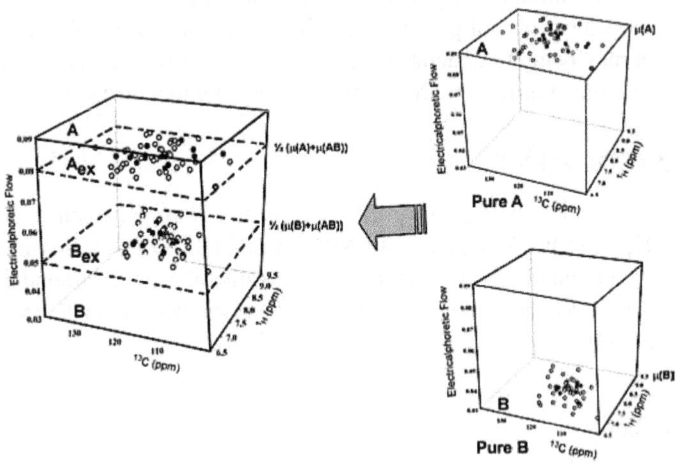

FIGURE 27.12 The effective electrophoretic migration rates of protein A and protein B changes when they are involved in a chemical reaction using $A + B \rightleftharpoons AB$ as a model system. The schematic 3D Exchange electrophoretic ^{1}H-^{15}N HSQC spectrum gives 2D-NMR spectral planes of A_{ex} and B_{ex}, which differ from that of the pure protein spectra A or B. The simulation plots a case when protein A, B, and their complex AB are equally populated. The protein resonances involved in the chemical exchange are represented with filled circles, while those not involved with open circles. (unpublished results)

27.6 CONCLUSIONS

Electrophoretic NMR proteomics is potentially a powerful tool for studying protein interactions in system biology and may be combined with the current in-cell NMR techniques.[32–34] to map biological signaling events across cell membranes mediated by interactions of cell protein receptors and other proteins, ligands, or drugs.[35] Without physical separation of the proteins in a free solution (or in gel matrix), ENMR can map unique proteomic profiles of coexisting proteins and protein conformations, retaining information of protein interactions. Thus, the protein markers of diseases identified in a proteomic sample can be characterized at the structural level to define their chemical identities and functional roles in cellular biochemistry and organ physiology. Clinically, the abnormal protein expression profiles in different biological signaling pathways of cancerous cells in a specific physiological microenvironment can be identified with detailed 3D structural characterizations. High-resolution ENMR proteomic profiles from spontaneous human cancer cells can be compared to that from their normal cell origin to obtain a picture of the altered cellular biochemical events in pathological conditions. Such information will be extremely valuable for effective drug designs and therapeutic interventions. As demonstrated in the preliminary experiments of 3D electrophoretic COSY and HSQC methods, the structures of coexisting proteins can be studied simultaneously in solution. In addition, high-throughput structural determination of protein–protein interactions can be performed to map 3D conformational changes of the protein active pockets. Ultimately, through proteomic characterization of proteins, protein interactions, and biological signaling pathways, we may begin to understand what makes an individual special and how to discover personalized drugs and therapies to manage human diseases.

ACKNOWLEDGMENTS

The work was supported in part by grants from the National Institutes of Health (RR12774-01, GM/OD55209, R21CA80906, and R21EB001756), Susan G. Komen Breast Cancer Foundation (IMG0100117), the National Science Foundation (NSF MCB-9707550), and the American Chemical Society Petroleum Research Funds (PRF# 32308-G4).

REFERENCES

1. Bouwmeester, T., Bauch, A., Ruffner, H., Angrand, P.-O., Bergamini, G., Croughton, K., Cruciat, C., Eberhard, D., Gagneur, J., Ghidelli, S., Hopf, C., Huhse, B., Mangano, R., Michon, A.-M., Schirle, M., Schlegl, J., Schwab, M., Stein, M.A., Bauer, A., Casari, G., Drewes, G., Gavin, A.-C., Jackson, D.B., Joberty, G.; Neubauer, G., Rick, J., Kuster, B., Superti-Furga, G., A physical and functional map of the human TNF-α/NF-κB signal transduction pathway, *Nat. Cell. Biol.,* 6, 97–105, 2004.
2. Petricoin, E.F., Zoon, K.C., Kohn, E.C., Barrett, J.C., Liotta, L.A., Clinical proteomics: translating benchside promise into bedside reality, *Nat. Rev. Drug Discovery*, 1, 683–695, 2002.
3. Li, J., Zhang, Z., Rosenzweig, J., Wang, Y.Y., Chan, D.W., Proteomics and bioinformatics approaches for identification of serum biomarkers to detect breast cancer, *Clin. Chem.,* 48, 1296–1304, 2002.
4. Koopmann, J., Zhang, Z., White, N., Rosenzweig, J., Fedarko, N., Jagannath, S., Canto, M.I., Yeo, C.J., Chan, D.W., and Goggins, M., Serum diagnosis of pancreatic

adenocarcinoma using surface-enhanced laser desorption and ionization mass spectrometry, *Clin. Cancer Res.,* 10, 860–868, 2004.

5. Taylor, C.F., Paton, N.W., Garwood, K.L., Kirby, P.D., Stead, D.A., Yin, Z., Deutsch, E.W., Selway, L., Walker, J., Riba-Garcia, I., Mohammed, S., Deery, M.J., Howard, J.A., Dunkley, T., Aebersold, R., Kell, D.B., Lilley, K.S., Roepstorff, P., Yates, J.R., III, Brass, A., Brown, A.J.P., Cash, P., Gaskell, S.J., Hubbard, S.J., and Oliver, S.G., A systematic approach to modeling, capturing, and disseminating proteomics experimental data, *Nat. Biotechnol.,* 21, 247–254, 2003.
6. Jones, M.B., Krutzsch, H., Shu, H., Zhao, Y., Liotta, L.A., Kohn, E.C., Petricoin E.F., III, Proteomic analysis and identification of new biomarkers and therapeutic targets for invasive ovarian cancer, *Proteomics,* 2, 76–84, 2002.
7. Chaiken, I., Cocklin, S., Zurawski, J., and Sergi, M., Proteins, recognition networks and developing interfaces for macromolecular biosensing, *J. Mol. Recognit.,* 17, 198–208, 2004.
8. Eustace, B.K., Sakurai, T., Stewart, J.K., Yimlamai, D., Unger, C., Zehetmeier, C., Lain, B., Torella, C., Henning, S.W., Beste, G., Scroggins, B.T., Neckers, L., Ilag, L.L., and Jay, D.G., Functional proteomic screens reveal an essential extracellular role for hsp90 in cancer cell invasiveness, *Nat. Cell Biol.,* 6, 507–514, 2004.
9. Colland, F., Jacq, X., Trouplin, V., Mougin, C., Groizeleau, C., Hamburger, A., Meil, A., Wojcik, J., Legrain, P., and Gauthier, J.-M., Functional proteomics mapping of a human signaling pathway, *Genome Res.,* 14, 1324–1332, 2004.
10. Hanahan, D. and Weinberg, R.A., The hallmarks of cancer, *Cell,* 100, 57–70, 2000.
11. Lenz, G.R., Nash, H.M., and Jindal, S., Chemical ligands, genomics, and drug discovery, *DDT,* 5, 145–156, 2000.
12. Jeffery, D.A., and Bogyo, M., Chemical proteomics and its application to drug discovery, *Curr. Opinion Biotechnol.,* 14, 87–95, 2003.
13. Steitz, T.A., *Structural Studies of Protein–Nucleic Acid Interaction*, Cambridge University Press, New York, 1993.
14. Wuthrich, K., *NMR of Proteins and Nucleic Acids*, John Wiley & Son, Inc., New York, 1986.
15. Ernst, R.R., Bodenhausen, G., and Wokaun, A., *Principle of Nuclear Magnetic Resonance in One and Two dimensions*, Oxford Science Publications, Oxford, 1987, pp. 482–485.
16. Kuznetsov, Y.G., Victoria, J.G., Robinson W.E., Jr., and McPherson, A., Atomic force microscopy investigation of human immunodeficiency virus (HIV) and HIV-infected lymphocytes, *J. Virol.,* 77, 11,896–11,909 2003.
17. Zuiderweg, E.R.P., Mapping protein–protein interactions in solution by NMR spectroscopy, *Biochemistry,* 41, 1–7, 2002.
18. Tran, T.-K., Sailasuta, N., Hurd, R., and Jue, T., Spatial distribution of deoxymyoglobin in human muscle: An index of local tissue oxygenation, *NMR Biomed.,* 12, 26–30, 1999.
19. Zhou, J., Payen, F., Wilson, D.A., Traystman, R.J., and van Zijl, P.C.M., Using the amide proton signals of intracellular proteins and peptides to detect pH effects in MRI, *Nat. Med.,* 9, 1085–1090, 2003.
20. Shulman, R.G., and Rothman, D.L., ^{13}C NMR of intermediary metabolism: Implications for systemic physiology, *Annu. Rev. Physiol.,* 63, 15–48, 2001.
21. He, Q., Liu, Y., and Nixon, T., High-field electrophoretic NMR of mixed proteins in solution, *J. Am. Chem. Soc.,* 120, 1341–1342, 1998.

22. Morris, K.F., and Johnson, C.S., Jr., Mobility-ordered 2D NMR spectroscopy for the analysis of ionic mixtures, *J. Magn. Reson. Ser. A,* 100, 67–73, 1993.
23. Holz, M., Seiferling, D., and Mao, X., Design of a new electrophoretic NMR probe and its application to $^{7}Li^{+}$ and $^{133}Cs^{+}$ mobility studies, *J. Magn. Reson. Ser. A,* 105, 90–94, 1993.
24. He, Q., Liu, Y., Sun, H., and Li, E., Capillary array electrophoretic NMR of proteins in biological buffer solutions, *J. Magn. Reson.,* 141, 355–359, 1999.
25. Li, E., and He, Q., Constant-time multidimensional electrophoretic NMR, *J. Magn. Reson.,* 156, 1–6, 2002.
26. He, Q., Electrophoretic Nuclear Magnetic Resonance, Ph.D. thesis, University of North Carolina at Chapel Hill, 1990.
27. Kay, L.E., Clore, G.M., Bax, A., and Gronenborn, A.M., Four-dimensional heteronuclear triple-resonance NMR spectroscopy of Interleukin-1β in solution, *Science,* 262, 411–414, 1990.
28. He, Q., Lin, W., Liu, Y., and Li, E., Three-dimensional electrophoretic NMR correlation spectroscopy, *J. Magn. Reson.,* 147, 361–365, 2000.
29. He, Q., Electroosmosis-enhanced electrophoretic NMR, *Concepts Magn. Reson.* (in revision).
30. He, Q., and Wei, Z., Convection compensated electrophoretic NMR, *J. Magn. Reson.* 150, 126–131, 2001.
31. Johnson, C.S., Jr., Effects of chemical exchange in diffusion-ordered 2D NMR spectra, *J. Magn. Reson. Ser. A,* 102, 214–218, 1993.
32. Serber, Z., and Dotsch, V., In-cell NMR spectroscopy, *Biochemistry,* 40, 14,317–14,323, 2001.
33. Serber, Z., Ledwidge, R., Miller, S.M., and Dotsch, V., Evaluation of parameters critical to observing proteins inside living *Escherichia coli* by in-cell NMR spectroscopy, *J. Am. Chem. Soc.,* 123, 8895–8901, 2001.
34. Serber, Z., Keatinge-Clay, A.T., Ledwidge, R., Kelly, A.E., Miller, S.M., and Dotsch, V., High-resolution macromolecular NMR spectroscopy inside living cells, *J. Am. Chem. Soc.,* 123, 2446–2447, 2001.
35. Sem, D., Villar, H., and Kelly, M., NMR on target: Technological innovations are creating a powerful tool for target identification and chemical proteomics, *Mod. Drug Discovery,* 6, 26–31, 2003

Index

Milton Keynes UK
Ingram Content Group UK Ltd.
UKHW021923071024
449327UK00022B/1695

9 780367 391577